Principles, Techniques and Practices of Biotechnology

Principles, Techniques and Practices of Biotechnology

Edited by **Lydell Norris**

SYRAWOOD
PUBLISHING HOUSE

New York

Published by Syrawood Publishing House,
750 Third Avenue, 9th Floor,
New York, NY 10017, USA
www.syrawoodpublishinghouse.com

Principles, Techniques and Practices of Biotechnology
Edited by Lydell Norris

International Standard Book Number: 978-1-68286-190-5 (Hardback)

Contents

Preface

Biotechnology is a fast developing field of research and its applications are finding more and more relevance with the passage of time. From stem cell research to applications in agriculture and medical science, the scope of biotechnology is vast. The topics included in this book like – food and bioprocessing, bioengineering, immobilization, biotechnology, etc. are of utmost significance and are bound to provide incredible insights to readers. This text elucidates the concepts and innovative models around prospective developments with respect to biotechnology. It will benefit biotechnologists, researchers, scholars and scientists in this field.

All of the data presented henceforth, was collaborated in the wake of recent advancements in the field. The aim of this book is to present the diversified developments from across the globe in a comprehensible manner. The opinions expressed in each chapter belong solely to the contributing authors. Their interpretations of the topics are the integral part of this book, which I have carefully compiled for a better understanding of the readers.

At the end, I would like to thank all those who dedicated their time and efforts for the successful completion of this book. I also wish to convey my gratitude towards my friends and family who supported me at every step.

Editor

Antibacterial Compounds in Predominant Trees in Finland: Review

Sari Metsämuuronen[1] and Heli Siren[2]*

[1]*Lappeenranta University of Technology, PO Box 20, FI-53851 Lappeenranta, Finland*
[2]*Department of Chemistry, University of Helsinki, PO Box 55, FI-00014 University of Helsinki, Finland*

Abstract

The extracts of Scots pine, Norway spruce, silver and white birches stem, bark, roots, leaves and needles contain several useful bioactive compounds that exhibit antibacterial activity against pathogens. Both phenolic extracts and essential oils are bacteriostatic against several bacteria. The main individual antibacterial phenolic compounds in Scots pine are pinosylvins that effectively inhibit growth of pathogens such as *Bacillus cereus*, *Staphylococcus aureus* and *Listeria monocytogenes*. From other phenolic compounds lignans appeared to be the least bacteriostatic and flavonoids tend to occur as glycosylated forms which have lower antibacterial activity than their aglycones. Gram-positive bacteria are generally more susceptible to plants bioactive compounds than gram-negative bacteria.

Keywords: Antibacterial compounds; Norway spuce; Extract; Hydrolysis; Fermentation

Introduction

Plants synthesise low molecular mass compounds, phytoalexins, which protect them against attacks by fungi, bacteria and insects [1,2]. Several studies of these plants used as traditional folk medicine have recently been published [3-10]. Interest in natural bioactive compounds has arisen for their multiple biological effects, including antioxidant, antifungal and antibacterial activity. The potential use of these compounds in food preservation and pharmaceutical applications as oxidants and for cancer chemoprevention has been investigated [1,11,12]. Their potential use against pathogenic microorganisms and infections that are currently difficult to treat because of the resistance that microorganisms have built against antibiotics would be one interesting application. Essential oils and phenolic extracts have been tested against multi-drug-resistant human pathogens and intestinal bacteria, like methicillin-resistant *Staphylococcus aureus* (MRSA) [8,13-19].

Essential oils are very complex natural mixtures which can contain about 20-60 components at quite different concentrations [20]. They are characterised by two or three major components at fairly high concentrations (20-70%) compared to others components present in trace amounts. The main group is composed of terpenes and terpenoids.

The main groups of bioactive phenolic compounds in plants are simple phenols and phenolic acids, stilbenes, flavonoids and lignans, which are derivatives of phenylpropanoid metabolism via the shikimate and acetate pathways [11,12,21,22]. These secondary metabolites are often bound to a mono- or oligosaccharide or to uronic acid [2]. The saccharide or uronic acid part is called glycone and the other part the aglycone. Flavonoids, phenolic acids, stilbenes, tannins and lignans are especially common in leaves, flowering tissues and woody parts such as stems and barks. In bark and knots they are especially important defence against microbial attack after injury. Hence, the bark and knot extracts have been observed to be more active against bacteria than the wood extracts [23]. The effectiveness of the defence varies among plant species. The long-lived and slow growing plants have been observed to be more active than the fast growing ones [24].

The aim of this literature review is to clarify the antibacterial compounds present in the predominant tree species in Finland, Scots pine (*Pinus sylvestris*), Norway spruce (*Picea abies*), silver birch (*Betula pendula*) and white birch (*Betula pubescens*). The extraction of these valuable compounds from forest biomass is of special interest as they are available in different wood harvesting and industrial residues, such as bark, knots, stump and roots.

Phenolic Compounds

The bioactive compounds are present in wood and, thus, they can be solubilised by different solvents [12]. In published studies phenolic compounds have been most often obtained through ethanol, methanol or acetone extraction as alone or after hexane extraction (Table 1). The most frequently used method to determine the total phenols in the extracts is colorimetric measurement with the Folin-Ciocalteu reagent. However, this reagent may react with any reducing substances other than phenols and therefore measures the total reducing capacity of a sample. By this method total phenol content is expressed in terms of gallic acid (GAE) or tannic acid (TAE) equivalents. The individual components have been identified by using gas (liquid) chromatography (GC, GLC) and mass spectrometry (MS) or high-performance liquid chromatography (HPLC).

The evaluation methods for antibacterial activity can be divided into diffusion, dilution and optical density methods. The most commonly used screen to evaluate antimicrobial activity is the agar diffusion technique. In this method, the diameter of inhibition zone is measured at the end of incubation time. The usefulness of diffusion method is limited to the generation of preliminary data only [25]. With respect to that, the activity values are not comparable, since the studies are made with different procedures and chemicals. Thus, extracts, extraction methods, assess of antimicrobial activity and strains of test organisms vary in the publications.

The minimal inhibiting (MIC) and bactericidal (MBC) concentrations are defined as the lowest concentrations of tested

***Corresponding author:** Heli Siren, University of Helsinki, Department of Chemistry, PO Box 55, FI-00014 University of Helsinki, Finland

Tree species	Part	Drying method, Solvent extraction	Compounds	Identification	Ref.
Scots pine	heartwood	Oven-dried 60˚C, 48h, 80% acetone	Total phenolics 6.7-13.6 mg TAE^{g-1}	Folin-Ciocalteau	[26]
	heartwood	Oven-dried 60˚C 48h, acetone	Total pinosylvins 4.7-7.5 mg^{g-1} Pinosylvin, PMME, PDME	GLC-MS	[26]
	heartwood	Oven-dried 60˚C 48h, 80% acetone	Total phenolics 4.55-4.66 mg TAE cm^{-1}	Folin-Ciocalteu	[27]
	heartwood	Oven-dried 60˚C 48h, acetone	Pinosylvin, PMME and PDME 2.07-3.16 mg cm^{-1}	GLC-MS	[27]
	knotwood	Freeze-dried 1. hexane 2. acetone:water	pinosylvin, PMME, DHPMME chromatographic purification, chrystallization	GC after	[28]
	knotwood	Freeze-dried, 1. Hexane 2. Acetone	Nortrachelogenin fractionated by flash chromatography	GC-MS	[29,30]
	knotwood	N.a., 1. Hexane 2. Acetone/water (95:5)	Lingnans, stilbenes and flavonoids	GC-FID	[30]
	knotwood	N.a., 1. hexane 2. acetone	Pinosylvin, PSME Fractionation by flash chromatography	GC-MS	[30]
	knotwood, bark	Freeze-dried, 1. Hexane 2. Acetone	Pinosylvin Fractionation using flash chromatography	GC-FID HPSEC	[31]
	knotwood	Resin Methyl-tert-butyl ether	Pinoresinol Purification with flash chromatography	GC-FID HPSEC	[31]
	phloem	Freeze/vacuum dried, 80% methanol			[32]
	phloem, bark	Air-dried 1. 70% acetone 2. Chloroform	22 phenolic compounds	HPLC-DAD HPLC-ESI-MS	[33]
	needle, cork, bark	Air dried, 80% methanol	Phenolics	Folin-Ciocalteau	[34]
	needles	Distillation	Essential oil (mainly terpenes and terpenoids)	GC	[35]
Norway spruce	knotwood bark	1. Hexane 2. Acetone/water (95:5)	Lingnans, stilbenes and flavonoids	GC-FID	[30]
	knotwood	Freeze-dried, 1.Hexane 2. Acetone/water (95:5)	Hydroxymatairesinol	GC-MS	[30]
	phloem	Freeze-dried, 80% ethanol	Isorhapontin, astringin, piceid	GC-MS	[37]
	phloem	Freeze-dried, 80% methanol	Isorhapontin, astringin, piceid, taxifolin gulcoside, (+)-catechin	HPLC	[38]
Silver birch	bark	1. Hexane 2. Acetone/water (95:5)	Lingnans, stilbenes and flavonoids	GC-FID	[30]
	phloem, bark	Air dried, 80% methanol	Phenolics	Folin-Ciocalteau	[34]

ASE = accelerated solvent extractor, DHPMME = dihydropinosylvin monomethyl ether, FID = flame ionization detector, PDME = pinosylvin dimethyl ether, PMME = pinosylvin monomethyl ether, TAE = tannic acid equivalent

Table 1: Extraction and identification methods of antibacterial substances.

compounds which completely inhibited bacterial growth and which results in more than 99.9% killing of the bacteria being tested, respectively. The MIC and MBC have been determined by the liquid dilution method.

Scots pine

Antioxidant, antifungal and antibacterial [26-34] activity of phenolic extracts of Scots pine growing in Finland have been investigated during the last decades. In most of these studies activity of knotwood extracts were detected and heartwood extracts were detected only against brown-rot fungus. The total phenolic concentration has been observed to vary a lot between the different parts of the tree: 76.0, 17.5 and 1.1 mg GAE g^{-1} for dried bark, needles and cork, respectively [35-38]. For the heartwood the total phenolic concentration of 4.55-4.66 mg TAE cm^{-3} of wood [27] and 6.7-13.6 mg TAE g^{-1} wood [26] have been reported. Willför et al. [39] reported that in knotwood, the amount of extractable phenolic compounds can be as large as 10%

(w/w), which was several times more than that they observed in the stem wood.

Knotwood extracts containing lignans, stilbenes and flavonoids have been observed to show antibacterial activity against some pathogenic bacteria [30] (Table 2). The strongest inhibition has been observed against the gram-positive pathogenic bacteria *Bacillus cereus*, *S. aureus* and *Listeria monocytogenes*, while inhibition against the gram-negative bacteria *Escherichia coli*, and *Pseudomonas fluorescens* was only slight. The extract did not show any inhibitory effect on *Lactobacillus plantarum* and *Salmonella infantis*. Also the phoem extracts of Scots pine have shown to be clearly active against *S. aureus*, but not against *E. coli* [32]. The hydrophilic extracts from knotwood of several pines, rich in stilbenes, proved to be efficient antibacterial agents when tested against paper mill bacteria *Burkholderia multivorans*, *Bacillus coagulans* and *Alcaligenes xylosoxydans* [28].

The main components in hydrophilic extracts of Scots pine have

Tree species	Bacteria	Activity	Method	Ref.
Scots pine, knotwood	Escherichia coli Salmonella infantis Pseudomonas fluorescens Bacillus cereus Staphylococcus aureus Listeria monocytogenes Lactobacillus plantarum	5 ± 13% 0 ± 4% 13 ± 2% 79 ± 3% 30 ± 23% 47 ± 20% 0 ± 3%	Well microplates and turbidity reader	[30]
Scots pine, knotwood extract	Burkholderia multivorans Bacillus coagulans Alcaligenes xylosoxydans	strong, > 20 mm strong, > 20 mm moderate, 16-19 mm	Agar diffusion	[28]
Scots pine, phloem	Escherichia coli Staphylococcus aureus	slight, 1-3 mm clear, 4-10 mm	Cylinder diffusion	[32]
Norway spruce, knotwood	Burkholderia multivorans Bacillus coagulans Alcaligenes xylosoxydans	0 0 small, 11-15 mm	Agar diffusion	[28]
Norway spruce, knotwood and bark	Escherichia coli Salmonella infantis Pseudomonas fluorescens Bacillus cereus Staphylococcus aureus Listeria monocytogenes Lactobacillus plantarum	0 0 knotwood 38%, bark 0% knotwood 15%, bark 0% 0 0 0	Well microplates and turbidity reader	[30]
Norway spruce, needles	Escherichia coli Staphylococcus aureus	0 slight, 1-3 mm	Cylinder diffusion	[32]
Silver birch, knotwood	Burkholderia multivorans Bacillus coagulans Alcaligenes xylosoxydans	0 0 small, 11-15 mm		[28]
Silver birch, knotwood and bark	Escherichia coli Salmonella infantis Pseudomonas fluorescens Bacillus cereus Staphylococcus aureus Listeria monocytogenes Lactobacillus plantarum	0 0 knotwood 20%, bark 22% knotwood 14%, bark 0% 0 0 0	Well microplates and turbidity reader	[30]
White birch, leaf	Escherichia coli Staphylococcus aureus	slight, 1-3 mm clear, 4-10 mm	Cylinder diffusion	[32]

MBC = minimal bactericidal concentration, MIC = minimal inhibiting concentration

Table 2: Antibacterial activity of wood extracts and essential oils against selected bacteria.

been observed to be stilbenes, lignans (31%) nortrachelogenin being the most abundant lignan (30%) and oligolignans (6%) [31]. In the heartwood of the brown-rot fungus resistant and susceptible trees, the average total stilbenes concentration 6.4-7.5 mg g^{-1} and 4.7-5.0 mg g^{-1} of dry weight, respectively, have been measured [26].

Stilbenes (Figure 1) are 1,2-diarylethenes, the A ring usually having two hydroxyl groups in the m-position, while B ring is substituted by hydroxy and methoxy groups in the o-, m- and/or p-positions [11]. Stilbenes are synthesised mainly by forest trees [12], in monomeric form and as dimeric, trimeric and polymeric stilbenes, the so-called viniferins. They are commonly found in the roots, barks, rhizomes and leaves [11]. The most abundant stilbenes in Scots pine extracts are pinosylvins: 38% in knotwoods [31] and 6-7% in heartwoods [27], whereas in the bark extracts they have not been found [40]. Pinosylvin and pinosylvin monomethyl ether (PMME) are the main pinosylvins, pinosylvin dimethyl ether (PDME) being less abundant [26,28,31,41]. Dihydropinosylvin monomethyl ether (DHPMME) has been isolated from Pinus strobus knotwood [30]. The pinosylvin-3-O-methyltransferase enzyme catalyses the conversion of pinosylvin to the monomethyl ether that plays a role in the resistance of the plant to stress including ozone and infection [41]. Hence, a high concentration of PMME relative to pinosylvin may be an indication of high stress levels of the trees.

The highest antimicrobial activities of the pure compounds present in Scots pine have been observed with pinosylvin and PMME, followed by DHPMME and flavanone pinocembrin [30] (Table 3). Very strong inhibition effects (62-100%) have been observed against human pathogens B. cereus, S. aureus and L. monocytogenes. Pinosylvin, DHPMME, PMME and flavonoid pinocembrin (from P. cembra) have shown a very similar activity against bacteria as the Pinus extracts where they have been isolated [30]. Both Välimaa et al. [30] and Lindberg et al. [28] have observed that the antibacterial activity correlate with the stilbene content of the extracts and, hence, stilbenes have been concluded to be the main antibacterial compounds of hydrophilic extracts of Scots pine.

The precise mechanism of antibacterial action of stilbenes is unclear. One possibility is that they destroy the membrane structure resulting in burst of the cell [42]. Välimaa et al. [30] suggested that two hydroxyl groups in meta position in one of the aromatic rings and the double bond in the carbon chain between the rings may play an important role. From phenolic acids chlorogenic acid has shown stronger activity against E. coli than ferulic acid [43].

Flavonoids consist of a central three-ring structure (Table 4). Their activity is proposed to be due to their ability to complex with extracellular and soluble proteins and to complex with bacterial cell walls [1]. Flavonoids and oligomers of flavonoids and proanthocyanidins, frequently occur as glycosides [2]. Different flavonol glycosides are typical in pine needles and the sugar residues are found to be

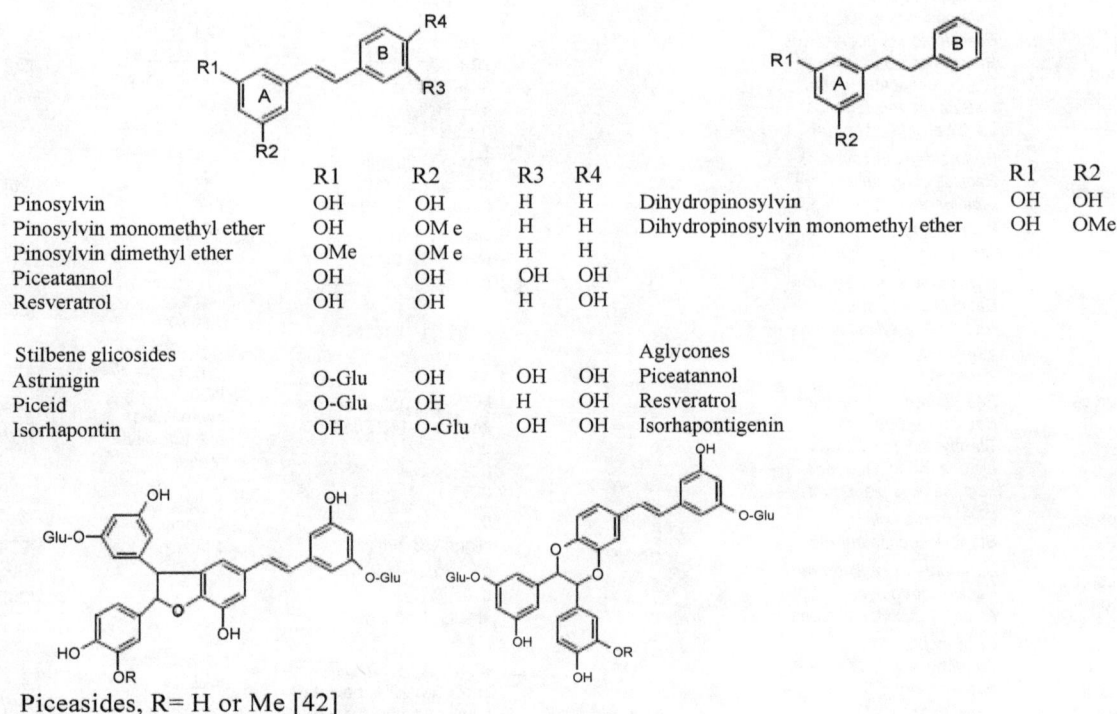

Figure 1: Chemical structures of stilbenes and stilbenes glucosides found in conifers.

bonded mainly at the 3-position [28,44-48] (Appendix 1). However, the glycoside contents on other parts than needles are not available. Dihydro-flavonol type taxifolin and flavanone type pinocembrin were the main flavonoids in knotwood [29]. Pinocembrin has been observed to inhibit growth of several bacteria, the strongest activity being against *B. cereus* (Table 3).

Lignans isolated from knotwoods of conifers are strong antioxidants [29], but their antibacterial activity is observed to be low. Purified lignans (Figure 2), matairesinol, hydroxymatairesinol, lariciresinol and secoisolaricinol, have not shown activity against any of the tested bacteria and isolariciresinol and nortrachelogenin have shown slight activity only against *B. cereus* [30] (Table 3).

Norway spruce

The average amount of extractable phenolic compound in Norway spruce knotwood is around 15% (w/w), but as high values as nearly 30% (w/w) have been detected [49]. The amount of phenolic compounds in the stem wood has been observed to be much lower, usually 0.15-0.3%. Malá et al. [50] observed that in Norway spruce cells the soluble glycoside-bound forms of phenolic acids accounted for ~ 85% of the total content, followed by the methanol-insoluble cell wall-bound phenolic esters (7-8%). The amount of methanol soluble esters and free phenolic acids were low, accounting for ~2 and 4-5% of total phenolic contents, respectively [50]. Free, ester-bound (released after alkaline hydrolysis) and glycoside-bound (released after acid hydrolysis) phenolic acids were obtained from a methanol extract [50]. Two cinnamic acid derivatives, p-coumaric and ferulic acids and five benzoic acid derivatives (anisic, p-hydroxybenzoic, vanillic and syringic acids) were found in the Norway spruce cells. p-Hydroxybenzoic acid

glucoside and native ferulic acid have been reported in the extracts of roots [51].

Several stilbenes and stilbenes glucosides have been detected in different parts of Norway spruce. Astringin and isorhapontin are the main constitutive stilbenes glycosides in Norway spruce [36,37,50]. Zeneli et al. [36] detected the contents of astringin and isorhapontin of 20.2 and 71.8% in sapwood phenolics and 38.8 and 46.5% in bark phenolics, respectively, in trees growing in Norway. Wood phenolics contained also 5.1% piceid and bark phenolics 7.7% piceid and 0.4% piceatannol. Viiri et al. [37] have detected stilbene glycoside concentration of ~7-8 µg mg^{-1} in fresh phloem tissue. Over half of it was isorhapontin and rest astringin and piceid, while resveratrol was the most abundant aglycone (~0.5 µg mg^{-1}). These stilbenes aglycones and glucosides have been detected also in bark extracts [31]. In healthy phloem stilbenes typically occur as glycosides. Piceoside, piceatannol and its glucoside, isorhapontin have been detected in roots [51]. On the contrary, Willför et al. [49] have not found stilbenes in the hydrophilic knotwood extractives of Norway spruce. They reported that more than a half of the knotwood extracts are lignans, the rest being mainly oligolignans. The most abundant lignan was hydroxymatairesinol [29,31]. Its two isomers constitute over 70 % of the lignans [29].

Shan et al. [45] evaluated antibacterial activity of resveratrol and its glucoside piceid against five bacteria (Table 3). In general, efficacy of aglycone and glucoside appeared to be almost the same. The MIC values of both compounds were 313-625 mg L^{-1} and in the case of *L. monocytogenes* MIC was also bacteriostatic concentration.

Needles have considerably high content of phenolic substances: 155.3 mg GAE g^{-1} dry weight of the original sample [34]. Five types of flavonoids (flavones, flavonols, flavanones, dihydro-flavonols

Substance class	Substance	Bacteria	Inhibitory effect, %	Method	Ref.
Hydroxy-cinnamic acid	Ferulic acid (commercial)	*Escherichia coli* *Salmonella enterica* *Enterococcus faecalis*	Slight Slight No inhibition	Agar well diffusion	[43]
		Escherichia coli *Bacillus licheniformis* *Micrococcus luteus*	MIC 375 mg L⁻¹ MIC 375 mg L⁻¹ MIC 375 mg L⁻¹	Not announced	[44]
	Chlorogenic acid (commercial)	*Escherichia coli* *Salmonella enterica* *Enterococcus faecalis*	Clear No inhibition No inhibition	Agar well diffusion	[43]
Stilbene	Pinosylvin (Scots pine knotwood)	*Escherichia coli* *Salmonella infantis* *Pseudomonas fluorescens* *Bacillus cereus* *Staphylococcus aureus* *Listeria monocytogenes* *Lactobacillus plantarum*	54 ± 8 42 ± 20 50 ± 15 101 ± 6 76 ± 2 62 ± 15 0 ± 0	Well microplates and turbidity reader	[30]
		Burkholderia multivorans *Bacillus coagulans* *Alcaligenes xylosoxydans*	29-34 mm 21 mm 23 mm	Disc diffusion and liquid culture	[28]
	Pinosylvin monomethyl ether (Scots pine knotwood)	*Escherichia coli* *Salmonella infantis* *Pseudomonas fluorescens* *Bacillus cereus* *Staphylococcus aureus* *Listeria monocytogenes* *Lactobacillus plantarum*	17 ± 7 40 ± 7 35 ± 2 92 ± 4 105 ± 12 100 ± 7 2 ± 11	Well microplates and turbidity reader	[30]
		Burkholderia multivorans *Bacillus coagulans* *Alcaligenes xylosoxydans*	15-17 mm 18 mm 13 mm	Disc diffusion and liquid culture	[28]
	Dihydropinosylvin monomethyl ether (Scots pine knotwood)	*Escherichia coli* *Salmonella infantis* *Pseudomonas fluorescens* *Bacillus cereus* *Staphylococcus aureus* *Listeria monocytogenes* *Lactobacillus plantarum*	18 ± 2 14 ± 2 22 ± 6 82 ± 3 76 ± 4 64 ± 12 1 ± 1	Well microplates and turbidity reader	[30]
		Burkholderia multivorans *Bacillus coagulans* *Alcaligenes xylosoxydans*	20-23 mm 19 mm 13 mm	Disc diffusion and liquid culture	[28]
	Resveratrol *(C. baiaessi)*	MRSA	MIC > 128 mg L⁻¹	Liquid microdilution	[15]
	Resveratrol (commercial)	*Escherichia coli*	MIC 313 mg L⁻¹ MBC 625 mg L⁻¹	Agar well diffusion	[45]
		Salmonella anatum	MIC 313 mg L⁻¹ MBC 625 mg L⁻¹		
		Bacillus cereus	MIC 313 mg L⁻¹ MBC 2500 mg L⁻¹		
		Listeria monocytogenes	MIC 625 mg L⁻¹ MBC 625 mg L⁻¹		
		Staphylococcus aureus	MIC 313 mg L⁻¹ MBC 625 mg L⁻¹		
Stilbene glucoside	Piceid (commercial)	*Escherichia coli*	MIC 625 mg L⁻¹ MBC 1250 mg L⁻¹	Agar well diffusion	[45]
		Salmonella anatum	MIC 313 mg L⁻¹ MBC 313 mg L⁻¹		
		Bacillus cereus	MIC >2500 mg L⁻¹ MBC 2500 mg L⁻¹		
		Listeria monocytogenes	MIC 625 mg L⁻¹ MBC 625 mg L⁻¹		
		Staphylococcus aureus	MIC 313 mg L⁻¹ MBC 625 mg L⁻¹		
Flavone	Apigenin	*Escherichia coli* *Salmonella enterica* *Enterococcus faecalis*	No inhibition No inhibition No inhibition	Agar well diffusion	[43]
Flavanone	Pinocembrin (Pinus cembra knotwood)	*Escherichia coli* *Salmonella infantis* *Pseudomonas fluorescens* *Bacillus cereus* *Staphylococcus aureus* *Listeria monocytogenes* *Lactobacillus plantarum*	15 ± 2 14 ± 8 7 ± 3 74 ± 6 30 ± 9 55 ± 16 4 ± 5	Well microplates and turbidity reader	[30]

Flavonol		Escherichia coli	Slight-moderate		
	Quercetin, 500 μg (commercial)	Pseudomonas aeruginosa Micrococcus luteus Bacillus subtilis Staphylococcus aureus Staphylococcus epidermis	Slight Moderate Slight Clear Clear	Hole-plate diffusion	[32]
	Quercetin	Staphylococcus aureus	MIC 2 mM	Hole-in-plate	[46]
	Quercetin	Escherichia coli Salmonella enterica	No - clear	Agar well diffusion	[43]
	Isoguercitrin (quercetin-3-glucoside)	Escherichia coli Salmonella enterica Enterococcus faecalis	No inhibition No inhibition No inhibition	Agar well diffusion	[43]
	Kaempferol, (commercial)	Escherichia coli Micrococcus luteus Bacillus subtilis Staphylococcus aureus Staphylococcus epidermis	No activity No activity Slight Clear No activity	Hole-plate diffusion	[32]
		MRSA	MIC 25->400 mg L^{-1}	Agar dilution	[19]
	Myricetin	Escherichia coli Salmonella enterica Enterococcus faecalis	Strong No inhibition No inhibition	Agar well diffusion	[43]
Flavanone		MRSA	MIC 200-400 mg L^{-1}	Agar dilution	[19]
	Naringenin (commercial)	Escherichia coli Pseudomonas aeruginosa Micrococcus luteus Bacillus subtilis Staphylococcus aureus Staphylococcus epidermis	Moderate-clear Slight Strong Clear-strong Strong Strong	Hole-plate diffusion	[32]
Flavan	Catechin	Escherichia coli Salmonella enterica Enterococcus faecalis	No inhibition No inhibition No inhibition	Agar well diffusion	[43]
Tannins	Gallic acid	Escherichia coli Proteus mirabilis Staphylococcus aureus Staphylococcus epidermis Staphylococcus haemolyticus	MIC 125 mg L^{-1} MIC 62 mg L^{-1} MIC 31 mg L^{-1} MIC 31 mg L^{-1} MIC 62 mg L^{-1}	Disc diffusion	[47]
Lignans	Matairesinol (synthesized), Hydroxymatairesinol (Norway spruce), Lariciresinol (balsam fir), Secoisolariciresinol (Brazilian pine)	Escherichia coli Salmonella infantis Pseudomonas fluorescens Bacillus cereus Staphylococcus aureus Listeria monocytogenes Lactobacillus plantarum	0 0 0 0 0 0 0	Well microplates and turbidity reader	[30]
	Isolariciresinol, Nortrachelogenin (Scots pine knotwood)	Escherichia coli Salmonella infantis Pseudomonas fluorescens Bacillus cereus Staphylococcus aureus Listeria monocytogenes Lactobacillus plantarum	0 0 0 10-13 0 0 0	Well microplates and turbidity reader	[30]

MIC = minimal inhibiting concentration, MRSA = methicillin-resistant *Staphylococcus aureus*

Table 3: Pure antibacterial substances and their activity against selected bacteria.

and flavans [22]) occur in Norway spruce (Appendix 1). Several glycosides of quercetin, isorhamnetin, kaempferol, myricetin, lericitrin and syringetin have been found in needles of Norway spruce [52]. Glucose at the 3- or 7-position is the most common glycone part of the glycosides. However, majority of the antibacterial activity data concerns aglycones and only limited data on glyciosides are available. From these flavonoids aglycones quercetin [32,43,46], kaempferol [19,32] and myricetin [43] have been observed to have antibacterial activity (Table 3), whereas quercetin-3-glucoside (quercitrin) [43] have been observed to be inactive. Silva [47] have reported that quercetin and myricetin-3-rhamnosides are inactive against *Proteus mirabilis* and *E. coli* at a concentration of 500 mg L^{-1} and against *S. aureus*, *Staphylococcus epidermidis* and *Staphylococcus haemolyticus* at a concentration of 350 mg L^{-1}. It may be that glycosylation of flavonoids reduces their antibacterial activity when compared to corresponding aglycones.

Betula genera

Phloem of silver birch contains phenolics 85.5 mg GAE g^{-1}, but bark only 2.0 mg GAE g^{-1} [34]. Kähkönen et al. [34] have found three different types of phenols in silver birch inner bark (phloem): arylbutanoid glycosides, benzoic acid derivatives and catechins, whereas Willför et al. [29] have found stilbene-derived compounds. Main phenolics in white birch and silver birch leaves have been identified by Ossipov et al. [53]. Chlorogenic acid constituted almost 50 % of the phenolic content in white birch leaves, but in silver birch leaves only 5 % of the total phenolic contents. (+)-Catechin and several quercetin, kaempferol and myricetin glycosides were the next abundant compounds identified (Appendix 1).

Structure	Compound	R1	R2	R3	R4	Trees
	Flavanones					
	Pinocembrin	H	H			P
	Naringenin	H	OH			B
	Eriodictyol	OH	OH			S
	Flavones					
	Apigenin	H	H	H		S, B
	Luteolin	OH	H	H		B
	Flavonols					
	Kaempferol	H	H	OH		P,S,B
	Quercetin	OH	H	OH		P,S,B
	Isorhamnetin	OMe	H	OH		P,S
	Myricetin	OH	OH	OH		S
	Laricitrin	OMe	OH	OH		S
	Syringetin	OMe	OMe	OH		S
	Chrysin	H	H	H	H	
	Acacetin	H	OMe	H	H	B
	Dihydro-flavanols					
	Aromadendrin	H	H	OH		S
	Taxifolin	H	OH	OH		P,S
	Ampelopsin	OH	OH	OH		S
	Flavans					
	Catechin	OH	H			P,S,B
	Epicatechin	H	OH			S
	3′-Me-Catechin	OMe	H			S
	Gallocatechin	OH	OH			S
	Anthocyanins					
	Pelargonidin	H	H			S
	Cyanidin	OH	H			S
	Peonidin	OMe	H			S
	Delphinidin	OH	OH			S

Table 4: Flavonoids found in extracts of Scots pine (S), Norway spruce (S) and silver birch (B).

The leaf extracts of white birch have been observed to inhibit clearly the growth of gram-positive bacteria *S. aureus* [54]. Omar et al. [23] found that the bark extract of *Betula papyrifera* was active against gram-positive bacteria *S. aureus*, *Bacillus subtilis*, *Enterococcus faecalis* and *Mycobacterium phlei*, whilst the wood extract showed activity only against S. aureus. None of the extracts inhibited growth of gram-negative pathogens *E. coli*, *Pseudomonas aeruginosa*, *Salmonella typhimurium* and *Klebsiella pneumonia*.

The growth of *S. aureus* was inhibited very effectively by quercetin, kaempferol and naringenin [32,54], whereas +/- -catechin and (+)-catechin were inactive against it. In addition, Tsuchiya et al. [19] observed the antibacterial activity of kaempferol and naringenin against methicillin-resistant *S. aureus* (MRSA).

Essential Oils

The primary constituents of the essential oils of conifers are terpenes [55] and when they contain additional elements, usually oxygen, they are termed terpenoids [1,20]. Monoterpenes are the most representative molecules constituting 90% of the essential oils [20]. Terpenes are derived through isoprenoid pathway in plants [21] and

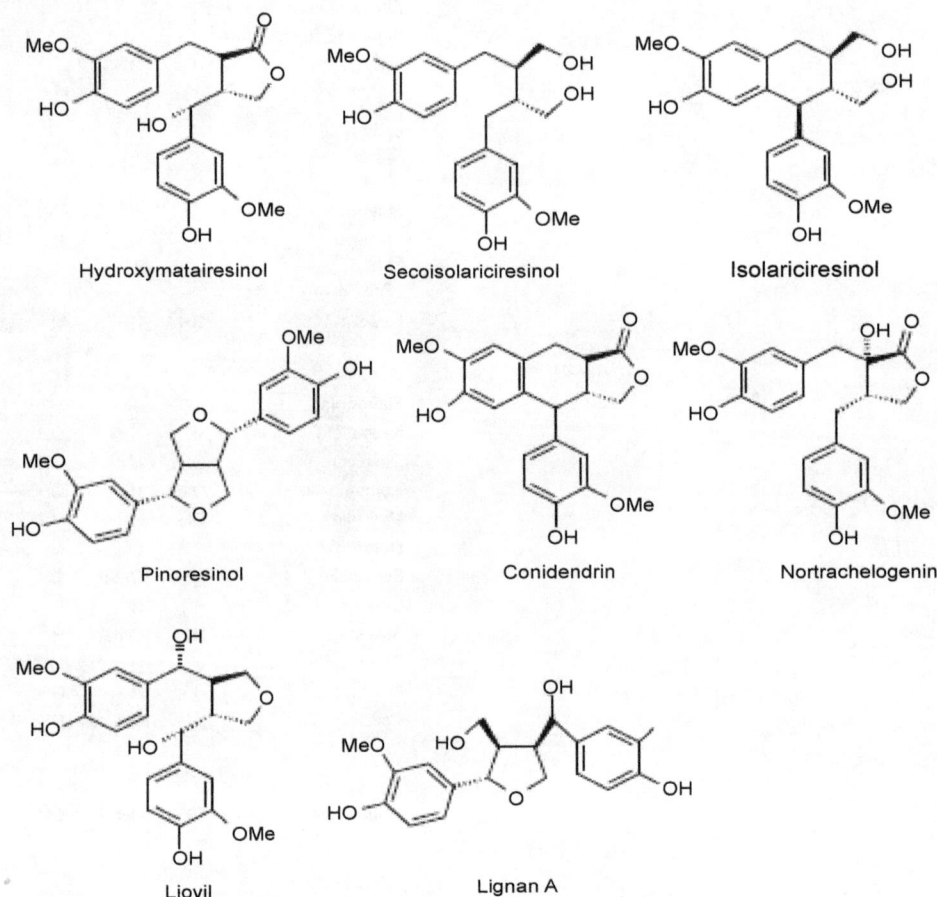

Figure 2: Lignans and oligolignans found in hydrophilic knotwood extracts [29,31].

The leaf extracts of white birch have been observed to inhibit clearly the growth of gram-positive bacteria S. aureus [54]. Omar et al. [23] found that the bark extract of Betula papyrifera was active against gram-positive bacteria S. aureus, Bacillus subtilis, Enterococcus faecalis and Mycobacterium phlei, whilst the wood extract showed activity only against S. aureus. None of the extracts inhibited growth of gram-negative pathogens E. coli, Pseudomonas aeruginosa, Salmonella typhimurium and Klebsiella pneumonia.

The growth of S. aureus was inhibited very effectively by quercetin, kaempferol and naringenin [32,54], whereas +/- -catechin and (+)-catechin were inactive against it. In addition, Tsuchiya et al. [19] observed the antibacterial activity of kaempferol and naringenin against methicillin-resistant S. aureus (MRSA).

Essential Oils

The primary constituents of the essential oils of conifers are terpenes [55] and when they contain additional elements, usually oxygen, they are termed terpenoids [1,20]. Monoterpenes are the most representative molecules constituting 90% of the essential oils [20]. Terpenes are derived through isoprenoid pathway in plants [21] and they are based on an isoprene structure (Figure 3). Different conifer species often contain the same terpenes but in different portions

[55] (Appendix 2). They can be obtained by expression, fermentation, extraction or by stream distillation that is the most commonly used method for commercial production of essential oils [56]. However, greater antibacterial activity has been observed with essential oils extracted by hexane than the corresponding steam distilled essential oils.

Essential oils of pine needles and spruce have been reported to be inactive against gram-negative bacteria but to have significant activities against gram-positive bacteria S. aureus, E. faecalis and B. subtilis, L. monocytogenes and Listeria ivanovii) [35,57] (Table 5). On the contrary, Hammer et al. [25] have noticed stronger activity against gram-negative E. coli than S. aureus.

Few studies have been published at the antimicrobial activity of terpenes present in conifer extracts. β-Pinene (from nutmeg) has been found to be particularly effective against E. coli O157:H7. Mourey and Canillac [55] studied activity of commercial terpenes, α-pinene, β-pinene, 1,8-cineole, R-limonene, S-limonene and borneol, against L. monocytogenes serovars, which is one of the most dangerous food pathogens. The terpenes studied had a significant anti-Listeria activity (Table 5). α-Pinene was the most active compound with an average MIC of 0.019-0.025% against L. monocytogenes, while 1,8-cineole was the least inhibitory and had the lowest activity against bacteria being 0.375-0.417%, although this concentration was directly bactericidal.

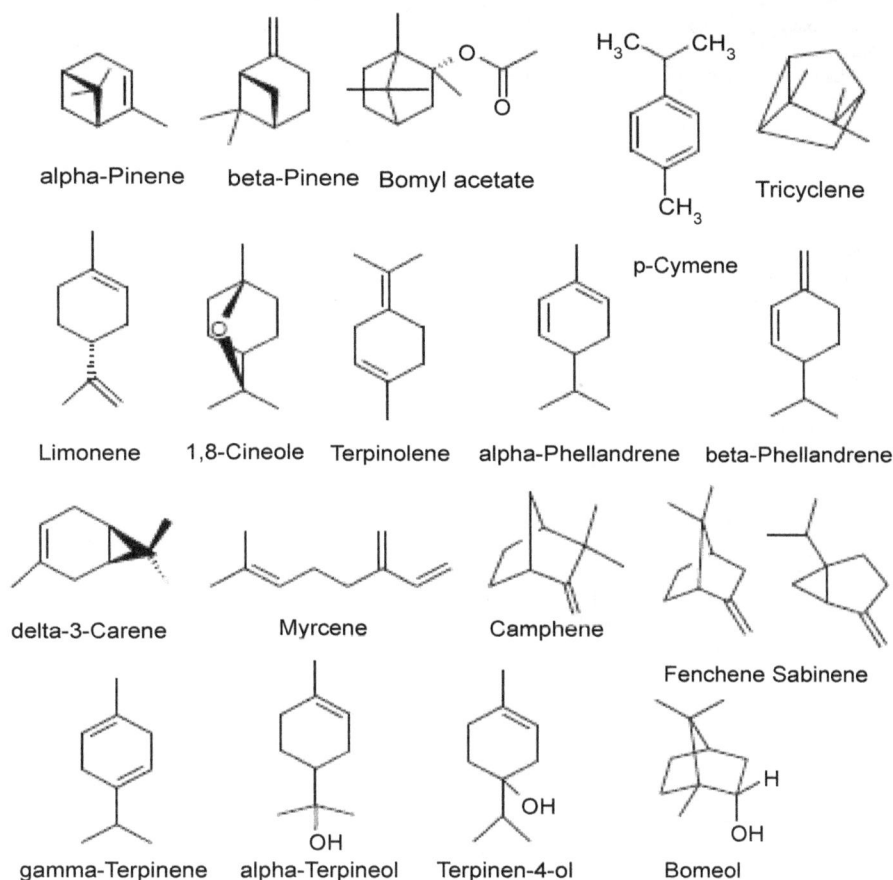

Figure 3: Structures of monoterpenes found in Scots pine and Norway spruce.

Furthermore, 1,8-cineole has exhibited low antibacterial activity against MRSA and vancomycin-resistant enterococci (VRE) *E. faecalis* [18].

Non-oxygenated monoterpene hydrocarbons, α-pinene, p-cymene and γ-terpinene have shown the least antibacterial activity among essential oil components [58-60]. Furthermore, these compounds may produce antagonistic effects and therefore, lower the antimicrobial activity of essential oil. Terpinen-4-ol from tea tree oil has been active on its own against *P. aeruginosa* and *S. aureus*, but reduced efficacy has been observed in combination with either γ-terpinene or p-cymene due to lowered aqueous solubility [61]. Also minor components in essential oil may play a role in antibacterial activity of the main component as interactions between components may lead to additive, synergistic or antagonistic effects [20,56].

Mechanisms of Activity

In general, the extracts were more active against gram-positive bacteria than gram-negative bacteria [23,28,58-66]. The main difference between gram-positive and gram-negative bacteria is the structure of their cell walls. Therefore it seems that the main target of the antibacterial activity is to destroy the cell walls of the bacteria. The gram-negative cell envelope is made up of lipopolysaccharide that renders the surface highly hydrophilic whereas the lipophilic

structure of the cell membrane of gram-positive bacteria may facilitate penetration by hydrophobic compounds [13,57,66]. Thus, flavonoids and stilbenes with lower hydroxylation should be more active against bacteria than those with the several hydroxyl groups. However, there is no clear comparability between the degree of hydroxylation and toxicity to bacteria. Either the mechanism of action of terpenes is not fully understood but is speculated to involve membrane disruption by the lipophilic compounds [1,67,68].

Conclusions

The extracts of Scots pine, Norway spruce, silver and white birches stem, bark, roots, leaves and needles contain several useful bioactive compounds that exhibit antibacterial activity against pathogens. Both phenolic extracts and essential oils are bacteriostatic against several bacteria. The main individual antibacterial phenolic compounds in Scots pine are pinosylvins that effectively inhibit growth of pathogens such as *B. cereus, S. aureus* and *L. monocytogenes*. From other phenolic compounds lignans appeared to be the least bacteriostatic and flavonoids tend to occur as glycosylated forms which have lower antibacterial activity than their aglycones. Gram-positive bacteria are generally more susceptible to plants bioactive compounds than gram-negative bacteria.

Substance	Bacteria	Inhibitory effect, %		Ref.
Scots pine, needles	*Escherichia coli*	MIC > 100 mg L^{-1}	Agar diffusion and liquid dilution methods	[35]
	Pseudomonas aeruginosa	MIC > 100 mg L^{-1}		
	Proteus mirabilis	MIC > 100 mg L^{-1}		
	Klebsiella pneumoniae	MIC > 100 mg L^{-1}		
	Staphylococcus aureus	MIC 25 mg L^{-1}, MBC >50 mg L^{-1}		
	Staphylococcus aureus	MIC > 100 mg L^{-1}		
	Staphylococcus aureus	MIC 12.5 mg L^{-1}, MBC > 50 mg L^{-1}		
	Enterococcus faecalis	MIC 3.1 mg L^{-1}		
	Bacillus subtilis	MIC 50 mg L^{-1}		
Scots pine, needles	*Escherichia coli*	MIC 2.0 % (v/v)	Agar dilution method	[25]
	Pseudomonas aeruginosa	MIC > 2.0 % (v/v)		
	Klebsiella pneumoniae	MIC > 2.0 % (v/v)		
	Staphylococcus aureus	MIC > 2.0 % (v/v)		
	Acinetobacter humanii	MIC 2.0 % (v/v)		
	Aeromonas sobria	MIC 2.0 % (v/v)		
	Enterococcus faecalis	MIC > 2.0 % (v/v)		
	Salmonella typhimurium	MIC > 2.0 % (v/v)		
	Serratia marcescens	MIC > 2.0 % (v/v)		
Norway spruce, essential oil commercial	*Escherichia coli*	>16 %	Dilution	[57]
	Klebsiella pneumoniae	≥16 %		
	Klebsiella oxytoca	≥16 %		
	Enterobacter cloacae	≥16 %		
	Staphylococcus aureus	MIC 0.022-0.061%, MBC 0.25%		
	Listeria monocytogenes	MIC 0.015-0.067%, MBC 0.20%		
	Listeria ivanovii	MIC 0.025%, MBC 0.27 %		
α-Pinene	*Escherichia coli*	MBC > 900 mg L^{-1}	Broth dilution, visible growth	[58]
	Pseudomonas aeruginosa	MBC > 900 mg L^{-1}		
	Salmonella typhimurium	MBC > 900 mg L^{-1}		
	Yersinia enterocolitica	MBC > 900 mg L^{-1}		
	Staphylococcus aureus	MBC > 900 mg L^{-1}		
	Staphylococcus epidermis	MBC > 900 mg L^{-1}		
	Enterococcus faecalis	MBC > 900 mg L^{-1}		
	Bacillus cereus	MBC > 900 mg L^{-1}		
	Listeria monocytogenes	MIC 0.019-0.025%, MBC 0.192-0.354%	Broth dilution, visible growth	[55]
	Several bacteria	No activity	Vapour diffusion test, Agar test	[59, 60]
β-Pinene	*Listeria monocytogenes*	MIC 0.041-0.060%, MBC 0.55-1.167%	Broth dilution, visible growth	[55]
R Limonene	*Listeria monocytogenes*	MIC 0.047-0.062%, MBC 0.208-0.45%	Broth dilution, visible growth	[55]
S Limonene	*Listeria monocytogenes*	MIC 0.028-0.062%, MBC 0.15-0.45%	Broth dilution, visible growth	[55]
Borneol	*Listeria monocytogenes*	MIC 0.039-0.094%, MBC 0.039-0.156%	Broth dilution, visible growth	[55]
1,8-Cineole	*Listeria monocytogenes*	MIC 0.375-0.417%, MBC 0.375-0.417%	Broth dilution, visible growth	[55]
	Escherichia coli	MIC > 8 mg mL^{-1}	Broth microdilution	[18]
	Pseudomonas aeruginosa	MIC > 8 mg mL^{-1}		
	Klebsiella pneumoniae	MIC > 8 mg mL^{-1}		
	Acinetobacter baumanii	MIC 8 mg mL^{-1}		
	Bacillus subtilis	MIC 32 mg mL^{-1}		
	Staphylococcus saprophyticus	MIC > 8 mg mL^{-1}		
	Staphylococcus aureus, MRSA	MIC 64 mg mL^{-1}		
	Staphylococcus epidermis	MIC > 8 mg mL^{-1}		
	Staphylococcus agalactiae	MIC > 8 mg mL^{-1}		
	Staphylococcus pyogenes	MIC 16 mg mL^{-1}		
	Enterococcus faecalis, VRE	MIC > 8 mg mL^{-1}		

	Bacillus acereus	Not active	Vapour diffusion	[59]
	Escherichia coli	MIC 1% v/v	N.a.	[61]
	Pseudomonas aeruginosa	MIC >8% v/v		
	Staphylococcus aureus	MIC 0.5% v/v		
p-Cymene	Escherichia coli	MBC > 900 mg L^{-1}	Broth	[58]
	Pseudomonas aeruginosa	MBC > 900 mg L^{-1}	microdilution	
	Salmonella typhimurium	MBC > 900 mg L^{-1}		
	Yersinia enterocolitica	MBC > 900 mg L^{-1}		
	Staphylococcus aureus	MBC > 900 mg L^{-1}		
	Staphylococcus epidermis	MBC > 900 mg L^{-1}		
	Enterococcus faecalis	MBC > 900 mg L^{-1}		
	Listeria monocytogenes	MBC > 900 mg L^{-1}		
	Bacillus cereus	MBC > 900 mg L^{-1}		
	Escherichia coli	MIC >8% v/v	N.a.	[61]
	Pseudomonas aeruginosa	MIC >8% v/v		
	Staphylococcus aureus	MIC >8% v/v		
Terpinen-4-ol	Escherichia coli	MIC >8% v/v	N.a.	[61]
	Pseudomonas aeruginosa	MIC >8% v/v		
	Staphylococcus aureus	MIC 0.25% v/v		
γ-Terpinene	Escherichia coli	MBC > 900 mg L^{-1}	Broth	[58]
	Pseudomonas aeruginosa	MBC > 900 mg L^{-1}	microdilution	
	Salmonella typhimurium	MBC > 900 mg L^{-1}		
	Yersinia enterocolitica	MBC > 900 mg L^{-1}		
	Staphylococcus aureus	MBC > 900 mg L^{-1}		
	Staphylococcus epidermis	MBC > 900 mg L^{-1}		
	Enterococcus faecalis	MBC > 900 mg L^{-1}		
	Listeria monocytogenes	MBC > 900 mg L^{-1}		
	Bacillus acereus	MBC > 900 mg L^{-1}		
	Staphylococcus, Micrococcus, Bacillus, Enteropacter sp.	MIC > 1 mg mL^{-1}	Agar	[60]
	Escherichia coli	MIC >8% v/v	N.a.	[61]
	Pseudomonas aeruginosa	MIC >8% v/v		
	Staphylococcus aureus	MIC 0.25% v/v		
α-Terpineol	Escherichia coli	MBC 450 mg L^{-1}	Broth dilution,	[58]
	Escherichia coli O157:H7	MBC > 900 mg L^{-1}	visible growth	
	Pseudomonas aeruginosa	MBC > 900 mg L^{-1}		
	Salmonella typhimurium	MBC 225 mg L^{-1}		
	Yersinia enterocolitica	MBC > 900 mg L^{-1}		
	Staphylococcus aureus	MBC 900 mg L^{-1}		
	Staphylococcus epidermis	MBC > 900 mg L^{-1}		
	Enterococcus faecalis	MBC 900 mg L^{-1}		
	Listeria monocytogenes	MBC > 900 mg L^{-1}		
	Bacillus acereus	MBC 900 mg L^{-1}		

Na. = not announced

Table 5: Essential oils of conifers and pure monoterpenes and -terpenoids with their activity against selected bacteria.

References

1. Cowan MM (1999) Plant Products as Antimicrobial Agents. Clin Microbiol Rev 12: 564-582.

2. Bernhoft A (2008) A brief review on bioactive compounds in plants. Bioactive compounds in plants - benefits and risks for man and animals. The Notwegian Academy of Science and Letters, Oslo, Norway.

3. Pesewu GA, Cutler RR, Humber DP (2008) Antibacterial activity of plants used in traditional medicines of Ghana with particular reference to MRSA. J Ethnopharmacol 116: 102-111.

4. Ao C, Li A, Elzaawely AA, Xuan TD, Tawata S (2008) Evaluation of antioxidant and antibacterial activities of Ficus microcarpa L. fil. Extract. Food Control 19: 940-948.

5. Aremu AO, Ndhlala AR, Fawole OA, Light ME, Finnie JF, et al., (2010) In vitro pharmacological evaluation and phenolic content of ten South African medicinal plants used as anthelmintics. S Afr J Bot 76: 558-566.

6. Kumar VP, Chauhan NS, Padh H, Rajani M (2006) Search for antibacterial and antifungal agents from selected Indian medicinal plants. J Ethnopharmacol 107: 182-188.

7. Navarro V, Villarreal ML, Rojas G, Lozoya X (1996) Antimicrobial evaluation of some plants used in Mexican traditional medicine for the treatment of infectious diseases. J Ethnopharmacol 53: 143-147.

8. Ahmad I, Beg AZ (2001) Antimicrobial and phytochemical studies on 45 Indian medicinal plants against multi-drug resistant human pathogens. J Ethnopharmacol 74: 113-123.

9. Maregesi SM, Pieters L, Ngassapa OD, Apers S, Vingerhoets R, et al., (2008) Screening of some Tanzanian medicinal plants from Bunda district for antibacterial, antifungal and antiviral activities. J Ethnopharmacol 119: 58-66.

10. Jagtap UB, Bapat VA (2010) Artocarpus: A review of its traditional uses, phytochemistry and pharmacology. J Ethnopharmacol 129: 142-166.

11. Cassidy A, Hanley B, Lamuela-Raventos RM (2000) Isoflavones, lignans and stilbenes – origins, metabolism and potential importance to human health. J Sci Food Agric 80: 1044-1062.

12. Stevanovic T, Diouf PN, Garcia-Perez ME (2009) Bioactive Polyphenols from Healthy Diets and Forest Biomass. Current Nutrition & Food Science 5: 264-295.

13. Takahashi T, Kokubo R, Sakaino M (2004) Antimicrobial activities of eucalyptus leaf extracts and flavonoids from Eucalyptus maculate. Lett Appl Microbiol 39: 60-64.

14. Tohidpour, Sattari M, Omidbaigi R, Yadegar A, J Nazemi (2010) Antibacterial effect of essential oils from two medicinal plants against Methicillin-resistant Staphylococcus aureus (MRSA). Phytomedicine 17: 142-145.

15. Adugna B, Terefe G, Kebede N (2014) Potential In vitro Anti-Bacterial Action of Selected Medicinal Plants Against Escherichia coli and Three Salmonella Species. International Journal of Microbiological Research 5: 85-89.

16. Essawi T, Srour M (2000) Screening of some Palestinian medicinal plants for antibacterial activity. J Ethnopharmacol 70: 343-349.

17. Chang S, Chen P, Chang S (2001) Antibacterial activity of leaf essential oils and their constituents from Cinnamomum osmophloeum. J Ethnopharmacol 77: 123-157.

18. Mulyaningsih S, Sporer F, Zimmermann S, Reichling J, Wink M (2010) Synergistic properties of the terpenoids aromadendrene and 1,8-cineole from the essential oil of Eucalyptus globulus against antibiotic-susceptible and antibiotic-resistant pathogens. Phytomedicine 17: 1061-1066.

19. Tsuchiya H, Sato M, Miyazaki T, Fujiwara S, Tanigaki S, et al., (1996) Comparative study on the antibacterial activity of phytochemical flavanones against methicillin-resistant Staphylococcus aureus. J Ethnopharmacol 50: 27-34.

20. Bakkali F, Averbeck S, Averbeck D, Idaomar M (2008) Biological effects of essential oils – A review. Food and Chemical Toxicology 46: 446-475.

21. Iriti M, Faoro F (2009) Chemical Diversity and Defence Metabolism: How Plants Cope with Pathogens and Ozone Pollution. Int J Mol Sci 10: 3371-3399.

22. Rice-Evans CA, Miller NJ, Paganga G (1996) Structure-antioxidant activity relationships of flavonoids and phenolic acids. Free Radic Biol Med 20: 933-956.

23. Omar S, Lemonnier B, Jones N, Ficker C, Smith ML, et al., (2000) Antimicrobial activity of extracts of eastern North American hardwood trees and relation to traditional medicine. J Ethnopharmacol 73: 161-170.

24. Omar S, Lalonde M, Marcotte M, Cook M, Proulx J, et al., (2000) Insect growth-reducing and antifeedant activity in Eastern North America hardwood species and bioassay-guided isolation of active principles from Prunus serotina. Agricult for Entomol 2: 253-257.

25. Hammer KA, Carson CF, Riley TV (1999) Antimicrobial activity of essential oils and other plant extracts. J Appl Microbiol 86: 985-990.

26. Venäläinen M, Harju AM, Saranpää P, Kainulainen P, Tiitta M, et al., (2004) The concentration of phenolics in brown-rot decay resistant and susceptible Scots pine heartwood. Wood Sci Technol 38: 109-118.

27. Harju, Venäläinen M, Anttonen S, Viitanen H, Kainulainen P, et al., (2003) Chemical factors affecting the brown-rot decay resistance of Scots pine heartwood. Trees 17: 263-268.

28. Lindberg LE, Willför SM, Holmbom BR (2004) Antibacterial effects of knotwood extractives on paper mill bacteria. Journal of Industrial Microbiology & Biotechnology 31: 137-147.

29. Willför SM, Ahotupa MO, Hemming JE, Reunanen MHT, Eklund PC, et al., (2003) Antioxidant Activity of Knotwood Extractives and Phenolic Compounds of Selected Tree Species. J Agric Food Chem 51: 7600-7606.

30. Välimaa, Honkalampi-Hämäläinen U, Pietarinen S, Willför S, Holmbom B, et al., (2007) Antimicrobial and cytotoxic knotwood extracts and related pure compounds and their effects on food-associated microorganisms. Int J Food Microbiol 115: 235-243.

31. Pietarinen SP, Willför SM, Ahotupa MO, Hemming JE, Holmbom BR (2006) Knotwood and bark extracts: strong antioxidants from waste materials. Journal of Wood Science 52: 436-444.

32. Rauha J, Remes S, Heinonen M, Hopia A, Kähkönen M, et al., (2000) Antimicrobial effects of Finnish plant extracts containing flavonoids and other phenolic compounds. Int J Food Microbiol 56: 3-12.

33. Karonen M, Hämäläinen M, Nieminen R, Klika KD, Loponen J, et al., (2004) Phenolic Extractives from the Bark of Pinus sylvestris L. and Their Effects on Inflammatory Mediators Nitric Oxide and Prostaglandin E2. J Agric Food Chem 52: 7532-7540.

34. Kähkönen MP, Hopia AI, Vuorela HJ, Rauha J, Pihlaja K, et al., (1999) Antioxidant Activity of Plant Extracts Containing Phenolic Compounds. J Agric Food Chem 47: 3954-3962.

35. Kartnig T, Still F, Reinthaler F (1991) Antimicrobial activity of the essential oil of young pine shoots (Picea abies L.). J Ethnopharmacol 35: 155-157.

36. Zeneli G, Krokene P, Christiansen E, Krekling T, Gershenzon J (2006) Methyl jasmonate treatment of mature Norway spruce (Picea abies) trees increases the accumulation of terpenoid resin components and protects against infection by Ceratocystis polonica, a bark beetle-associated fungus. Tree Physiol 26: 977-988.

37. Viiri H, Annila E, Kitunen V, Niemelä P (2001) Induced responses in stilbenes and terpenes in fertilized Norway spruce after inoculation with blue-stain fungus, Ceratocystis polonica. Trees 15: 112-122.

38. Lieutier F, Brignolas F, Sauvard D, Yart A, Galet C, et al., (2003) Intra- and inter-provenance variability in phloem phenols of Picea abies and relationship to a bark beetle-associated fungus. Tree Physiol 23: 247-256.

39. http://www.academia.edu/741832/Phenolic_compounds_in_different_parts_of_fir

40. Pan H, Lundgren LN (1995) Phenolic extractives from root bark of Picea abies. Phytochemistry 42: 1423-1428.

41. Hovelstad H, Leirset I, Oyaas K, Fiksdahl A (2006) Screening Analyses of Pinosylvin Stilbenes, Resin Acids and Lignans in Norwegian Conifers. Molecules 11: 103-114.

42. Li SH, Niu XM, Zahn S, Gershenzon J, Weston J, et al., (2008) Diastereomeric stilbene glucoside dimers from the bark of Norway spruce (Picea abies). Phytochemistry 69: 772-782.

43. Puupponen-Pimiä R, Nohynek L, Meier C, Kähkönen M, Heinonen M, et al., (2001) Antimicrobial properties of phenolic compounds from berries. J Appl Microbiol 90: 494-507.

44. Baurhoo B, Ruiz-Feria CA, Zhao X (2008) Purified lignin: Nutritional and health impacts on farm animals—A review. Anim Feed Sci Technol 144: 175-184.

45. Shan B, Cai YZ, Brooks JD, Corke H (2007) The in vitro antibacterial activity of dietary spice and medicinal herb extracts. Food Chem 117: 112-119.

46. Ibewuike JC, Ogungbamila FO, Ogundaini AO, Okeke IN, Bohlin L (1997) Antiinflammatory and antibacterial activities of C-methylflavonols from Piliostigma thonningii. Phytother Res 11: 281-284.

47. Ji Y, Zhou Y (2013) Mathematical Modelling of Protein Precipitation Based on the Phase Equilibrium for an Antibody Fragment from E. coli Lysis. J Bioprocess Biotech 3: 129.

48. Beninger CW, Abou-Zaid MM, Kistner ALE, Hallett RH, Iqbal MJ, et al., (1997) A Flavanone and Two Phenolic Acids from Chrysanthemum morifolium with Phytotoxic and Insect Growth Regulating Activity. Journal of Chemical Ecology 30: 589-606.

49. Eklund PC, Willför SM, Smeds AI, Sundell FJ, Sjöholm RE, et al., (2004) A New Lariciresinol-Type Butyrolactone Lignan Derived from Hydroxymatairesinol and Its Identification in Spruce Wood. J Nat Prod 67: 927-931.

50. Malá J, Hrubcová M, Máchová P, Cvrcková H, Martincová O, et al., (2011) Changes in phenolic acids and stilbenes induced in embryogenic cell cultures of Norway spruce by two fractions of Sirococcus strobilinus mycelia. J For Sci 57: 1-7.

51. Münzenberger B, Heilemann J, Strack D, Kottke I, Oberwinkler F (1990) Phenolics of mycorrhizas and non-mycorrhizal roots of Norway spruce. Planta 182: 142-148.

52. Slimestad R, Hostettmann K (1996) Characterisation of Phenolic Constituents from Juvenile and Mature Needles of Norway Spruce by Means of High Performance Liquid Chromatography–Mass Spectrometry. Phytochem Anal 7: 42-48.

53. Ossipov V, Nurmi K, Loponen J, Prokopiev N, Haukioja E, et al. (1995) HPLC isolation and identification of flavonoids from white birch Betula pubescens leaves. Biochem Syst Ecol 23: 213-222.

54. Chu X, Zhen Z, Tang Z, Zhuang Y, Chu J, et al.(2012) Introduction of Extra Copy of Oxytetracycline Resistance Gene otrB Enhances the Biosynthesis of Oxytetracycline in Streptomyces rimosus. J Bioprocess Biotechniq 2: 117.

55. Mourey A, Canillac N (2002) Anti-Listeria monocytogenes activity of essential oils components of conifers. Food Control 13: 289-292.

56. Burt C (2004) Essential oils: their antibacterial properties and potential applications in foods—a review Int. J Food Microbiol 94: 223-253.

57. Canillac N, Mourey A (2001) Antibacterial activity of the essential oil of Picea excelsa on Listeria, Staphylococcus aureus and coliform bacteria. Food Microbiol 18: 261-268.

58. Cosentino S, Tuberoso CIG, Pisano B, Satta M, Mascia V, et al. (1999) In-vitro antimicrobial activity and chemical composition of Sardinian Thymus essential oils. Lett Appl Microbiol 29: 130-135.

59. López P, Sánchez C, Batlle R, Nerín C (2007) Vapor-Phase Activities of Cinnamon, Thyme, and Oregano Essential Oils and Key Constituents against Foodborne Microorganisms. J Agric Food Chem 55: 4348-4356.

60. Moleyar V, Narasimham P (1992) Antibacterial activity of essential oil components. Int J Food Microbiol 16: 337-342.

61. Cox SD, Mann CM, Markham JL (2001) Interactions between components of the essential oil of Melaleuca alternifolia. J Appl Microbiol 91: 492-497.

62. Takikawa A, Keiko A, Yamamoto M, Ishimaru S, Yasui M, et al. (2002) Antimicrobial activity of Nutmeg against Escherichia coli O157. Journal of Bioscience and Bioengineering 94: 315-320.

63. Cimanga K, Kambu K, Tona L, Apers S, De Bruyne T, et al., (2002) Correlation between chemical composition and antibacterial activity of essential oils of some aromatic medicinal plants growing in the Democratic Republic of Congo. J Ethnopharmacol 79: 213-220.

64. Eyles A, Davies NW, Mohammed C (2003) Novel Detection of Formylated Phloroglucinol Compounds (FPCs) in the Wound Wood of Eucalyptus globulus and E. nitens. Journal of Chemical Ecology 29: 881-898.

65. Gilles M, Zhao J, Min A, Agboola S (2010) Chemical composition and antimicrobial properties of essential oils of three Australian Eucalyptus species. Food Chem 119: 731-737.

66. Tyagi AK, Malik A (2011) Antimicrobial potential and chemical composition of Eucalyptusglobulus oil in liquid and vapour phase against food spoilage microorganisms. Food Chem 126: 228-235.

67. Cox SD, Mann CM, Markham JL, Bell HC, Gustafson JE, et al., (2000) The mode of antimicrobial action of the essential oil of Melaleuca alternifolia (tea tree oil). J Appl Microbiol 88: 170-175.

68. Lambert RJW, Skandamis PN, Coote PJ, Nychas GE (2001) A study of the minimum inhibitory concentration and mode of action of oregano essential oil, thymol and carvacrol. J Appl Microbiol 91: 453-462.

Perspectives on Computational Structural Bio-Systems

Stefano Piccoli[1] and Alejandro Giorgetti[1,2]*

[1]*Applied Bioinformatics Group, Dept. of Biotechnology, University of Verona, Italy*
[2]*German Research School for Simulation Sciences, Jülich, Germany*

Abstract

Non-molecular systems biology is aimed at the prediction of the functional features of bio-systems on the basis of known cell proteomes and interactomes. Understanding the interactions between all the involved molecules is therefore the key for gaining a deep understanding of such processes. Albeit many thousands of interactions are known, accurate molecular insights are available for only a small fraction of them. The difficulties found in the resolution of atomic level structures for interacting pairs, make the predictive power of molecular computational biology methods essential for the advancement of the field. Indeed, bridging the gap formed due to the lack of structural details can therefore transform systems biology into models that more accurately reflect biological reality.

Introduction

Molecular biology, in the genome era, does not refer to studies involving just single macromolecules, it actually involves the study of complete cellular pathways, and why not, even entire organisms. Indeed, the world-wide genome-sequencing projects revolutionized the field and are producing unimaginable amount of biological data, providing a near complete list of the components that are present in an organism. Furthermore, the post-genomic projects that have the main scope of offering the scientific community with the relationships between all these components, are giving more and more data, not always of high quality.

A natural consequence of the outcomes of these large efforts was the birth, several years ago, of the "Systems Biology" field, mainly devoted to the unraveling and understanding of metabolic and signaling pathways or gene-regulatory networks. Systems biology relies on a detailed knowledge of protein–metabolite, protein–protein and protein-nucleic-acid interactions at the cellular level. In this regard, it is also important to take into account that, at the very bases, all of these processes involve molecular recognition: for example the process by which two or more molecules interact to form a specific complex. Albeit far from being completely characterized [1], the complex formation processes are surely dominated by short-range, often transient, interactions at the contact surface of the molecules. Furthermore, conformational changes and assemblies of very large macromolecular complexes, which can be propagated through long distances (tens of angstroms), are the effect of local interactions between small molecules (like messengers) or macromolecules with their cellular targets. In conclusion, small molecules control an enormous amount of cellular functions by binding to their target macromolecules, firing complex cellular pathways characterized by reactions, environmental changes, intermolecular interactions, and allosteric modifications.

In short, a deep understanding of the molecular basis of ligand-target interactions requires the integration of biological complexes into cellular pathways, that is "systems biology". In the other hand, systems biology (see, e.g., [2,3]), needs to be accompanied by a quantitative molecular description of pathways, so far most lacking. This novel area of investigation will impact strongly on pharmaceutical sciences and toxicology, as drugs target (and mutations affect) pathways, rather than a single biomolecule. It is also crucial in areas such as nanobiotechnology and in bioprocessing techniques.

World-wide projects such as the structural genomics, that are pushing forward the entire field of structural biology, will be able in a near future to produce such important results that it will be difficult to find a single protein for which no structural information is available or for which structural information is not readily accessible by straightforward [4]. It is probable that a near-complete structural picture will be available for most of the proteins in any given organism soon. However, structural biology remains limited in terms of the outcomes that it can offer to the community and still find difficulties when dealing with big macromolecular complexes. Large protein complexes or whole systems still require years of study for a detailed structural understanding to be reached. Thus, computational biology complemented with X-ray crystallography and carefully designed molecular biology experiments may be the key to face these difficulties.

Results and Discussion

Known structures, predicted interactions and predicted binding sites can greatly illuminate the understanding of a pathway. Very recently [5] we have tried to give structural detail to different proteins and complexes present on the pathway fundamental for the degradation of polycyclic aromatic hydrocarbons (PAHs). PAHs are widespread in the environment and persist over long periods of time: many polycyclic aromatic hydrocarbons (PAHs) are largely suspected to be mutagenic or carcinogenic [6] and their contamination in soil and aquifer is of great environmental concern. Among the PAHs, dibenzothiophene (DBT) represents the prevailing compound, and is generally considered as a model chemical structure in studies dealing with either biodegradation of organo-sulfur contaminants by petroleum biodesulfurisation through the "4-S pathway" [7] or through the "Kodama pathway" [8-10] Figure 1. When degraded through the latter pathway, the molecule is transformed into 3-hydroxy-2-formylbenzothiophene (HFBT). Di Gregorio and collaborators, very recently have identified a novel genotype for the initial steps of the oxidative degradation of dibenzothiophene, found in the bacterial strain *Burkholderia fungorum* DBT [11].

*****Corresponding author:** Alejandro Giorgetti, Ca'Vignal1, strada'LeGrazie'15, 37134Verona, Italy

Burkholderia fungorum DBT1 is a NON-PATHOGENIC strain capable of transforming DBT completely through the "Kodama pathway" with higher efficiency than other microorganisms. The first step of the Kodama pathway was shown to be catalyzed by the Rieske Oxygenase (RO) systems [12], forming *cis*-dihydrodiols from a large variety of substrates and molecular oxygen (dioxygen).

Detail of the initial step: RO systems (in the box, composed by reductase, ferredoxin and an oxigenease) use electrons from NAD (P) H to activate molecular oxygen, which is then used to oxidize the substrate. The reductase component liberates electrons from NAD (P) H and transfers the electrons to the ferredoxin, that shuttles the electrons to the oxygenase, where they are used in catalysis. This component consists of an alpha subunit, which contains both a Rieske binding domain and a catalytic domain. A beta subunit is present, which is believed to primarily function as a stabilizer for the alpha subunits. Rieske Oxygenase (RO) systems of *Burkholderia fungorum* DBT1 is called dibenzothiophene dioxygenase [11]. Homology models were produced for each of the proteins shown. This structural information covers only a small fraction of the possible interactions between these families of ligands and receptors.

While structural studies have been performed on a number of ROs, no crystal structure exists for the DBT1 enzymes. Dibenzothiophene dioxygenase is able to degrade a wide spectra of molecules, including naphthalene, phenanthrene and DBT [13]. For its particular characteristics, Burkholderia fungorum DBT1 might be interestingly exploited in bioremediation protocols of PHA-contaminated sites. Our initial studies comprised the structural modeling of all the players of the initial step in the Kodama pathway for the degradation of PAHs by the *Burkholderia fungorum* DBT1 strain. The availability of several structural evolutionary correlated templates, covering the entire RO complex gave us the possibility of building not only the structural models of the isolated components but also to gain insight into the big protein complexes involved in the process. The possibility of modeling one of the most important proteins complexes and its validation with experiments extracted from literature, prompted us to hypothesize that more refined models will offer more a important overview of the system under study and may allow the full characterization of the entire pathway Figure1.

Combining pathways with 3D details ultimately makes them more useful for the complete characterization of bio-systems. Indeed, if the nature of an interaction is known, then it may be easier to estimate the affinity of the association. Crystal structures or models can give clues about the interaction regions, indeed, they can also give insights into the order of interaction events in a pathway, by indicating which interactions cannot occur simultaneously or which of them may compete for a similar binding region. It may also provide elements for an intelligent design of inhibitors, providing a rational basis for deciding how to interfere with a pathway.

The previous example illustrates how known structures, when combined with modeling, can provide insights into the interacting components of a well-studied pathway. However, a more interesting possibility is to use interaction modeling/prediction as a means to propose new pathway elements, or indeed pathways that are completely new. Joining this information with methods for prediction of protein interactions can be used to give insights into the molecular details of the interactions, and therefore to give some guidance regarding the order of events (see above). Such approaches might also allow us to determine whether clusters of interacting proteins correspond to a single large complex or to a set of proteins that belong to a pathway.

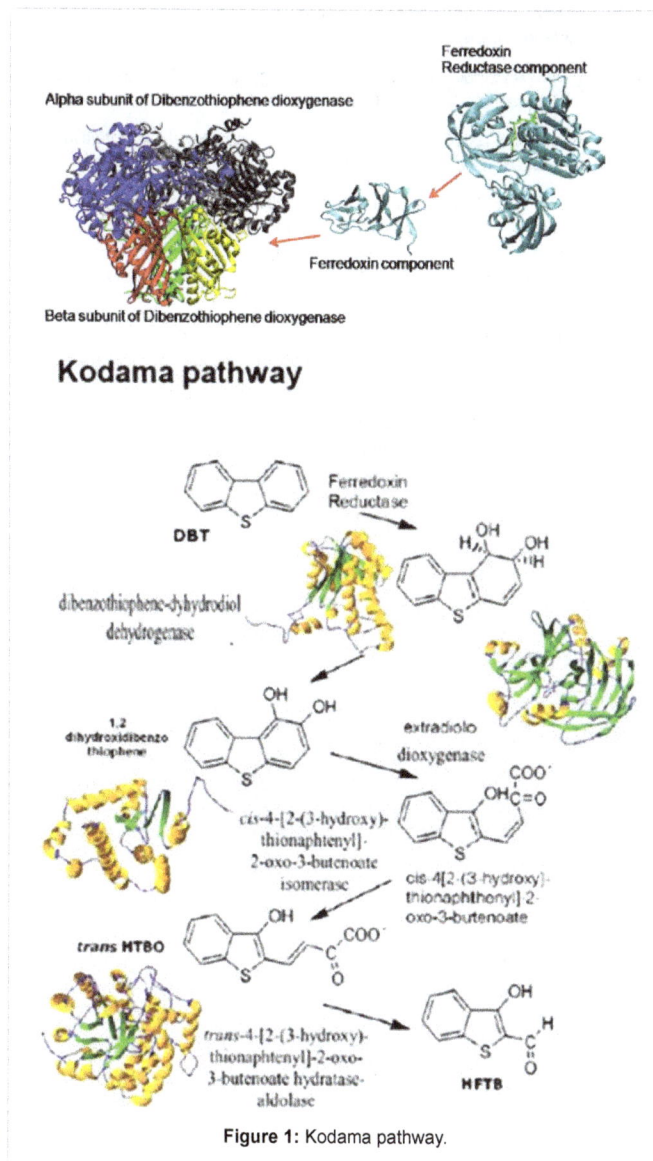

Figure 1: Kodama pathway.

A last word

Proteomes and interactomes, albeit extremely useful for the characterization at a macro level of the functioning of a whole cell, provide a rather abstract network of macromolecules. However, the distance to a real physical picture is very large. Indeed, a more concrete description of the cell networks will arrive when complete interaction pathways will be complemented with all-atom three dimensional structures of protein complexes. This kind of information will confer an extremely important novel role to experimental/computational structural biology in the field of systems biology. Indeed, structural information for interacting cellular components will produce a more and more complete all-atom-detail scaffold for the characterization of entire cellular pathways, which will be of immense benefit to anybody studying or modeling biological systems.

References

1. Whitesides GM, Snyder PW, Moustakas DT, Mirica KA (2008) Designing ligands to bind tightly to proteins. In Physical Biology: From Atoms to Medicine, Zewail A H Imperial College Press, London, UK.

2. Segrè D, Vitkup D,Church GM (2002) Analysis of optimality in natural and perturbed metabolic networks. Proc. Natl. Acad. Sci. U.S.A. 99: 15112–15117.

3. Alon U (2006) An Introduction to Systems Biology: Design Principles of Biological Circuits. Chapman & Hall/CRC, Boca Raton, FL.

4. Aloy P, Russell RB (2006) Structural systems biology: modelling protein interactions. Nat Rev Mol Cell Biol. 7: 188-97.

5. Piccoli S, Lampis S, Vallini G Giorgetti A (2011) Structural Characterization of the Rieske Oxygenase Complex from Burkholderia fungorum DBT1 strain: Insights from bioinformatics. Proceedings of BIOTECHNO 2011: The Third International Conference on Bioinformatics, Biocomputational Systems and Biotechnologies.

6. Fujikawa K, Fort FL, Samejima K, Sakamoto Y (1993) Genotoxic potency in Drosophila melanogaster of selected aromatic amines and polycyclic aromatic hydrocarbons as assayed in the DNA repair test. Mutat Res

7. GallagherJR, Olson ES, Stanley DC (1993) Microbial desulfurization of dibenzothiophene: a sulfur specific pathway. Fems Microbiol Lett 107: 31-36.

8. Kodama K, Nakatani S, Umehara K, Shimizu K, Minoda Y, et al. (1973) Identification of microbial products from dibenzothiophene and its proposed oxidation pathway. Agric Biol Chem 37: 45-50.

9. Kodama K, Nakatani S, Umehara K, Shimizu K, Minoda Y, et al. (1970) Microbial conversion of petrosulfur compounds Part III Isolation and identification of products from dibenzothiophene .Agric Biol Chem 34: 1320-1324.

10. Kropp KG , Fedorak PM (1998) A review of occurrence, toxicity, and biodegradation of condensed thiophenes found in petroleum. Can J Microbiol 44: 605-622.

11. Di Gregorio S, Zocca C, Sidler S, Toffanin A, Lizzari D, et al. (2004) Identification of two new sets of genes for dibenzothiophene transformation in Burkholderia sp. DBT1. Biodegradation 15:111–123.

12. Habe H, Omori T (2003) Genetic of polycyclic aromatic hydrocarbon metabolism in diverse aerobic bacteria. Biosci Biotechnol Biochem 67: 225–243.

13. Andreolli M, Lampis S, Zocca C, Vallini G (2008) Biodegradative potential of Burkholderia fungorum DBT1 in the abatement of polycyclic aromatic hydrocarbons. Proceedings of the 4th European Bioremediation Conference, Chania, Crete, Greece.

A Cost Effective Strategy for Production of Bio-surfactant from Locally Isolated *Penicillium chrysogenum* SNP5 and Its Applications

Gunjan Gautam, Vishwas Mishra, Payal Verma, Ajay Kumar Pandey and Sangeeta Negi*

Department of Biotechnology, Motilal Nehru National Institute of Technology, Allahabad, India

Abstract

The current work is enlightened about a cost effective bioprocess using one factor at a time approach for the production of bio-surfactant through solid state fermentation. A fungal strain *Penicillium chrysogenum* SNP5 isolated from grease contaminated soil was reconnoitered for the production of bio-surfactant. Various physiochemical parameters i.e., substrate composition, nitrogen supplements, extraction media and pH were optimized in order to optimized the production in terms of emulsification index and oil displacement assay. Maximum oil displacement area produced using grease waste and wheat bran (1:1 w/w), waste cooking oil and wheat bran (1:1 v/w) as a substrate were 3.5 cm and 5 cm, respectively. Whereas, considered values for emulsification activity with oil and diesel were 43% and 22% during optimization of substrate composition. Variable ratios of grease waste and wheat bran were capable to enhance the emulsification activity with oil and diesel up to 45% and 24% in presence of grease and wheat bran (1.5:1). The strain also showed enhancement of emulsification activity 45% and 23% with oil and diesel respectively to utilized yeast extract as a nitrogen source and the highest emulsification activity 38% in diesel, 47% in oil and oil displacement 5.5 cm was found at pH 8 with grease and wheat bran as a substrate. Preliminary characterizations by thin layer chromatography showed that the bio-surfactant was lipopeptide in nature and was also confirmed through FTIR analysis. Metabolization of industrial grease waste through solid state fermentation has never been reported before for the production of bio-surfactants therefore would be applicable in petroleum and biodiesel industry. The partially purified biosurfactants was further investigated for antimicrobial activity and enhanced oil recovery. It displayed effective zones of inhibition against both gram +ve (1.67 cm) and gram –ve (1.93 cm) as well as 16.5% enhanced recovery of oil. Both results also give a positive support to its role in pharmaceuticals as well as in petroleum and oil industry.

Keywords: Biosurfactant; Lipopeptides; Solid state fermentation; FTIR; Production and extraction

Introduction

Science, mysteries, miracles, inventions and Eureka!" all are required for invigoration of a new approach to develop a product. Biosurfactant (BS) word implies to "A surface active substance which synthesized from microbes to metabolize water insoluble substance such as hydrocarbons and lipids etc". BSs are amphiphilic compounds contain hydrophobic and hydrophilic moieties that reduce surface and interfacial tensions between individual molecules at the surface and interface, respectively. BS are generally characterized as anionic (usually due to sulphonate or sulphur group), cationic (positively charged quaternary ammonium group) and nonionic (lack their ionic constituents). They may be characterized in terms of molecular weight as high mass containing BS (polymeric and particulate surfactants) or low mass BSs (i.e., lipopeptides, glycolipids and phospholipids etc.) [1,2].

Unlike petroleum based surfactant which are usually non biodegradable thus remain toxic to the environment, biosurfactants have many advantages in terms of biodegradability, biocompatibility, digestibility, low toxicity, availability of raw material for production and surface activeness (can lower down surface tension), which allows their application in cosmetic, pharmaceuticals, as functional food additives, agriculture, medicine, petroleum and industry etc.

Due to this increasing awareness on the need to protect the ecosystem and mankind, necessitate an increased interest in surfactants of microbial origin as possible alternatives to chemically synthesized ones [3,4]. According to a new market report of transparency market research global market of BSs was worth USD 1,735.5 million in 2011 and is expected to attain USD 2,210.5 million in 2018, growing at a CAGR of 3.5% from 2011 to 2018 with leading position of European region sharing 53.3% of global BSs market revenue share [5].

To meet such boost in demand of BS production must be cost effective and at industrial scale. Although scale up process has lacunae due to its high production cost with low rate of production compare to synthetic surfactant. So it is very imperative that our target should be relying on development of successive effort and strategies in such a way that our needs replenish the constancy of nature without any major changes in next generation [6] and environmental sustainability emphasizing, the production of eco-friendly natural bioactive compounds from renewable substrates those have the potential to replace chemically synthesized surfactants [7].

In this concern various microorganisms have been explored for the production of BS and major of studies used submerged fermentation. However, it leads to a main disadvantage of foaming which further has a tendency to accumulate micro-organisms on foam and finally lead to washing out of cells from media. Instead of that, solid state fermentation represents a promising future of production with two main advantages firstly, use of low cost substrate for production and secondly it evades foaming problems [8,9].

The present work is centered on achieving a cost effective and

*Corresponding author: Sangeeta Negi, Assistant Professor, Department of Biotechnology, National Institute of Technology, Allahabad, 211004, India

optimum amount of BS production by *P. chrysogenum* SNP5 isolated from grease contaminated soil, using one at a time approach to optimize variable features like substrates compositions, nitrogen sources, substrate ratio, solvent system etc. isolated surfactant were identified through thin layer chromatography method and FTIR. In this study grease waste along with wheat bran has been taken as substrate for the production of BS, which acted as stimulator of the synthesis of hydrophobic moiety as well as protein structure of BS, which has been never reported earlier.

Material and Methods

Organism and maintenance of inoculums

A fungal strain of *Penicillium chrysogenum* SNP5 was isolated from soil contaminated grease waste. The strain *P. chrysogenum* SNP5 was reported as good producer of lipases (46U/ml at 30°C) using Solid State Fermentation (SSF) [10]. The microorganism was maintained on Malt Extract Agar (MEA) slope with 1% v/v glycerol at 4°C.

Media and inoculum preparation:

50 ml Czepek- dox media(NaNO₃ 2.5 g/l,KH₂PO₄ 1.0 g/l, MgSO₄·7H₂O 0.5 g/l, KCl 0.5 g/l) supplemented with 1% glucose and 5% glycerol as carbon sources incubated with 10^7 spore/ml of a culture of *P. chrysogenum* and allowed to grow at 30°C for next 5-6 days. The next of 6th day the culture was vortexed for the separation of spores and mycelium. Spores containing mycelium free supernatant was further used as inoculate for production of BSs.

SSF for BSs production:

The substrate was prepared with wheat bran, grease waste at different ratios and supplemented with Czepek-dox media (NaNO₃ 2.5 g/l, KH₂PO₄ 1.0 g/l, MgSO₄·7H₂O 0.5 g/l, KCl 0.5 g/l) in 1:2 (w/v) ratios in 250 ml Erlenmeyer flasks to provide the basic micronutrients. The culture medium was autoclaved at 121°C for 20 min. after the sterilization medium was inoculated with 2 ml of spore suspension (10⁷spores/ml) and incubated for 10 days at 30°C.

Optimization of process parameters:

The production optimization was regulated in a series of experiments, changing one parameter at a time while keeping other variables constant at a particular set of conditions. Five factors were considered to achieve high production of BSs: the substrate sources (C), nitrogen sources (N), C/N ratio, solvent systems, pH. The substrate sources used were grease waste, waste cooking oil, soyabean oil and oil cake with NaNO₃ as a nitrogen source. In case of nitrogen sources NaNO₃, yeast extract, peptone and malt extract were used at a concentration of 1% with optimum carbon source. Variation in C/N ratio was applied through changes in optimum carbon sources from 1, 1.5, 2, 2.5 and3 by keeping constant nitrogen source. pH ranges were varied from 3-12 for optimization of higher production (Table 1).

Analytical examinations

Collection of extract: 5 g of fermented bran were extracted with 1:3 ratios of distilled water at 30°C followed by extraction with appropriate filtration and centrifugation at 10000 for 10 min at 4°C for the separation of solids and spores. Recovery of BS was expressed in amount of crude extract obtained per gm. of dry substrate (gds) [6].

Determination of Emulsification Index (EI): Emulsification activity was observed with respect of oil and hydrocarbon. By adding

Substrate	Ratio of substrate (w/w)	EI (oil) (%)	EI(diesel) (%)	ODA(cm)	DCA
Grease waste + wheat bran	1:1	43 ± 0.12	22 ± 0.01	3 ± 0.15	++
waste cooking oil+ wheat bran	1:1	40 ± 0.15	20 ± 0.05	5 ± 0.90	++
soybeans oil+ wheat bran	1:1	53 ± 0.20	37 ± 0.04	5 ± 0.05	+++
oil cake+ wheat bran	1:1	25 ± 0.62	10 ± 0.90	2 ± 0.01	+

Table 1: Influence of substrate sources for biosurfactant production in terms of EI: Emulsification Index; ODA: Oil Displacement Assay; DCA: Drop Collapse Assay; (+: less dispersed drop; ++: moderate dispersed drop; +++ highly dispersed drop of surfactant).

2 ml of oil or hydrocarbon to the same amount of extract, subjected to vortexed for 2 min and keeping stand for 24 hours. After 24 hours the emulsification activity was determined by the measurement of percentage of height of emulsified layer (mm) divided by total height of the liquid (mm) [11].

$$EI = \frac{emulsion\ height}{total\ height} \times 100 \quad EI = \frac{emulsion\ height}{total\ height} \times 100$$

Drop collapse assay: Drop collapse assay depends on destabilization of a liquid on a solid surface due to presence of surfactant. Drop of collected extract was placed on an oil coated glass slide and the destabilization of liquid drop was observed against distilled water used as a control [12].

Oil displacement assay: 40 ml of distilled water was added in 90 mm petri plate and then 50 μl of oil was added on surface to make a thin layer after that it was followed by addition of 10 μl of collected extract on middle of surface. Diameter of displaced oil was observed after 30 sec under visual light 10 μl distilled water used as control [13] (Figure 1).

Foam height analysis: 10 ml of collected extract was shacked vigorously for 10 min to attain a good frothing and left to stand for 2 minutes. Foaming capacity was calculated according to the following equation [14]:

Foaming = emulsion height / total height × 100

Optimization of BS extraction

For estimation of the most appropriate solvent system for the extraction of BSs: various buffer from range of 3-10 (sodium citrate buffer; 3-6, sodium phosphate buffer; 7-8, sodium carbonate buffer; 9-10 and water) were employed. BS production was further studied with emulsification assay, oil displacement assay and drop collapse assy.

Partial purification of surfactant

Partial purified extraction of surfactant was done through ethanol precipitation method at -20°C for16 h. After 16 hour precipitate was collected by centrifugation at 4000×g for 15 min at 4°C [15] resulted pellet was treated with hexane to remove excess grease traces and BS was collected in form of pellet after air drying.

Preliminary compositional analysis

Thin layer chromatography (TLC): A part of crude BS was separated on a TLC Sheet using CHCl₃: CH₃OH: H₂O (65:25:4. v/v/v)

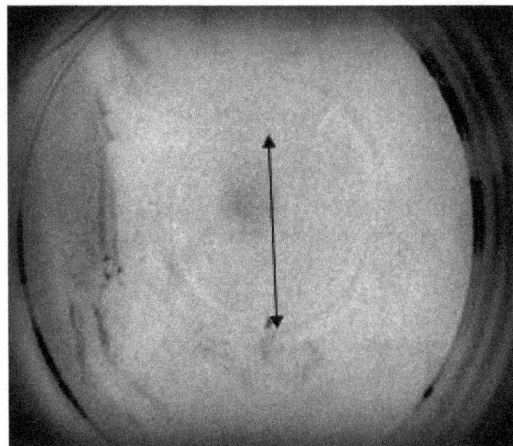

Figure 1: Oil displacement assay. The arrow shows the displaced area of oil (grease and wheat bran ration 1.5:1).

as a solvent system with different color developing reagents. Ninhydrin reagent (0.5 g ninhydrin in 100 ml anhydrous acetone) was used to detect lipo-pepetide kind of BS as a red spot [16,17] and iodine vapors was used to develop lipid moiety of surfactant as a yellow spot.

Structural characterization

Lowery assay: Presence of protein concentration was assayed by lowery method with the help of standard curve BSA protein [18].

Fourier transform Infra-red spectra analysis: For detection of functional group IR analysis was done by KBr pellet method, using resolution of 2 cm^{-1} and 4000-400 cm^{-1} spectral region [19].

Antimicrobial activity

Anti microbial activity of partially purified BS evaluated using agar diffusion method. 20 ml Muller Hinton Agar media was prepared each for petri plates on each of which 2 wells were made and were named as A and B respectively. The plates were swabbed with *P. aeruginosa and S. aureus*. To the wells A, 50 µl of partially purified BS was added, to and to the wells B, distilled water (control) was added. The plates were kept in incubation at 37°C for 24 hours. The presence of clear zone marked the antimicrobial activity of BS. Three readings of the clear zone diameter were taken for each well and the mean was calculated to determine the actual zone diameter [20].

Application of BS in oil recovery

To evaluate the efficiency of crude BS produced by *P. chrysogenum SNP5* in oil recovery process, sand pack column as a laboratory scale technique was employed. A glass column height 17 cm was packed with 17 g of acid pretreated sand. The brine solution (5% NaCl, w/v) was then passed through the column and Pore Volume (PV) was determined by measuring the volume required to make the sand matrix wet in brine solution. After saturating column with brine solution, two stroke engine oil (Honda) was passed through the column until the column got saturated with oil. Once, oil entered in the column, discharge of brine solution was observed from the matrix of sand. The discharged volume of brine from sand pack column was collected and measured to calculate initial oil saturation (Soi). Then again oil saturated column was washed with brine solution until no further oil was discharged in the effluent. The oil retained i.e. residual oil saturation (Sor) after brine solution wash was calculated on the basis of oil loaded and oil

discharged in the effluent from column. BSs was then injected and kept it for 24 hours. After 24 hours residual oil was calculated with following equations.

$$\text{Initial oil saturation (Soi \%)} = \frac{OOIP}{PV} \times 100$$

Pore volume (PV) (mL) = Volume of brine required to saturate the column.

Original oil in place (OOIP) (mL) = Amount of brine solution discharged upon displacement by oil sand pack column.

$$\text{Initial water saturation (Swi \%)} = \frac{X}{PV} \times 100$$

Where X = pore volume- volume of brine displaced after injection of oil in sand pack coloumn.

$$\text{Residual oil saturation (Sor\%)} = \frac{Xi}{00IP} \times 100$$

Where Xi = OOIP- Volume of oil displaced after water flooding

$$\text{Oil recovery after water flooding (Orecwf)} = \frac{Sorwf}{00IP} \times 100$$

Sorwf (mL) = Oil retained after brine flooding.

Sorbf (mL) = Oil released after the feeding of sand pack column saturated with residual oil

Additional oil recovery after bio surfactant flooding =

$$\frac{oil\,re\cos ered\,u\sin g\,biosurfac\tan t}{oil\,in\,column\,after\,water\,flooding} \times 100$$

Results

Effect of substrate sources

In order to optimize the production of BSs from *P. chrysogenum* SNP5, different substrates such as grease waste, waste cooking oil, soya bean oil and oil cake were used for solid state fermentation and mapped in term of Emulsification Index (EI), Oil Displacement Area in mm (ODA) and drop collapse assay (Figure 1). Although soya bean oil as a substrate had shown high emulsification activity 56% with oil and 37% with diesel but grease waste was also found promising alternative due to its low cost and good emulsification activity 43% with oil and 22% with diesel.

Growing on waste cooking oil where displaced oil on a water surface was 5 cm without displaying very effective emulsification. Bio-surfactant produced over grease waste and wheat bran displaced the oil (3.5 cm) and shown a good destabilization of surfactant drop on an oil coated glass surface.

Effect of nitrogen sources

Different nitrogen sources like NaNO$_3$, yeast extract, peptone and malt extract were used and observed that yeast and NaNO$_3$ were growth as well as production promoting (Table 2). Other complex sources like peptone and malt extract were good for growth but not suitable for BS production.

Effect of substrate ratio

Grease waste in different ratio with wheat bran (NaNO$_3$ as a constant nitrogen source) shows an effective change in emulsification index and oil displacement activity as observed in Table 3. The best value of emulsification index and oil Displacement activity was detected at 1.5:1 ratio of grease and wheat bran respectively. A significant decrease in activity of BS was observed as the ratio of grease waste was increased

Nitrogen (1%)	EI (oil) (%)	EI(diesel) (%)	ODA (cm)	DCA
NaNO$_3$	43 ± 0.15	22 ± 0.09	3 ± 0.15	++
Peptone	40 ± 0.10	20 ± 0.12	3 ± 0.01	++
Yeast	45 ± 0.61	23 ± 0.13	4 ± 0.75	+++
malt	37 ± 0.23	17 ± 0.21	2 ± 0.93	+

Table 2: Influence of nitrogen content (1%) for biosurfactant production in terms of EI: Emulsification Index; ODA: Oil Displacement Assay; DCA: Drop Collapse Assay; (+: less dispersed drop; ++: moderate dispersed drop; +++ highly dispersed drop of surfactant).

Substrate concentration (Wheat bran and grease ratio)	EI (oil) (%)	EI(diesel) (%)	ODA (cm)	DCA
1:1	43 ± 0.30	22 ± 0.07	3 ± 0.17	++
1.5:1	45 ± 0.06	24 ± 0.01	4 ± 0.30	+++
2:1	43 ± 0.10	18 ± 0.50	2 ± 0.2	++
2.5:1	40 ± 0.22	18 ± 0.03	1 ± 0.72	+
3:1	37 ± 0.13	15 ± 0.04	1 ± 0.51	+

Table 3: Effects of substrate ratio for biosurfactant production in terms of EI: emulsification index; ODA: Oil Displacement Assay; DCA: Drop Collapse Assay; (+: less dispersed drop; ++: moderate dispersed drop; +++ highly dispersed drop of surfactant).

with constant ratio of wheat bran. Oil displacement results were similar for NaNO$_3$ with Grease and wheat bran (1.5:1) which previously found using grease waste and wheat bran (1:1) ratio with 1% yeast extract as a nitrogen source.

Effect of pH on production of BS

The production of BS was tested over a wide range of pH (3-12) of medium with wheat bran and grease waste (1:1) and 1% NaNO$_3$ as nitrogen source. Growth was observed at all condition but performance of surfactants was observed with significant variations under different pH ranges (Table 4).

Surfactant shows the changes in the activity of emulsion formation in different pH ranges a positive increase was observed from 2 to 8 (maximum activity with E24, 47.61% in oil and 38.9% in diesel (Figure 2) and further some significant reduction was observed from pH range 8-12.

Foam height analysis:

Foaming ability of BS was determined 23% in 10 ml crude surfactant. To improve this observation further stability of foam was continuously monitored after every 5 minutes for 1 hr and after 1 hour 3% reduction was found in initial foam height (2 m) (Figure 3).

Effect of solvent system for extraction of BS:

Another aspect which was fundamental to analysis of suitable solvent for collection of cell free extracts (a source of crude BS) from fermented bran was application of various buffers (pH 3-10) and their comparative study with water as a solvent system. It was observed that all buffer system was depicted as week solvent comparative to water for BS extraction. Obtained results were shown the dependency of emulsification activity with various buffer ranges. It is clearly observed that emulsification activity was gradually increased from pH (3-7) then it was again expressed reduction but overall it cumulative effects was very low compare to water as a solvent with maximum activity 43% in case of oil and 22.06% in case of hydrocarbon (Table 5).

Analysis through thin layer chromatography:

Characterization of BS collected by precipitation method was determined by TLC method. A red color spot was observed after the use of ninhydrin reagent and a yellow color spot was developed by

iodine vapors with a Rf value 0.69, 0.69, 0.73 respectively in soya bean oil, grease and waste cooking oil. Presence of red color in ninhydrin with Rf values 0.62, 0.68, 0.66 respectively in soya bean oil, grease and waste cooking oil, confirms presence of protein and yellow spot with iodine confirmed the lipid moiety of surfactant (Figure 4).

Lowery assay

Protein concentration of surfactant was determined through lowery assay [18]. In 1 mg of partially purified BS the protein concentration was 13% (0.13 mg/ml)

pH range	EI (oil) (%)	EI (diesel) (%)	ODA (cm)	DCA
3	37 ± 0.09	14 ± 0.28	2.2cm	+
4	40 ± 0.05	19 ± 0.04	3.3 cm	+
5	4 ± 0.30	22 ± 0.07	3.9cm	++
6	45 ± 0.03	23 ± 0.80	4.1m	++
7	45 ± 0.09	23 ± 0.9	4.8cm	+++
8	47 ± 0.61	38 ± 0.09	5.5	+++
9	45 ± 0.07	30 ± 0.45	5.3	+++
10	44 ± 0.10	25 ± 0.03	4.7	++
11	43 ± 0.01	18 ± 0.18	4.1	++
12	38 ± 0.09	15 ± 0.03	3.4	+

Table 4: Influence of different pH values for production of biosurfactant in terms of EI: Emulsification Index; ODA: Oil Displacement Assay; DCA: Drop Collapse Assay; (+: less dispersed drop; ++: moderate dispersed drop; +++ highly dispersed drop of surfactant).

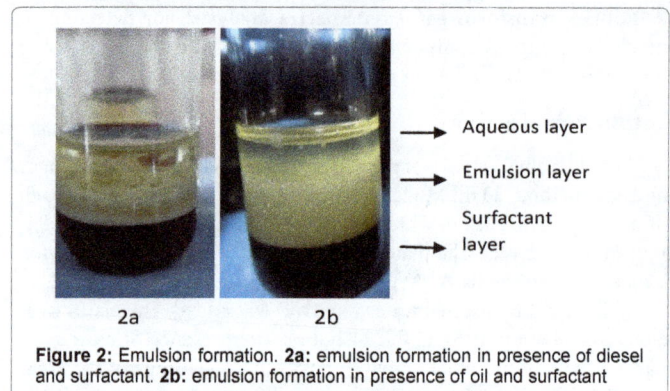

Figure 2: Emulsion formation. **2a:** emulsion formation in presence of diesel and surfactant. **2b:** emulsion formation in presence of oil and surfactant

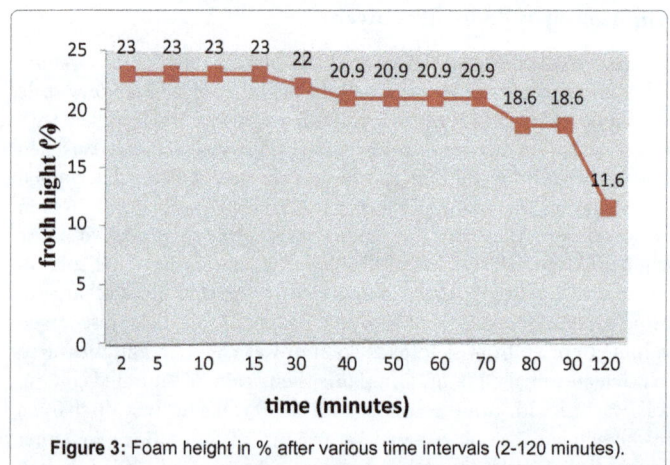

Figure 3: Foam height in % after various time intervals (2-120 minutes).

Solvent system	EI (oil) (%)	EI(diesel) (%)	ODA (in cm)	DCA
Sodium citrate buffer 3	27 ± 0.11	9 ± 0.17	1 ± 0.40	+
Sodium citrate buffer- 4	28 ± 0.20	10 ± 0.30	1 ± 0.1	+
Sodium citrate buffer 5	30 ± 0.05	12 ± 0.51	0 ± 0.95	+
Sodium citrate buffer 6	33 ± 0.12	13 ± 0.61	1 ± 0.70	+
sodium carbonate buffer 7	35 ± 0.19	15 ± 0.21	2 ± 0.4	+
sodium carbonate buffer 8	29 ± 0.05	12 ± 0.05	1 ± 0.05	+
sodium carbonate buffer9	19 ± 0.01	10 ± 0.02	nd	nd
sodium carbonate buffer 10	13 ± 0.08	8 ± 0.14	nd	nd
Water	43 ± 0.31	22 ± 0.06	3 ± 0.19	++

Table 5: Optimization of different solvent system for extraction of biosurfactant from fermented substrate in terms of EI: Emulsification Index; ODA: Oil Displacement Assay; DCA: Drop Collapse Assay; (+: less dispersed drop; ++: moderate dispersed drop; +++ highly dispersed drop of surfactant).

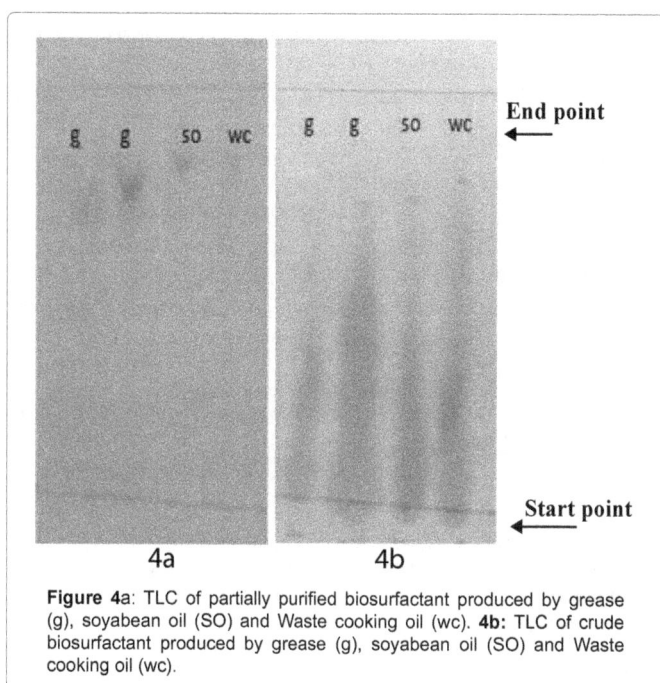

Figure 4a: TLC of partially purified biosurfactant produced by grease (g), soyabean oil (SO) and Waste cooking oil (wc). **4b:** TLC of crude biosurfactant produced by grease (g), soyabean oil (SO) and Waste cooking oil (wc).

FTIR analysis

The FTIR spectra of partially purified BS are shown in figure 5. Characterization absorption peak were found at 3398 cm^{-1} ranging from 3100 cm^{-1} to 3600cm^{-1}. This feature typically confirmed presence of carbon and amino groups and caused due to stretching vibrations of C-H and N-H bonds characterized to carbon containing amino group. Two other sharp bands at 2926 cm^{-1} and 2850 cm^{-1} showed presence of long alkyl chain (-CH$_2$- and -CH$_3$-). The medium band near 1637 cm^{-1} attributed to CO- NH bend which confirms presence of peptide group in the 286 molecule. Other significant peaks observed at 1384cm^{-1} correspond to C-H vibrations. A medium stretch peak around 1084 cm^{-1} signifies to presence of C-O-C stretch of ester. And at last a sharp peak at the region around 669 cm^{-1} and 770 cm^{-1} also represented -CH-bending of alkenes.

Antimicrobial activity of BS

The results of antimicrobial activity of lipopeptide produced by *P. chrysogenum* SNP5 against pathogenic bacteria *S. aureus* and *P. aeruginosa* are shown in table 6. Presence of clear halos indicating that antimicrobial activity of the compound against gram +ve (*S. aureus*)

and gram-ve (*P. aeruginosa*) bacteria with a hallo diameter of mean values 1.67 cm and 1.93 cm respectively (Table 6).

Application of BS for oil recovery

While applying 0.6 pore volume of cell free BS to the oil saturated sand column, to recover oil trapped (43.2%) in the column, 16.5 ± 2 AOR was achieved upon 24 h of incubation (Table 7).

Discussion

On the basis of results obtained in current study it can be stated that production of BS through solid state fermentation is much advantageous than submerged fermentation. Especially in case of BS where SSF emphasize to less energy and cost requirement and very important foaming nuisance in fermentation system, which generally found as drawback in Submerged Fermentation (SmF). Various types of strains of bacteria and fungus have been reported for the efficient production of secondary metabolites under SSF by many scientists [21-24].

In present work production and effectiveness of BS was tried to enhance through optimizing various physiochemical conditions and selecting appropriate media for SSF. However results were promising using wheat bran and soybean oil (emulsification activity 56% with oil and 37% with diesel) but for further study wheat bran and grease waste (emulsification activity 43% with oil and 22% with diesel) was considered in order to find out good alternative tool for remediation of oil spills and hydrocarbon contaminated soil sites etc. By using hydrocarbon based grease waste as substrate for production, BS produced might be specific for petroleum products removal or

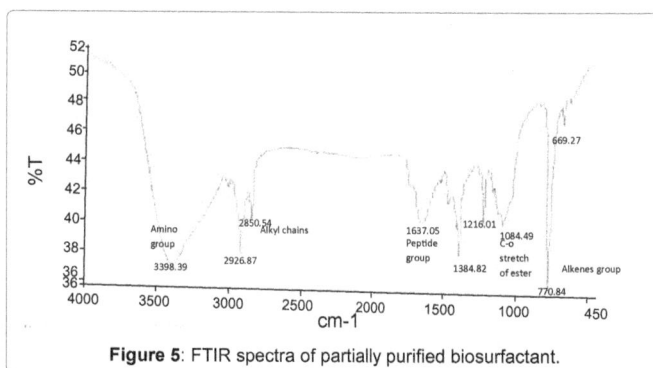

Figure 5: FTIR spectra of partially purified biosurfactant.

S.No.	Name of strain	Zone of inhibition (cm)			Mean value
1	*S. aureus*	1.8	1.6	1.6	1.67
2	*P. aeruginosa*	2	1.8	2.0	1.93

Table 6: Antimicrobial activity with partially purified biosurfactant.

Parameters	Sand Pack Column (SPC)		
	SPC1	SPC2	SPC3
PV (ml)	5.5	5.4	5.45 ± 0.70
OOIP (ml)	4	3.8	3.9 ± 0.14
Sorwf (ML)	1.7	1.6	1.45 ± 0.70
Sorfb(ml)	0.6	0.5	0.55 ± 0.70
Swi (%)	72	70.3	71.15 ± 1.2
Soi (%)	27	29	28 ± 1.4
Orecwf (%)	42	44	43 ± 1.4
AOR (%)	18	15	16.5 ± 2.1

Table 7: Parameters for oil recovery using water flood and lipopeptide bio surfactant from sand pack glass column.

recovery, whereas soybean oil is lipid based. Yeast extract as nitrogen source has shown good impact (displacement of oil is 4 cm and destabilization of surfactant drop 1.4 cm compare to water was 0.4 cm on an oil coated glass surface) on BS production, however $NaNO_3$ was also found enhancer (displacement of oil 3.5 cm and destabilization of surfactant drop 1.1 cm in compare to water 0.4 cm on an oil coated glass surface). Yeast extract being complex and mix kind of nitrogen source might have served as fortifying rather than $NaNO_3$. Selection of substrates and nitrogen sources has major effect on the production of BS [6,11]. *P. chrysogenum* showed different preference on complex and simple nitrogen sources and substrates for BS production [4].

While optimizing the ratio of Wheat Bran (WB) and grease waste for production it was observed that with increase ratio of Grease Waste (GW), emulsification activity reached up to 45% with oil and 24% with diesel at a ratio of 1.5:1 w/w of grease waste and wheat bran respectively without use of other complex nitrogen sources. But with more than this ratio reduction in emulsification index and oil displacement activity from 22-15% and 3-1 cm respectively was observed due to its complexity in nature and hard to utilize completely.

Use of various buffer as a solvent systems showed high impact over recovery of crude BS and found water as best solvent in compare to other solvents. Similar findings were also observed by other scientist [22]. However, in this study the optimization of BS production was confirmed through some indirect methods like Emulsification index, Oil displacement assay and Drop collapse. These are very sensitive and effective methods to confirm the production of BS [11,23]. Abouseoud, et al. [11] had optimized the carbon source, nitrogen source and C/N ratio for bio surfactants production on the basis of emulsification index [11]. In similar manner bio surfactant production optimization was reported by Kiran et al. [23], where optimization of BS production by *Nocardiopsis lucentensis* MSA04 with various carbon sources in solid-state cultivation had been performed by using Emulsification index reported of about E24=25% with wheat bran as substrate [23]. Colla et al. [24] have studied simultaneous production of lipases and BSs by submerged and solid-state bioprocesses with a fungal strain *Aspergillus* sp. and they reported that the production of BS was very low in SSF (emulsification activity after 24 hours was 2.85%) compare to SmF (emulsification activity after 24 hours was 42.67%). Therefore this current study with *P. chrysogenum* SNP5 by using grease waste as substrate for BS production with their optimized emulsification value 43% would be a good benchmark among already reported BSs and also can be accentuate Solid state cultivation of BSs production. And using grease waste as substrate for production make the process cost effective, sustainable and innovative.

BS obtained after extraction was precipitated out and characterized through thin layer chromatography. Results confirmed that BS produced by *P. chrysogenum* SNP5 by using grease waste as substrate was lipopeptide in nature. Presence of peptide and lipid was confirmed on the basis of Rf value 0.68 and 0.69 obtained with ninhydrin and iodine respectively during TLC.

Earlier study confirmed the TLC data with the Rf value of 0.68 and 0.70 for lipopeptide in case of iodine treatment [25] and in case of lipopeptide isolated from *Kocuria marina* BS-15 also confirmed Rf value 0.68 [26]. It was again verified from FTIR spectrum of BS which has shown vibration at 2926 cm^{-1} and 2850 cm^{-1} and 1384cm^{-1} suggested the presence of aliphatic chain of fatty acids, at 3398 cm^{-1} indicating presence of C-H and N-H bonds stretching of peptides. A vibration near 1647 cm^{-1} showed the presence of CO-NH bond and vibration near 1084 cm^{-1} signified presence of C-O-C stretch of ester.

The FTIR analysis was quite similar to result obtained by Thavasi et al. [27] They found spectra at 2,852, 2,923, 1,421, and 1,465 cm^{-1} for the C-H stretching mode. 3,383 and 1,647 cm^{-1} for C-O and N-H vibration and the C-O ester bonds were observed at 1,058 cm^{-1} [27].

Surfactant produced by *P. chrysogenum* SNP5 was lipopeptide in nature, hence pH, solvent and buffer have shown high impact on its activity, stability and emulsification efficiency etc. due to presence of free NH, C=O and COOH group. It can be characterized further for its molecular structure in detail.

Lipopeptides are commercially known for their antimicrobial activity against Multi Drug Resistance (MDR) strains [20]. The partially purified lipopetide was shown good antimicrobial activities of 1.67cm and 1.93 cm against gram +ve (*S. aureus*) and gram –ve (*P. aeruginosa*) respectively. This is little contrasting to previous reports where lipopeptides have been reported mostly against gram +ve bacteria [28]. The current scenario of pathogenic strain against drug resistance depends upon indiscriminate use of regular common antibiotics. Both strain are a very common example of MDR and generally show resistance against Methicillin and Streptomycin. Hence these new lipopeptide produced from *P. chrysogenum* SNP5 can gives an alternative to already existing antimicrobial drugs [6]. So finally BS produced due to its light molecular weight and lipopeptide in nature would be most desirable and useful in food, pharmaceutical like, medicine, agriculture industries and for bioremediation purpose like oil tanks clean up [29-31].

BS produced as crude extract was also tested for it application in oil recovery which is very challenging and costly efforts in petroleum and oil industries. And it was found very efficient even in crude form for recovery of trapped oil in column. Hence could be very useful in petroleum and oil industries, also it can be applied for pipeline washing due to its both the properties i.e., antimicrobial property to check microbial corrosion and surface activity for washing out retained oil from the surface.

Conclusion

In present study SSF using grease waste as substrate for BS production had been found promising due to its cost effectiveness and single step downstream processing. BS produced had shown an effective emulsification and oil displacement activity against crude oil as well as it has shown a potential antimicrobial activity also. Therefore, BS produced may be utilized towards the solid waste management, in oil recovery and in food processing. The BS produced by *P. chrysogenum* SNP5 was characterized as lipopeptide derivative, which is different from the liposaccharide BS produced by *P. citrium*, reported by Demorais et al. [15] and finally create a potential alternative over the various other chemical surfactants. Now a day's solid waste management is a critical issue and this strain or its surface active compounds can be used as a good alternative for biostimulation or bioaugmentation respectively. Further study may open new horizon for industries.

Acknowledgement

Authors are thankful to, Technical Education Quality Improvement Programme, India and Department of Science and Technology, Government of India for providing facility and resources for current work. Author declared no conflict of interest.

References

1. Rahman PKSM, Gakpe E (2008) Production, characterization and applications of biosurfactants-Review. Biotechnology 7: 360-370.

2. Ron EZ, Rosenberg E (2002) Biosurfactants and oil bioremediation. CurrOpin Biotechnol 13: 249-252.

3. Kosaric N (1992) Biosurfactants in industry. Pure & Appl Chem 64: 1731-1737.

4. Makkar RS, Cameotra SS (2002) An update on the use of unconventional substrates for biosurfactant production and their new applications. Appl Microbiol Biotechnol 58: 428-434.

5. Transparency market research (2012) Biosurfactant Market- Global scenario, raw material and consumption trends industry analysis, size, share and forecast, 2011-2018.

6. Das P, Mukherjee S, Sen R (2008) Antimicrobial potential of a lipopeptidebiosurfactant derived from a marine *Bacillus circulans*. J Appl Microbiol 104: 1675-1684.

7. Shaligram NS, Singhal RS (2010) Surfactin – A Review on Biosynthesis, Fermentation, Purification and Applications. Food Technol Biotechnol 48: 119-134.

8. Krieger N, Camilios Neto D, Mitchell DA (2010) Production of microbial biosurfactants by solid-state cultivation. Adv Exp Med Biol 672: 203-210.

9. Satpute SK, Bhuyan SS, Pardesi KR, Mujumdar SS, Dhakephalkar PK, et al. (2010) Molecular genetics of biosurfactant synthesis in microorganisms. Adv Exp Med Biol 672: 14-41.

10. Kumar S, Katiyar N, Ingle P, Negi S (2011) Use of evolutionary operation (EVOP) factorial design technique to develop a bioprocess using grease waste as a substrate for lipase production. Bioresour Technol 102: 4909-4912.

11. Abouseoud M, Maachi R, Amrane A, Boudergua S, Nabi A (2008) Evaluation of different carbon and nitrogen sources in production of biosurfactant by *Pseudomonas fluorescens*. Desalination 223: 143-151.

12. Walter V, Syldatk C, Hausmann R (2010) Screening concepts for the isolation of biosurfactant producing microorganisms. Adv Exp Med Biol 672: 1-13.

13. Violeta O, Oana S, Matilda C, Maria CD, Catalina V, et al. (2011) Production of biosurfactants and antifungal compounds by new strains of *Bacillus* Spp. isolated from different sources. Rom Biotechnol Lett 16: 84-91.

14. El- Shesthawy HS, Doheim MM (2014) Selection of *Pseudomonas aeruginosa* for biosurfactant production and studies of its antimicrobial activity. Egyptian Journal of Petroleum 23: 1-6.

15. De-Morais MMC, Ramos SAF, Pimental MCB, Melo EHM, Morais MA, et al. (2006) Liposaccharide extracellular emulsifier produced by *Penicilliumcitrinum*. Journal of Biological Sciences 6: 511-515.

16. Carrillo PG, Mardaraz C, Pitta-Alvarez SI, Giulietti AM (1996) Isolation and selection of biosurfactant-producing bacteria. World J Microbiol Biotechnol 12: 82-84.

17. Symth TJP, Permuo A, McClean S, Marchant R, Banat IM (2010) Isolation and Analysis of Lipopeptides and High Molecular Weight Biosurfactants: Handbook of Hydrocarbon and Lipid Microbiology. (5th edition), springer, Berlin, Heidelberg.

18. Lowry OH, Rosebrough NJ, Farr AL, Randall RJ (1951) Protein measurement with the Folin phenol reagent. J Biol Chem 193: 265-275.

19. Mukherjee S, Das P, Sen R (2006) Towards commercial production of microbial surfactants. Trends Biotechnol 24: 509-515.

20. Rodrigues L, Banat IM, Teixeira J, Oliveira R (2006) Biosurfactants: potential applications in medicine. J Antimicrob Chemother 57: 609-618.

21. Negi S, Kumar S (2012) Evaluation of techniques used for parameters estimation: an application to bioremediation of grease waste. Appl Biochem Biotechnol 167: 1613-1621.

22. Kitamoto D, Ikegami T, Suzuki GT, Sasaki A, Takeyama Y, et al.(2001) Microbial conversion of n-alkanes into glycolipid biosurfactants, mannosylerythritol lipids, by Pseudozyma (*Candida antarctica*). Biotechnol Lett 23: 1709-1714.

23. Kiran GS, Thomas TA, Selvin J (2010) Production of a new glycolipid biosurfactant from marine *Nocardiopsis lucentensis* MSA04 in solid-state cultivation. Colloids Surf B Biointerfaces 78: 8-16.

24. Colla LM, Rizzardi J, Pinto MH, Reinehr CO, Bertolin TE, et al. (2010) Simultaneous production of lipases and biosurfactants by submerged and solid-state bioprocesses. Bioresour Technol 101: 8308-8314.

25. Anyanwu CU, Obi SKC, Okolo BN (2011) Lipopeptide biosurfactant production by *Serratia marcescens* NSK-1 strain isolated from petroleum-contaminated soil. Journal of Applied Sciences Research 7: 79-87.

26. Sarafin Y, Donio MBS, Velmurugan S, Michaelbabu M, Citarasu T (2014) *Kocuria marina* BS-15 a biosurfactant producing halophilic bacteria isolated from solar salt works in India. Saudi Journal of Biological Sciences. In Press.

27. Thavasi R, Subramanyam Nambaru VR, Jayalakshmi S, Balasubramanian T, Banat IM (2009) Biosurfactant production by *Azotobacter chroococcum* isolated from the marine environment. Mar Biotechnol (NY) 11: 551-556.

28. Singh P, Cameotra SS (2004) Potential applications of microbial surfactants in biomedical sciences. Trends Biotechnol 22: 142-146.

29. Banat IM, Samarah N, Murad M, Horne R, Banerjee S (1991) Biosurfactant production and use in oil tank clean-up. World J Microbiol Biotechnol 7: 80-88.

30. Bhadauriya SS, Madoriya N, Shukla K, Parihar MS (2013) Biosurfactants: A new pharmaceutical additive for solubility enhancement and pharmaceutical development. Biochem Pharmacol 2: 113.

31. Colin VL (2012) Technology of biosurfactants for the development of environmental remediation processes. Ferment Technol 1: e109.

Smoking Related Changes in Neurotransmitters in African Americans

Sudhish Mishra, Anita Mandal and Prabir K. Mandal*

Department of Biology, Edward Waters College, 1658 Kings Road, Jacksonville, FL 32209, USA

Abstract

Smoking is a most common way of tobacco consumption that provides a pleasurable effect by modulating neurotransmitters in the brain. African Americans (AA) have shown higher susceptibility to nicotine consumption and highest rate of mortality due to cancer. In this study we have examined the levels of 6 neurotransmitters in AA smokers and compared them with nonsmokers of same population. We observed decrease in plasma levels of glutamate and increase in serotonin, epinephrine and dopamine. The level of gamma amino butyric acid (GABA) and norepinephrine was not changed significantly. The results, in part, explain the basis of higher nicotine susceptibility in AAs and will be helpful to develop population specific strategies for smoking cessation.

Keywords: African Americans; Neurotransmitter; Smoking; Nicotine; Tobacco

Introduction

Smoking is the most common way of tobacco consumption and it's the most difficult addiction to remove. It gives a pleasurable effect to the individuals through stimulation of dopanergic pathway [1]. However, Dopamine alone cannot be accounted for the development of addiction. Other neurotransmitters implicated in the development of addiction include serotonin, gamma amino butyric acid (GABA), glutamate and noradrenaline [2].

A range of cellular changes occurs following the repeated use of stimulants. These changes include synthesis of new proteins including transcription factors of FOS family. These transcription factors increase responsiveness to the effect of drugs [3]. These cellular changes coincide with depression, irritability and anxiety during withdrawal. Changes in neurotransmitters can also play a role in establishing the cycle of adaptation. One reason may be the down regulation of dopamine receptor. Other adaptations include changes in noradrenaline and glutanergic system [2]. These neuroadaptive changes develop responses in addicts, which force them to seek new drug supplies and begin a fresh addiction cycle.

The rate of smoking is historically higher in African Americans when compared to the general US population. Smoking related diseases kill approximately 45,000 African American each year [4]. According to epidemiological surveys smoking rate declined dramatically in African American teens since 1976 but recent trends indicate an increase of smoking among younger generation. If this trend continues, an estimated 1.6 million African Americans who are under the age of 18 will become regular smokers. About 500,000 of these smokers will die of smoking-related diseases [5]. Smoking is responsible for almost 90 percent of all lung cancer cases and also a major cause of heart disease and stroke. Heart diseases account for more deaths (27%) in African Americans when compared to deaths due to all forms of cancer (21%) [6].

Tobacco is obtained from Plant Nicotiana tabacum and contains an alkaloid, called Nicotine. Nicotine has been accounted as main substance responsible for forming dependence on tobacco. Nicotine is metabolized in the liver by cytochrome P450 enzymes (CYP2A6 and CYP2B6). It increases level of several neurotransmitters by binding to nicotinic acetyl choline receptors (nAChR). At high concentrations it can induce muscle contractions and respiratory paralysis [7]. Tobacco smoke contains monoamine oxidase inhibitors in addition to nicotine. Monoamine oxidase enzyme is responsible for breakdown of monoam-

inergic transmitters such as dopamine, norepinephrine and serotonin. Chronic smoking also up-regulates nAChR [8]. Nicotine also activates the sympathetic nervous system, acting via splanchnic nerves to the adrenal medulla, stimulates the release of epinephrine [9].

By binding to ganglion type nicotinic receptors, Nicotine may also cause cell depolarization and influx of calcium through voltage gated calcium channels. Calcium triggers the release of epinephrine in blood stream, which causes an increase in heart rate, blood pressure and respiration [10]. Nicotine is also a vasoconstrictor, which makes harder for the heart to pump through the constricted arteries. It causes the body to release its stored fat and cholesterol into the blood. It is also speculated that nicotine also increases the risk of blood clot by increasing plasminogen activator inhibitor-1. The elevated levels of plasma fibrinogen and Factor XIII have been reported among smokers.

In the current study we have analyzed the plasma levels of 6 neurotransmitters (GABA, glutamate, serotonin, dopamine, epinephrine and norepinephrine) in African American smokers and compared the results with non-smokers of same population. Since these neurotransmitters play an important role in tobacco addiction, these results can have implications in tobacco cessation studies.

Materials and Methods

Materials

Blood samples were collected in sterile condition from 58 African American volunteers of either sex as per the institutional guidelines. Plasma were separated from collected samples by centrifugation of blood for 10 min and stored at -20°C in aliquots until further use. ELISA kits for GABA and Glutamate were purchased from Alpco, Salem (NH). ELISA kits for Serotonin and 3-CAT (epinephrine, norepineph-

*Corresponding author: Prabir K Mandal, Department of Biology, Edward Waters College, 1658 Kings Road, Jacksonville-FL 32209, USA

rine and Dopamine) were purchased from Genway, San Diego (CA). Microplate reader was purchased from Bio Tek (Winooski, VT).

Experimental procedure

Plasma concentration of neurotransmitters: The plasma levels of neurotransmitters were measured by using Enzyme-linked immuno-sorbent assays (ELISA).

ELISAs: The ELISAs were conducted according to the manufacturers' instructions to quantify their levels in plasma of African American volunteers. Plates were read in an EPOCH ELISA reader from Biotek Instruments assay following the manufacturer's protocol and data were analyzed using GEN5 software (Biotek). For Serotonin, adrenaline, noradrenaline and dopamine, samples were acylated before ELISA. For serotonin, biotin-antiserum system was used and results were measured at 405nm after conducting competitive ELISA. For adrenaline, noradrenaline and dopamine, enzyme conjugate and antiserum system was used for separate solid phase competitive ELISA and the results were measured at 450 nm.

Statistical analysis: All the results were compared using Student's paired t-test. To determine whether significant differences in changes were present between groups, ANOVA was performed with α set to 0.05. In all instances, if significance was attained by overall ANOVA, pairwise comparisons were performed using the Student-Newman-Keuls test. For all pairwise comparisons, probability values of ≤ 0.05 were considered significant. All data are reported as means $\pm$ SE.

Results

Regulation of GABA and glutamate

Plasma concentration of GABA was not changed significantly when volunteer smokers were compared with non-smoker volunteers (115.1 ± 21.3 vs. 126.9 ± 19.9; ng/ml; Figure1A).

Plasma concentration of Glutamate was found significantly decreased (12.1 ± 2.5 vs. 19.7 ± 2.5; µg/ml) in smokers when compared to non-smokers (Figure1B).

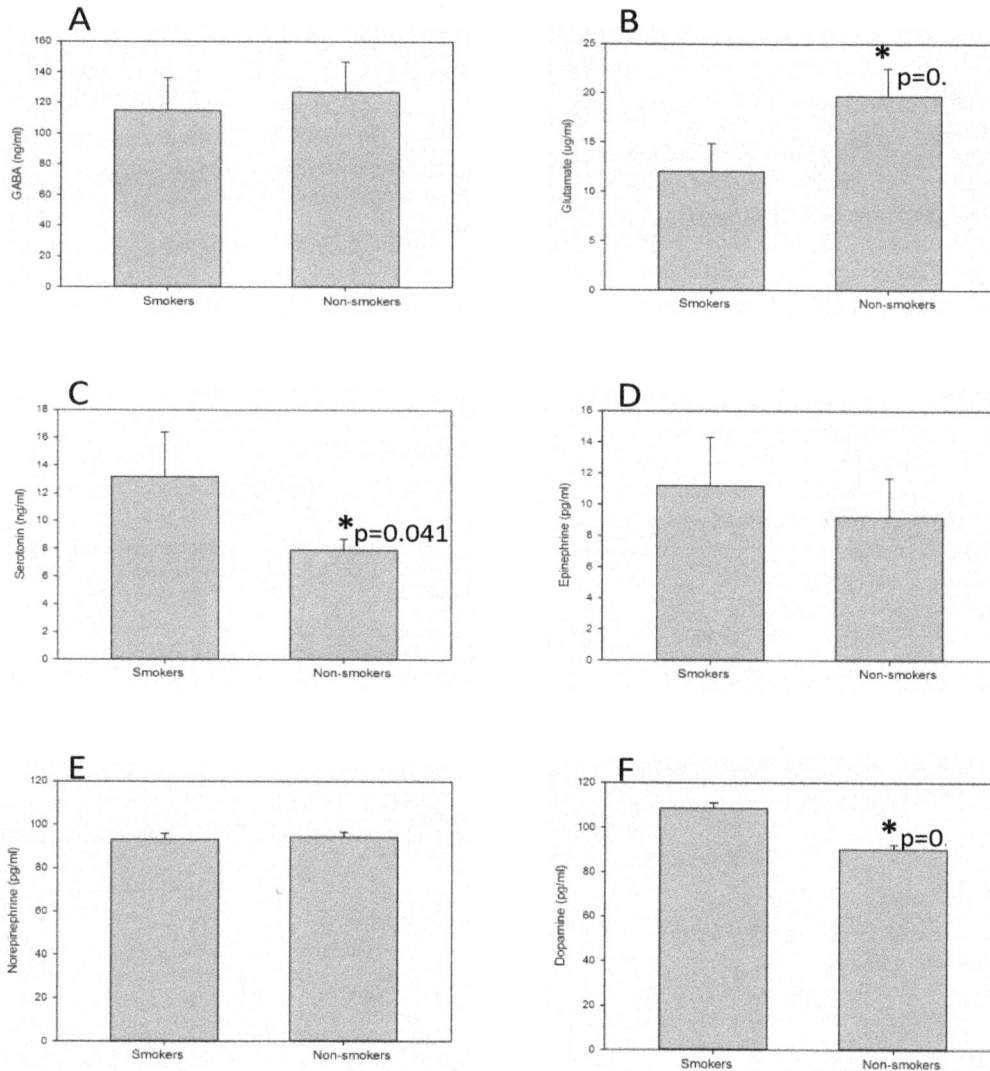

Figure1: Plasma concentrations of neurotransmitters in smoker and non-smoker African American volunteers. A: GABA; B: Glutamate; C:Serotonin; D: Epinephrine; E: Norepinephrine; F: Dopamine; * p<0.05 vs smokers.

Up-regulation of serotonin

Plasma concentration of serotonin significantly increased (13.2±2.8 vs. 7.9±0.8; ng/ml) in smokers when compared to non-smokers (Figure1C).

Regulation of epinephrine, norepinephrine and dopamine

Plasma concentration of epinephrine was slightly increased (11.2±3.1 vs. 9.2±2.5; pg/ml; Figure1D) in smoker volunteers when compared to non-smokers but concentration of norepinephrine was not changed significantly (93.1± 2.7 vs. 94.3±2.4; pg/ml; Figure1E). The concentration of Dopamine was found significantly increased in smokers (108.5±2.6 vs. 90.2±2.1; pg/ml) when compared to non-smokers (Figure1F).

Discussion

In the present study we have identified the alteration in neurotransmitter levels in smokers of African American population. During this study we measured the plasma levels of GABA, Glutamate, serotonin, epinephrine, norepinephrine and dopamine. The results were analyzed by comparing with non-smokers.

Gamma-aminobutyric acid (GABA) is a multifunctional mediator that functions as a neurotransmitter in the central nervous system and [11,12] is responsible for controlling dopamine release through receptors present on dopamine releasing neurons. Because of its role in nicotine addiction pathway, GABA has been considered a very useful target for nicotine addiction therapies [13]. Nicotine and/or cigarette smoking modulates brain gamma aminobutyric acid (GABA) concentrations in animals and humans [14,15]. Our results show slight decrease in GABA concentration among smokers but it was not found statistically significant. These findings of reduced GABA are in accord with investigations for tobacco specific carcinogen that have revealed a reduction of GABA in the lungs and small airway-derived pulmonary adenocarcinoma of smokers [16].

Glutamate is an amino acid neurotransmitter responsible for creating and storing memories in the brain. Each time a nicotine dependent individual smokes or chews a memory of alertness (from the acetylcholine) and pleasure (from the dopamine) reinforces the desire for continued nicotine consumption. Glutamate also synthesizes GABA through the action of glutamic acid carboxylase [17]. We have observed significant reduction in glutamate levels among smokers [18]. These results are in contrast to the results reported for hippocampus and cingulate cortex region of brain, where no change in glutamate concentration was found among smokers [19].

Serotonin is the "feel good" neurotransmitter of the brain and is responsible for us feeling happy, relaxed, calm, motivated, and at peace with our lives and our role in the world. Low level of serotonin is associated with stress, food craving, obesity, poor memory, fatigue, insulin resistance, insomnia, food intolerance, heart disease and smoking addiction [20].

Smoking causes decrease in the serotonin receptors in the brain which stop it being flooded with the body's natural stress-busting hormone, serotonin. It makes smokers feel deficient in serotonin and therefore they are less able to cope with the everyday pressures of life and suffer from high levels of stress [21]. Smokers enjoy short term relief from their anxiety because nicotine temporarily increases levels of dopamine. We observed significant increase in plasma serotonin level in smokers [22]. These results are in agreement with earlier reports

where investigators observed higher levels of platelet [23] and plasma serotonin [24] in smokers.

Epinephrine and norepinephrine are catecholamine, secreted by adrenal gland and is responsible for fight or flight response of sympathetic nervous system. These neurotransmitters exert their tissue specific effect by binding to their specific receptors expressed in target tissue [25].In our study, we observed slight increase in plasma epinephrine level in African American smokers but norepinephrine levels were not changed. These results are in agreement with earlier studies to measure nicotine effect on these catecholamines in plasma [26-28] and serum [29], where they found increased level of epinephrine but not of norepinephrine. Some other studies have shown increased plasma level of both catecholamines after smoking [30,31].

Dopamine, another catecholamine neurotransmitter of brain and precursor of norepinephrine and epinephrine has widely implicated for smoking addiction. Smoking increases the level of dopamine in the body, which stimulates reward centers in the brain. This process reinforces the smoking behavior, so that each cigarette makes smokers desire yet another [32].Our results of increased dopamine level in plasma of smokers are in agreement of earlier studies for its concentration in basal ganglia of brain in smokers [33,34].

African Americans show higher susceptibility to various types of cancer and cardiac abnormalities. Smoking behavior further increases the chances of disease manifestation. Present population specific study of neurotransmitters has provided some information to understand the pathophysiology of AA smokers and will be helpful to control smoking behavior.

Grant

This study was supported by James & Esther King Biomedical Research Program of Florida Department of Health Grant HBC-01 to Drs. Prabir K. Mandal (PI), Sudhish Mishra (Co-PI) and Anita Mandal (Co-PI) of Biology Department, EWC.

References

1. Koob GF, Le Moal M (1997) Drug abuse: Hedonic homeostatic dysregulation. Science 278: 52-58.

2. Delfs JM, Zhu Y, Druhan JP, Aston-Jones G (2000) Noradrenaline in the ventral forebrain is critical for opiate withdrawal-induced aversion. Nature 403: 430-434.

3. Kelz MB, Chen J, Carlezon WA Jr, Whisler K, Gilden L, et al. (1999) Expression of the transcription factor deltaFosB in the brain controls sensitivity to cocaine. Nature 401: 272-276.

4. CDC (1995) Office on Smoking and Health, Unpublished Data.

5. CDC (1998) At A Glance. Tobacco use among US racial/Ethnic Minority Groups-African Americans, American Indians and Alaska Natives and Hispanics, Atlanta.

6. American Cancer Society (2005-2006)Cancer Facts and Figures for African Americans.

7. Katzung BG (2006) Basic and Clinical Pharmacology. New York: McGraw-Hill Medical Pp. 99-105.

8. Walsh H, Govind AP, Mastro R, Hoda JC, Bertrand D, et al. (2008) Up-regulation of nicotinic receptors by nicotine varies with receptor subtype. J Biol Chem 283: 6022-6032.

9. Yoshida T, Sakane N, Umekawa T, Kondo M (1994) Effect of nicotine on sympathetic nervous system activity of mice subjected to immobilization stress. Physiol Behav 55: 53-57.

10. Marieb EN, Hoehn K (2007) Human Anatomy & Physiology (7th Ed.). Pearson Benzamin Cummings.

11. Owens DF, Kriegstein AR (2002) Is there more to GABA than synaptic inhibition? Nat Rev Neurosci 3: 715-727.

12. Watanabe M, Maemura K, Kanbara K, Tamayama T, Hayasaki H (2002) GABA and GABA receptors in the central nervous system and other organs. Int Rev Cytol 213: 1-47.

13. Mason G (2007) Nicotine has significant effects on brain GABA. American College of Neuropsychopharmacology (ACNP) Annual meeting.

14. Epperson C N, O'Malley S, Czarkowski K A, Gueorguieva R, Jatlow P, et al. (2005) Sex, GABA, and nicotine: the impact of smoking on cortical GABA levels across the menstrual cycle as measured with proton magnetic resonance spectroscopy. Biological Psychiatry 57: 44–48.

15. Zhu PJ, Chiappinelli VA (1999) Nicotine modulates evoked GABAergic transmission in the brain. Journal of Neurophysiology 82: 3041–3045.

16. Schuller HM, Al-Wadei HA, Majidi M (2008) Gamma-aminobutyric acid, a potential tumor suppressor for small airway-derived lung adenocarcinoma Carcinogenesis 29: 1979–1985.

17. How Nicotine Addiction Really Works.

18. Wang G, Wang R, Ferris B, Salit J, Strulovici-Barel Y, et al. (2010) Smoking-mediated up-regulation of GAD67 expression in the human airway epithelium Respiratory Research 11: 150-164.

19. Gallinat J, Schubert F (2007) Regional cerebral glutamate concentrations and chronic tobacco consumption. Pharmacopsychiatry 40: 64-67.

20. Berger M, Gray JA, Roth BL (2009) The expanded biology of serotonin. Annu Rev Med 60: 355–366.

21. Xu Z, Seidler FJ, Cousins MM, Slikker W Jr, Slotkin TA (2002) Adolescent nicotine administration alters serotonin receptors and cell signaling mediated through adenylyl cyclase. Brain Res 951: 280-292.

22. Stokes C, Watson J (2004) Smokers Have Serotonin Deficiency Causing High Stress Levels. Scotsman.com

23. Racké K, Schwörer H, Simson G (1992) Effects of cigarette smoking or ingestion of nicotine on platelet 5-hydroxytryptamine (5-HT) levels in smokers and non-smokers. Clin Investig 70: 201-204.

24. Sugiura T, Dohi Y, Yamashita S, Hirowatari Y, Yatomi Y, et al. (2010) Increase of plasma serotonin mediates impaired endothelial function in habitual smokers. Eur Heart J 31: Abstract Supplement: 682.

25. Berecek Kh, BM (1982) Evidence for a neurotransmitter role for epinephrine derived from the adrenal medulla. Am J Physiol 242: H593–H601.

26. Trap-Jensen J, Carlsen JE, Svendsen TL, Christensen NJ (1979) Cardiovascular and adrenergic effects of cigarette smoking during immediate non-selective and selective beta adrenoceptor blockade in humans. European Journal of Clin Invest 9: 181–183.

27. Myers MG, Benowitz NL, Dubbin JD, Haynes RB, Sole MJ (1988) Cardiovascular Effects of Smoking in Patients with Ischemic Heart Disease. Chest 93: 14-19.

28. Wolk R, Shamsuzzaman ASM, Svatikova A, Huyber CM, Huck C, et al. (2005) Hemodynamic and autonomic effects of smokeless tobacco in healthy young men. J Am Coll Cardiol 45: 910-914.

29. Hill P, Wynder EL (1974) Smoking and cardiovascular disease. Effect of nicotine on the serum epinephrine and corticoids. Am Heart J 87: 491-496.

30. Walker JF, Collins LC, Rowell PP, Goldsmith LJ, Stamford BA, et al. (1999) The effect of smoking on energy expenditure and plasma catecholamine and nicotine levels during light physical activity. Nicotine Tob Res 1: 365-370.

31. Siess W, Lorenz R, Roth P, Weber PC (1982) Plasma Catecholamines, Platelet Aggregation and Associated Thromboxane Formation After Physical Exercise, Smoking or Norepinephrine Infusion. Circulation 66: 44-48.

32. Arias-Carrión O, Pöppel E (2007) Dopamine, learning and reward-seeking behavior. Act Neurobiol Exp 67: 481–488.

33. Court JA, Lloyd S, Thomas N, Piggott MA, Marshall EF, et al. (1998) Dopamine and nicotinic receptor binding and the levels of dopamine and homovanillic acid in human brain related to tobacco use. Neuroscience 87: 63-78.

34. Salokangas RKR, Vilkman H, Ilonen T, Taiminen T, Bergman J, et al. (2000) High Levels of Dopamine Activity in the Basal Ganglia of Cigarette Smokers. Am J Psychiatry 157: 632–634.

Cytotoxicity of Chitosan Oligomers Produced by Crude Enzyme Extract from the Fungus *Metarhizium Anisopliae* in Hepg2 and Hela Cells

Cristiane Fernandes de Assis[1], Raniere Fagundes Melo-Silveira[2], Ruth Medeiros de Oliveira[2], Leandro Silva Costa[2], Hugo Alexandre de Oliveira Rocha[2], Gorete Ribeiro de Macedo[1]* and Everaldo Silvino dos Santos[1]

[1]Department of Chemical Engineering, Federal University of Rio Grande do Norte, Natal, Rio Grande do Norte, Brazil
[2]Department of Biochemistry, Federal University of Rio Grande do Norte, Natal, Rio Grande do Norte, Brazil

Abstract

Chitooligosaccharides exhibit biological activities, including antitumor, antimicrobial and antioxidant. In this study we used a mixture of chitooligosaccharides produced by enzymatic hydrolysis in two tumor cell lines and assessed the cell proliferation and cytotoxicity of these compounds. The proliferation of HeLa cells was inhibited around by 60%.

Keywords: Chitooligosaccharides; HepG2 cells; HeLa cells; Cytotoxicity

Introduction

In recent years, the research for more efficient alternatives in the treatment of infectious and neoplastic diseases has mobilized professionals from a host of different areas. Promising results have emerged from the use of substances produced by microorganisms. Chitosanolytic enzymes from different microorganisms, including fungi [1] and bacteria [2], have been reported and used with excellent results in the production of chitooligosaccharides (COS)[3]. Chitooligosaccharides are partially hydrolyzed products of chitosan, within which pentamers and hexamers can be obtained as intermediate reactions [4].

Two methods can be used to obtain chitooligosaccharides: chemical and enzymatic. Chemical hydrolysis is carried out using high temperatures under acidic conditions and produces a large amount of glucosamine (chitosan monomer), compromising control over the reaction progress. This method produces low pentamer and hexamer yields. Enzymatic hydrolysis has a number of advantages for the production of chitooligosaccharides and some chitosanases may catalyze hydrolysis under mild conditions and not produce monosaccharides [4].

COS is applied more widely than chitosan in health-care, food, medicine, pesticides and feedstuffs. The anti-tumor activity of COS has been known since 1986 [5], and several mechanisms have been proposed. These include the regulation of immunity [6], the direct killing of tumor cells, or causing tumor cell apoptosis and inhibiting tumor angiogenesis [7].

The aim of this study was to quantify and analyze the chitooligosaccharides produced during 20 minutes of enzymatic hydrolysis of chitosan produced by fungus *Metarhizium anisopliae* and evaluates cell viability of these compounds in hepatocarcinome (HepG2) and uterine carcinome (HeLa) cells. The 3T3 cell lines, which are fibroblast cells, were used as standard non-toxic concentrations.

Materials and Methods

Fermentation conditions

The fungus *Metarhizium anisopliae* was kindly gifted by Embrapa Recursos Genéticos e Biotecnologia (Brasília/DF-Brazil). Ten milliliters of spore suspension (10^7 spores/mL) from a 5-day culture in PDA medium were transferred using 2 mL of sterile water. Next, this spore suspension was once again transferred to a 250 mL Erlenmeyer containing 90 mL of culture medium consisting of: 0.2% chitosan, 0.1% K_2HPO_4, 0.05% $MgSO_4$, 0.5 % KCl, 0.3% yeast extract, 0.5% peptone, 0.2% $NaNO_3$, 0.001% $FeSO_4$ (pH 5.5); growth took place in a rotating incubator (110 rpm) for 2 days at 25°C. From this suspension, 10 mL was transferred to 90 mL of the same medium. After 48 hours of culture the broth was centrifuged at 13400 $x\,g$ for 15 minutes and the supernatant was used to determine the enzymatic activity assay.

Enzyme activity

Enzymatic activity was assessed by determining the reducing sugars generated by chitosan hydrolysis. In this case, 500 lL of the fermented broth was mixed with 500 lL of chitosan solution solubilized in hydrochloric acid (0.1 N). The reaction was carried out for 30 min at 55°C. To terminate the reaction, 2.5 mL of dinitrosalicylic acid was added and then cooled in an ice bath, and quantification of the reducing sugars was performed using a spectrophotometer (Thermo Spectronic) at 600 nm [8] and a standard curve with D-glucosamine. One unit (U) of chitosanase was defined as the amount of enzyme that is capable of releasing 1 lmol of reduced sugar equivalent to chitosan D-glucosamine/min.

Chitooligosaccharide production

The hydrolysis reaction of chitosan to obtain the chitooligosaccharides consisted of mixing a solution of 1% (m/v) soluble chitosan in chloridric acid (pH adjusted with NaOH to 5.5) to the broth fermented at a ratio of 1:1; the mixture was maintained at 55°C for 20 minutes. The reaction was finalized by the thermal inactivation of the enzyme at 100°C for 10 minutes, followed by centrifugation at 13400 x g for 15 minutes and filtration in a 0.22 μm filtering membrane. Detection of glucosamine and of chitosan oligosaccharides from dimer to hexamer was determined by high-performance liquid chromatography using a CLC-NH_2 Shim-Pack column (Shimadzu Co, Japan). Oligomer analy-

*Corresponding author: Gorete Ribeiro de Macedo, Departamento de Engenharia Química ,Centro de Tecnologia, Av. Senador Salgado Filho, s/n, Lagoa Nova, 50072-970 ,NATAL - Rio Grande do Norte, Brazil

sis was performed in HPLC with 60% acetonitrile as mobile phase and flow rate of 0.8 mL/min and using an RI detector. The peaks of $(GlcN)_{n=1-6}$ were identified and estimated using a standard calibration curve (1-10 mg/mL) according to Liang's equation [9].

Cell culture

Embryo 3T3 fibroblast cells (ATCC CCL-164), HepG2 hepato-carcinome cells (ATCC HB-8065) and cervical adenocarcinome HeLa cells (ATCC CCL-2) were used for viability and cytotoxicity. The cells grew in DMEM medium (Dulbecco's Modified Eagle Medium) supplemented with 10% newborn calf serum (CUTILAB, Campinas-SP, Brazil) and penicillin/streptomycin (1μg/mL) (Sigma-Aldrich, St. Louis, USA). The cells were incubated in oven at 37°C under moisture atmosphere containing 5% CO_2. The viability assays were conducted *in vitro* as follows: 100μL ($5x10^3$ cells) of cell suspension was plated in 96-well polystyrene plates. Cell adhesion was conducted for 12h. Before the addition of chitosan oligomers, the medium with the non-adhered cells was removed and the medium containing the oligomers (from 0.1 to 1 mg/mL) was used. A control without oligomers was also used. After treatment, the cytotoxic effect of the chitosan oligomers in HepG2, HeLa 3T3 cells was determined using the MIT method described by Mosmann [10] with some modifications. The cells were washed with PBS at 37°C and then 100 μL of serum-free medium containing 0.5mg/mL of MTT was added to each well. After 4 hours of incubation, the culture medium was removed and 100 μL of isopropyl alcohol was added to each well for solubilization of the formazan formed. The plates were agitated for 10 minutes and the mean absorbance of the plate was measured in the spectrophotometer at 570 nm [11]. The absorbance of the treated cells was compared with that of the control. The control cells were considered 100% viable, whereas the percentage of cell growth inhibition was calculated for those treated with the hydrolysate. All the analyses were conducted in triplicate.

Statistic analysis

All data were expressed as mean ± standard deviation. Statistical analysis was done by one-way Anova using the SIGMAStat version 2.01 computer software. Student-Newmans-Keuls post-tests were performed for multiple group comparison. In all cases statistical significance was set at $p < 0.05$.

Results and Discussion

The products of chitosan hydrolysis were analyzed by HPLC. Figure 1 shows the chromatogram obtained for a hydrolysis time of 20 minutes using the crude enzyme extract. This figure also demonstrates the formation of monomers (GlcN), trimers $(GlcN)_3$, tetramers $(GlcN)_4$ and pentamers $(GlcN)_5$ at concentrations of 6.532; 0.792; 0.395 and 0.352 (mg/mL), respectively. The monomer (glucosamine) shows the highest concentration. Such a polymer profile suggests the presence of enzymes with exo and endochitosanase activity [12,13].

Figure 2 shows the cytotoxicity of the tested compounds in tumor and normal cells. In the MTT assays we used different hydrolysate concentrations containing a mixture of chitooligosaccharides (0.1 – 1 mg/mL) in the different cell lines.

The cells were treated with different concentrations of the supernatant, for 72 hours. In the absence of these compounds the reduction of MTT was considered as being 100%. The experiment was carried out in 96-well plates. The results represent the mean ± SD of three experiments in triplicate ($p < 0.05$).

The 3T3 cells were used as control to assess the toxicity of the com-

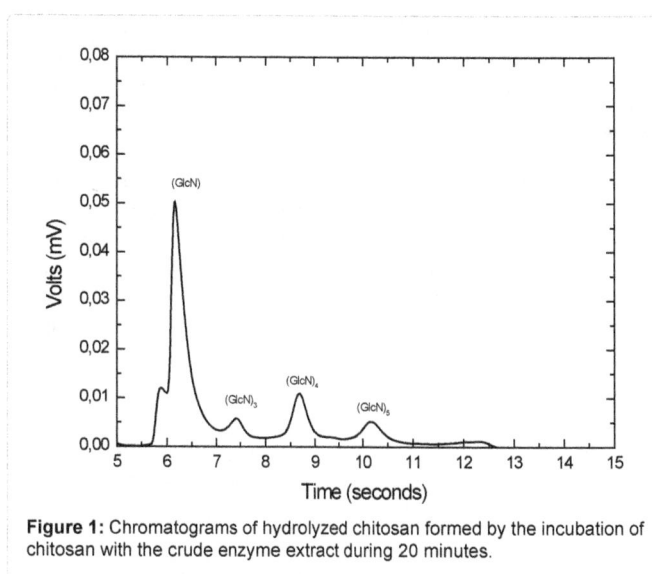

Figure 1: Chromatograms of hydrolyzed chitosan formed by the incubation of chitosan with the crude enzyme extract during 20 minutes.

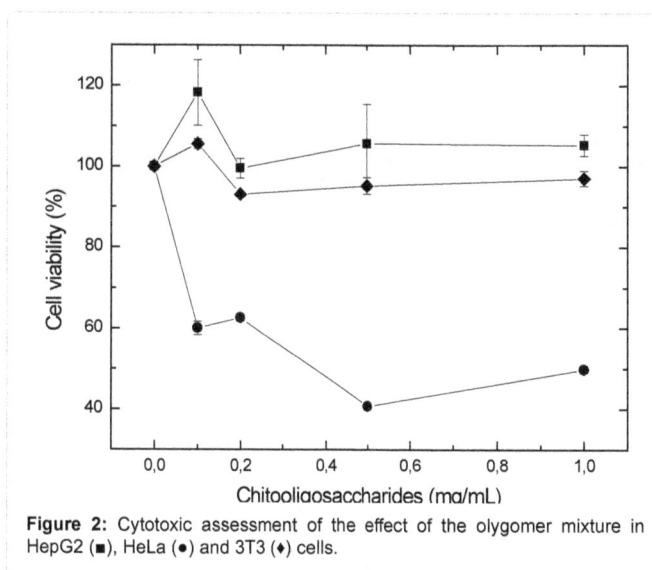

Figure 2: Cytotoxic assessment of the effect of the olygomer mixture in HepG2 (■), HeLa (●) and 3T3 (♦) cells.

pound. When these cells were treated with the hydrolysate containing the chitosan oligomers, no alteration in cell proliferation was detected. Thus, the tumor cells (HepG2 and HeLa) were treated with concentrations that were non-toxic to normal cells.

The HepG2 cells, when treated with the chitooligosaccharides at a concentration of 0.2 mg/mL, induced cell viability of around 20% and at higher concentrations, it remained practically unaltered. These results corroborate literature data, in that chitooligosaccharides did not show any activity against Hep3B cells [14,15].

In the HeLa cells, diminished cell viability was directly proportional to the increase in concentration, where cell viability decreased by 60% at a concentration of 0.5 mg/mL. Similar results were found by Jeon and Kim [16], where chitooligosaccharides inhibited uterine tumors in rats by 73.6%. The modified chitooligosaccharides used in HeLa cells showed an IC_{50} value of 0.45 mg/mL [14] and similar results were found in this study, where IC50 was 0.67 mg/mL. Kim and Rajapakse [12] observed that a mixture of chitosan oligomers from tetramer to pentamer could inhibit tumor cell growth in rats. It was observed that HeLa cells were more sensitive then the other ones showing that the biological functions of chitooligosaccharides depend not only on the degree of polymerization, but also on its molecular weight [12-16].

Conclusion

According to the results obtained, chitooligosaccharides may reduce the viability of HeLa tumor cells. The effect of the antitumor activity of chitooligosaccharides is mainly due to stimulation in the immune system, increasing *in vivo* macrophage activity. However, according to our findings, chitosan oligomers act directly on HeLa tumor cells, decreasing cell proliferation and viability. Moreover, it is interesting to observe the non-toxicity of the compound in 3T3 fibroblast cells.

Acknowledgements

The authors thank Financiadora de Estudos e Projetos (FINEP), Conselho Nacional de Desenvolvimentos Científico e Tecnológico (CNPq) and (Coordenação de Aperfeiçoamento de Pessoal de Nível Superior (CAPES) for financial support that made this work possible.

References

1. Kim SY, Shon DH, Lee KH (1998) Purification and characteristics of two types of chitosanases from Aspergillus fumigatus KH-94. J Microbiol Biotechol 8: 568 - 574.

2. Lee HW, Choi JW, Han DP, Lee NW, Park SL, et al. (1996) Identification and production of constitutive chitosanase from Bacillus sp. HW-002. J Microbiol Biotech 6: 12-18.

3. Zhang H, Du Y, Yu X, Mitsutomi M, Aiba SI (1999) Preparation of chitooligosaccharides from chitosan by a complex enzyme. Carbohyd Res 320: 257-260.

4. Ming M, Kuroiwa T, Ichikawa S, Sato S, Mukataka S (2006) Production of chitosan oligosaccharides by chitosanase directly immobilized on an agar gel-coated multidisk impeller. Biochem Engin J 28: 289-294.

5. Suzuki K, Mikami T, Okawa Y, Tokoro A, Suzuki S, et al. (1986) Antitumor effect of hexa-N-acetylchitohexaose and chitohexaose Carbohyd Res 151: 403-408.

6. Yu Z, Zhao L, Ke H (2004) Potential role of nuclear factor-kappaB in the induction of nitric oxide nd tumor necrosis factor-alpha by oligochitosan in macrophages. Int Immunopharmacol 4: 193-200.

7. Prashanth KVH, Tharanathan RN (2005) Depolymerized products of chitosan as potent inhibitors of tumor-induced angiogenesis. Biochimica et Biophysica Acta 1722: 22-29.

8. Miller GL (1959) Use of Dinitrosalicylic Acid Reagent for Determination of Reducing Sugar. Anal Chem 31: 426-428.

9. Liang TW, Chen YJ, Yen YH, Wang SL (2007) The antitumor activity of the hydrolysates of chitinous materials hydrolyzed by crude enzyme from Bacillus amyloliquefaciens V656. Proc Biochem 42: 527-534.

10. Mosmann T (1983) Rapid colorimetric assay for cellular growth and survival: Application to proliferation and cytotoxicity assays. J Immunol Methods 65: 55-63.

11. Denizot F, Lang R (1986) Rapid colorimetric assay for cell growth and survival: Modifications to the tetrazolium dye procedure giving improved sensitivity and reliability. J Immunol Methods 89: 271-277

12. Kim SK, Rajapakse N (2005) Enzymatic production and biological activities of chitosan oligosaccharides (COS): A review. Carbohydr Pol 62: 357-368.

13. Assis CF, Araújo NK, Pagnoncelli MGB, Pedrini MRS, Macedo GR, et al. (2010) Chitooligosaccharides enzymatic production by *Metarhizium anisopliae*. Bioprocess Biosys Eng 33: 893-899.

14. Assis CF, Costa LS, Melo-Silveira RF, Oliveira RM, Pagnoncelli MGB, et al. (2011) Chitooligosaccharides antagonize the cytotoxic effect of glucosamine. World J Microbiol Biotechnol 28: 1097-1105.

15. Jeon YJ, Kim SK (2002) Antitumor Activity of Chitosan Oligosaccharides Produced in Ultrafiltration Membrane Reactor System. J Microbiol Biotechnol 12: 503-507.

16. Shen KT, Chen MH, Chan HY, Jeng JH, Wang YJ (2009) Inhibitory effects of chitooligosaccharides on tumor growth and metastasis. Food Chem Toxic 47: 1864-1871.

6

Introduction of Extra Copy of Oxytetracycline Resistance Gene *otrB* Enhances the Biosynthesis of Oxytetracycline in *Streptomyces rimosus*

Xiaohe Chu[1,2], Zijing Zhen[1], Zhenyu Tang[1], Yingping Zhuang[1], Ju Chu[1], Siliang Zhang[1] and Meijin Guo[1*]

[1]*State Key Laboratory of Bioreactor Engineering, East China University of Science and Technology, Shanghai 200237, P.R. China*
[2]*Zhejiang Shenghua Biok Biology Co., Zhejiang 313220, P.R. China*

Abstract

The aromatic polyketide antibiotic oxytetracycline (OTC) is produced by *Streptomyces rimosus* as an important secondary metabolite. Enhancement of self-resistance is one effective way to improve antibiotic production in *Streptomyces* spp. In the present study, we aimed to improve the production of OTC by introducing extra copies of the OTC resistance genes, *otrA* and *otrB*, into the chromosome of the industrial strain of *S. rimosus* (SRI). First, *otrA* and *otrB* were cloned and ligated with pSET152 to generate the recombinant plasmids pSET152-*otrA/otrB*, the demethylated pSET152-*otrA/otrB* by *Escherichia coli* ET12567(pUZ8002) were then introduced into *SRI* to yield *otrA/otrB* knock-in mutants: *SRI-A* (*otrA*) and *SRI-B* (*otrB*). Ten selected mutants and the parent *SRI* strain were cultured in shake flasks. Production of OTC was increased by 67% in one *SRI-B* mutant compared with the parent strain, suggesting that the enhancement of resistance gene *otrB* in the antibiotic producer is an effective way to improve OTC biosynthesis. However, introduction of extra copy of *otrB* could retard growth of mutant cells.

Keywords: Antibiotic; Aromatic polyketide; Oxytetracycline (OTC) Biosynthesis; Self-Resistance

Introduction

Streptomyces spp. are filamentous Gram-positive aerobic soil-dwelling bacteria that belong to the family *Streptomycetaceae* and the order *Actinomycetales*. *Streptomyces* spp. and closely related genera have the ability to coordinate the production of a wide variety of secondary metabolites during morphological development [1]. Many of these secondary metabolites have antibiotic properties. Current methods employed to increase the antibiotic productivity of industrial microorganisms range from classical random mutagenesis studies performed in conjunction with the optimization of large-scale industrial fermentations. For more rational approaches, metabolic engineering is a common method used by researchers to regulate the production of many antibiotics. For example, genetic modifications of primary metabolic fluxes can lead to increases in the productivity of antibiotic synthesis [2,3], since the availability of biosynthetic precursors is a key factor that determines their production. To date, many studies have reported the improvement of antibiotic production by engineering the availability of certain precursors in the producer organisms [4-6]. Other methods, such as the optimization of fermentation conditions, are also available to regulate antibiotic production [7].

The aromatic polyketide antibiotic oxytetracycline (OTC) is produced by *Streptomyces rimosus* as an important secondary metabolite. There are three resistance genes in the OTC biosynthesis cluster,

namely *otrA*, *otrB* and *otrC*, of which *otrA* and *otrB* are located at either end of this cluster. *otrA* changes the conformation of the 30S ribosome non-covalently and prevents the binding of OTC . Furthermore, *otrA* may be a substitute for the regulatory elongation factor [8], and *otrB* encodes a membrane transport protein that aids the transportation of OTC out of the cell. The *otrB* sequence shares great similarity with other transport genes, including *tetA* from Tn10 [9]. However, the function of *otrC* remains to be elucidated. A traditional mutation program has resulted in the improvement of OTC production from 2 g•l^{-1} to 80 g•l^{-1}, and OTC production can also be improved by disrupting the zwf (coding glucose-6-phosphate dehydrogenase) gene [10]. Nevertheless, there is no information concerning the effects of *otrA*, *otrB* and *otrC* on OTC production.

In this present study, we aimed to investigate the influence of *otrA* and *otrB* on OTC production by introducing extra copies of these resistance genes into the genome of the industrial strain of *S. rimosus* (SRI). Herein, it is shown that production of OTC was increased in the *otrB* mutant, though an extra copy of *otrA* had no effect on OTC production.

Materials and Methods

Bacterial strains, media, and plasmids

The strains and plasmids used in this present study are listed in

*Corresponding author: Meijin Guo, State Key Laboratory of Bioreactor Engineering, P.O. box 329, East China University of Science and Technology, 130 Meilong Rd., Shanghai 200237, P.R. China

Strains or plasmids	Function	Source
E. coli Top10	plasmid amplification	our laboratory
E. coli ET12567 (pUZ8002)	plasmid demethylation	John Innes Center, UK
Industrial *S. rimosus* (SRI)	OTC producer	Shanxi Datong Antibiotic Company
pMD19T	gene amplification	TARAKA, Japan
pMD19T-*otrA*	gene amplification	this study
pMD19T-*otrB*	gene amplification	this study
pSET152	gene integration	
pSET152-*otrA*	gene integration	this study
pSET152-*otrB*	gene integration	this study

Table 1: Strains and plasmids used in this present study.

Primer	Sequence	Digestion site
P1(*otrA*-F)	5' CGCCATATGATGAACAAGCTGAATCTGGG 3'	*Nde*I
P2(*otrA*-R)	5' GGAAGCTTTCTAGATCACACGCGCTTGAGC 3'	*Xba*I
P3(*otrB*-F)	5' CGCCATATGGTGTGTCATCCGCAAATCCG 3'	*Nde*I
P4(*otrB*-R)	5' CCAAGCTTGCTCTAGATCAGGCGTCCGACGC 3'	*Xba*I
P5(*attB*-F)	5' GTTCACCAACAGCTGGAGGC 3'	
P6(*attB*-R)	5' CGTCATGCCCGCAGTGACC 3'	

Table 2: Primers used in this present study.

Table 1. Organisms were grown at 37°C in Luria-Bertani medium (1% tryptone, 1% NaCl, 0.5% yeast extract), and standard procedures were used for transformations [11]. *Escherichia coli* transformants were selected with ampicillin (100 mg•mL⁻¹), apramycin (50 mg•ml⁻¹), kanamycin (10 mg•ml⁻¹) or chloramphenicol (25 mg•ml⁻¹). *SRI* (S. rimosus M4018) was grown and manipulated as described previously [12].

Construction of recombinant plasmids and *SRI* mutants

Primers: All the primers used are listed in Table 2.

Construction of recombinant plasmids: The *otrA* (2.1 kb) and *otrB* (1.7 kb) gene fragments were amplified using primers P1 and P2 or P3 and P4. Genomic DNA of *SRI* was used as the template. The fragments were cloned into the pMD19T vector to yield pMD19T-*otrA/otrB*, and then transformed into competent cells of *E. coli* Top10. Recombinant clones were screened by white-blue plaque selection and recombinant plasmids were analyzed by both single and double restriction enzyme digestion. pMD19T-*otrA/otrB* were digested with *Nde*I and *Xba*I, and then cloned into pSET152, which was digested with the same restriction enzymes to give pSET152-*otrA/otrB*. Then, the plasmids were transferred into *E. coli* ET12567 for demethylation and stored at -20°C until later use.

Gene enhancement and identification of mutants: Demethylated pSET152-*otrA/otrB* were electroporated into *SRI* competent cells at 2 kV, 25 μF and 400Ω. Exconjugants were selected on tryptone soy agar plates containing apramycin 500 µg•mL⁻¹ and incubated at 30°C for 4–6 days. Mutants were confirmed by polymerase chain reactions (PCR) using *aprF* (P5) and *aprR* (P6) as primers.

Fermentation experiments

A spore suspension was inoculated into 30 ml of seed medium containing glucose (10 g•l⁻¹), yeast extract (0.5 g•l⁻¹), tryptone (15 g•l⁻¹), sucrose (2.8 g•l⁻¹) and calcium carbonate (0.1 g•l⁻¹). The first seed cultures were grown for 3 days at 260 rpm and 30°C. Then, 2 ml of the first seed culture was inoculated into 50 ml of fermentation medium [12] in a 500-ml shaking flask with a spring. The second cultures were grown for 8 days at 260 rpm and 30°C.

For determination of dry cell weight, 5-ml samples of each culture were collected every 24 h and dried at 105°C to constant weight. OTC production *in vivo* and *in vitro* was analyzed by high performance liquid chromatography according to reference 9.

Results

Validation of the introduction of extra *otrA* and *otrB* genes

The recombinant plasmids pSET152-*otrA* and pSET152-*otrB* were verified by *Xba*I and *Nde*I digestion (Figure 1A). As indicated in Figure 1B, the positive clones of *otrA* and *otrB* transformants showed strong

signals at 750 bp (apramycin resistance gene), while the wild-type pSET152 and pKC1139 showed no bands. After site-specific integration, the entire recombinant plasmid should be inserted into the genome of *SRI* as shown in Figure 1C. Then, positive clones were identified by cross-over sites (Figure 2) using the primers *otrA*-F and *attB*-R or *attB*-F and *apr*-R.

Screening of the high-productivity mutants

Eight *SRI-A* mutants (containing an extra copy of *otrA*) and 12 *SRI-B* mutants (containing an extra copy of *otrB*) were chosen for the shake flask experiments. After 8 days of fermentation, there were no obvious differences in the levels of OTC in the *SRI* parent strain, the *SRI-A* mutants and most of the *SRI-B* mutants (Figure 3). However,

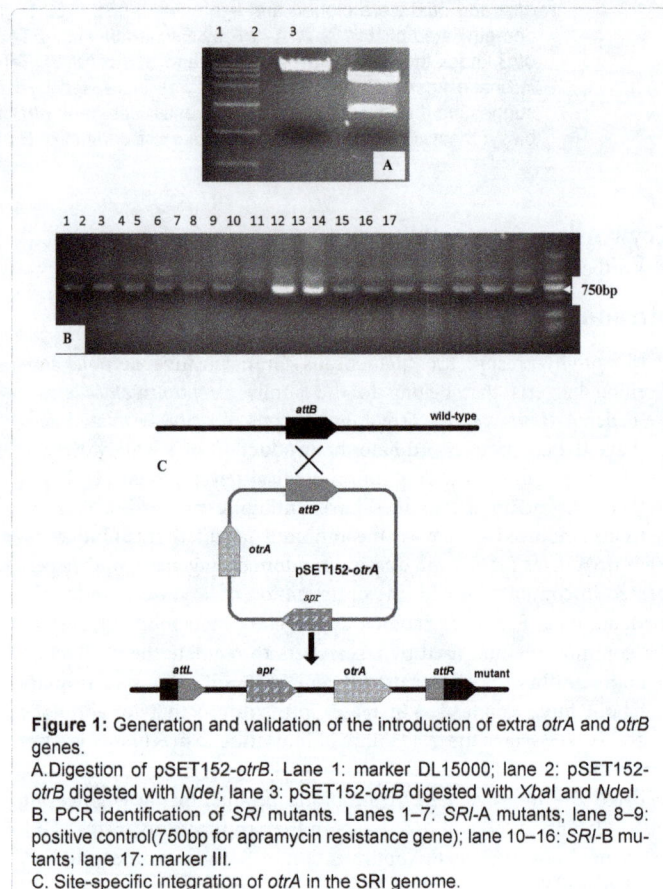

Figure 1: Generation and validation of the introduction of extra *otrA* and *otrB* genes.
A. Digestion of pSET152-*otrB*. Lane 1: marker DL15000; lane 2: pSET152-*otrB* digested with *Nde*I; lane 3: pSET152-*otrB* digested with *Xba*I and *Nde*I.
B. PCR identification of *SRI* mutants. Lanes 1–7: *SRI*-A mutants; lane 8–9: positive control(750bp for apramycin resistance gene); lane 10–16: *SRI*-B mutants; lane 17: marker III.
C. Site-specific integration of *otrA* in the SRI genome.

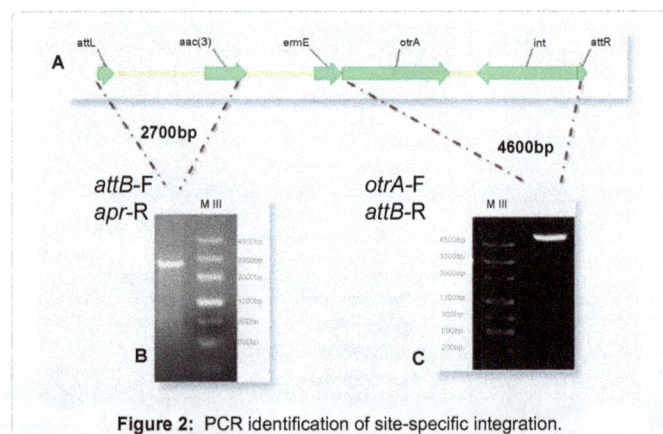

Figure 2: PCR identification of site-specific integration.

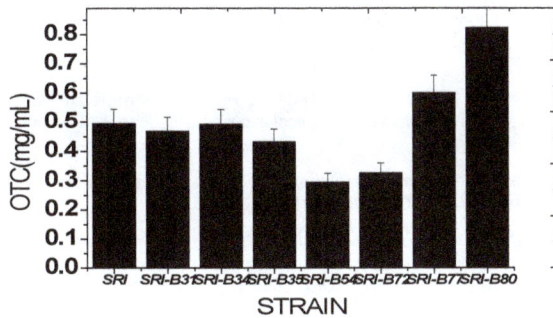

Figure 3: OTC levels in *SRI*, and *SRI-A* and *SRI-B* mutants.

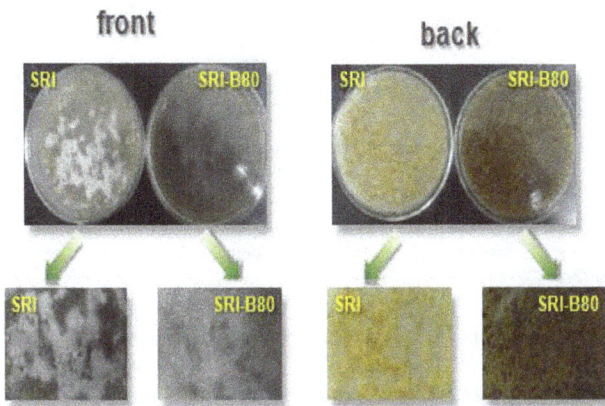

Figure 4: Pigment and spore formation by *SRI-B*80 and *SRI*.

OTC production was 67% higher in strain *SRI-B*80 compared with the control, indicating that *SRI-B*80 had high production of OTC.

Knowing that *Streptomyces* spp. spore formation is important for growth and the production of secondary metabolites [13], spore formation was compared between the *SRI* and *SRI-B*80 strains. After 6 days of incubation, mature spores were found on the surface of the *SRI* colony, but the *SRI-B*80 spores were malformed and immature. The onset of morphological differentiation generally coincides with the production of secondary metabolites and resistance genes are often co-regulated with these metabolites [14]. Indeed, spore formation is very closely related with the expression of resistance genes. *otrB* may be a negative regulator of spore formation and morphological differentiation. Moreover, the level of pigments produced by some *Streptomyces* spp. can reflect the magnitude of antibiotic production. Visual inspection of agar plates revealed that *SRI-B*80 produced more pigment than *SRI*, and this coincided with OTC production (Figure 4).

Growth characteristics of *SRI* and *SRI-B*80b

The physiological characteristics of *SRI-B*80 were further investigated and compared with the *SRI* parent. The growth curves of *SRI* and *SRI-B*80 are given in Figure 5A. The final biomass of *SRI-B*80 was markedly lower than *SRI*, indicating that the extra copy of *otrB* had a negative effect on growth. Furthermore, *SRI-B*80 had a slower growth rate, even during exponential growth phase. There are two possible reasons for this. First, the extra copy of *otrB* may increase the physiological burden of the cells. Second, re-distribution of cellular resources may occurred so that more materials could be directed into OTC biosyn-

thesis in the *SRI-B*80 strain because the specific OTC productivity of *SRI-B*80 was twice that of *SRI* after 5 days of fermentation (Figure 5B).

OTC efflux mediated by the extra copy of *otrB* in *SRI-B*80

The intracellular and extracellular concentrations of OTC were determined after fermentation for the *SRI* and *SRI-B*80 strains. As expected, *SRI-B*80 showed enhanced efflux of OTC to some extent. After 8 days of fermentation, the introcellular OTC concentration in *SRI-B*80 was 13.6% lower than found in *SRI* (Figure 6A), which may reduce the risk of cell suicide due to high concentrations of OTC. Concomitantly, OTC concentration outside of the cells was 21.2% greater for *SRI-B*80 than detected for *SRI*. This shows that the extra copy of *otrB* played an important role in transporting OTC out of the cells. The extra *otrB* gene in *SRI-B*80 enabled greater transport of OTC out of the cells and this ensured that OTC in the cells was maintained at sufficiently low concentrations to cause any harm (Figure 6B).

Discussion

In this present study, extra copies of *otrA* and *otrB* were introduced into the genome of *SRI*. The extra copy of *otrA* did not exert any effects on OTC production; however, the extra copy of *otrB* enhanced OTC production markedly. Nevertheless, the *SRI-B*80 mutant had a slower growth rate, and spore formation and morphological differentiation were also affected, which suggests that *otrB* may be a multi-functional regulator in *SRI*. Through comparison of OTC concentrations *in vivo* and *in vitro* for *SRI* and *SRI-B*80, the extra *otrB* gene was shown to facilitate the intracellular efflux of OTC. Thus, the lower concentrations of OTC *in vivo* decrease the toxicity of the excess OTC, which may in turn stimulate the cells to produce more of this metabolite.

Many microbes can synthesize potentially toxic secondary metabolites (mainly antibiotics), and so there must exist pathways or mech-

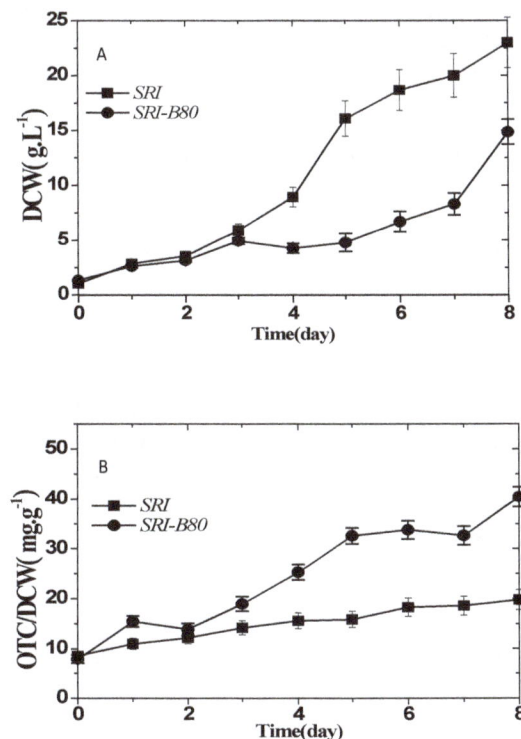

Figure 5: Growth and specific OTC productivities of *SRI* and *SRI-B*80.

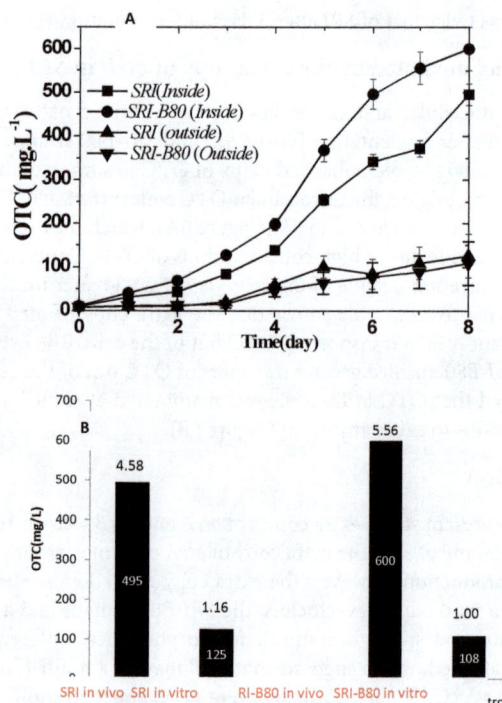

Figure 6: Intracellular and extracellular concentrations of OTC in *SRI* and *SRI-B*80.

anisms that protect the producer. Until now, various self-protection mechanisms have been reported [15], and by enhancing these self-protection mechanisms, the production of many antibiotics has been increased in various microorganisms. For example, Malla et al. [14] overexpressed the resistance genes *drrA*, *drrB* and *drrC* in *Streptomyces peucetius* ATCC 27952, such that the recombinant strains produced more doxorubicin (DXR) than the parental strain. Indeed, there was a 2.2-fold increase in DXR in the *drrAB* mutant, a 5.1-fold increase in the *drrC* mutant and a 2.4-fold increase in the *drrABC* mutant. Similarly, Olano et al. [16] demonstrated that improvements in self-resistance could increase the production of bioactive secondary metabolites in the actinomycetes. Thus, the introduction of extra resistance genes is an effective approach to enhance antibiotic production.

Although an extra copy of *otrB* improved OTC production, the additional copy of *otrA* had little or no effect on this parameter. This may be because *otrA* in *SRI* produces sufficient enzyme to modify the ribosome and provide a safe environment. Therefore, *otrA* may be saturating, but this assumption requires further investigation. Moreover, the observation that *SRI-B*80 has a slow rate of sporulation is consistent with the findings of Davies [17] and Den Hengst et al [18]. Specific growth states, such as biofilm formation, anaerobiosis and sporulation, can differentially impact on cell susceptibility to antimicrobials. It can be concluded that a series of genes that regulate *otrB* may have altered expression in *SRI-B*80, and these changes may have knock-on effects on growth and sporulation in these mutants. Since the regulation network in these cells is very complicated, further experiments are needed to understand the relations between all of these affected genes.

Acknowledgements

This work was supported by grants from the Fundamental Research Funds for the Central Universities (ECUST), the Outstanding Youth Talent Plan of East China University of Science and Technology (ECUST) and the Open Project Program of the State Key Laboratory of Bioreactor Engineering, ECUST (No.2060204).

References

1. Thompson CJ, Fink D, Nguyen LD (2002) Principles of microbial alchemy: insights from the *Streptomyces coelicolor* genome sequence. Genome Biol 3: 1020.

2. Nielsen J (1998) The role of metabolic engineering in the production of secondary metabolites. Curr Opin Microbiol 1: 330–336.

3. Butler MJ, Bruheim P, Jovetic S, Marinelli F, Postma PW, et al. (2002) Engineering of primary carbon metabolism for improved antibiotic production in *Streptomyces lividans*. Appl Environ Microbiol 68:4731-4739.

4. Reeves AR, Cernota WH, Brikun IA, Wesley RK, Weber JM (2004) Engineering precursor flow for increased erythromycin production in *Aeromicrobium erythreum*. Metab Eng 6: 300–312.

5. Wei X, Yunxiang L, Yinghua Z (2006) Enhancement and selective production of oligomycin through inactivation of avermectin's starter unit in *Streptomyces avermitilis*. Biotechnol Lett 28: 911–916.

6. Ryu YG, Butler MJ, Chater KF, Lee KJ (2006) Engineering of primary carbohydrate metabolism for increased production of actinorhodin in *Streptomyces coelicolor*. Appl Environ Microbiol 72: 7132–7139.

7. Avignone Rossa C, White J, Kuiper A, Postma PW, Bibb M, et al. (2002) Carbon flux distribution in antibiotic-producing chemostat cultures of *Streptomyces lividans*. Metab Eng 4: 138–150.

8. Doyle D, McDowall KJ, Butler MJ, Hunter IS (1991) Characterization of an oxytetracycline resistance gene, *otrA* of *Streptomyces rimosus*. Mol Microbiol 5: 2923-2933.

9. McMurry LM, Levy SB (1998) Revised sequence of *OtrB* (tet347) tetracycline efflux protein from *Streptomyces rimosus*. Antimicrob Agents Chemother 42: 3050.

10. Tang Z, Xiao C, Zhuang Y, Chu J, Zhang S, et al. (2011) Improved oxytetracycline production in *Streptomyces* rimosus M4018 by metabolic engineering of the G6PDH gene in the pentose phosphate pathway. Enzyme Microbial Technol 49: 17-24.

11. Sambrook J, Fritsch EF, Maniatis T (1989) Molecular cloning: a laboratory manual. (2ndedn), New York, Cold Spring Harbor Laboratory Press

12. Kieser T, Bibb MJ, Buttner MJ, Chater KF, Hopwood DA(2000) Practical *Streptomyces* genetics. John Innes Foundation, Norwich, United Kingdom.

13. Gverzdys TA (2011) The development of protocols to engineer and screen *Streptomyces* in high throughput to test for the activation of cryptic clusters by the heterologous expression of pleiotropic regulators. Open Access Dissertations and Theses. Paper 6090.

14. Malla S, Niraula NP, Liou K, Sohng JK (2010) Self-resistance mechanism in *Streptomyces peucetius*: Overexpression of *drrA*, *drrB* and *drrC* for doxorubicin enhancement. Microbiol Res 4: 259–267.

15. Cundliffe E, Demain AL (2010) Avoidance of suicide in antibiotic-producing microbes. J Ind Microbiol Biotechnol 37: 643–672.

16. Olano C , Lombó F, Méndez C, Salas JA (2008) Improving production of bioactive secondary metabolites in actinomycetes by metabolic engineering. Metab Eng 10: 281-292.

17. Crameri R, Davies JE (1986) Increased production of aminoglycosides associated with amplified antibiotic resistance genes. J Antibiot 39: 128-135.

18. Den Hengst CD, Tran NT, Bibb MJ, Chandra G, Leskiw BK, et al. (2010) Genes essential for morphological development and antibiotic production in *Streptomyces* coelicolor are targets of BldD during vegetative growth. Mol Microbiol 78: 361–379.

7

Cloning and Functional Expression of Z-Carotene Desaturase, A Novel Carotenoid Biosynthesis Gene From *Ficus Carica*

Araya-Garay JM[1], Feijoo-Siota L[1], Veiga-Crespo P[1,2], Sánchez-Pérez A[3] and González Villa T[1,2]*

[1]*Department of Microbiology, University of Santiago de Compostela, Spain*
[2]*School of Biotechnology, University of Santiago de Compostela, 15782 Santiago de Compostela, Spain*
[3]*Discipline of Physiology, Bosch Institute, University of Sydney, NSW 2006, Australia*

Abstract

Carotene desaturation, an essential step in the carotenoid biosynthesis pathway, is catalyzed by two enzymes, phytoene desaturase (PDS) and ζ-carotene desaturase (zeta carotene desaturase, ZDS). Here we describe cloning and *E. coli expression* of zdsfc, a novel *Ficus carica* ζ-carotene desaturase catalyzing dehydrogenation of ζ-carotene into neurosporene and finally lycopene. The ζ-carotene desaturase (ZDS) gene was amplified from the fig tree by rapid amplification of cDNA ends (RACE) and spanned a 1746 bp open reading frame (ORF), encoding a protein of 582 amino acid residues with a predicted molecular weight of 64kD. The N-terminal region of this polypeptide contained a putative transit sequence for plastid targeting. By phylogenetic and sequence analyses, zdsfc showed high homology with previously described ζ-carotene desaturases from higher plant species [1-4]. Additionally, sequence analysis revealed a high degree of conservation among plant ZDSs. The deduced ZDS protein, designated zdsfc, also contains an N-terminus dinucleotide-binding, followed by a conserved region identified in other carotene desaturase sequences. These data, taken together, confirm our cloned zdsfc as an integral part of the ZDS family of proteins.

Keywords: *Ficus carica*; Zeta-carotene desaturase; cDNA; Zeta-carotene; Neurosporene; Lycopene

Introduction

Carotenoids are pigments synthetized by plants, fungi, bacteria, and algae with the main function of protecting them from the action of singlet oxygen and other radicals [5]. In plants, carotenoids are either primary or secondary metabolites. As primary metabolites, carotenoids can function as regulators of plant growth and development, as accessory pigments in photosynthesis, as photoprotectors preventing photo-oxidative damage, or as precursors to the hormone abscisic acid (ABA) [6]. They are also responsible for the color of fruits and flowers, generating distinct yellow, orange, and red colors, thus substantially contributing to plant-animal communication [6,7]. In addition, the colors of many carotenoid-accumulating fruits and flowers also increase their appeal and hence their economic value [8,9]. These pigments also play an important protective role, in human and animal diets, as antioxidants.

The main carotenoid metabolic pathway is well known and may be shared by most of the carotenogenic species [10].The first step in this pathway is the condensation of two molecules of geranylgeranyl pyrophosphate (GGPP) to originate the first true carotenoid C_{40} molecule, phytoene [11]. This two-step reaction is catalyzed by a single soluble enzyme: phytoene synthase (PSY) (Figure 1). Two sequential desaturations of phytoene result in the formation of the first phytofluene, followed by ζ-carotene. Both of these reactions are catalyzed by phytoene desaturase (PDS). Two additional desaturations, catalyzed by ζ-carotene desaturase (ZDS), give rise to first neurosporene and finally lycopene, the symmetrical red carotenoid pigment (Figure 1) [12].

Since 1990, when cloned the carotenoid biosynthesis gene cluster from *Erwinia uredovora*, many carotenoid biosynthetic genes have been identified in plants and other organisms We have previously reported the production of β-carotene by expression of recombinant *Ficus carica* lycopene beta-cyclase in *E. coli* [13,14]. Here, we describe cloning and *Escherichia coli* expression of zdsfc, a novel *Ficus carica* ζ-carotene desaturase catalyzing dehydrogenation of ζ-carotene into

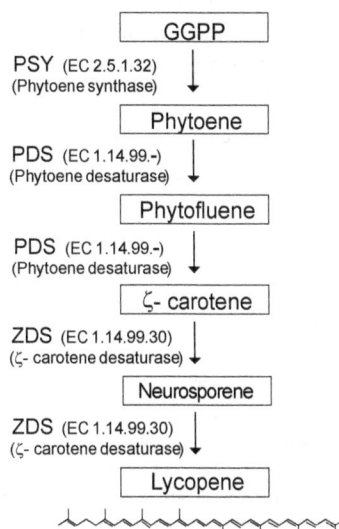

Figure 1: Schematic diagram of the early steps in the carotenoid biosynthetic pathway in plants, from geranyl geranyl pyrophosphate (GGPP) to lycopene, and the enzymes catalyzing those reactions.

neurosporene and finally lycopene. The *E. coli* culture containing zdsfc was able to convert ζ-carotene and neurosporene, two substrates of

*Corresponding author: González Villa T, Department of Microbiology, University of Santiago de Compostela, Spain

ζ-carotene desaturase, into neurosporene and lycopene, or lycopene, respectively; thus confirming, *in vivo*, the enzymatic activity of the recombinant ZDS we produced. The *F. carica* enzyme also shared the characteristic properties of other plant ZDSs, and this allowed us to construct a phylogenetic tree illustrating the evolutionary relationship between ZDS and other published carotenoid zeta carotene desaturases (ZDSs) from higher plants.

Materials and Methods

Materials

Plasmids pACCRT-EBI and pACCRT-EBR, encoding *Erwinia uredovora* crtE (GGPP synthase), crtB (phytoene synthase) and crtI (phytoene desaturase), and encoding *Erwinia uredovora* crtE, crtB plus Rodococcus crtR, respectively, were a present from Prof. Misawa (Research Institute for Bioresources and Biotechnology, Ishikawa Prefectural University, Japan). Whereas plasmid pACCRT-EBP, encoding *Erwinia uredovora* crtE, crtB, plus *Synechococcus* crtP was kindly donated by Prof. Gerhard Sandmann (J. W. Goethe Universität, Frankfurt, Germany) [15-17]. Plasmids pCR Blunt II Topo and pUC19, as well as *E. coli* strains TOP10 and BL21 (DE3) and Zero Blunt TOPO PCR Cloning Kit, were from Invitrogen (Carlsbad, CA, USA). The expression plasmid pET21a was purchased from Novagen, (Cambridge, UK). RNeasy Plant Mini Kit was from Qiagen (Valencia, CA, USA). The RT-PCR AMV kit was obtained from Roche Applied Science (Indianapolis, IN, USA). Accuzyme DNA polymerase was from Bioline (Taunton, MA, USA).

Cloning of *Ficus carica* ζ-carotene desaturase cDNA (zdsfc)

Total RNA was isolated, using the RNeasy Plant Mini Kit, from 50 mg of fresh *Ficus carica* leaf, following the instructions recommended by the manufacturer. RT-PCR was carried out using the RT-PCR AMV kit and degenerate primers ZDS DegF and ZDS DegR (Table 1). These oligonucleotide primers were designed using Primer Premier 5.0 (Biosoft International), according to the conserved motifs found on other published zeta-carotene desaturase sequences. Amplification of zdsfc conserved internal DNA fragment was achieved by PCR, using the above-described oligonucleotide primers and 2 μL of first-strand cDNA. The reaction mixture contained polymerase buffer, 0.2 mM of each primer, 1.5 mM $MgCl_2$, 0.2 mM of each deoxynucleotide, and 1 unit of Accuzyme DNA polymerase, following program: 94°C for 2 min, 35 cycles of 94°C for 45 s; 49°C for 1 min and 72°C for 2 min; and a final extension at 72°C for 2 min. The PCR product obtained was subcloned into pCR Blunt II TOPO, using the Zero Blunt TOPO PCR Cloning Kit. All DNA constructs were checked by sequence analysis.

RACE-PCR

The 5′and 3′ ends of zdsfc were obtained by RACE (rapid amplification of cDNA ends), using the 5′/3′ RACE *kit*, 2nd Generation (Roche Applied Science), primers ZDSFc1, ZDSFc2, ZDSFc3, ZDSFc4 and ZDSFc5 (Table 1), and 1 unit of Accuzyme DNA Polymerase. ZDSFc1 was the primer used for first strand cDNA synthesis, whereas the missing 5′ region of the gene was amplified with ZDSFc2 and Oligo dT-Anchor Primer, and incubated as follows: 1 cycle at 94°C for 2 min; 10 cycles of 94°C for 15 s, 55°C for 30 s and 72°C for 40 s; 25 cycles of 94°C for 15 s, 55°C for 30 s and 72°C for 40 s, increasing the elongation time by 20 s for each successive cycle (i.e. elongation time of cycle no.11 is 40 s, of cycle no. 12 is 60 s, of cycle no. 13 is 80 s etc.), and a final extension at 72°C for 7 min.

The PCR-amplified DNA product was then used as a template

Primer name	Primer sequence	Application
ZDS DegF	5′ GGBGAACTTGATTTYCGVTT 3′	RT-PCR
ZDS DegR	5′ ATCACANGCWGCMACATATGCATC 3′	RT-PCR
ZDS-Fc1	5′ CCCCACCTAAGATGAAACCTGCCC 3′	5′ RACE-PCR
ZDS-Fc2	5′ GCGTGCCACCTTTGGAGATGAAC 3′	5′ RACE-PCR
ZDS-Fc3	5′ GCCCCAACTGGAAACCGGAAAT 3′	5′ RACE-PCR
ZDS-Fc4	5′ GCCCAGTTGTAAAGGCTCTCGTTGA 3′	3′ RACE-PCR
ZDS-Fc5	5′ GGTTCATCTCCAAAGGTGGCACGC 3′	3′ RACE-PCR
ZDS-EcoRI F	5′ **CGGAATTC**ATGGCTTCTT-GGGTTCTTTTCC 3′	PCR (full ORF)
ZDS-XhoI R	5′ **CCGCTCGAG**CTAAACAAGACTA-AGCTTGTC 3′	PCR (full ORF)
	B=C, G, T; Y=C, T; V= A, C, G; N= A, T, C, G; W=A, T; M=C, A	

Table 1: PCR primers used in this study, their oligonucleotide sequence is shown 5' to 3', and the purpose the primers were generated for.
Note: F, forward; R, reverse; ORF, open reading frame.

(dilution 1:20) for a subsequent PCR, with ZDSFc3 and PCR Anchor Primer as primers. This second PCR was incubated as described above, with the exception of the primer annealing step, carried out at 60°C.

The 3′ end of zdsfc gene was amplified in two sequential PCR reactions, using ZDSFc4 and ZDSFc5, respectively, and the PCR Anchor as primers. The DNA sequence of the PCR-amplified fragments was confirmed by sequence analysis.

PCR primers ZDS-EcoRI F and ZDS-XhoI R were designed to amplify the complete zdsfc coding region (Table 1), using the above described zdsfc DNA fragments as template. These primers contained EcoRI and XhoI restriction sites at their 5′ and 3′ ends, respectively (in bold in Table 1). The PCR conditions were: 1 cycle at 95°C for 3 min; 35 cycles of 95°C for 10 s, 57°C for 10 s and 72°C for 2 min ; and a final extension at 72°C for 10 min. The resulting PCR product was cloned into a pET21a vector, using the restriction sites present in the primers, thus producing the expression plasmid pET21-zdsfc (Table 2).

Bacterial transformation and growth

Transformation of plasmid pACCRT-EBI into *E. coli* BL21 (DE3) generated *E. coli* BL-pEBI (Table 2), a bacterial strain expressing the enzymes GGPP synthase, phytoene synthase and phytoene desaturase, from *Erwinia uredovora*, and hence capable of producing lycopene. This recombinant *E. coli* was used as positive control in our zeta-carotene desaturase enzymatic assay (see below). Transformation of plasmid pACCRT-EBP into *E. coli* BL21 (DE3) resulted in *E. coli* BL-EBP (Table 2), a bacterial strain expressing Eu-crtE, Eu-crtB and Sc-crtP. *E. coli* BL-pEBR (Table 2) was generated by transforming *E. coli* BL21 (DE3) with plasmid pACCRT-EBR, encoding Eu-crtE, Eu-crtB and Rc-crtR, and thus capable of producing neurosporene. These recombinant *E. coli* strains were used for heterologous complementation assays (see below). All recombinant *E. coli* strains were grown in LB medium (1% tryptone, 0.5% yeast extract, 1% NaCl) for 48 h at 28°C, with agitation at 180 rpm, in the dark to maximize carotenoid production. Chloramphenicol (34 μg/ml) was used for selection of *E. coli* BL-EBI, BL-EBP and BL-EBR transformants. The culture broth was supplemented with 2% agarose for growth on solid medium. Plasmid pET21a, containing the complete coding area of the zdsfc gene (pET21-zdsfc), was transformed into *E. coli* BL21 (DE3) to produce *E. coli* BL-zdsfc (Table 2). *E. coli* BL-zdsfc was grown in LB medium for 48 h at 28°C, in the presence of ampicillin (50 μg/ml), 1 mM IPTG (Isopropyl-β-D thiogalactopyranoside) was added at the end of the logarithmic growth phase to induce ZDS protein expression.

Recombinant strain	Transformed plasmid	Genes encoded	DNA origin	Enzymatic activity	Enzymatic substrate	Enzymatic product	GenBank accession No.
E. coli BL-EBI	pACCRT-EBI	Eu-crtE	Erwinia uredovora	GGPP synthase	FPP/GPP	GGPP	D90087
		Eu-crtB	Erwinia uredovora	Phytoene synthase	GGPP	Phytoene	D90087
		Eu-crtI	Erwinia uredovora	Phytoene desaturase	Phytoene	Lycopene	D90087
					ζ-carotene	Lycopene	D90087
					Neurosporene	Lycopene	D90087
E. coli BL-EBP	pACCRT-EBP	Eu-crtE	Erwinia uredovora	GGPP synthase	FPP/GPP	GGPP	D90087
		Eu-crtB	Erwinia uredovora	Phytoene synthase	GGPP	Phytoene	D90087
		Sc-crtP	Synechoccocus sp.	Phytoene desaturase	Phytoene	ζ-carotene	X55289
E. coli BL-EBR	pACCRT-EBR	Eu-crtE	Erwinia uredovora	GGPP synthase	FPP/GPP	GGPP	D90087
		Eu-crtB	Erwinia uredovora	Phytoene synthase	GGPP	Phytoene	D90087
		Rc-crtR	Rhodobacter capsulatus	Phytoene desaturase	Phytoene	Neurosporene	CAA77540
E. coli BL-zdsFc	pET21-zdsFc	zds-Fc	Ficus carica	Zeta carotene desaturase	ζ-carotene	Lycopene	**JN896309**
					Neurosporene	Lycopene	**JN896309**
E. coli BL-EBPZ	pACCRT-EBP	Eu-crtE	Erwinia uredovora	GGPP synthase	FPP/GPP	GGPP	D90087
		Eu-crtB	Erwinia uredovora	Phytoene synthase	GGPP	Phytoene	D90087
		Sc-crtP	Synechoccocus sp.	Phytoene desaturase	Phytoene	ζ-carotene	X55289
	pET21-zdsFc	zds-Fc	Ficus carica	Zeta carotene desaturase	ζ-carotene	Lycopene	**JN896309**
					Neurosporene	Lycopene	**JN896309**
E. coli BL-EBRZ	pACCRT-EBR	Eu-crtE	Erwinia uredovora	GGPP synthase	FPP/GPP	GGPP	D90087
		Eu-crtB	Erwinia uredovora	Phytoene synthase	GGPP	Phytoene	D90087
		Rc-crtR	Rhodobacter capsulatus	Phytoene desaturase	Phytoene	Neurosporene	CAA77540
	pET21-zdsFc	zds-Fc	Ficus carica	Zeta carotene desaturase	Neurosporene	Lycopene	**JN896309**

Table 2: Recombinant E. coli strains generated in this study, and the DNA plasmids used to transform them. The table also includes the genes encoded by the recombinant plasmids, plant species where the genes originated from, as well as the enzyme names and enzymatic activity of the proteins encoded by those genes.

Protein electrophoresis (SDS-PAGE)

For protein analysis, recombinant E coli BL21 (DE3) cells, containing either pET21a or pET21-zdsfc, were grown at 37°C, with orbital shaking (180 rpm), until they reached the exponential growth phase. The cultures were then induced for recombinant protein expression, by addition of 1mM IPTG, and incubated for a further four hours before sample collection and centrifugation, as recommended by the pET21a system (Novagen). Prior to polyacrylamide gel electrophoresis, the bacterial cells were solubilized, at 85°C for 3 min, in a buffer containing 2.5% (w/v) SDS, 125 mM dithiothreitol, 25% (v/v) glycerol, and 112.5 mM Tris-HCl, pH 6.8. The protein samples were then loaded onto a 12% polyacrylamide gel, on a Mini Protean II cell (Bio-Rad, Hercules, CA), as described previously [18]. The gel was run, at 120 V for 1 h, using Tris-Glycine buffer (25 mM Tris, 192 mM Glycine, 0.1% SDS and pH: 8.3). The molecular weight of the proteins was estimated by comparing their gel migration patterns to those of Precision plus Protein Standard (BioRad), using the Quantity One software (Bio-Rad).

Bioinformatic analyses

To construct the phylogenetic tree (Figure 2), the inferred fig tree (Ficus carica) ZDS amino acid sequence was compared to 22 homologous amino acid sequences from three citrus trees (Citrus sinensis, C. unshiu and C. maxima), physic nut tree (Jatropha curcas), pawpaw (Carica papaya), strawberry (Fragaria x ananassa), two apple trees (Malus x domestica), peppers (Capsicum annuum), tomato (Solanum lycopersicum), sweet potato (Ipomea sp.), gentian (Gentiana lutea), chrysanthemum (Chrysanthemum x morifolium), marigold (Tagetes erecta), two wild carrots (Daucus carota), daffodil (Narcissus tazetta var. Chinensis), orchid (Oncidium Gower Ramsey), cress (Arabidopsis thaliana), turnip (Brassica rapa), sorghum (Sorghum bicolour), wheat (Triticum aestivum), and green algae (Dunaliella salina). Vector NTI Advance 11 software (Invitrogen) and BioEdit Sequence Alignment Editor version 7.0.5.3, were used to analyze the nucleotide and deduced amino acid sequences, and for sequence alignment, respectively. The NCBI database was searched for plant ZDS sequences using the BLAST software [19]. The ChloroP 1.1 Prediction Server program (Emanuelsson et al. 1999 [20]) and TargetP 1.1 Server were used to identify the ZDS signal/sorting peptide and for predicting its cleavage site [21,22]. Phylogenetic and molecular evolutionary analyses were conducted using MEGA, version 5.0, package program [23]. Data were analyzed by the neighbor-joining method [24] . The reliability of the neighbor-joining tree was estimated by calculating bootstrap confidence limits (BCL) based on 1000 replicates [24]. The evolutionary distances were computed using the Jones-Taylor-Thornton method and are in the units of the number of base substitutions per site. The rate variation among sites was modeled with a Gamma distribution (shape parameter=1) [25]. The GenBank accession numbers of the amino acid sequences used in the analysis are shown in Figure 2. Protein sorting was predicted by PSORT, a web server for analyzing and predicting protein-sorting signals, from the Institute for Molecular and Cellular Biology (Osaka, Japan) [26]. Finally, PSIpred v3.0 was used for hydrophobicity and protein secondary structure predictions [27].

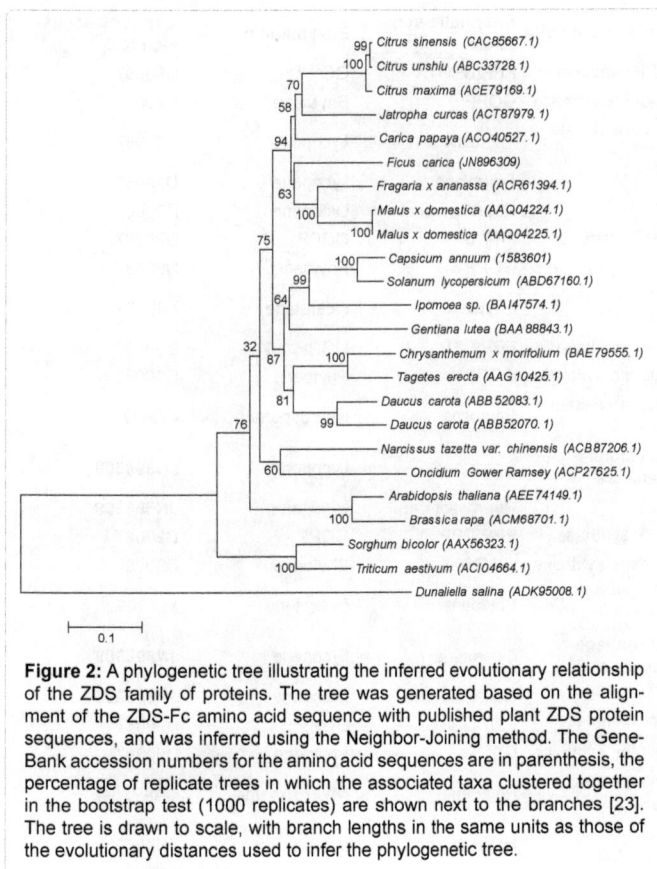

Figure 2: A phylogenetic tree illustrating the inferred evolutionary relationship of the ZDS family of proteins. The tree was generated based on the alignment of the ZDS-Fc amino acid sequence with published plant ZDS protein sequences, and was inferred using the Neighbor-Joining method. The Gene-Bank accession numbers for the amino acid sequences are in parenthesis, the percentage of replicate trees in which the associated taxa clustered together in the bootstrap test (1000 replicates) are shown next to the branches [23]. The tree is drawn to scale, with branch lengths in the same units as those of the evolutionary distances used to infer the phylogenetic tree.

HPLC analysis

For HPLC analysis, *E. coli* BL-EBPZ and BL-EBRZ (Table 2) were grown overnight in LB broth, supplemented with chloramphenicol (34 µg/ml) and ampicillin (50 µg/ml), at 37°C with agitation (200 rpm). Recombinant protein expression was then induced, by IPTG addition (1 mM), and the cultures incubated, in the dark, at 28°C for 48h with orbital shaking (200 rpm) After incubation, the cells were centrifuged (4000xg for 5 min), and the bacterial pellet washed twice with deionized water, re-suspended in acetone (Sigma) and homogenized by vortexing (10 min at 4°C). After a new centrifugation (13000xg, 2 min, 4°C), the cell supernatant was collected, dried down under N2 flow, and stored at -80°C for high performance liquid chromatography (HPLC) analysis. All sample manipulations were carried out on ice, under dim light, to prevent photodegradation, isomerization, or other structural changes in the recombinant terpenoids. For HPLC analysis, the dry residues were suspended in 2 ml of chlorophorm:metanol:acetone (3:2:1, v/v) and filtered through a 0.22 µm polycarbonate filter. The HPLC device was equipped with a photodiode array detector, set to scan from 250 to 540 nm throughout the elution procedure, and controlled by the Empower2 software program. Twenty µl of the terpenoid samples were loaded onto a C_{30} carotenoid column (250 mm x 4-6 mm, 5 µm; YMC Europa) and the flow rate was maintained at 1ml/min. The mobile phase was: A, methyl *tert*-butil ether; B, water; and C, methanol. The linear ternary gradient elution program was performed as follows: Initially A-B-C (5:5:90); followed by 0-12 min, A-B-C (5:0:95); 12-20 min, A-B-C (14:0:86); 20-30 min, A-B-C (25:0:75); 30-50 min, A-B-C (50:0:50); 50-70 min, A-B-C (75:0:25); and finally back to A-B-C (5:5:90) for column re-equilibration at 23°C. A Maxplot chromatogram, which plots each carotenoid peak and its corresponding maximum absorbance

wavelength, was obtained for each HPLC sample; and the recombinant proteins were identified by comparing their HPLC retention profiles to those of standards run in the same HPLC conditions, or to published data. The lycopene standard was obtained from Sigma-Aldrich (Madrid, Spain), and used as previously described, whereas ζ-carotene and neurosporene were produced by the recombinant *E. coli* cultures BL-EBP and BL-EBR, respectively [28,29]. Quantitative analyses were carried out by comparing the Maxplot chromatogram peak areas of the carotenoid samples to a calibration curve constructed with the lycopene standard.

Zeta-carotene desaturase activity determination

Zeta-carotene desaturase enzymatic activity was determined by functional complementation assays on *E. coli* strains BL-pEBP and BL-pEBR. These strains showed a light yellow and dark yellow coloration, due to ζ-carotene and neurosporene accumulation, respectively. Both *E. coli* BL-pEBP and BL-pEBR were -transformed with the plasmid pET21-zdsfc, to generate *E. coli* BL-EBPZ and *E. coli* BL-EBRZ, respectively. Transformed cells were plated onto LB agar plates containing chloramphenicol (34µg/ml) and ampicillin (50µg/ ml). These plates also contained 10µl of a 100 mM aqueous solution of IPTG, added one hour before the cells were plated. The cells were incubated at 28°C, for 48 h in the darkness, to maximize terpenoid production. After incubation, the desaturase enzymatic activity on the zeta-carotene and neurosporene substrates produced by the *E. coli* strains, was identified by the pigments present on the recombinant colonies, and confirmed by HPLC analysis. The negative control for our assay was an *E. coli* BL21 (DE3) containing the plasmid pET21a alone, whereas recombinant *E. coli* BL-pEBI was used as the positive control.

Results

Cloning the full-length zdsfc cDNA

The strategy used to construct the full-length coding region of *Ficus carica* zds (zdsfc) is outlined in Figure 3, and the primers used for this purpose are listed in Table 1. As a first step, we analyzed the published zds sequences, obtained from different plant species, and identified conserved regions in the gene. Based on these regions, we designed degenerate primers ZDS DegF and ZDS DegR. We used these primers to RT-PCR amplify the RNA we isolated from *Ficus carica* leaves, and this generated a DNA fragment of 567 bp (Figure 3A). By sequence analyses, the amplified DNA was shown to span to the central area of zdsfc, and displayed a high similarity (80% to 83%) to other plant zeta-carotene desaturase genes, such as those from *Citrus unshi* (AB072343.1), *Citrus maxima* (EU798286.1), *Malus x domestica* (AF429983.1), *Citrus sinensis* (AJ319762.1), *Jatropha curcas* (GQ337075.1), *Citrus x paradisi* (AF372617.1), *Fragaria x ananassa* (FJ795343.1), *Carica papaya* (FJ812088.1) and *Daucus carota* (DQ192189.1).

The missing 5'and 3' regions of zdsfc were then obtained by RACE-PCR, using the *Ficus carica* cDNA, generated above by RT, as template, and using primers designed from either the 5' or 3' area, respectively, of the above described central area of zdsfc. This resulted in a 727 bp encompassing the 5'-coding region of the gene (Figure 3B), and a 1318 bp spanning the missing 3'-coding region of zdsfc (Figure 3C). The DNA fragments also share a common area, at either the 5' or 3'-end, respectively, with the central part of the zdsfc gene, used to design the PCR primers (Figure 3E). Hence, the three DNA fragments together span the full-length coding region of zdsfc, and were used to

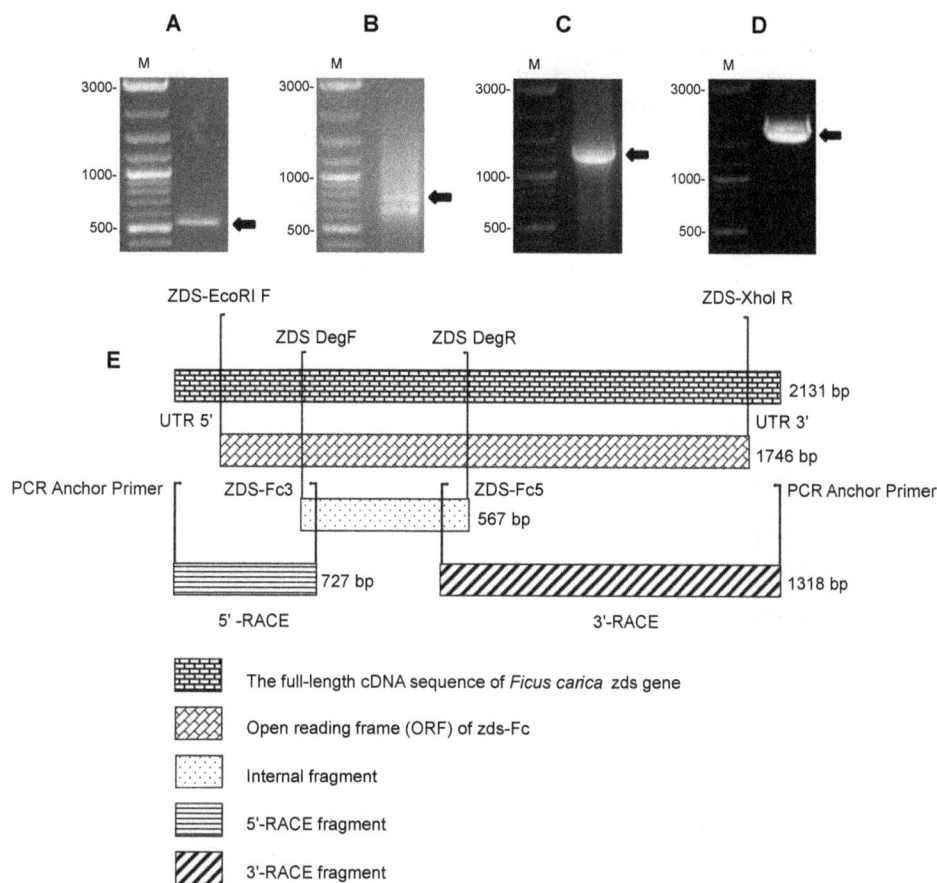

Figure 3: Isolation of the full-length zeta-carotene desaturase cDNA from *Ficus carica*. Agarose gels (1%) stained with ethidium bromide show RT-PCR amplified products: A) Internal zds-Fc cDNA fragment of 567 bp, amplified with primers ZDS DegF and ZDS DegR; B) DNA fragment of 727 bp, corresponding to the 5´-end of the zds-Fc cDNA, obtained by 5´-RACE; C) The 1318 bp cDNA fragment amplified by 3´-RACE, corresponding to the 3´-coding region of zds-Fc; D) A DNA fragment of 2131 bp, spanning the complete open reading frame of the zds-Fc gene from *Ficus carica*. E) Diagram of the strategy used to construct the full-length coding region of *Ficus carica* zds-Fc (1746 bp), indicating the length of the PCR-amplified fragments and the PCR primers used.

PCR amplify a 2131 bp DNA fragment. Apart from the zdsfc coding region, the PCR product also contained 197 nucleotides of 5´-UTR and 188-nucleotides of 3´-UTR (data not shown). Finally, we sequenced the zdsfc cDNA from *Ficus carica* and deposited it in as the GenBank data base (accession number of JN896309). The complete open reading frame (ORF) of the gene has a length of 1746 bp (Figure 3D).

Ficus carica ZDS protein characterization and recombinant expression

The zdsfc gene coded for a polypeptide of 582 amino acids, with an estimated molecular weight of 64 kD. The ZDSFc polypeptide sequence was compared to published zeta-carotene desaturase proteins from other plants. ChloroP 1.1 and TargetP 1.1 predicted a putative chloroplast transit peptide targeting sequence, encompassing the N-terminal amino acids 1 to 43 of ZDSFc; this correlates well with the transit peptides found in maize, *Arabidopsis, Narcissus,* and *Triticum,* which are 30, 34, 42 and 32 residues long, respectively [2,30].The PSORT program predicted a putative polypeptide cleavage site, located: between amino acids 33 and 34, whereas PSIpred found, in our *F. carica* protein (ZDSFc), a highly conserved motif characteristic of higher plants ZDS enzymes. Another conserved motif present in ZDSFc is a typical pyridine dinucleotide binding domain (FAD-binding domain),

with a secondary structure of β-sheet-α-helix loop-β-sheet, that is also characteristic of other plant ZDSs [3,31].

The degree of identity (and homology) between the ZDSFc polypeptide and other plant ZDSs was examined by BlastP search. This analysis revealed that our enzyme has an 84% amino acid identity (91% similarity) with *Jatropha curcas*, 85% (91%) with *Citrus maxima*, 85% (91%) with *Citrus unshiu*, 84% (91%) with *Citrus sinensis*, 84% (90%) with *Malus x domestica*, 83% (91%) with *Fragaria x ananassa*, and 83% (90%) with *Carica papaya*. These data, taken together, confirm the cloned zdsfc as an integral part of the ZDS family of proteins.

Expression of *Ficus carica* zds was confirmed by SDS/PAGE analyses of *E. coli* lysates transformed with the plasmid pET21a, either alone, or containing the complete coding area of the zdsfc gene (*E. coli* BL-zdsfc, Table 2). A protein with a molecular weight of 64 kD was observed only in the bacterial cells expressing zdsfc, and only after protein expression was induced by addition of IPTG (data not shown).

Phylogenetic analysis of plant ZDS proteins

We conducted a phylogenetic study of ZDSFc on the basis of its predicted polypeptide sequence and that of a variety of clone plant ZDS sequences, available from GenBank. The phylogenetic tree shown in Figure 2 was inferred using the Neighbor-Joining method [24]. This test showed that our expressed protein has a high homology to other plant zeta-carotene desaturases. The zeta-carotene desaturase from marine microalgae *Dunaliella salina* was included as an out-group and, hence, appears in an isolated branch. The phylogenetic analyses clearly demonstrated that *Ficus carica* ZDS (JN896309) clusters better with *Fragaria x ananassa* (FJ795343.1), *Malus x domestica* (AF429983.1), *Carica papaya* (FJ812088.1), *Jatropha curcas* (GQ337075.1), *Citrus maxima* (EU798286.1), *Citrus sinensis* (AJ319762.1), and *Citrus unshiu* (DQ309869.1), than with the other ZDSs included in Figure 2. Contrary to expectations, the monocot ZDS proteins from *Narcissus tazetta var. chinensis* and *Oncidium Gower Ramsey* were not more closely related to the grass carotenoid desaturase than the ZDSs from dicot plants, such as *Arabidopsis thaliana or Brassica rapa*. The Bootstrap values on the nodes (Figure 2) indicate the number of times that each group occurred with 1000 replicates. Models with the lowest BIC scores (Bayesian Information Criterion) are considered to describe the substitution pattern as best. So, the evolutionary distances were computed using the Jones-Taylor-Thornton method (Nei and Kumar 2000). The rate variation among sites was modeled with a Gamma distribution (shape parameter=1) with 5 rate categories, and by assuming that a certain fraction of the sites are evolutionarily invariable.

Carotenoid production and ZDSFc enzymatic activity determination

We constructed a recombinant *E. coli* strain (*E. coli* BL-pEBI, Table 2), encoding GGPP and phytoene synthases, as well as phytoene desaturase, hence producing lycopene. We used the cell extracts from this bacterial culture, expressing lycopene, as a positive control for our HPLC analyses. We also constructed two recombinant *E. coli* strains (*E. coli* BL-pEBP and *E. coli* BL-pEBR, Table 2) to use as negative controls. Induction of recombinant protein expression in *E. coli* BL-pEBP should result in the accumulation of ζ-carotene, which can be used as a substrate by zeta carotene desaturase. On the other hand, the recombinant *E. coli* BL-pEBR should produce neurosporene. HPLC analyses confirmed the carotenoids present in the bacterial extracts from our engineered controls as zeta-carotene (*E. coli* BL-pEBP), neurosporene (*E. coli* BL-pEBR) and lycopene (*E. coli* BL-pEBI).

In order to test the enzymatic activity of our recombinant zdsfc, we constructed two more recombinant *E. coli* strains (*E. coli* BL-EBPZ and *E. coli* BL-EBRZ, Table 2). The only difference between the negative control *E. coli* BL-pEBP and *E. coli* BL-EBPZ is the presence in the latter of an additional plasmid (pET21-zdsFc), encoding *Ficus carica* ZDS. Production of a functional ZDSFc enzyme by *E. coli* BL-EBPZ would result in the conversion of the zeta-carotene, produced by *E. coli* BL-pEBP, into neurosporene and lycopene (Figure 1). For its part, *E. coli* BL-EBRZ is only different from the negative control (*E. coli* BL-pEBR) by the presence in the former of an additional plasmid (pET21-zdsFc), encoding *Ficus carica* ZDS. As above, production of a functional ZDSFc enzyme by *E. coli* BL-EBRZ would result in the conversion of the carotenoid accumulated by *E. coli* BL-pEBR (neurosporene) into the product of ZDS activity, in this case lycopene (Figure 1). As expected, ZDS expression by *E. coli* BL-EBPZ resulted in the enzymatic conversion of part of the ζ-carotene, accumulated by the *E. coli* BL-EBP culture, into neurosporene and lycopene. This was confirmed by the HPLC detection of two new compounds, not present in the *E. coli* BL-pEBP culture. The HPLC elution profile for these compounds recorded maximum absorbance peaks at 416, 441 and 471 nm (identified as neurosporene) and 445, 472, and 504 nm (identified as lycopene). Accordingly, ZDS expression by *E. coli* BL-EBRZ resulted in the enzymatic conversion of part of the neurosporene accumulated

by the *E. coli* BL-EBR culture into lycopene, a compound not present in the *E. coli* BL-pEBR culture. The new pigment was confirmed as lycopene by HPLC analysis, and exhibited the expected maximum absorbance peaks at 445, 472 and 504 nm. These results, taken together, confirm the enzymatic ability of ZDS to act on both zeta-carotene and neurosporene, as expected from a plant zeta-carotene desaturase.

The neurosporene content of the recombinant *E. coli* BL-EBPZ culture was estimated at 34 μg per g, of cells (dry weight) and the lycopene content at 11 μg per gram of cells (Table 3). Whereas, the lycopene content of the recombinant *E. coli* BL-EBRZ culture was estimated at 13 μg per gram of cells (dry weight).

Carotenoid production was also apparent by the accumulation of pigment in the recombinant *E. coli* cultures and was found to be dependent on the temperature at which the bacteria were grown. Both *E. coli* BL-EBPZ and *E. coli* BL-EBRZ cultures accumulated most pigment when grown at 28°C in the dark.

Discussion

Carotenoids are widely distributed natural pigments responsible for the yellow, orange and red colors of fruits, roots and flowers, as they invariably occur in the chloroplasts of higher plants. But the importance of carotenoids in food goes far beyond their role as natural pigments. Carotenoids are not only the starting material for the synthesis of vitamin A, an essential vitamin, but also carry out a variety of biological functions independent of the provitamin A activity, attributed to the antioxidant property of carotenoids resulting in deactivation of free radicals [8,30]. They have been reported to enhance the immune system, as well as decrease the risk of degenerative diseases such as cancer, cardiovascular disease, age-related macular degeneration, and cataract formation [8,31].

With consumers these days looking for natural products that would offer them more health benefits, carotenoids emerge as a logical target for the food industry. Plant carotenoid production is insufficient to satisfy this market; hence it is essential to generate a sustainable alternative capable of producing large amounts of natural, high quality carotenoids. Carotenoid production can be rapidly increased by the use of recombinant DNA technology, but before this can be done, the carotenoid biosynthetic genes from many different plants and other organisms need to be cloned and characterized. Moreover, heterologous expression of biosynthetic pathways in *E. coli* continues to be a powerful approach for developing metabolic engineering applications in plants. The utility of the bacterial system lies in its inherent similarity to the biochemistry of the plant plastid, thus generating carotenoids that structurally identical to those produced by plants [32].

Here, we describe the cloning and characterization of a novel ζ-carotene desaturase from the fig tree *Ficus carica*. Our strategy was to analyze the published zds sequences, from other plant species, identify conserved regions in the gene and design degenerate PCR primer that were used to clone the central region of the zdsfc gene. The rest of the

E. coli strain	ζ-carotene	Neurosporene	Lycopene
E. coli BL-pEBP	201	n.d.	n.d.
E. coli BL-EBPZ	122	34	11
E. coli BL-pEBR	n.d.	215	n.d.
E. coli BL-EBRZ	n.d.	186	13

Table 3: Carotenoid production by four of the recombinant *E. coli* strains we generated to assess the enzymatic activity of recombinant zds-Fc. The figures indicate the μg of carotenoid per g of cells (dry weight) produced by the bacterial cultures, whereas n.d. stands for not determined.

gene was then obtained by RACE-PCR. The sequence amplified for the zeta-carotene desaturase gene from *Ficus carica* is 2131 bp long and spans a 1746 bp open reading frame, encoding a protein of 582 amino acid residues, with an estimated molecular weight of 64 kD.

The ZDSFc polypeptide sequence was compared to published zeta-carotene desaturase proteins from other plants and was found to share many characteristics common to other plant ZDS s so, as it is the case for other enzymes in the carotenoid biosynthesis pathway, ZDS appears to be highly conserved in higher plants [2]. Another common characteristic was the presence in ZDS of an N-terminal putative chloroplast transit peptide targeting sequence, which was predicated to be 43 residues long, this is also typical of other enzymes involved in carotenoid biosynthesis in plants [8,10,33]. Finally, the sequences also contained a typical pyridine dinucleotide binding domain (FAD-binding domain), with a secondary structure of β-sheet-α-helix loop-β-sheet, that is characteristic of other plant ZDSs [3,29] . These data, taken together, confirms the cloned zdsfc as an integral part of the ZDS family of proteins.

Based on the shared sequences between the *F. carica* enzyme and other plant ZDSs, we constructed a phylogenetic tree illustrating the evolutionary relationship between the fig tree ZDS and other published carotenoid zeta carotene desaturases from higher plants. *Ficus carica* ζ-carotene desaturase was further characterized as part of the ZDS family of plant enzymes, and it clustered better with the dicot plant trees ZDSs such as *Citrus unshiu, Jatropha curcas, Carica papaya* and *Malus x domestica.*

Recombinant expression of *Ficus carica* zds in *E. coli* produced a protein with a molecular weight of 64 kD, consistent with that estimated from the amino acid sequence of the polypeptide. In order to test the *in vivo* enzymatic activity of our recombinant zdsfc, we engineered four recombinant *E. coli* strains (*E. coli* BL-EBP, BL-EBR, BL-EBPZ, and BL-EBRZ, Table 2). Whereas the first two bacterial cultures only accumulated ζ-carotene and neurosporene, respectively, the last two recombinant strains (*E. coli* BL-EBPZ, and BL-EBRZ), expressing our recombinant zdsFc, were able to convert the ζ-carotene and neurosporene into neurosporene and lycopene, or lycopene, respectively. These results demonstrated, *in vivo*, that ZDS, a single desaturase plant enzyme, efficiently catalyzed the last two desaturation steps leading to the production of Lycopene (Table 3). Hence, as described for other plant enzymes the *Ficus carica* zds gene we cloned encodes a zeta-carotene desaturase which is capable of catalyzing the conversion of ζ-carotene into first neurosporene and then lycopene [1-3].

In summary, this study, represents the first time that enzymes the Ficus carica zds was, not only cloned and characterized, but was also shown, by recombinant expression in *E. coli*, to catalyze the desaturation of ζ-carotene to produce first neurosporene and finally lycopene.

Acknowledgements

J.M. A-G is the recipient of an AECID scholarship from the Spanish Foreign Affairs Ministry.

The authors wish to thank Dr. Norihiko Misawa (Research Institute for Bioresources and Biotechnology, Ishikawa Prefectural University, Japan) for the gift of plasmids pACCRT-EBR and pACCRT-EBI; and Dr. Gerhard Sandmann (J. W. Goethe-Universität, Frankfurt, Germany) for the gift of plasmid pACCRT-EBP. Thanks are also due to both the Faculty of Pharmacy and the School of Biotechnology for their support throughout this project.

References

1. Al-Babili S, Oelschlegel J, Beyer P (1998) A cDNA encoding for beta carotene desaturase (Accession No. AJ224683) from *Narcissus pseudonarcissus* L. Plant Physiol 117: 719.

2. Cong L, Wang C, Li Z, Chen L, Yang G, et al. (2010) cDNA cloning and expression analysis of wheat (*Triticum aestivum* L.) phytoene and zeta-carotene desaturase genes. Mol Biol Rep 37: 3351-3361.

3. Matthews PD, Luo R, Wurtzel ET (2003) Maize phytoene desaturase and zeta-carotene desaturase catalyse a poly-Z desaturation pathway: implications for genetic engineering of carotenoid content among cereal crops. J Exp Bot 54: 2215-2230.

4. Yan P, Gao XZ, Shen WT, Zhou P (2011) Cloning and expression analysis of phytoene desaturase and ζ-carotene desaturase genes in *Carica papaya*. Mol Biol Rep 38: 785-791.

5. Krinsky NI (1989) Antioxidant functions of carotenoids. Free Radic Biol Med 7: 617-635.

6. Hirschberg J (2001) Carotenoid biosynthesis in flowering plants. Curr Opin Plant Biol 4: 210-218.

7. DellaPenna D, Pogson BJ (2006) Vitamin synthesis in plants: Tocopherols and carotenoids. Annu Rev Plant Biol 57: 711-738.

8. Fraser PD, Bramley PM (2004) The biosynthesis and nutritional uses of carotenoids. Prog Lipid Res 43: 228-265.

9. Rodríguez Concepción M (2006) Early steps in isoprenoid biosynthesis: Multilevel regulation of the supply of common precursors in plant cells. Phytochem Rev 5: 1-15.

10. Sandmann G, Romer S, Fraser PD (2006) Understanding carotenoid metabolism as a necessity for genetic engineering of crop plants. Metab Eng 8: 291-302.

11. Park H, Kreunen SS, Cuttriss AJ, Dellapenna D, Pogson BJ (2002) Identification of the carotenoid isomerase provides insight into carotenoid biosynthesis, prolamellar body formation, and photomorphogenesis. Plant Cell 14: 321-332.

12. Sandmann G (1994) Phytoene desaturase - Genes, enzymes and phylogenetic aspects. J Plant Physiol 143: 444-447.

13. Misawa N, Nakagawa M, Kobayashi K, Yamano S, Izawa Y, et al., (1990) Elucidation of the *Erwinia uredovora* carotenoid biosynthetic pathway by functional analysis of gene products expressed in *Escherichia coli*. J Bacteriol 172: 6704-6712.

14. Araya-Garay JM, Feijoo-Siota L, Veiga-Crespo P, Villa TG (2011) cDNA cloning of a novel gene codifying for the enzyme lycopene β-cyclase from *Ficus carica* and its expression in the *Escherichia coli*. Appl Microbiol Biotechnol 92: 769-777.

15. Misawa N, Shimada H (1997) Metabolic engineering for the production of carotenoids in non-carotenogenic bacteria and yeasts. J Biotechnol 59: 169-181.

16. Takaichi S, Sandmann G, Schnurr G, Satomi Y, Suzuki A, et al. (1996) The carotenoid 7,8-dihydro-ψ end group can be cyclized by the lycopene cyclases from the bacterium *Erwinia uredovora* and the higher plant *Capsicum annuum*. Eur J Biochem 241:291-296.

17. Linden H, Vioque A, Sandmann G (1993) Isolation of a carotenoid biosynthesis gene coding for zeta-carotene desaturase from *Anabaena* PCC 7120 by heterologous complementation. FEMS Microbiol Lett 106:99-104.

18. Laemmli UK (1970) Cleavage of structural proteins during the assembly of the head of bacteriophage T4. Nature 227: 680-685.

19. Altschul SF, Gish W, Miller W, Myers EW, Lipman DJ (1990) Basic local alignment search tool. J Mol Biol 215: 403-410.

20. Emanuelsson O, Nielsen H, Von Heijne G (1999) ChloroP, a neural network-based method for predicting chloroplast transit peptides and their cleavage sites. Protein Sci 8: 978-984.

21. Emanuelsson O, Nielsen H, Brunak S, von Heijne G (2000) Predicting subcellular localization of proteins based on their N-terminal amino acid sequence. J Mol Biol 300: 1005-1016.

22. Nielsen H, Engelbrecht J, Brunak S, von Heijne G (1997) Identification of

prokaryotic and eukaryotic signal peptides and prediction of their cleavage sites. Protein Eng 10: 1-6.

23. Tamura K, Peterson D, Peterson N, Stecher G, Nei M, et al. (2011) MEGA5: Molecular evolutionary genetics analysis using maximum likelihood, evolutionary distance, and maximum parsimony methods. Mol Biol Evol.

24. Saitou N, Nei M (1987) The neighbor-joining method: a new method for reconstructing phylogenetic trees. Mol Biol Evol 4: 406-425.

25. Nei M, Kumar S (2000) Molecular Evolution and Phylogenetics. Oxford University Press, New York.

26. Nakai K, Horton P (1999) PSORT: A program for detecting sorting signals in proteins and predicting their subcellular localization. Trends Biochem Sci 24: 34-36.

27. Jones DT (1999) Protein secondary structure prediction based on position-specific scoring matrices. J.Mol.Biol 292: 195-202.

28. Melendez-Martinez AJ, Vicario IM, Heredia FJ (2003) A routine high-perfomance liquid chromatography method for carotenoid determination in ultrafrozen orange juices. J Agric Food Chem 51: 4219-4224.

29. Wirenga RK, Terpstra P, Hol WGJ (1986) Prediction of the occurrence of the ADP-binding $\beta\alpha\beta$-fold in proteins, using an amino acid sequence fingerprint. GBB-University of Groningen.

30. Richmond A (2000) Microalgal biotechnology at the turn of the millennium: A personal view. J Appl Psychol 12: 441-451.

31. van den Berg H, Faulks R, Fernando Granado H, Hirschberg J, Olmedilla B, et al. (2000) The potential for the improvement of carotenoid levels in food and the likely systemic effects. J Sci Food Agric 80: 880-912.

32. Gallagher CE, Cervantes-Cervantes M, Wurtzel ET (2003) Surrogate biochemistry: use of *Escherichia coli* to identify plant cDNAs that impact metabolic engineering of carotenoid accumulation. Appl Microbiol Biotechnol 60: 713-719.

33. Bouvier F, Rahier A, Camara B (2005) Biogenesis, molecular regulation and function of plant isoprenoids. Prog.Lipid Res 44: 357-429.

8

Optimization of Process Conditions for Biotransformation of Caffeine to Theobromine using Induced Whole Cells of *Pseudomonas sp*

Sreeahila Retnadhas and Sathyanarayana N Gummadi*

Applied and Industrial Microbiology Laboratory, Department of Biotechnology, Bhupat and Jyoti Mehta School of Biosciences, Indian Institute of Technology Madras, Chennai, India

Abstract

The obromine is a metabolic intermediate produced in caffeine degradation pathway by many bacterial species, which has potential applications in food and pharmaceutical industries. Conventional methods of Theobromine production from xanthine involve harsh physical and chemical conditions which are harmful to the environment. To overcome this, we employed biotechnological route to convert caffeine to theobromine by single demethylation reaction using induced cells of *Pseudomonas sp*. Initially we screened various divalent metal ions for the production of Theobromine by *Pseudomonas sp*. from caffeine. Co^{2+} and Ni^{2+} accumulates 400 and 100 mg/l of theobromine under initial reaction conditions (2 g/l caffeine, 8 g/l cell loading, pH 7.0, 30°C). Co^{2+} was chosen for further optimization of reaction conditions for Theobromine production using response surface methodology. Data were fitted into a quadratic model and the optimal condition for theobromine production was found to be 3.2 g/l caffeine, 11.3 g/l initial cell loading and pH 7.0. Quadratic regression models were validated at the optimized conditions and the experimental theobromine produced 689.7 mg/l corresponds to the model predicted theobromine 729.4 mg/l. Theobromine production was further improved to 1.08 ± 0.10 g/l by optimizing the reaction temperature. This study reports highest production of theobromine from caffeine using induced cells of *Pseudomonas sp*. Induced cells are better suited for metabolite production as it is metabolically very active and can be re-used several times. Optimization of reactor parameters will enable us to make microbial production of theobromine feasible in industries at reduced cost.

Keywords: Caffeine; Theobromine; Central composite design; Resting cells; *Pseudomonas sp*; Co^{2+}

Introduction

Theobromine (3, 7-dimethylxanthine) is a metabolite formed in the first step of caffeine metabolism by bacteria [1] and is naturally present in cocoa bean. It has potential applications in food and pharmaceutical industries. Even though theobromine is similar to caffeine in its structure, its effect on human physiology is milder and beneficial. While caffeine increases blood pressure, theobromine reduces blood pressure by dilating blood vessels. Theobromine sodium salicylate, a derivative of theobromine was found to act as a vasodilator early in 1935 [2]. According to the U.S. National Institutes of Health, an antitussive drug BC1036which contains theobromine is under phase III clinical trials. The antitussive action of this drug was found to be because of direct inhibition of cough reflex by theobromine [3]. It is also an intermediate used for the production of a vasotherapeutic agent, Trental and Hextol [4]. Conventionally theobromine has been produced by extraction from cocoa beans or by synthesis from 3-methyl urea. It is also produced by methylation of 3-methylxanthine using dimethyl sulfate as methylating agent in methanolic solution [4]. Exposure of dimethyl sulfate to human has toxic effects like genital and mucous membrane lesions, conjunctivitis and keratitis in eyes, ulceration of lips, etc. [5]. Hence, there is a need to develop alternative methods which are less toxic to the environment and human health. Biotechnological route has been used as an eco-friendly and cheap alternative to chemical methods to produce a number of industrially important molecules. Since theobromine is a metabolite formed in the caffeine metabolic pathway by certain bacterial species, it can be effectively produced from caffeine using such bacteria in an eco-friendly method by single demethylation reaction. Caffeine (1,3,7-trimethylxanthine) is a naturally occurring purine alkaloid abundantly present in tea leaves, coffee beans, cocoa beans and cola nuts. It has been used as a psycho-stimulant [6] and as food ingredients by humans. Acute intake of caffeine causes elevated blood pressure [7], osteoporosis [8], sterility [9], nervousness, mood

change and many types of cancer [10,11]. Hence, there is a need to degrade caffeine in food products and as well as in effluents from tea and coffee industries. Instead of degrading completely, it will be advantageous to convert caffeine into less harmful and value added molecule such as theobromine.

Microbial production of theobromine can be used as a cheaper and ecofriendly alternative to the conventional chemical methods available. Very few studies are reported in literature on the production of theobromine from caffeine using microbes [12-14]. A new strain of *Pseudomonas sp* that is capable of growing on very high concentration of caffeine as the sole carbon and nitrogen source was isolated from the soil of coffee plantation area [15]. It was reported that the growth of *Pseudomonas sp*. gets inhibited by caffeine concentration greater than 2.5 g/l while it was able to grow on medium containing as high as 10 g/l of caffeine with low specific growth rate [16]. After optimization of all other media components, caffeine degradation rate was as high as 0.29 g/l h which is the highest reported so far [17,18]. It was also reported that the enzyme system required for caffeine degradation in *Pseudomonas sp*. is inducible [19] and certain divalent metal ions (Co^{2+}, Ni^{2+}, Cu^{2+}, Zn^{2+}) strongly inhibit further degradation of theobromine accumulated from demethylation of caffeine [20]. Metal ions have been known to

*Corresponding author: Sathyanarayana N Gummadi, Applied and Industrial Microbiology Laboratory, Department of Biotechnology, Bhupat and Jyoti Mehta School of Biosciences, Indian Institute of Technology Madras, Chennai 600036, India

inhibit enzymes thereby resulting in the accumulation of metabolites. Certain heavy metals have been shown to accumulate metabolite in plants [21]. Asano et al. reported that the strain *Pseudomonas putida* No. 352 accumulates theobromine due to the inhibition of a monooxygenase specific for theobromine demethylation by Zn^{2+} [22]. Even at low concentration (8 g/l) of induced cells of *Pseudomonas sp.* as high as 10 g/l of caffeine was found to be degraded in 25 h [23]. Since *Pseudomonas sp.* was shown to be very effective in caffeine degradation, in the present study we attempted to exploit its ability to accumulate theobromine from caffeine in the presence of certain metal ions.

Materials and Methods

Chemicals

Pure caffeine was obtained from Merck India. 3, 7 dimethyl xanthine, 7 methyl xanthine and xanthine were obtained from Sigma India. Solvents used for HPLC analysis were of HPLC grade. All other reagents were of analytical grade.

Bacterial strain

Pseudomonas sp. was previously isolated in our laboratory from soil of coffee plantation area of Ooty, India during August 2003 [15]. The isolate was maintained on Caffeine Added Sucrose (CAS) agar with sub-culturing every 36 h. CAS agar has the following composition -Na_2HPO_4 (120 mg/l), KH_2PO_4 (1.3 g/l), $CaCl_2$(300 mg/l), $MgSO_4$ (300 mg/l), sucrose (5 g/l) and agar (2.5%). pH adjusted to 6.0.

Production of induced cells of *Pseudomonas sp.*

Pseudomonas sp. was maintained on CAS agar at 30°C. After 36 h, three loops full of cells were inoculated in 25 ml seed culture medium and incubated at 30°C and 180 rpm. The seed medium has the following composition -beef extract (1 g/l), yeast extract (2 g/l), peptone (5 g/l), NaCl (5 g/l). Once A_{600} reaches 1.6-1.8, production media was inoculated with 6% of seed culture and incubated at30°C and 180 rpm till 90-95% of initial caffeine gets degraded. The production medium has the following composition - KH_2PO_4 (3.4 g/l), Na_2HPO_4 (0.36 g/l), caffeine (6.4 g/l), sucrose (5 g/l), $CaCl_2$ (0.3 g/l), $MgSO_4.7H_2O$ (0.3 g/l), $FeSO_4$(10.05 g/l). Sucrose and trace elements were prepared and autoclaved separately. Production medium was prepared just before inoculation by mixing sterilized media components under aseptic conditions. Initial pH was adjusted to 7.8 with 3N NaOH. Once 90-95% of caffeine gets degraded, cells were harvested by spinning at 10,000 rpm, 4°C for 5 minutes under sterile conditions. At this point, the enzymes required for caffeine metabolism are induced and expressed at high levels and therefore these induced cells are metabolically very active in utilizing caffeine.

Screening of metal ions for theobromine accumulation by the induced cells of *Pseudomonas sp.*

Harvested induced cells were washed with 10 mM potassium phosphate buffer pH 8.0 and resuspended in 10 mM potassium phosphate buffer pH 8.0, so that the induced cell concentration in the buffer was 100 g/l (catalyst stock). Resuspended induced cells were inoculated in the reaction media containing 2 g/l caffeine such that initial concentration of induced cells in reaction media was 8 g/l. Metal ions were added at a concentration of 1 mM and incubated at 30°C, 180 rpm. Samples were collected every hour till caffeine was completely degraded in control flask, where no metal ion was added and the collected sample was centrifuged at 10,000 rpm for 10 min. The supernatant was analyzed by reverse phase High Performance

Liquid Chromatography (HPLC) using JASCO HPLC system (Multiwavelength detector MD-2010 plus) at 254 nm to determine the concentration of caffeine, theobromine and other methylxanthine derivatives remaining in the medium. C-18 column was used with 30% methanol in water as mobile phase. Retention time of caffeine and theobromine was 9.2 min and 4.5 min respectively under the above mentioned stationary and mobile phases.

Optimization of reaction media parameters for maximized theobromine production using Central Composite Design (CCD)

CCD was chosen to show the statistical significance of the effects of pH, initial cell loading and initial caffeine concentration on theobromine production by *Pseudomonas sp.* The software 'Design Expert 9', a trial version from Stat-Ease, Inc., was used to do a two-level factorial design in which three independent variables (X_1, X_2 and X_3) were represented as three dimensionless variables x_1, x_2 and x_3 with coded levels at -1, 0 and 1.Statistical relationship between coded and actual values is given below,

$$x_i = (X_i - X_0)/\delta X$$

Where X_0 is the corresponding center point value and δX is the step change. The coded values and its corresponding actual values are given in table 1. Total number of experiments that need to be performed according to this design is given by $2^k + 2k + n_0$ where 'k' is the number of independent variables and n_0 the number of repetitions of the experiments at the center point. For three independent variables (caffeine concentration, cell concentration and pH), 20 runs were performed in three blocks with two center points in each block. pH of the reaction media (X_3) was varied from 6.32 to 9.68. 10 mM potassium phosphate buffer was used for pH ranges 6.32 to 8.0 and 10 mM carbonate buffer was used for pH 9.68. Appropriate amount of caffeine (X_1) (ranges from 0.32 g/l to 3.68 g/l) was dissolved in appropriate buffers according to the experimental design in table 2. Induced cells (X_2) (ranges from 1.27 to 14.73 g/l) were resuspended in corresponding buffers to make a stock of 100 g/l and inoculated in reaction media according to the design. Theobromine produced after 4 h, molar yield of theobromine and productivity were measured as the responses.

The dependence of these independent variables (x_1, x_2, x_3) on the responses can be approximated by the quadratic equation (Equation 1) as given below,

$$Y = \beta_0 + \beta_1\chi_1 + \beta_2\chi_2 + \beta_3\chi_3 + \beta_{11}\chi_1^2 +$$
$$\beta_{22}\chi_2^2 + \beta_{33}\chi_3^2 + \beta_{12}\chi_1\chi_2 + \beta_{13}\chi_1\chi_3 + \beta_{23}\chi_2\chi_3 \quad (1)$$

where Y is the predicted response, β_0 is the intercept term, β_1,β_2 and β_3 are the linear coefficients, β_{11}, β_{22} and β_{33} are the quadratic coefficients and β_{12}, β_{13} and β_{23} are the interaction coefficients.

Effect of reaction temperature

Medium components	Coded values				
	-1.68	-1	0	1	1.68
Caffeine concentration X_1 (g/l)	0.32	1	2	2	3.68
Initial cell loading X_2 (g/l)	1.27	4	8	12	14.73
pH X_3	6.32	7	8	9	9.68

Table 1: Coded and actual values of the medium components chosen for optimization using CCD.

The effect of temperature on conversion of caffeine to theobromine was studied by incubating reaction media with induced cell sat three different temperatures 25°C, 30°C and 35°C.The reaction conditions were 3.2 g/l caffeine in 10 mM potassium phosphate buffer, pH 7.0 and induced cell concentration 11.3 g/l. Samples were collected at regular intervals of time and the supernatants were analyzed for theobromine and caffeine concentration by HPLC.

Results

Screening of metal ions for theobromine accumulation from caffeine by *Pseudomonas sp.*

Divalent metal ions like Co^{2+}, Cu^{2+}, Ca^{2+}, Fe^{2+}, Mg^{2+}, Mn^{2+}, Ni^{2+} and Zn^{2+}were tested for their ability to accumulate theobromine from caffeine by *Pseudomonas sp.*Cu^{2+} completely inhibits caffeine degradation by *Pseudomonas sp.* whereas Ca^{2+}, Fe^{2+} and Mg^{2+} slightly enhanced caffeine degradation and Mn^{2+} has no effect. Out of all the metal ions screened, only $Ni^{2+}$and Co^{2+} accumulated significant amount of theobromine in reaction media (Figure 1A). After 4 h of adding induced cells to the reaction media, $Ni^{2+}$and Co^{2+} accumulated 106.67 ± 4.73 mg/l and 384.53 ± 54.01 mg/l with molar yields of 7.31 ± 1.10% and 29.89 ± 4.23% respectively. We also observed that when these metal

ions were added, caffeine degradation by *Pseudomonas sp.* was less compared to control till fourth hour and completely stopped there after (Figure 1B). Experiments were carried out to check synergism between Ni^{2+} and Co^{2+} on theobromine accumulation and caffeine degradation (Figures 1C and 1D) by adding 1 mM each in the media and it was observed that they produce only around 182.89 ± 36.75 mg/l with a molar yield of 13.59 ± 3.04%; thus confirming that they do not act synergistically in accumulating theobromine from caffeine. Amount of theobromine accumulated by synergistic experiment was less than that accumulated by cobalt ions but more than what was produced in the presence of nickel ions. Hence, $Co^{2+}$was chosen for further studies.

Optimization of reaction parameters for maximum theobromine accumulation by CCD

From the preliminary experiments, the upper and lower values for each parameter were fixed and the experimental design for CCD is given in table 1. The design required 20 experimental runs which were performed in three blocks with two center point values in each block. The experiments were carried out in random run order so as to avoid systematic error. Theobromine production (mg/l), molar yield (%) and productivity (mg/l) were considered as the responses after 4 h reaction time. Analysis of the data from CCD experiments

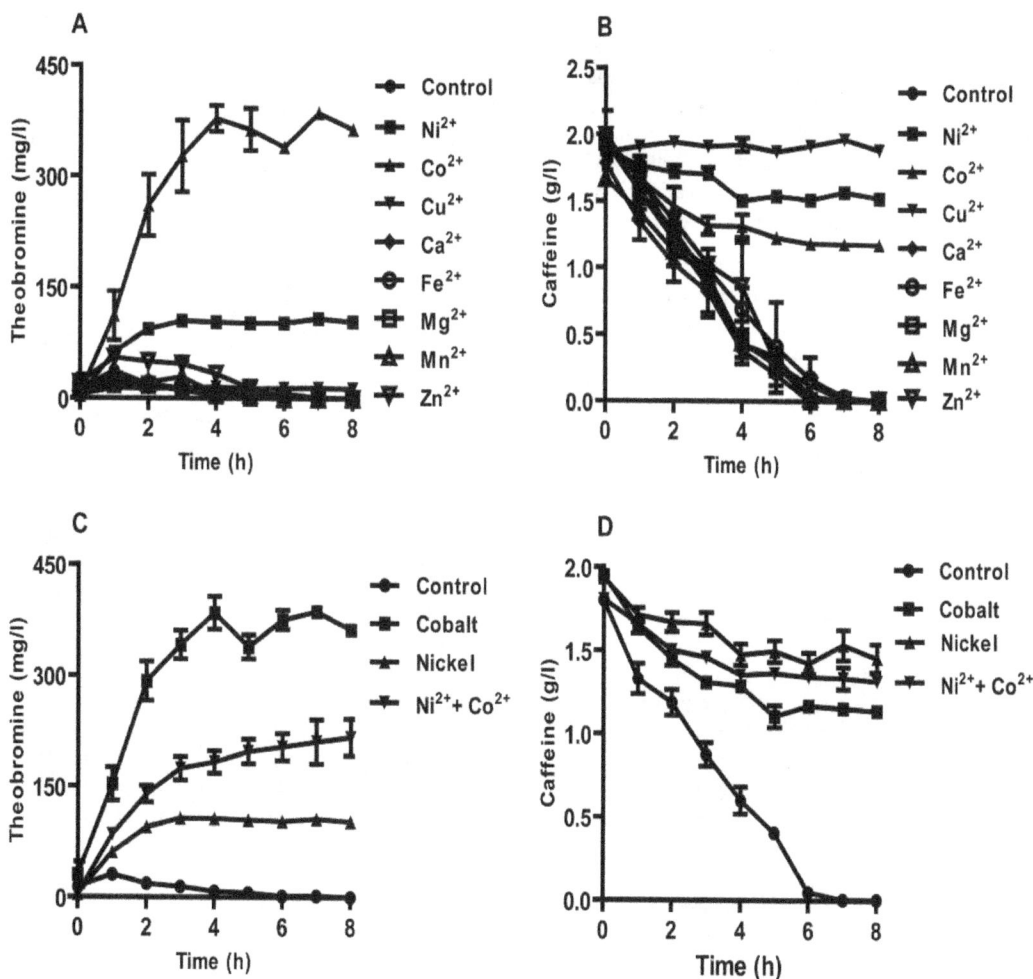

Figure 1: Screening of metal ions for theobromine production from caffeine using *Pseudomonas sp*. Effect of various divalent metal ions on (a) caffeine degradation, (b) theobromine accumulation. Synergistic effect of Ni^{2+} and Co^{2+} on (c) caffeine degradation, (d) theobromine accumulation.

showed that theobromine production was enhanced when pH was lower while caffeine and cell concentration were higher than that used in the preliminary experiments (center point values). When caffeine concentration and cell loading were maintained at +1 with pH at -1 level, maximum theobromine accumulation was observed (643.09 mg/l, run 6, table 2). Similarly, when we compared the runs where caffeine concentration and cell loading were maintained at +1 with varying pH (-1 and +1), it was found that lower levels of pH favored theobromine production [643.09 (run 6, table 2) and 256.90 (run 11, Table 2) mg/l respectively]. Reduction in theobromine production (487.05 mg/l, run 17) observed at -1.68 level of pH with other two parameters at 0 levels, shows that the dependence of theobromine production on pH is not linear and the optimal value lies somewhere between 0 and -1.68. Huge difference in theobromine accumulation was observed when only the caffeine concentration was changed from level +1 to -1 [643.09 (run 6) to 298.82 (run 9) mg/l] with cell loading and pH maintained at +1 and -1 respectively. This trend was also observed for cell loading and thus at higher concentrations of cell loading and caffeine, theobromine accumulation was favored (run 6, 13 and 14, Table 2). Marginal increase in theobromine production at +1.68 levels of cell loading (572.56 mg/l, run 14) and caffeine concentration (637.4 mg/l, run 13) when compared to center point values (556.46 mg/l, run 2) suggests that the optimal value of cell loading and caffeine concentration lies within the range of values (-1.68 to 1.68) selected for optimization. Productivity followed the same pattern as that of theobromine produced after 4 h, which can be explained from the relation that productivity is proportional to the amount of theobromine produced at a particular time. Molar yield was higher at higher levels of caffeine concentration, cell loading and pH (+1, +1 and +1 respectively) (run 11, Table 2). Yield was found to have reduced in experiments where the pH level was +1 and cell loading was maintained at -1 or less (run 1, 7,16). Otherwise not much variation in theobromine yield was observed between the CCD experiments.

Using 'Design Expert9' software, the response data were fitted into a quadratic equation to explain the effects of the three variables on theobromine production and yield. The quadratic regression models

predicting theobromine accumulation, molar yield and productivity are given below

$$T = 546.85 + 118.56x_1 + 130.71x_2 - 126.27x_3 +$$
$$30.34x_1x_2 - 55.52x_1x_3 - 6.36x_2x_3 - 81.78x_1^2 \qquad (2)$$

$$Y = 63.27 + 5.57\chi_1 + 15.26\chi_2 - 9.45\chi_3 +$$
$$3.32\chi_1\chi_2 + 2.13\chi_1\chi_3 + 11.58\chi_2\chi_3 - 5.06x_1^2 - 11.18x_2^2 \qquad (3)$$

$$P = 136.71 + 29.64\chi_1 + 32.68x_2 - 31.37x_3 -$$
$$7.58x_1x_2 - 13.88x_1x_3 - 1.59x_2x_3 - 40.45\chi_1^2 - 25.9\chi_2^2 \qquad (4)$$

Where, T is the predicted theobromine produced (mg/l), Y is the predicted molar yield, P is the predicted productivity (g/l) and x_1, x_2 and x_3 are the coded values of X_1, X_2 and X_3 respectively. The co-efficient for the interaction term x_1x_3 and x_2x_3 in equation (2) is negative indicating that the variable x_3 interacts negatively with the other variables x_1 and x_2. It shows that the effect of x_3 on theobromine production pattern is different from the effect of other two variables (x_2 and x_3). Similar pattern was observed in equation (4), which represents the model equation for theobromine productivity. Even the coefficient for the factor x_3 is negative in equation (1) and equation (3), which represents the model equation for molar yield of theobromine produced after 4 h. It was reflected in experimental data which showed increase in theobromine production and theobromine productivity as pH was lowered from the center point values (Table 2).

Low p-value (0.0001) for theobromine production (Equation 2) in ANOVA analysis indicates that the quadratic regression model is very significant (Table 3). R^2 value is coefficient of determination which examines how well the data fits into the model. R^2 value closer to 1 indicates that the model significantly fits the experimental data. R^2 value of 0.9598 and adjusted R^2 value of 0.9146 obtained for the model equation (2) indicates that the quadratic regression model can predict theobromine production at any given point of x_1, x_2 and x_3 with good precision. Since productivity is directly related to theobromine produced at a particular time, quadratic model predicting theobromine

Run no	Caffeine concentration (g/l)	Cell loading (g/l)	pH	Theobromine produced (mg/l)		Molar yield %		Productivity (mg/l/h)	
				Experimental	Predicted	Experimental	Predicted	Experimental	Predicted
1	1 (≡3)	-1 (≡4)	1 (≡9)	7.38	35.33	4.80	12.03	1.84	8.83
2	0 (≡2)	0 (≡8)	0 (≡8)	556.46	546.85	66.97	63.27	139.11	136.71
3	-1 (≡1)	1 (≡12)	1 (≡9)	149.79	157.95	48.40	50.31	37.44	39.49
4	0 (≡2)	0 (≡8)	0 (≡8)	556.46	546.85	66.97	63.27	139.11	136.71
5	-1 (≡1)	-1 (≡4)	-1 (≡7)	145.19	98.71	53.59	49.59	36.29	24.67
6	1 (≡3)	1 (≡12)	-1 (≡7)	643.09	721.01	61.60	63.83	160.77	180.25
7	-1 (≡1)	-1 (≡4)	1 (≡9)	6.55	-30.07	10.77	3.27	1.63	-7.53
8	0 (≡2)	0 (≡8)	0 (≡8)	546.26	546.85	60.06	63.27	136.56	136.71
9	-1 (≡1)	1 (≡12)	-1 (≡7)	298.82	312.17	62.85	50.31	74.70	78.05
10	1 (≡3)	-1 (≡4)	-1 (≡7)	353.05	386.19	57.05	49.83	88.26	96.55
11	1 (≡3)	1 (≡12)	1 (≡9)	256.90	344.71	73.66	72.35	64.22	86.17
12	0 (≡2)	0 (≡8)	0 (≡8)	539.56	546.85	62.25	63.27	134.89	136.71
13	1.68 (≡3.68)	0 (≡8)	0 (≡8)	637.40	515.21	60.19	58.34	159.35	128.78
14	0 (≡2)	1.68 (≡14.73)	0 (≡8)	572.56	474.01	52.85	57.35	143.14	118.51
15	-1.68 (≡0.32)	0 (≡8)	0 (≡8)	67.09	116.85	27.77	39.63	16.77	29.19
16	0 (≡2)	-1.68 (≡1.27)	0 (≡8)	8.45	34.82	0.47	6.07	2.11	8.70
17	0 (≡2)	0 (≡8)	-1.68 (≡6.32)	487.05	453.57	58.96	70.42	121.76	113.40
18	0 (≡2)	0 (≡8)	1.68 (9.68≡)	67.92	29.30	40.16	38.67	16.98	7.32
19	0 (≡2)	0 (≡8)	0 (≡8)	530.29	546.85	60.06	63.27	132.57	136.71
20	0 (≡2)	0 (≡8)	0 (≡8)	548.26	546.85	63.56	63.27	137.06	136.71

Experimental response values are the average responses from three independent experiments

Table 2: Central Composite Design for optimization of reaction conditions for theobromine production by Pseudomonas sp.

productivity (Equation 4) was found to be proportional to the model which predicts theobromine production (Equation 2). For the model which predicts theobromine yield (Equation 3), low p-value (0.0004) indicates that the model is significant. High value of R^2 (0.9431) and adjusted R^2 (0.8792) confirm the significance of the model to predict theobromine yield. Hence, the model developed using 'Design Expert 9' for the responses of theobromine production, theobromine yield and molar yield were found to be significant after statistical analysis (Table 3).

Response surface plots showing the two factor interaction effect on theobromine production (Figures 2A-2C), theobromine yield (Figures 2D-2F) and theobromine productivity (Figures 2G-2I) when the other factor was maintained at optimal value, are given in figure 2. The two dimensional response surface plots are the graphical representations of the quadratic regression model equation. Optimum values of the variables for maximum response can be predicted from contour plots. The contour lines represent the interaction of the two variables when the other is maintained at optimal value predicted by solving the model equation. Area enclosed by the smallest ellipse is the optimal response area and the optimal point is the intersection point of the major and minor axes of the ellipse. Figure 2A shows the effect of caffeine concentration (x_1) and cell loading (x_2) on theobromine production when pH is maintained at optimal value $(x_3 = -0.913)$. Elliptical contours in Figure 2A indicate that there is interaction between two variable x_1 (caffeine concentration) and x_2 (initial cell loading) in determining the response (theobromine production). Similarly, perfect elliptical contours in Figure 2B, Figure 2D, Figure 2E, Figure 2G and Figure 2H indicates that the variables represented in the plots have interaction effects on the corresponding responses. Almost circular contour plots in Figure 2C and Figure 2I indicates that the interaction between the variables represented is less. Parallel contour lines in Figure 2F show that the interaction between pH (x_1) and initial cell loading (x_2) is not significant in determining the molar yield of theobromine. In all the contour plots in Figure 2, the intersection point between the major and minor axes of the smallest ellipse lies in the contour which indicates that the optimum point lies within the ranges of X_1, X_2 and X_3 taken for optimization.

All three model equations were solved for the values of x_1, x_2 and x_3 at which the responses were maxima. Model predicted values of theobromine production and productivity at optimal values of X_1, X_2 and X_3 (3.2 g/l, 11.33 g/l and 7.0) are 729.43 mg/l and 182.35 mg/l, which correspond to the responses obtained from experimental values of 689.72 ± 22.86 mg/l and 172.46 mg/l respectively with <4% error. Predicted quadratic regression model for molar yield was completely different from that obtained for theobromine production

and productivity, and thus the optimal values of X_1, X_2 and X_3 were also different. The maximum experimental yield obtained at optimal conditions of X_1, X_2 and X_3 (2.28 g/l, 8.17 g/l and 6.64 respectively) was 64.46 ± 3.87% and it was comparable with model predicted yield of 70.4%. At points X_1, X_2 and X_3, where we got maximum theobromine production (3.2 g/l caffeine, 11.3 g/l induced cell and pH 7.0), experimental molar yield was only 43 ± 3.9% which is 33% less than the molar yield obtained at optimized point for molar yield (2.28 g/l caffeine, 8.17 g/l induced cells and pH 6.64). Figure 3 shows the distribution of experimental and model predicted theobromine production (Figure 3A), molar yield (Figure 3B) and productivity (Figure 3C). Almost all the points fall within ± 1σ for theobromine production, productivity and theobromine yield.

Effect of reaction temperature

All the preliminary experiments and the optimization of reaction media parameters were carried out at 30°C which is the optimal temperature for *Pseudomonas sp.* to grow on CAS medium. In an effort to understand the effect of reaction temperature on theobromine production, reactions were carried out at three different temperatures 25°C, 30°C and 35°C. It was observed that after 4 h, theobromine production was maximum (880.57 ± 18.46 mg/l) at 25°C (Figure 4b) which is 27% more than that observed at 30°C (689.72 ± 22.86 mg/l) with molar yield of 57.77 ± 9.8%. At 35°C, theobromine production was only around 525.93 ± 19.32 mg/l but the yield was higher (79.6 ± 5.18%). We also observed that at 30°C and 35°C, there is no significant conversion of caffeine to theobromine after 4 h whereas at 25°C, *Pseudomonas sp.* converted caffeine to theobromine (Figure 4A and Figure 4B) even after 4 hours with maximum theobromine concentration at 6th h (1.08 ± 0.10 g/l) which is 57% more than that observed at 30°C with a molar yield of 57.92 ± 11.9%. Initial conditions used for converting caffeine to theobromine was 2 g/l caffeine, 8 g/l induced cells and pH 8.0 at 30°C which were optimized to 3.2 g/l caffeine, 11.33 g/l induced cells, pH 7.0 and at 25°C. The comparison in theobromine production between optimized and unoptimized conditions is given in Figure 4C. Theobromine production was increased from 0.384 ± 0.054 g/l to 1.08 ± 0.1 g/l thereby obtaining a 181% increase after optimization of reaction parameters. Molar yield was also increased from 62.70 ± 9.9 % to 68.23 ± 17.6 %.

Discussion

It was reported that a caffeine degrading strain, *Pseudomonas putida* No. 352 accumulated theobromine when zinc ions were added to the growth medium [13]. About 20 g/l of theobromine was produced from caffeine with a yield of 92%. Later they showed that zinc specifically inhibited a theobromine specific monooxygenase [22]. However, this study was performed under growing conditions of *Pseudomonas putida* No. 352 in the presence of zinc in a medium which contained complex components like tryptone and expensive micronutrients like vitamins and amino acids which increases the cost of production. In addition, downstream processing to recover theobromine will be tedious as the growth medium has several compounds. In another study, a mutant strain *Pseudomonas putida IF-3-9C-21* developed from a wild strain *Pseudomonas putida IF3* was found to degrade caffeine at a faster rate and accumulated high theobromine in the media. It was cultured at 30°C in a medium containing caffeine, glucose and sodium glutamate with 1 vv m aeration. Caffeine was continuously added into the medium after 9 h culture time so as to maintain caffeine concentration between 0.5 g/l to 2.5 g/l. It was observed that when 165 g of caffeine was introduced 133 g of theobromine was produced with final caffeine

Response variable	Source	DF	SS	MS	F-value	p-value
Theobromine accumulated[a]	Model	9	1027000	114100	21.23	0.0001
	Error	8	42973.56	5371.69		
	Lack of fit	5	42789.67	8557.93	139.62	0.0009
	Pure error	3	183.89	61.3		
	Total	19	1073000			
Molar yield[b]	Model	9	8086.74	898.53	14.74	0.0004
	Error	8	487.52	60.94		
	Lack of fit	5	479.01	95.80	33.76	0.0077
	Pure error	3	8.51	2.84		
	Total	19	8852.86			

[a]R^2 = 0.9598; adjusted R^2= 0.9146
[b]R^2 = 0.9431; adjusted R^2= 0.8792

Table 3: ANOVA for the quadratic models predicted for each response variable.

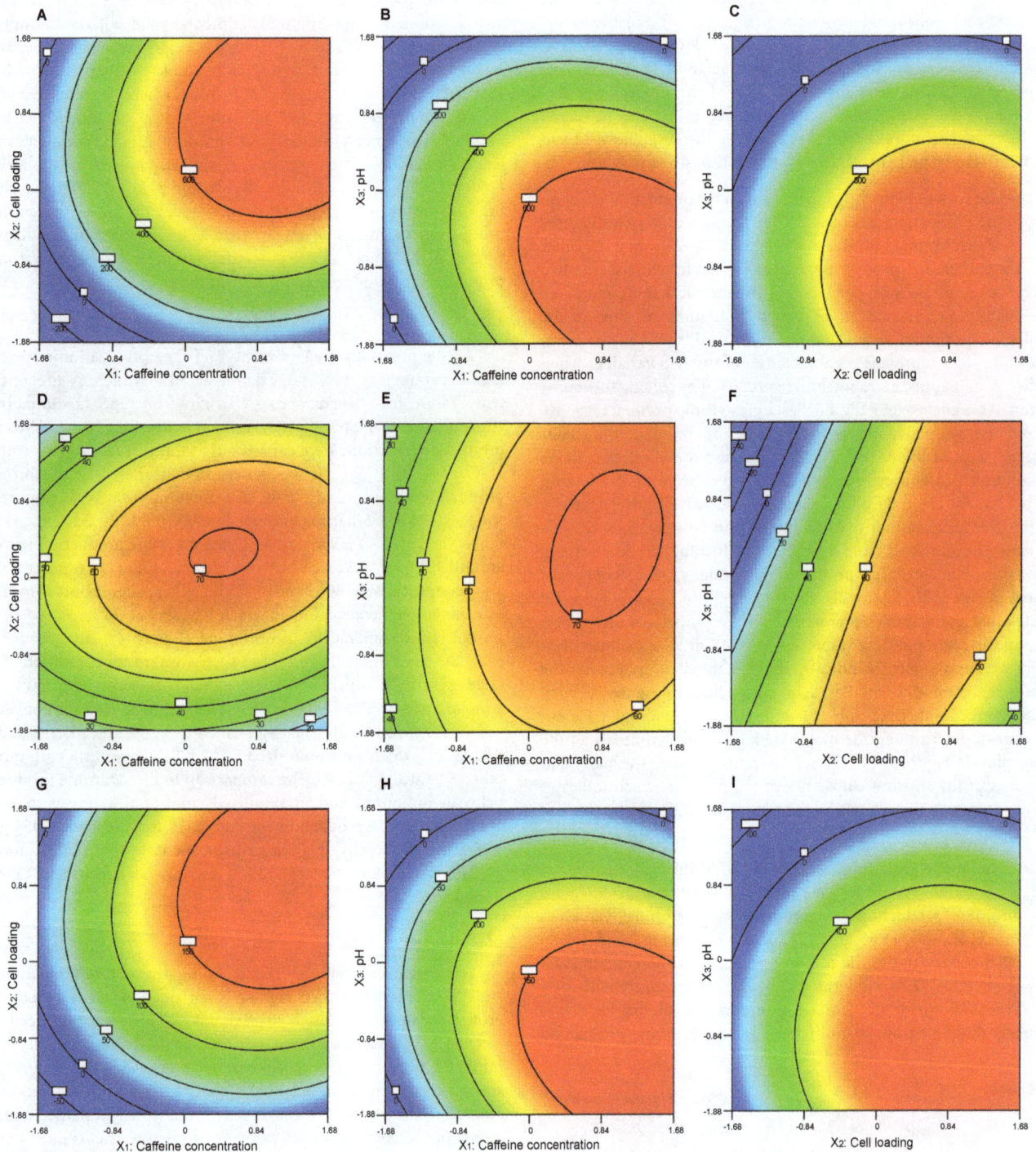

Figure 2: Contour plots showing the effect reaction parameters on responses. (a) theobromine production at optimal pH, (b) theobromine production at optimal cell loading, (c) theobromine production at optimal caffeine concentration, (d) theobromine yield at optimal pH, (e) theobromine yield at optimal cell loading, (f) theobromine yield at optimal caffeine concentration, (g) theobromine productivity at optimal pH, (h) theobromine productivity at optimal cell loading, (i) theobromine productivity at optimal caffeine concentration.

and theobromine concentration of 0.3 g/l and 39.8 g/l respectively [14]. Even though a process for theobromine production from caffeine using a *Pseudomonas* strain was developed with very good conversion (88%), it was done in the growing conditions in the presence of many other complex nutrients which need tedious downstream processing. As reported by Glück et al. [12], there is also a possibility of the mutant strain reverting to its wild type. All these studies were performed to produce theobromine in the growth medium with complex components.

However, we used induced cells of *Pseudomonas sp.* as the catalyst to convert pure caffeine in 10 mM potassium phosphate

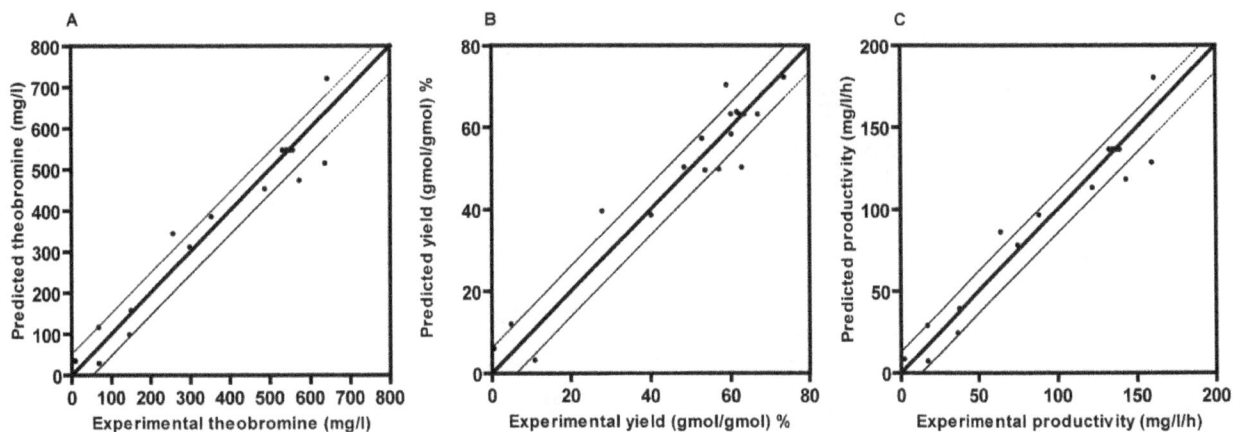

Figure 3: Parity plots for the quadratic models. (a) Parity plot showing the distribution of experimental and predicted theobromine production (σ = 51.48), (b) Parity plot showing the distribution of experimental and predicted theobromine yield (σ = 6.17), (C) Parity plot showing the distribution of experimental and predicted theobromine productivity (σ = 12.8).

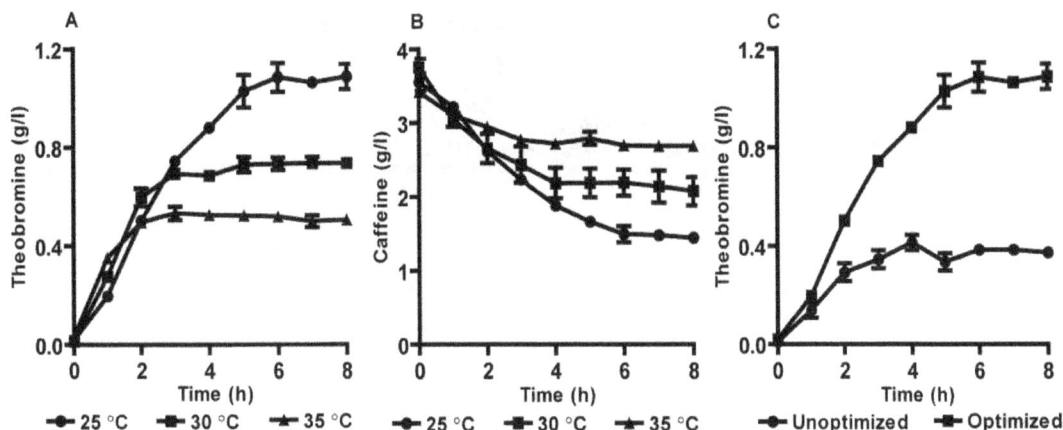

Figure 4: Optimization of reaction temperature at optimized caffeine concentration, cell loading and pH. Effect of temperature on (a) caffeine degradation, (b) theobromine production, (c) plot showing increased theobromine production under optimal conditions.

buffer to theobromine. Whole cell catalyst ensures reusability and low production cost. We report highest theobromine production (~1 g/l) from caffeine using resting cells of *Pseudomonas sp.* as compared to already reported 225 mg/l by Glück et al. They reported accumulation of theobromine and 7-methylxanthine, using resting cells of *Pseudomonas sp. H8* which is a mutant of *Pseudomonas putida WS* [12]. But very less amount of theobromine was accumulated (around 100 mg/l) and they shifted their focus to metabolite production during growth of *Pseudomonas H8* in a medium containing caffeine. Even during growth, only around 225 mg/l of theobromine was reported along with 425 mg/l of 7-methylxanthine which is very less for industrial production. Major advantages of using induced whole cell biocatalyst for theobromine production as demonstrated in this study are: 1. Induced cells are metabolically active as it has enough caffeine degrading enzymes synthesized already during growth in production media. Therefore, conversion of caffeine to theobromine is really fast. Productivity of theobromine after 6 h under optimized conditions is 220.14 ± 4.61 mg/l/h which can be increased by applying continuous or fed-batch strategy and 2. Reaction media contains only caffeine and theobromine which can be easily separated by preparative HPLC.

Conclusions

Theobromine is a commercially important molecule which is currently produced in industries using toxic chemicals and tedious procedures. Microbial production of theobromine is practically a safer and economical approach compared to the chemical methods available. We optimized the reaction parameters for theobromine production from caffeine by *Pseudomonas sp.* as 3.2 g/l caffeine; 11.3 g/l induced *Pseudomonas sp.*, pH 7.0 at 25°C. Under this condition, up to 1 g/l of theobromine was produced from caffeine in 6 h with a molar yield of 57%. More research is needed on increasing the yield and production of theobromine from caffeine by employing bioprocess strategies so that microbial production of theobromine in industry will become feasible in future.

Acknowledgement

The authors would like to acknowledge Department of Biotechnology (DBT), India for funding the project.SR acknowledges UGC for JRF. Authors also acknowledge Anchitha Krishna from Humanities and Social Sciences, IIT Madras for her help in editing the manuscript. Study design, data collection, interpretation of data and manuscript preparation was done by the authors and the funding agency has no part in it.

References

1. Dash SS, Gummadi SN (2006) Catabolic pathways and biotechnological applications of microbial caffeine degradation. Biotechnol Lett 28: 1993-2002.

2. Mc Govern T, Mc Devitt E, Wright IS (1936) Theobromine sodium salicylate as a vasodilator. J Clin Invest 15: 11-16.

3. Usmani OS, Belvisi MG, Patel HJ, Crispino N, Birrell MA, et al. (2005) Theobromine inhibits sensory nerve activation and cough. FASEB J 19: 231-233.

4. Christ C (2008) Production-integrated environmental protection and waste management in the chemical industry. WILEY-VCH, USA.

5. Littler TR, McConnell RB (1955) Dimethyl sulphate poisoning. Br J Ind Med 12: 54-56.

6. Nehlig A, Daval JL, Debry G (1992) Caffeine and the central nervous system: mechanisms of action, biochemical, metabolic and psychostimulant effects. Brain Res Brain Res Rev 17: 139-170.

7. Nurminen ML, Niittynen L, Korpela R, Vapaatalo H (1999) Coffee, caffeine and blood pressure: a critical review. Eur J Clin Nutr 53: 831-839.

8. Rapuri PB, Gallagher JC, Kinyamu HK, Ryschon KL (2001) Caffeine intake increases the rate of bone loss in elderly women and interacts with vitamin D receptor genotypes. Am J Clin Nutr 74: 694-700.

9. Hartley-Asp B, Kihlman BA (1971) Caffeine, caffeine derivatives and chromosomal aberrations. IV. Synergism between Mitomycin C and caffeine in Chinese hamster cells. Hereditas 69: 326-328.

10. Slattery ML, West DW, Robison LM, French TK, Ford MH, et al. (1990) Tobacco, alcohol, coffee, and caffeine as risk factors for colon cancer in a low-risk population. Epidemiology 1: 141-145.

11. Smith SJ, Deacon JM, Chilvers CE (1994) Alcohol, smoking, passive smoking and caffeine in relation to breast cancer risk in young women. UK National Case-Control Study Group. Br J Cancer 70: 112-119.

12. Glück M, Lingens F (1987) Studies on the microbial production of theobromine and heteroxanthine from caffeine. ApplMicrob Biotechnol 25: 334-340.

13. Asano Y, Komeda T, Yamada H (1993) Microbial production of theobromine from caffeine. Biosci Biotechnol Biochem 57: 1286-1289.

14. Yoshinao K (1992) Novel bacterial strain and method for producing theobromine using the same. European Patent No. EP0509834A2.

15. Gokulakrishnan S, Chandraraj K, Gummadi SN (2007) A preliminary study of caffeine degradation by Pseudomonas sp. GSC 1182. Int J Food Microbiol 113: 346-350.

16. Gokulakrishnan S, Gummadi SN (2006) Kinetics of cell growth and caffeine utilization by Pseudomonas sp. GSC 1182. Process Biochem 41: 1417-1421.

17. Dash SS, Gummadi SN (2007) Enhanced biodegradation of caffeine by Pseudomonas sp. using response surface methodology. Biochem Eng J 36: 288-293.

18. Dash SS, Gummadi SN (2007) Optimization of physical parameters for biodegradation of caffeine by Pseudomonassp.: a statistical approach. American Journal of Food Technology 2: 21-29.

19. Dash SS, Gummadi SN (2008) Inducible nature of the enzymes involved in catabolism of caffeine and related methylxanthines. J Basic Microbiol 48: 227-233.

20. Dash SS, Gummadi SN (2007) Degradation kinetics of caffeine and related methylxanthines by induced cells of Pseudomonas sp.Curr Microbiol 55: 56-60.

21. Jahangir M, Abdel-Farid IB, Choi YH, Verpoorte R (2008) Metal ion-inducing metabolite accumulation in Brassicarapa. J Plant Physiol 165: 1429-1437.

22. Asano Y, Komeda T, Yamada H (1994) Enzymes involved in theobromine production from caffeine by Pseudomonas putida No. 352. Biosci Biotechnol Biochem 58: 2303-2304.

23. Gummadi SN, Santhosh D (2006) How induced cells of Pseudomonas sp. increase the degradation of caffeine. Cent Eur J Biol 1: 561-571.

9

Process Development to Recover Pectinases Produced by Solid-State Fermentation

Daniel E. Rodríguez Fernández[1]*, José A. Rodríguez León[2], Julio C. de Carvalho[1], Susan G. Karp[1], José L. Parada[2] and Carlos R. Soccol[1]

[1]Department of Bioprocess Engineering and Biotechnology, Federal University of Paraná, P.O. Box 19011, Zip Code 81.531-970, Curitiba, PR, Brazil
[2]Department of Biology Science, Positivo University, Prof. Pedro Viriato Parigot de Souza Street, 5300, Zip Code 81.280-330, Curitiba, PR, Brazil

Abstract

Leaching, or solid-liquid extraction, is the first step that must be done in the recovery process of a metabolite produced by Solid State Fermentation (SSF). In this work, the leaching of a Polygalacturonase (PG) produced by a strain of *Aspergillus niger* by SSF of citrus dried pulp was performed. A fractionated factor design 2^{4-1} was developed to establish the influences of five factors: solid-liquid ratio, temperature, pH, agitation and surfactant addition (Tween 80). Agitation and surfactant addition effects were confounded by second order effects according to the fractionated factor design. Results showed that pH and surfactant addition did not influence the recovery process. The leaching process was characterized through the corresponding kinetic parameters for PG and total protein recovery, corresponding to first order kinetics. In the case of PG leaching, the theoretical C_s was 17993 U L^{-1} and the k was 0.107 min^{-1}, and in the case of total protein extraction, values were 13159 mg L^{-1} and 0.177 min^{-1}, respectively. The process was improved by using a system of six successive steps. The results showed that when a single step was performed the concentration obtained was 137 U g^{-1} (d.b.), while after successive extractions the PG extracted achieved a concentration of 537 U g^{-1} (d.b.), improving the process by 74%.

Keywords: Enzyme Recovery; Leaching kinetics; Solid state fermentation; Pectinase; Aspergillus niger

Abbreviations and Symbols: PG: Polymethylgalacturonase; SSF: Solid state fermentation; (d.b.): Dry basis; ANOVA: Variance analysis; p: Probability; C_{aPG}: PMG concentration at time t (IU/mL); C_p: Total protein concentration at time t (mg/mL); PG_{act}: IU PG/g (d.b.); X: Stage number; Y: mg protein /g (d.b.); r^2: Regression coefficient

Introduction

Annually, about 34 million tons of citrus residues (peel, seeds) are produced from the processing of citrus fruits [1-3]. Although a part of these residues is applied in the diet of monogastric animals, its accumulation in the biosphere causes several environmental problems. At the same time, citrus peel can be an important and economical raw material to produce enzymes, mainly pectinase, by fermentative processes. This material contains almost all nutrients needed for microorganism growth, and several processes have been studied for the production of many important metabolites including enzymes [4].

Polygalacturonases, including endoPGases and exoPGases, comprise a family of enzymes named pectinases that degrade pectin and pectic acid. These enzymes are extensively used in the feed and drink industries, chiefly in juice clarification because they are capable of reducing the viscosity of liquors during the clarification process [5,6].

The production of PG by SSF employing agro-industrial residues and different microorganisms has been previously studied [7-13]. In SSF, it is necessary to consider that the produced metabolites remain in a solid matrix and, therefore, must be extracted by solid-liquid extraction or leaching. Therefore, this is the first step in any recovery and purification process that is intended for any desirable metabolite produced by SSF [14,15].

The study of leaching involves establishment of the factors and the corresponding parameters that characterise the process to reach acceptable levels in yield during the purification process. However, there are only a few reports regarding the leaching of metabolites from a solid state fermented matrix [16,17] and, in particular, from matrixes containing PG. These reports mainly discuss how different factors affect enzyme extraction, but there is a lack of complete representation of the leaching process through the corresponding kinetic pattern. However, several mechanisms were postulated for their study from biological material [18].

The aim of this work was to analyse a leaching process to improve the recovery of PG produced by *Aspergillus niger F3* growing on citrus wastes by SSF, as part a strategy to apply the enzyme PG in juice clarification within a global project to reuse the solid waste generated during citrus juice production. This study was done by statistically determining the significance of different factors that may influence the process. Once the significance of these factors was established, the kinetic patterns of the process were determined. The process was considered a proper procedure by using successive extraction steps.

Materials and Methods

Microorganism and inoculum preparation

The strain *A. niger F3* with low sporulation was employed. This strain was obtained from the Federal University of Paraná (UFPR) and maintained on Potato Dextrose Agar (PDA) slants. Mycelial biomass was produced by growing the *A. niger F3* on Czapek liquid medium for 72 hours at pH 5, 30°C and 120 rpm.

*Corresponding author: Daniel E. Rodríguez-Fernández, Department of Bioprocess Engineering and Biotechnology, Federal University of Paraná, P.O. Box 19011, Zip Code 81.531-970, Curitiba, PR, Brazil

Solid state fermentation

Dried citrus peel (75% of the peel had a particle size ranging from 0.8 mm to 2 mm, and the other 25% had a particle size between 2 mm and 3 mm) was used and supplemented with the following substances in dry basis: NH_4NO_3 0.43%; Na_2SO_4 0.021%; $MgSO_4.7H_2O$ 0.077%; $ZnSO_4.7H_2O$ 0.042%; KCl 0.162%; and $Ca(OH)_2$ 0.011%. Water was added to obtain 60% of initial humidity. Inoculation was made in a 1:10 (v/w) ratio in 1 kg of wet solid media. Fermentations were carried out in a cylindrical bioreactor (25 cm diameter and 50 cm height) loaded with 2 kg for 96 h at controlled room temperature (30°C). Aeration intensity was set at 1 VkgM (1 LAir kg^{-1} min^{-1}) [19].

Moisture content

Moisture content was determined in an infrared balance (Sartorius model MA-50) at 105°C.

Polygalacturonase (PG) activity

PG was assayed by measuring the release of reducing sugars by the di-nitro-salicylic (DNS) method [20]. In this case, galacturonic acid (Sigma) was considered as simple sugar formed after pectin degradation. In a test tube, 0.9 mL of 0.5% (w/v) pectin (Sigma) in 0.1 M citrate buffer (pH 4.0) was added to 0.1 mL of the diluted enzyme solution. After incubation for 15 min at 50°C, the reaction was stopped by the addition of 1 mL DNS, and the mixture was heated in boiling water for 5 min. Distilled water (5 mL) was then added to each sample. Absorbances of the samples were read at 540 nm. One unit of PG activity was defined as the amount of enzyme that liberates 1 μmol of D-galacturonic acid per minute at 50°C and pH 4.0 [19].

Protein concentration

Protein was determined with a Folin-phenol reagent [21] using Bovine Serum Albumin (BSA) as the protein standard. O

Enzyme leaching

Extractions were carried out in a stirred and jacketed tank (Heidolph RZR 2021; Germany) with 1 L of effective volume and a paddle agitator. Different solid-liquid relations were studied to determine the best condition to extract the enzymes. Enzymatic extracts were prepared by adding different masses of fermented solid matrix in 1L of pure water to obtain the desired solid-liquid ratio. Mass of solids was added considering the moisture content to express the solid concentration in dry basis. Mass of solids was calculated by the following equation:

$$M_S = \frac{Sol/liq_{(d.b.)} \cdot V}{\left(1 - \dfrac{\%Moist}{100}\right)}$$

Where:

M_S: Mass of solids (g).

$Sol/liq_{(d.b.)}$: Solid-liquid relation in dry basis (g dry solids mL^{-1}).

V: Total working volume (mL)

%Moist: Moisture content of the fermented solid matrix.

Experimental design

Four independent factors were analysed as follows: temperature, pH, agitation and addition of a non-anionic detergent as surfactant (Tween 80). After pH adjustment and temperature stabilisation, solids were agitated for 30 min to obtain enzymatic extracts. Extracts were centrifuged at 4,000 rpm (Sigma centrifuge; B. Braun Biotech International) for 15 min to analyse the enzymatic activity.

The selected factors considered in the experimental design that may influence the leaching study were the following:

Temperature (A): This factor influences both enzyme stability and enzyme diffusion into the solvent or liquid phase.

pH (B): This parameter may have an important effect on enzyme characteristics and enzyme-surface ratios but a lower impact on diffusion in the process.

Surfactant addition (C): This factor may avoid solute attachment to the solid matrix and increase cell permeability to increase extract yield. The surfactant selected was Tween 80 due to its non-anionic character.

Agitation (D): The most important influence of this factor is related to the gradients that may eventually form during the process. In addition, aeration influences the liquid film formed around the solid particles, which conducts the mass transfer from the solute to the solvent.

A two level fractionated factorial design (2^{4-1}) with three replicates at the central point was carried out to evaluate the influence of the selected factors and their possible interactions in the leaching process [22]. This procedure allowed the statistical analysis of these factors with a minimal number of experiments. Table 1 shows the levels and values for the independent factors analysed in the 2^{4-1} plan developed for the study.

The factorial plan was generated by using a full three factorial experimental design involving three factors (A, B and C) and then confounding the remaining factor (D) with interactions generated by the following plan generators: D = A * B. The defined relation allowed the establishment of an alias that determined which effects were confounded with each other through the coefficients of the predicted response equation as follows:

$$Y = \beta 0 + \beta_1 A + \beta_2 B + \beta_3 C + \beta_{12} AB + \beta_{13} AC + \beta_{14} AD \qquad (1)$$

The experimental design matrix is shown in Table 2. Runs were carried out in a random sequence with central point repetition.

In order to optimize the extraction a central composite design

Factors	Corresponding variable	Coded level of variable		
		- 1	0	1
A	Temperature (°C)	30	40	50
B	pH	3	5	7
C	Tween 80 (% v/v)	0	0.75	1.5
D	Agitation (rpm)	100	300	500

Table 1: Experimental range and levels of independent variables studied using a two level fractionated factorial design (2^{4-1}) in terms of actual and coded factors.

Run	A	B	C	D = A*B
1	-1	1	1	1
2	1	-1	-1	-1
3	-1	1	-1	-1
4	1	1	-1	1
5	-1	-1	1	1
6	1	-1	1	-1
7	-1	1	1	-1
8	1	1	1	1
Central Point	0	0	0	0
Central Point	0	0	0	0
Central Point	0	0	0	0

Table 2: Experimental design matrix for the 24-1 plan developed for independent factors analyzed in the PG extraction.

was developed after determining the significant factors for the PG extraction. In Table 3, the corresponding experimental design matrix is shown. All data were processed using the Design Expert Version 5.07 program.

Effect of ions on the recovery of PG

The effect of anions (Cl-, SO_4^{-2}, NO_3^{-1}, citrate anion, acetate anion and EDTA) and cations (Na^+, K^+, Ca^{+2}, Mg^{+2}, Mn^{+2} and Zn^{+2}) on enzyme activity was investigated by the addition of acid solutions containing these anions to the solvent for 30 min at room temperature (approximately 30°C). For all runs the ionic concentration was 30mM. The EDTA was added at a concentration of 50 mM. Distilled water was considered as the control.

Results and Discussion

PG from *Aspergillus niger F3* was produced by SSF employing citrus peel during 96 hours. PG concentration at the end of the fermentation was 265 U g^{-1}(d.b.) and the final moisture was 61.23%. Based on this result the extraction of PG from the solid matrix for further downstream processing was analysed. The PG concentration obtained is higher than that reported in a review by Favela-Torres et al. [23] which is 71.2 - 81 U g^{-1} for medium composed by orange bagasse.

Solid-liquid relation

Different quantities of fermented solid matrix were processed in 1L of distilled water in order to obtain the best solid-liquid relation. Table 4 shows the results of concentration, the activity expressed as U mL^{-1}. It is important to point out that when working with a 1:10 solid-liquid relation, all water was absorbed and retained by the solid and it was impossible to obtain an extract. According to the results showed in Table 3, the higher values of enzyme concentration were obtained at 1:20 g mL^{-1} in dry basis, but it was also the condition with lower yield, whereas in the condition of 1:50 g mL^{-1} enzyme was more diluted and the yield of the extraction was higher. There is an apparent contradiction, the more concentrated the activity was, the lower yield was obtained in the extraction. To explain this it is necessary to point out that when the yield was higher the amount of solid employed was lower and it was easier to extract the enzyme present in the solid matrix. But, at the same time, this condition represented more volume used to extract, so more steps required for concentration and more liquid residues generated, influencing negatively on the economy of the process. Applying an engineering criterion for extractive processes, "*the best condition for recovery is where the product is more concentrated and not where more is extracted*", it is possible to economize time and energy during the

concentration steps and in all recovery process. It was considered as the best condition where the PG was recovered more concentrated and it was decided to go on working in this research considering 1:20 g mL^{-1} in dry basis as the solid liquid relation.

Statistical analysis for fractionated factorial design 2^{4-1}

The statistical analysis was carried out by searching for possible factor interactions. It was considered that the confounded effects in the coefficients may be deduced from the defining relations of the plan, and confounded effects higher than second order were overlooked.

The response matrix (Y) obtained from the experiments developed in the fractionated factorial plan 2^{4-1} is shown in Table 5. The highest activities obtained in the developed experiments corresponded to runs 5, 7 and 1. These runs coincided with the lowest values of the temperature, while for the rest of the independent factors analysed, the values appeared at higher or lower level in different cases. It is possible to think that temperature is the only factor with significant influence on the PG activity in the extract.

In order to explain this behaviour, analysis of variance (ANOVA) was performed to obtain the significant coefficients corresponding to Equation (1) in terms of coded factors and to validate the regression model. In Table 6, the results for the values and discrimination of coefficients ($\beta_0, \beta_1, \beta_2, \beta_3, \beta_{12}=\beta_4, \beta_{13}=\beta_4, \beta_{13}=\beta_{24}, \beta_{14}=\beta_{23}$) are shown, while independent factor (β_4) is associated with the interaction between temperature and pH (β_{12}).

In Table 6 coefficients obtained from ANOVA of the fractionated factorial design developed are shown. The β_1, β_{12} and β_{13} coefficients were statistically significant at $\alpha = 0.05$ confidence level and the rest of the coefficients had no significant influence on the process. Once the significant coefficients were established, it was important to analyse which effects represented each coefficient through the confounding effects determined by the alias pattern and the defining relations. The β_1

Factor	Units	Coded level of variable				
		-α	-1	0	1	α
Temperature	(°C)	25.86	30	40	50	54.14
Agitation	(rpm)	88	150	300	450	512

Table 3: Experimental range and levels of independent variables studied using a central composite design in terms of actual and coded factors.

Solid-liquid relation (g mL^{-1}) (d.b.)	Mass of solids added (g)	PG concentration (U L^{-1})	Yield PG Extraction (%)
1:10	257.9	-	-
1:20	129.0	11348	37.40
1:30	86.0	9276	42.14
1:50	64.5	6329	44.40
1:60	51.6	4983	47.58

Table 4: Concentration obtained for different solid-liquid relations considering dry basis.

Run	Temperature (°C)	pH	Tween 80 (%)	Agitation (rpm)	PG Concentration (U L^{-1})
1	30	3	0	500	11982
2	50	3	0	100	7785
3	30	7	0	100	10457
4	50	7	0	500	8884
5	30	7	1	500	12007
6	50	3	1	100	7792
7	30	3	1	100	12162
8	50	7	1	500	7831
PC1	40	5	0.75	300	10267
PC2	40	5	0.75	300	10354
PC1	40	5	0.75	300	10199

Table 5: Concentration of PG extracted from fermented solid matrix according to the fractionated factorial design 24-1.

Coefficient (confounded effects)	Value	p
β_0 (Block effect)	9862.65	-
β_1 (A) *	-1789.61	0.002165
β_2 (B)	-29.03	0.006773
β_3 (C)	85.50	0.054505
β_{12} (D; AB) *	313.53	0.006769
β_{13} (AC; BD) *	-346.81	0.006223
β_{14} (AD; BC)	77.45	0.105698
Lack of fit*	-	0.010336

Table 6: Factor coefficients discrimination corresponding to the fractionated factorial design 24-1.

coefficient showed the largest influence on the process, which means that the temperature is the most important independent factor analyzed. At the same time, second order interactions of temperature and pH, and temperature and Tween 80 are statistically significant. However second order interaction is confounded with the agitation and it is not possible to determine which of them has the real significant influence on the extraction. On the other hand, the fit of the model obtained was 0.86514 and its lack of fit was statically significant too. This behaviour suggested that a non-linear model may adjust better in the range analysed and that there may be a zone of optimal conditions.

Optimisation of the solid liquid conditions

A central composite design based on the results of the fractionated factorial design was carried out to optimise the conditions for lixiviation. The design and results are shown in Table 7. Analysis of variance was performed to validate the regression model. The regression coefficients and p-values are presented in Table 8. Linear and quadratic coefficients of temperature had significant influence in the extraction of PG, also the quadratic coefficient of the agitation formed a curve in the surface obtained, therefore it was possible to optimize the extraction of PG. Non-significance of the linear coefficient of agitation and interaction are in accordance with the results showed in the fractionated factorial design, and at the same time it is possible to affirm that interaction between temperature and pH had significance influence on the extraction and not the linear coefficient of agitation The fit of the model was 0.9525, therefore, the statistical model accounted for 95.25% of the response. Statistical model was obtained considering only significant coefficients and is expressed by Equation 2:

$$Y = 10225.55 - 2893.9X_1 - 229.2X_1^2 - 140.68X_1^2 \qquad (2)$$

Decoded values for temperature and agitation were 25°C and 300 rpm respectively.

Run	Temperature (°C)	Agitation (rpm)	PG Extracted (U L^{-1})
1	3.5	150	12194.08
2	6.5	150	7129.85
3	3.5	450	12252.49
4	6.5	450	7185.91
PC1	5.0	300	10235.53
PC2	5.0	300	10296.22
PC3	5.0	300	10162.26
PC4	5.0	300	10209.74
Ax1	2.9	300	14790.63
Ax2	7.1	300	5076.88
Ax3	5.0	90	10295.16

Table 7: Response matrix (Y) of the Central Composite Design developed for optimizing PG extraction.

Coefficient (linear and quadratic effects)	Value	p
β_0 (Block effect)	10225.55	-
β_1 (L) *	-2983.90	0.000001
β_{11}(Q) *	-229.20	0.001910
β_2 (L)	-51.16	0.081112
β_{22} (Q) *	-140.68	0.007836
β_{12}	-0.50	0.986844

Table 8: Factor coefficients discrimination corresponding to the Central Composite Design.

Effect of ions on the recovery of PG

The control sample had an activity of 13682 U mL^{-1}. Figures 1 and 2 show that none of the anions and cations used in the study had an activating effect when used at a concentration of 30 mM. It is possible to see that the inhibition effect was stronger for cations than for anions. Zn^{2+} and Mn^{2+} were the strongest inhibitors of PG activity with 46% and 51% of inhibition, respectively. For anions the strongest inhibitory effect was given by NO_3^-, when 85% of the PG activity was achieved compared with the control.

EDTA addition improved the PG activity because it tends to chelate metal ions, which inhibit enzyme activity. Furthermore, the growth medium contained different ions that are considered to be inhibitors of enzyme activity. Therefore, the enzyme stability was tested in solution at room temperature (± 25°C) after 24 hours. The sample with EDTA conserved approximately 90% of the initial activity whereas the control sample lost nearly 33% of its initial activity.

Kinetic pattern of the PG leaching process

The leaching kinetic study was performed considering the optimal conditions determined for the extraction of PG from a solid state fermented dried citrus pulp by A. niger F3. These optimal conditions were 1 dried gram of solid fermented medium per 20 mL of solvent in agitated system at 300 rpm and 25°C of temperature. EDTA was added to get a concentration of 50 μM in the solvent employed.

Figure 3 illustrates the extraction kinetics. The highest concentra-

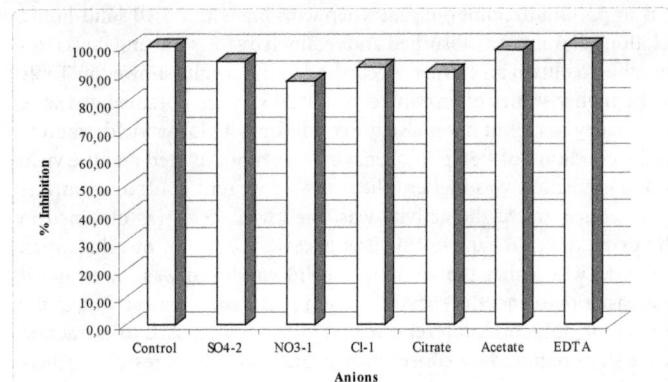

Figure 1: Effect of different anions on the PG activity considering the control to be 11.65 IU/mL leached from a solid fermented (by A. niger F3) dried citrus pulp.

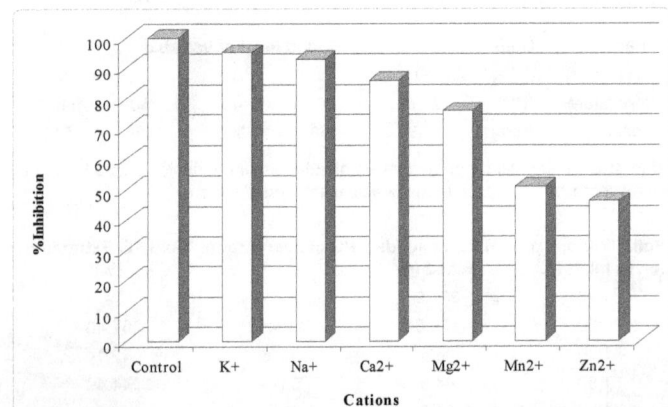

Figure 2: Effect of different anions on the PG activity considering the control to be 11.65 IU/mL leached from a solid fermented (by A. niger F3) dried citrus pulp.

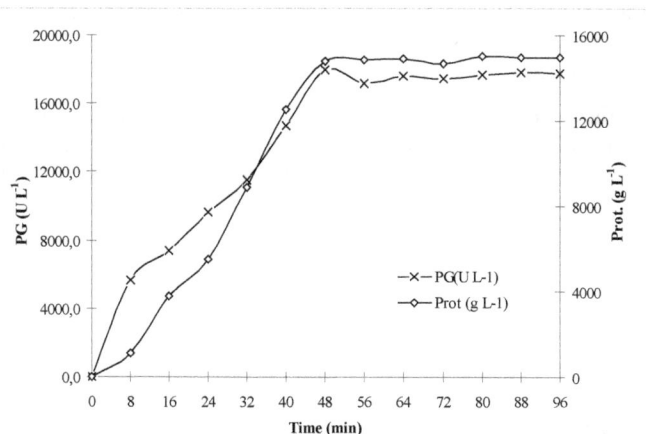

Figure 3: Kinetics of PG and total proteins leaching from a solid fermented (by *A. niger F3*) dried citrus pulp matrix.

$$\ln\left(\frac{C_s - C}{C_s}\right) = -0.1077t + 0.1548 \qquad (4)$$

$R^2 = 0.964$

For the total protein extraction, the linear model is as follows:

$$\ln\left(\frac{C_s - C}{C_s}\right) = -0.177t + 0.1493 \qquad (5)$$

$R^2 = 0.9506$

From the former equations, the following can be deduced:

$$C_{PG} = 17.9\left(1 - e^{-0.107t}\right) \qquad (6)$$

$$C_P = 13.1\left(1 - e^{-0.177t}\right) \qquad (7)$$

Where C_{PG} is the PG concentration at time t (IU/mL), C_p is the total protein concentration at time t (mg/mL) and (t) is the time (min).

tion was 16 IU/mL at 40 min of extraction. At this time, the extraction reached the equilibrium point and the activity remained constant for the rest of the time studied. The concentration obtained was similar to the concentration of PG normally produced in submerged fermentation by several species of *Aspergillus niger*.

The mechanism that controls the leaching process depends on the characteristics of the original matrix from which the product is extracted. If this mechanism depends on the following factors, then a basic equation for the leaching process is represented by a first order kinetic process: migration of the extracted solute from the particle surface into the bulk solution without interferences by the entry of the solvent into the particles; redistribution of the solvent in the cell compartments; expansion of the solid matrix; and the dissolution and/or degradation of solutes. The basic equation reported in scientific literature [24] is as follows:

$$C = C_s (1 - e^{-kt}) \qquad (2)$$

Where:

k is the leaching specific rate constant (min^{-1}),

C is the enzyme concentration (IU/mL),

C_s is the saturated enzyme concentration (IU/mL)

t is the time (min).

From Equation (2), the following is deduced:

$$\ln\left(\frac{C_s - C}{C_s}\right) = -kt \qquad (3)$$

Equation (3) represents a linear form of Equation (2). Therefore, k may be deduced from Equation (3) by linear correlation if the process is controlled only by the migration of the extracted solute from the particle surface into the bulk solution.

Because *Aspergillus spp.* have the general capacity to synthesise several enzymes, the leaching of other metabolites that may be present in the fermented solid matrix, mainly other undetermined enzymes, was considered in the leaching study. This assumption was made considering the total protein extracted from the solid matrix. Figure 4 shows the linear models obtained for the leaching of PG (IU/mL) and total protein (mg/mL) considering the points obtained until 40 min.

The following linear model is for PG leaching from the values reported in Figure 5:

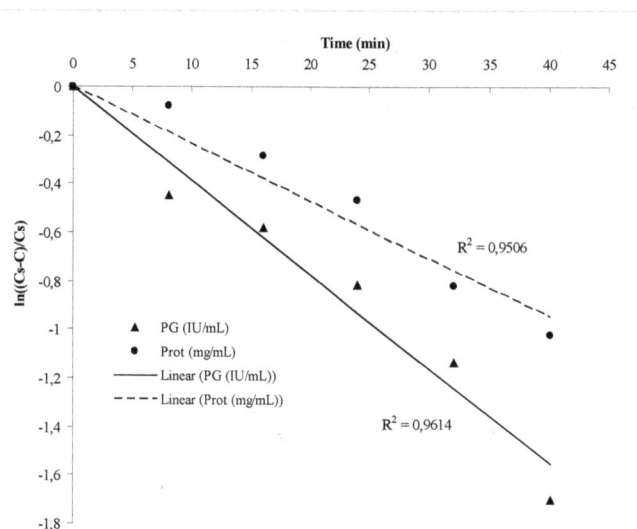

Figure 4: Representation of Equation (5) relating the kinetic parameters of lixiviation of PG and total proteins produced by SSF of citrus pulp by *A. niger F3* (dry basis).

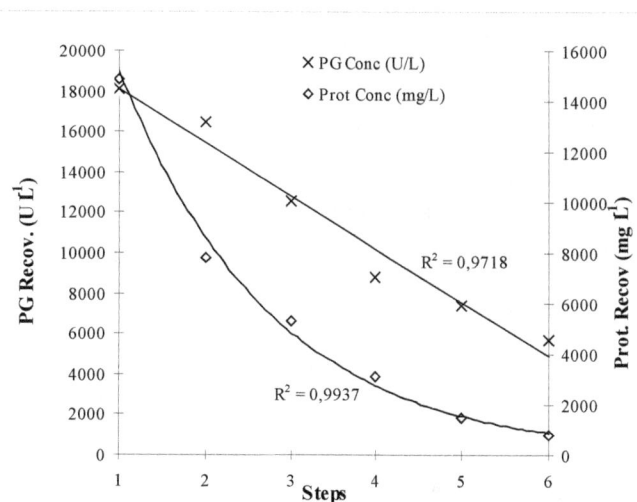

Figure 5: PG and Protein concentration obtained during successive leaching steps.

Step	PG Conc. (U L^{-1})	PG Recovery (Ug^{-1} d. b.)	Protein (mg g^{-1} d. b.)	Specific Activity (Ug^{-1}prot.)	Yield Total PG Recovery (%)	Yield Total Protein Recovery (%)
1	17684	137.09	115.44	1.19	29.10	44.45
2	13491	104.58	60.78	1.72	22.20	23.40
3	10586	82.06	41.26	1.99	17.42	15.88
4	7653	59.33	24.03	2.47	12.59	9.25
5	6022	46.68	11.83	3.95	9.91	4.55
6	5329	41.31	6.38	6.48	8.77	2.46

Table 9: Results of PG leaching from citrus pulp fermented by *A. niger F3* at six steps with a solid-liquid ratio of 1:25 in dry basis.

In the case of PG leaching, the C_s was 15.9 IU/mL and k was 0.107 min^{-1}. The corresponding values for total protein extraction were 13.1 mg/mL and 0.177 min^{-1}, respectively. These results demonstrated that the leaching process of PG was accompanied by the extraction of other proteins, and these proteins were easier to extract than PG according to their respective leaching specific rate constants. The kinetic constant for the total protein extraction was almost the double of the kinetic constant for the PG leaching, which has to be taken into account when the PG specific activity is considered in a further purification, because the total protein extracted does not only correspond to PG. Unfortunately, previous reports do not contain any values related to this topic.

The C_s indicated that the water employed as solvent was saturated at a concentration equivalent to 15.9 IU/mL. This concentration constant also indicated that not all of the enzyme may be extracted in a specific volume of solvent, in one step. Therefore, the solid-liquid ratio needs to be greater, which may be undesired because it results in more liquid disposal to handle within the recovery process. This is contrary to the features of a solid state process in which the disposal problem should impact less when compared with submerged fermentations. Thus, successive extractions or a multistage process is required to determine the complete extraction of the initial enzyme.

Study of the leaching process by a multistage process

As it was observed in the study of solid-liquid relation, working at lowest levels implies in treating a high amount of solids and it is possible to achieve a high concentration of enzyme, but at the same time, part of the enzyme that is not extracted and remains in the solid matrix. In the present case it was decided to proceed with successive leaching steps at a low solid-liquid ratio (1/20) with the residual solid from the preceding step. Each step lasted for 45 min at 300 rpm and 25°C and there were 6 successive steps. Although a countercurrent process could be better at operation, our purpose was to establish the relationship between PG and total protein extracted relative to the quantity extracted and remaining, compared to a single stage. As in the previous kinetic study, the PG and the total protein extracted were considered. Table 9 reports the values obtained for each step.

From Table 9 it is observed that in the one step leaching process for PG at a solid-liquid ratio of 1/20, there was an extraction of 140.29 U g^{-1} (d.b.) of PG with the residual PG activity as high as 395.98 IU g^{-1} (d.b.), that is almost three times the value obtained in a one step extraction. It can also be observed that the concentration in each step diminished with successive steps, meaning that the kinetic pattern established former no longer remained and the transport of the solute to the exterior of the particles, or internal diffusion, started to control the process. The increase of the specific activity with the steps indicates that at the same time a purification of the PG took place. This is expected when considering the values for the specific kinetic constants obtained before, which established that proteins not related to PG are easier and faster

to extract. This increase in the specific activity demonstrates that in the preceding steps, protein not related to PG was not as strongly attached to the solid fermented matrix as the PG protein.

These results can be better observed by modeling the values reported in Table 9. The following models were determined from the data reported in Table 9, considering the step number as the independent variable:

$$PG_{Rec} = -2651(\#step) + 20808 \tag{6}$$

$$R^2 = 0.9718$$

and

$$Prot_{Rec} = 26881e^{-0.57(\#Step)} \tag{7}$$

$$R^2 = 0.9937$$

In the case of total protein extraction, an exponential pattern is observed, while in the case of PG it is linear. This fact shows that protein extraction is easier than PG extraction. It can be inferred that the PG protein was more attached to the porous surface of the fermented solid (citric pulp) than the other existing proteins. The high regression coefficients obtained in both models allow for this conclusion. The yields for each step were determined from the total amount of PG and total protein recovery in the 6 steps. In 3 steps 70% of the PG is recovered and also 83% of the total protein. It is remarkable that the yield for the leaching of total protein attached was 45% in the first step, which does not occur in the case of the PG extraction (26%only). After 3 stages it is noticeable that the total protein yields dropped, indicaating that the remaining proteins are related mainly to residual PG so the specific activity increased. The increase of PG specific activity means that the system works as a purification procedure. The PG concentrations obtained in the 3 earlier steps are higher than 10 U mL^{-1}, values comparable with concentrations reported for submerged fermentations [9,24,25]. The former conclusion indicates that 3 liquid extractions from a solid state fermented substrate are equivalent to 3 submerged fermentations, emphasizing one of the advantages of this type of fermentation. Another point that is very important to emphasize is that the whole content of PG in the solid fermented substrate after six steps was calculated as 536.27 U g^{-1} (d.b.), a value that is overlooked when considered at merely one stage. In the case of a single step, only 26% of the total enzyme is obtained. Comparing this value with those reported it can be concluded that the process for production of PG from citrus pulp employing the strain *Aspergillus niger F3* is quite appropriate.

Conclusions

In this study, PG was recovered from a fermented solid matrix. The results indicated that the best conditions for extraction were room temperature and 1:20 solid-liquid ratio. None of the anions and cations assayed improved the enzyme activity when compared to the control, but addition of EDTA helped in the stabilization of PG, whose activity was measured 24 hours after extraction. Kinetic studies showed that the saturation concentrations for PG and protein were reached after 48 minutes. In three consecutive extraction steps it was possible to recover 70% of all PG produced by solid state fermentation.

Acknowledgement

Authors thank UFPR, CNPq, CAPES, for their financial and material support. Also authors thank American Journal Experts for their fast and professional review.

References

1. Huang YS, Ho SC (2010) Polymethoxy flavones are responsible for the anti-inflammatory activity of citrus fruit peel. Food Chem 119: 868-873.

2. Martín MA, Siles JA, Chica AF, Martín A (2010) Biomethanization of orange peel waste. Bioresour Technol 101: 8993-8999.

3. Anwar F, Naseer R, Bhanger MI, Ashraf S, Talpur FN, Aladedunye FA (2008) Physico-chemical characteristics of citrus seeds and seed oils from Pakistan. J Amer Oil Chem Soc 85: 321–330.

4. Palit S, Banerjee R (2001) Optimization of extraction parameters for recovery of α- amylase from the fermented bran of Bacillus circulans GRS313. Braz Arch Biol Technol 44: 107–111.

5. Oszmianski J, Wojdyło A, Kolniak J (2011) Effect of pectinase treatment on extraction of antioxidant phenols from pomace, for the production of puree-enriched cloudy apple juices. Food Chem 127: 623–631.

6. Pedrolli DB, Gomes E, Monti R, Cano-Carmona E (2008) Studies on productivity and characterization of polygalacturonase from Aspergillus giganteus submerged culture using citrus pectin and orange waste. Appl Biochem Biotechnol 144: 191–200.

7. Kumar YS, Varakumar S, Reddy OVS (2010) Production and optimization of polygalacturonase from mango (Mangifera indica L.) peel using Fusarium moniliforme in solid state fermentation. World J Microbiol Biotechnol 26: 1973–1980.

8. Mamma D, Kourtoglou E, Christakopoulos P (2008) Fungal multienzyme production on industrial by-products of the citrus-processing industry. Bioresour Technol 99: 2373-2383.

9. Tari C, Gögus N, Tokatli F (2007) Optimization of biomass, pellet size and polygalacturonase production by Aspergillus sojae ATCC 20235 using response surface methodology. Enz Microb Technol 40: 1108–1116.

10. Botella C, Diaz A, de Ory I, Webb C, Blandino A (2007) Xylanase and pectinase production by Aspergillus awamori on grape pomace in solid state fermentation. Process Biochem 42: 98–101.

11. Kuhad RC, Kapoor M, Rustagi R (2004) Enhanced production of an alkaline pectinase from Streptomyces sp. RCK-SC by whole-cell immobilization and solid-state cultivation. World J Microbiol Biotechnol 20: 257–263.

12. de Gregorio A, Mandalari G, Arena N, Nucita F, Tripodo MM, et al. (2002) SCP and crude pectinase production by slurry-state fermentation of lemon pulps. Bioresour Technol 83: 89–94.

13. Silva D, Martins ES, Silva R, Gomes E (2002) Pectinase production by Penicillium viridicatum rfc3 by solid state fermentation using agricultural wastes and agro-industrial by-products. Braz J Microb 33: 318-324.

14. Castilho LR, Alves TLM, Medronho RA (1999) Recovery of pectolytic enzymes produced by solid state culture of Aspergillus niger. Process Biochem 34: 181–186.

15. Gupta S, Kapoor M, Sharma KK, Nair LM, Kuhad RC (2008) Production and recovery of an alkaline exo-polygalacturonase from Bacillus subtilis RCK under solid-state fermentation using statistical approach. Bioresour Technol 99: 937-945.

16. Díaz AB, Caro I, de Ory I, Blandino A (2007) Evaluation of the conditions for the extraction of hydrolitic enzymes obtained by solid state fermentation from grape pomace. Enzyme Microb Technol 41: 302–306.

17. Ikasari L, Mitchell DA (1996) Leaching and characterization of Rhizopus oligosporus acid protease from solid-state fermentation. Enzyme Microb Technol 19: 171-175.

18. Bai Y, Nikolov ZL, Glatz CE (2002) Aqueous Extraction of β-Glucuronidase from Transgenic Canola: Kinetics and Microstructure. Biotechnol Prog 18: 1301-1305.

19. Rodríguez-Fernández DE, Rodríguez-León JA, de Carvalho JC, Sturm W, Soccol CR (2011) The behavior of kinetic parameters in production of pectinase and xylanase by solid-state fermentation. Bioresour Technol 102: 10657-10662.

20. Miller GL (1959) Use of dinitrisosalicilic acid reagent for determination of reducing sugars. Anal Chem 31: 426-428.

21. Lowry OH, Rosebroug NJ, Farr AL, Randall RJ (1951) Protein measurement with folin-phenol reagent. J Biol Chem 193: 265-271.

22. Montgomery DC (2001) Design and analysis of experiments. 5th Edn. John Willey & Sons Inc, New York.

23. Favela-Torres E, Volke-Sepúlveda T, Viniegra-González G (2006) Production of Hydrolytic Depolymerising Pectinases. Food Technol Biotechnol 44: 221–227.

24. Coulson JM, Richardson JF (2002) Leaching. In: Chemical Engineering, Vol. 2, Chapter 10, Butterworth-Heinemann 502-541.

25. Ustok FI , Tari C, Gogus N (2007) Solid-state production of polygalacturonase by Aspergillus sojae ATCC 20235. J Biotechnol 127: 322–334.

26. Gögus N, Tari C, Oncü S, Unluturk S, Tokatli F (2006) Relationship between morphology, rheology and polygalacturonase production by Aspergillus sojae ATCC 20235 in submerged cultures. Biochem Eng J 32: 171–178.

10

Mathematical Modelling of Protein Precipitation Based on the Phase Equilibrium for an Antibody Fragment from *E. coli* Lysis

Yu Ji and Yuhong Zhou*

Department of Biochemical Engineering, University College London, Torrington Place, London WC1E 7JE, UK

Abstract

Precipitation is an important operation in biopharmaceutical purification yet the mechanism of protein precipitation in multi-component solutions is not well understood. Existing models lack fundamental understanding of the process. In this paper, a new model describing how the protein solubility changes in the protein precipitation is proposed and is based on the phase equilibrium of the light liquid phase and dense solid phase. The model structure is generic and robust. It adequately reflects the non-linearity of protein precipitation kinetics and thus provides new fundamental insights into the protein precipitation in multi-component, complex protein solution.

Two feed stocks of a pure fragment antigen-binding (Fab') solution obtained by chromatographic purification and a clarified Fab' homogenate solution from *E. coli* were used to examine the effect of ammonium sulphate concentrations and pH conditions on precipitation. It was found that the model can describe pure Fab' precipitation well, and identify the non-ideal behavior of Fab' precipitation in multi-component homogenates. Through statistical analysis, the model parameters have been further reduced from 8 to 4. The quality of the model is such that errors were within the acceptable statistical confidence limits, even when applied to multi-component impurity precipitation. The new model with fewer parameters is better than existing empirical models in reflecting the salting-in and salting-out effect of the protein precipitation. This demonstrated that the structure of the model is sound and over-fitting in the parameter estimation is avoided. The model can be applied directly to industrial processes for protein precipitation process design after appropriate calibration with the required operating conditions of pH and salt concentration.

Keywords: Precipitation; Model; Multi-component protein solution; Antibody fragment; Impurity

Nomenclature

a_1, b_1, c_1, d_1	constants in Equation (9)
a_2, b_2, c_2, d_2	constants in Equation (18)
$a_3, b_3, c_3, d_3, e_3, f_3$	constants in Equation (19)
$a_4, b_4, c_4, d_4, e_4, f_4, g_4, h_4, i_4$	constants in Equation (20)
$a_5, b_5, c_5, d_5, e_5, f_5$	constants in Equation (21)
A, B	constants in Equation (12)
C_d	protein molar concentration in the dense phase ($\mathrm{mol \cdot L^{-1}}$)
C_i	other component molar concentration in the solution ($\mathrm{mol \cdot L^{-1}}$)
C_l	protein molar concentration in the light phase ($\mathrm{mol \cdot L^{-1}}$)
C_s	salt molar concentration ($\mathrm{mol \cdot L^{-1}}$)
C_T	the maximum protein concentration in the solution ($\mathrm{mol \cdot L^{-1}}$)
I	ionic strength ($\mathrm{mol \cdot L^{-1}}$)
k_s	salt activity coefficient (-)
k_i	components activity coefficient (-)
m_3	the salt mole concentration ($\mathrm{mol \cdot L^{-1}}$)
r_l	protein activity coefficient in the light phase (-)
r_d	protein activity coefficient in the dense phase (-)
Q_i	molar concentration of ion i ($\mathrm{mol \cdot L^{-1}}$)
R_g	ideal gas constant ($\mathrm{J \cdot mol^{-1} \cdot K^{-1}}$)
R^2	coefficient of determination
S	Fab' concentration in the supernatant ($\mathrm{mol \cdot L^{-1}}$)
S_0	Fab' concentration in the feedstock ($\mathrm{mol \cdot L^{-1}}$)
T	the absolute temperature (K)
V_l	light liquid phase volume (L)
V_d	dense phase volume (L)
V_T	total solution volume (L)

w_1, w_2 and w_3	constants in Equation (11)
Y	Fab's concentration in the supernatant
Z	net charge of the protein (-)
Z_i	charge number of ion i
$\alpha, \beta, \chi, \delta$	lumped constants in Equation (9)
θ, λ	constants in Equation (17)
K	lumped constants in Equation (15)
λ_{scaled}	scaled value
λ_{real}	real value
λ_L	real value at low limit
λ_U	real value at upper limit
μ_l	chemical potential for liquid phase ($\mathrm{J \cdot mol^{-1}}$)
μ_d	chemical potential for dense phase ($\mathrm{J \cdot mol^{-1}}$)
μ_l°	protein standard chemical potential in the light phase ($\mathrm{J \cdot mol^{-1}}$)
μ_d°	protein standard chemical potential in the dense phase ($\mathrm{J \cdot mol^{-1}}$)
μ_2°	protein standard chemical potential in the solution ($\mathrm{J \cdot mol^{-1}}$)
$\mu_{2,w}^\circ$	protein standard chemical potential in the water ($\mathrm{J \cdot mol^{-1}}$)
σ, ν, ρ, ξ	constants in Equation (16)
$\phi, \varphi, \gamma, \eta$	lumped constants in Equation (14)

***Corresponding author:** Yuhong Zhou, University College London, Department of Biochemical Engineering, Torrington Place, London WC1E 7JE, UK

Introduction

Protein precipitation is a technique that utilizes the differences of protein solubility to precipitate proteins into the solid phase from the liquid phase. It has been used extensively to separate and purify proteins for sample preparation [1]. Ammonium sulphate is usually used to separate protein from complex solutions because it does not denature protein and has a very high salting-out effect [2-4]. Currently, with advanced fermentation technology, higher protein titres can be achieved upstream and it is now possible to produce multi-kilogram quantities of therapeutic monoclonal antibodies in a single batch [5]. However, this creates problems at the downstream purification stages. The high concentration of target protein plus impurities in the feedstock changes the physical properties of the protein solution. If such a complex biological material is applied directly to the chromatographic columns, they are susceptible to fouling and blockages [6-8] so significantly increasing the chromatographic processing time and cost. Therefore, a primary separation, such as protein precipitation, may be beneficial in the preparation of a relatively clearer and less contaminated solution for expensive high resolution steps.

During precipitation, the solubility of a protein depends primarily on process conditions including pH, salt concentration and temperature [9]. In order to optimize the precipitation process operation, a good understanding of the impact of these conditions on the behavior of the protein is needed. For industrial scale process engineering and design purpose, a protein precipitation model that directly links protein solubility with operating conditions would help support industrial process development e.g. scale-up, predict process optimal conditions and provide information for process control [10]

The first attempt to model the protein solubility was by Cohn [11] His log-linear equation, discussed later, gives a simple empirical relationship between the soluble protein concentration and ionic strength in the solution over a narrow salt concentration range. Melander and Horvath [12] then improved Cohn's empirical equation by linking the hydrophobic effect with thermodynamic parameters such as the hydrophobic surface. Unfortunately the improved model often sheds little light on the bioprocess operation and design as the linkage between the operating conditions and the hydrophobic surface cannot be established. The universal quasi chemical (UNIQUAC) model describes protein solubility by protein activity coefficients and a polynomial relationship between protein activity coefficients and osmotic second virial coefficients can be used to model protein behavior [13,14] where experiments to obtain protein activity data are required. The theoretical thermodynamic equations to predict protein solubility with molecular radius and surface parameters [15-17] worked quite well in a simple and defined system in which all physical properties are known. However, such thermodynamic-based models are of limited use for process design and control because the thermodynamic properties for complex multi-component processing materials are unknown.

Modified empirical exponential models that describe the traditional sigmoid shape of the precipitation curve directly link predictions with process conditions [18,19], but these models provide little fundamental understanding. Despite pH having been reported as a strong factor on protein precipitation, pH was not considered in these models. Temperature is another variable that often strongly influences protein precipitation. However, as most proteins are sensitive to temperature, a fixed temperature is applied during the industrial precipitation process, typically a low temperature (~ 4°C), to prevent protein denature.

The goal of this paper is to develop and validate a protein precipitation model that uses bioprocessing conditions as inputs to predict the protein solubility for complex multi-component materials. The model will be based on theoretical phase equilibrium to achieve an improved process understanding.

Antibody fragments expressed intracellularly in *E. coli* as next generation therapeutics is cheaper to produce by fermentation than antibodies from mammalian cell culture because of shorter culture time and less expensive media. It also has better selectivity than antibodies but is more difficult to purify due to high level of impurities.

Two different feed stocks, a pure fragment of antibody (Fab') solution and a clarified Fab' solution from *E. coli* homogenate, will be used to examine the generality of the model. The model will be validated by experimental data, and then statistical tests will be used to evaluate the quality of the model. The predictions of the model will be compared with four existing models where pH will be introduced as an extra variable [11,18,19].

Materials and Methods

Sodium monobasic phosphate, sodium dibasic phosphate, sodium acetate and ammonium sulphate were purchased from Sigma Chemical Co. Ltd. (Dorset, UK). All chemicals were reagent grade. Fab' producing strain *E. coli* W3110 was kindly provided by UCB (Slough, UK) and the cell paste was provided by the Fermentation Group, Department of Biochemical Engineering in University College London.

Precipitation material preparation

E. coli cells were suspended in a 10 mM pH 7.0 phosphate buffer at 40% (wt) and homogenized in an APV Lab 40 Homogenizer at 750 bar. The homogenized solution was centrifuged in an Eppendorf Centrifuge 5810R at 12,000 rpm for 2 hours with supernatant collected as the stock solution for further study. Pure Fab' solution was prepared from the collected supernatant. An AKTA Basic HPLC (GE Healthcare, Sweden) and Hitrap Mabselect 5 ml HPLC column (GE Healthcare, Sweden) were used to purify Fab'. The eluate was buffer exchanged to 10 mM pH 7.0 phosphate buffer and stored at 4°C.

Fab' and impurity concentration analysis

Fab' concentration and total protein concentration were analyzed by HPLC Agilent 1200 (Agilent Technologies, UK) with a Hitrap Protein G 1ml HPLC column (GE Healthcare, Sweden). Fab' concentration was calculated from the peak area according to a calibration curve, which was obtained using pure Fab' after Protein A and size exclusion chromatography. The impurity concentration analysis method was the same as Fab' except that the feedstock was used as the standard and the impurity area monitored.

Fab' precipitation by microwell scale high throughput experimentation

The Fab' precipitation was carried out in the ABgene's 96 deep microwell plates by a Packard MultiPROBE II HT EX (Packard BioScience Company, Meriden, U.S.A.). The experimental conditions were selected as follows: pH from 4.5 to 8.0 with intervals of 0.5, ammonium sulphate concentration from 0 to 3.0 mol/L, with intervals of 0.2 mol/L for pure Fab' solution and with intervals of 0.3 mol/L for clarified homogenized solution. The total volume of precipitate supernatant was 1.8 ml. The precipitation plate was shaken on an Eppendorf thermomixer at 600 rpm for 2 hours and then centrifuged at 4000 rpm for 15 minutes. The clear supernatant was transferred to an Agilent 96 HPLC microwell plate and analyzed in triplicate.

Hybrid model derivation and parameter estimation

Phase equilibrium based protein precipitation model: Protein precipitation has been thermodynamically regarded as a pure crystallisation process because the solution has only protein and salt [15-17]. However, for proteins in a real fermentation broth or with other complex biological materials, the precipitation will not form pure crystal but an amorphous mixture [20,21]. The precipitation can be treated as a distribution between a light liquid phase (supernatant) and a dense liquid phase (precipitate). Therefore, the proposed model in this paper is based on the phase equilibrium for the target protein in a multi-component solution:

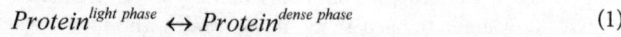

$$Protein^{light\ phase} \leftrightarrow Protein^{dense\ phase} \qquad (1)$$

When the two phases are in equilibrium, the chemical potentials of the proteins must be equal:

$$u_l = \mu_d \qquad (2)$$

and

$$\mu_l^o + RT \ln C_l r_l = \mu_d^o + RT \ln C_d r_d \qquad (3)$$

where μ_l is the chemical potential for the liquid phase and μ_d the chemical potential for the dense phase, C_l the protein mole concentration in the light phase, C_d the protein mole concentration in the dense phase, r_l the protein activity coefficient in the light phase, r_d the protein activity coefficient in the dense phase, R the ideal gas constant, T the absolute temperature, μ_l^o the protein standard chemical potential in the light phase and μ_d^o the protein standard chemical potential in the dense phase.

Equation (3) can be rearranged to

$$C_l/C_d = (r_d/r_l)\exp\left(-\left(\mu_l^o - \mu_d^o\right)/RT\right) \qquad (4)$$

Suppose V_l is the light liquid phase volume, V_d the dense phase volume, C_T the maximum protein concentration in the solution and V_T the total solution volume with the approximation that there is no volume change during precipitation. Then,

$$V_T = V_l + V_d \qquad (5)$$

$$C_l V_l + C_d V_d = C_T V_T \qquad (6)$$

Introducing equations (5) and (6) into equation (4) we obtain

$$C_l/C_T = 1/\left(V_l/V_T + (V_d/V_T)(r_l/r_d)\exp\left(-\left(\mu_l^o - \mu_d^o\right)/RT\right)\right) \qquad (7)$$

The dense phase volume in protein precipitation cases will increase with salt concentration as more proteins will be precipitated and reach a nearly constant level at high salt concentration. It is often very small compared to the total solution volume because the total protein concentration is low in biopharmaceutical processing material, so it is reasonable to assume that $V_l/V_T \approx 1$. As we know, V_d/V_T depends on the protein properties and pH probably with an apparent isoelectric point (pI). Our preliminary experimental results shown in Figure 1 illustrate that the Fab' concentration increased linearly with pH in the range 4-8. Hence pH having a linear effect was approximated and the effects of salt on V_d/V_T followed similarly to a Michaelis–Menten relationship. Thus, the empirical equation to represent the effect of pH and salt concentration on V_d/V_T is proposed as:

$$V_d/V_T = a_1 + b_1\left(pH - c_1\right)C_s/\left(d_1 + C_s\right) \qquad (8)$$

or

$$V_d/V_T = \left(\alpha + \beta C_s + \chi pHC_s\right)/\left(\delta + C_s\right) \qquad (9)$$

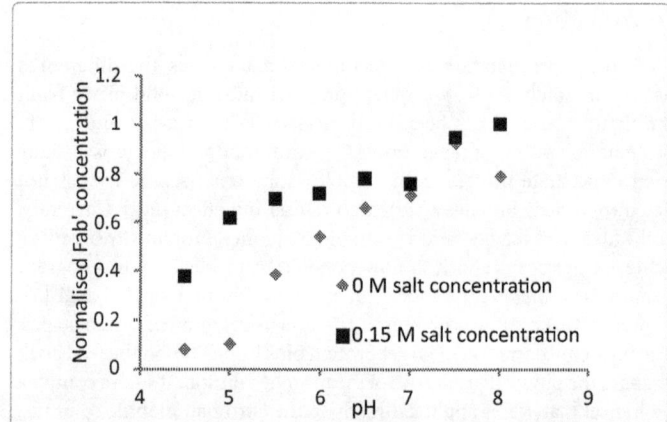

Figure 1: Preliminary result showing the linear dependence between Fab' protein precipitation and pH.

where C_s is the salt concentration, a_1, b_1, c_1, d_1 are constants, and $\alpha = a_1 d_1$, $\beta = a_1 - b_1 c_1$, $\chi = b_1$, $\delta = d_1$, are the lumped constants. With different protein solutions, these lumped parameters may vary and hence should be estimated from real experimental data.

In 1943, Kirkwood [22] defined the protein activity coefficient in a multi-component solution as a simple function of the concentrations of all solute species. Long and Mcdevit [23] assumed that the protein activity coefficient can be represented by a log-linear function based on fundamental chemical thermodynamics:

$$\log r_p = k_s C_s + \sum k_i C_i \quad i = 1,\ldots\ldots,n, \qquad (10)$$

where k_s is the salt activity coefficient, k_i the components activity coefficient and C_i the other component concentrations in the solution.

In a multi-component solution containing biomolecules and salt, the protein activity coefficient is dominated by the salt concentration; the other effects caused by biomolecules can be regarded as constant due to their very low concentrations. Therefore in the liquid phase, the second part of equation (10) can be represented by a constant. In the dense phase, the concentration of salt is considered as not significantly changing while the other molecules still have no or little effect; hence the overall protein activity coefficient can be regarded as a constant. Therefore

$$r_l/r_d = \exp\left(k_s C_s + w_1\right)/\exp\left(w_2\right) = \exp\left(k_s C_s + w_3\right) \qquad (11)$$

where w_1, w_2 and w_3 are constants.

In some cases, the protein property and its main interaction with salt will depend on the type of salt, so equation (11) may need a second order or even a higher order term of the salt concentration [23]. In this study, only the first order term was used in the model. A higher order model needs only to be considered if the first order model fails.

In 1985, Arakawa and Timasheff [20] published a theoretical protein precipitation model, which gave the following theoretical chemical potential equation:

$$\left(\mu_2^o - \mu_{2,w}^o\right)/RT = AZ^2\left(I\right)^{1/2}\left(1 + B\left(I\right)^{1/2}\right) - \left(1/2.303RT\right)\int_0^m \left(\partial\mu_2/\partial m_3\right)_{T,m_2} dm_3 \qquad (12)$$

where μ_2^o is the protein standard chemical potential in the solution, $\mu_{2,w}^o$ the protein standard chemical potential in the water, Z the net charge of the protein, I the ionic strength, m_3 the salt mole concentration with A and B the coefficients. The second differential term can be approximated by a first order term of salt concentration because $(\partial\mu_2/\partial m_3)_{T,m_2}$ is an empirical constant over a wide range of salt concentrations [20]. The temperature was regarded as a constant in this study.

The ionic strength is defined as:

$$I = (1/2)\sum Q_j \left(Z_j\right)^2 \qquad (13)$$

where Q_i is the molar concentration of ion i, and Z_i is the charge number of that ion. For a neutral salt such as ammonium sulphate the ionic strength is linearly proportional to salt concentration. As there is no general mathematical model for protein surface net charge as a function of pH, we will approximate Z^2 in equation (12) by a second order pH polynomial equation. Therefore, equation (12) can be simplified into a function of salt concentration and pH by:

$$\left(\mu_2^0 - \mu_{2,w}^0\right)/RT = \varphi\left(pH - \gamma\right)^2 \left(C_s\right)^{1/2}/\left(\eta + \left(C_s\right)^{1/2}\right) + \phi C_s \qquad (14)$$

where $\phi, \varphi, \gamma, \eta$ are lumped constants.

Under the assumption that the salt concentration in the dense phase is very small and does not change significantly, the value of equation (14) for dense phase protein will be considered as a constant, so

$$\left(\mu_l^0 - \mu_d^0\right)/RT = \varphi\left(pH - \gamma\right)^2 \left(C_s\right)^{1/2}/\left(\eta + \left(C_s\right)^{1/2}\right) + \phi C_s + \tau \qquad (15)$$

where τ is a lumped constant.

The second term essentially describes the protein salting-in effect at low salt concentration. To simplify the calculation, the second term is approximated by the pH effect for the low concentration range because from our experiments the pH effect dominated at low salt concentration. It was then described by a simplified second order polynomial function, while the salt effect was separated from this term and lumped into the first term on the right hand side of equation (15) to give:

$$\left(\mu_l^0 - \mu_d^0\right)/RT \approx \sigma pH^2 + vpH + \rho + \xi C_s \qquad (16)$$

where σ, v, ρ, ξ are constants. At high salt concentration the salting-in phenomena does not occur or the effect is very small compared to the first term in equation (15). Therefore, the coefficients in equation (16) were relatively insignificant for the overall model prediction at a high concentration range. It also kept the model mathematically consistent throughout the salt range.

Combining equations (9), (11) and (16), equation (7) becomes:

$$C_l/C_T = 1/\left(1 + \left(\left(\alpha + \beta C_s + \chi pHC_s\right)/\left(\delta + C_s\right)\right)\exp\left(k_s C_s + w_3 + \sigma pH^2 + vpH + \rho + \xi C_s\right)\right) \qquad (17)$$

or

$$C_l/C_T = 1/\left(1 + \left(\left(\alpha + \beta C_s + \chi pHC_s\right)/\left(\delta + C_s\right)\right)\exp\left(\theta C_s + vpH + \sigma pH^2 + \lambda\right)\right) \qquad (18)$$

where $\theta = k_s + \xi$, $\lambda = w_3 + \rho$.

This model is able to describe the strong non-linearity of the precipitation surface due to its sigmoid structure. All the parameters in equation (18) are lumped and so it is difficult to predict their values or limit their ranges. However, according to the modelling assumptions, parameters θ and δ should have positive values and the term $\alpha + \beta C_s + \chi pH C_s$ will be positive. At low salt concentrations, the exponential expression in the model is not a dominant effect and thus the decrease of dense phase volume caused by the salting-in effect, which makes the value of V_d/V_T smaller, explains the protein concentration increase. At high salt concentration, the second term which contains an exponential expression will have a value much larger than 1.0, so we can neglect the value of 1.0 in the denominator. Thus this model is similar to the exponential structure of Cohn's equation, discussed below.

This model has a thermodynamic base assisted by empirical relationships. The model structure involves eight parameters to allow a proper expression of the solubility surface. This is different from models from purely experimental observation which are too simplified and so loose the necessary details.

Model comparison: In order to evaluate the capability of this new model, it is useful to compare it with three published models, Cohn's [11], Niktari's [18] and Habib's [19] models plus a polynomial model. For process design purposes, all the selected models were modified to contain a pH factor by introducing a second order polynomial expression of pH to substitute for the model coefficients without changing the model structure, in order to link protein solubility directly to operating variables.

Cohn's equation is an empirical equation. In Cohn's papers, it has been shown that the second constant is only associated with the salt effect and the first constant is associated with all of the other effects, therefore the second constant d is not related to pH. Hence, the Cohn's equation will only be expanded in the first constant:

$$\ln\left(S/S_0\right) = a_2 + b_2 pH + c_2 pH^2 - d_2 C_s \qquad (19)$$

where S/S_0 is the ratio of Fab' concentration in the supernatant to the Fab' concentration in the feedstock. The expanded Niktari's equation is:

$$y = 1/\left(1 + \left(C_s/\left(a_3 + b_3 pH + c_3 pH^2\right)\right)^{\wedge}\left(d_3 + e_3 pH + f_3 pH^2\right)\right) \qquad (20)$$

The expanded Habib's sigmoid model is:

$$y = 1 - \left(a_4 + b_4 pH + c_4 pH^2\right)\left(1 - \exp\left(-\left(C_s/\left(d_4 + e_4 pH + f_4 pH^2\right)\right)^{\wedge}\left(g_4 + h_4 pH + i_4 pH^2\right)\right)\right) \qquad (21)$$

A second order polynomial equation with interaction terms is used to represent the conventional two factors polynomial model:

$$y = a_5 + b_5 C_s + c_5 pH + d_5 C_s^2 + e_5 pH + f_5 C_s pH^2 \qquad (22)$$

where y is Fab' concentration in the supernatant, and others are parameters.

Although the number of parameters involved in these models is similar, the structure of our model is very different.

Data analysis and parameter estimation: In order to eliminate the errors in the model parameter estimation caused by different orders of variables, the experimental conditions of pH and salt concentration were normalized to a 0-1 range by the scaling:

$$\lambda_{scaled} = \left(\lambda_{real} - \lambda_L\right)/\left(\lambda_U - \lambda_L\right) \qquad (23)$$

where λ_{scaled} is the scaled value, λ_{real} the real value, λ_L the real value at low limit and λ_U the real value at upper limit.

The Fab' concentration and the impurity concentration in both feed stocks were also normalized. The initial concentration was not preferred in the normalization because there was a salting-in effect. The maximum concentration of Fab' and impurities during salting-in were regarded as the true maximum protein concentration in the solution. In reality, this concentration was difficult to obtain so the maximum concentration from the experimental data was used as the approximation to the true maximum concentration, and also as the upper limit for equation (22). The model parameters in equations (18) - (22) were estimated by the nonlinear least-squares regression method in the Matlab Tool box (MathWorks, USA).

Results and Discussion

The Fab' solubility in pure Fab' solution and clarified homogenate was described by equation (18) with regards to pH and ammonium sulphate concentration. As the focus of this study is on the development of an accurate precipitation model, a large data set is needed to test the model accuracy and regression fitting. A nonlinear least squares method was used to estimate the parameters in the model. The accuracies of the model, equation (18), and the other models, equations (19), (20), (21) and (22), were measured by R^2. In addition, the statistical F-test was used to evaluate the new model. The Wilcoxon test and paired t-test for individual parameters in the model were used to validate the model.

Protein precipitation modelling for pure Fab', Fab' in clarified homogenate and impurities

119 experiments for pure Fab' solution and 79 experiments for clarified homogenate were carried out. The pure Fab' concentration, as shown in Figure 2a, slightly increased during the salting-in phase and then gradually decreased with salt concentration, similar to many previous pure protein precipitation curves[11, 20]. For the clarified homogenate shown in Figure 2b, when the salt concentration was low, the Fab' concentration was significantly affected by other components

in the solution and its concentration was altered compared to that of pure Fab' solution. Under low pH and at low salt concentration, the Fab' concentration was only 40% of the highest concentration. The same phenomena occurred with the impurity solubility, shown in Figure 2c. An explanation may be that the low Fab' concentration at certain conditions, while it was not a pure solution, was probably caused by the co-precipitation between the Fab' molecule and other impurity proteins, the solubility of which were significantly changed by pH at low ionic strength. The effect of salting-out dominated when the salt concentration increased and thus the solubility was nearly the same as that of pure Fab' solution.

These experimental data sets were then used to develop the pure Fab' precipitation model based on equation (18). The estimated parameters are shown in Table 1 and the R^2 value was 0.975. The F-test value of the model fitting was 624.81; indicating 95% confidence model accuracy was achieved. It is clear that the model prediction satisfactorily described both the salting-in and salting-out features of the concentrations without the cost of losing accuracy for any phases.

The generality of the model structure was then assessed by applying the model to Fab' precipitation in a clarified homogenate where multi-components exist. The parameters of the pure Fab' model were used as the initial guess for the parameter estimation. The R^2 value for Fab' in clarified homogenate was 0.972 as shown in Table 1, which was very similar to that of the pure Fab' model. The F-test value of this model was 320.36, which was smaller than that in the pure Fab' model. Nevertheless, this suggests that the model was accurate to within 95% confidence. The predicted Fab' concentration surface is shown in Figure 2b. It can be seen that the Fab' solubility in the clarified homogenate was in general predicted well by the model. Nevertheless, there is a slight discrepancy between predicted solubility and experimental data at a low pH range as well as very low salt concentrations.

We also assessed the model by applying it to impurity precipitation, where a mixture of proteins was treated as an assumed pseudo-single molecule with average characteristics of all the proteins in the solution, e.g. average electronic charges and hydrophobic behaviour. 79 experimental data points from an impurity precipitation in a clarified homogenate were used for parameter estimation. The results are given in Table 1 and the R^2 value was 0.945, which is slightly lower than that of the Fab' models. The F-test value of the model was 172.60. These measures showed that the model was accurate to 95% confidence. The predicted impurity concentration surface is shown in Figure 2c. The geometrical pattern for the real data points and the model predicted surface were slightly different, especially in the high salt concentration region. We believe the difference was caused by the simplification of a mixture of proteins into a pseudo-single protein. Although the impurities were regarded as a pseudo-single molecule with an average value of all the mixture, in reality, different proteins will have a different sensitivity to pH and salt concentrations. Conditions may significantly affect one protein with no great effect on other molecules. Thus, the

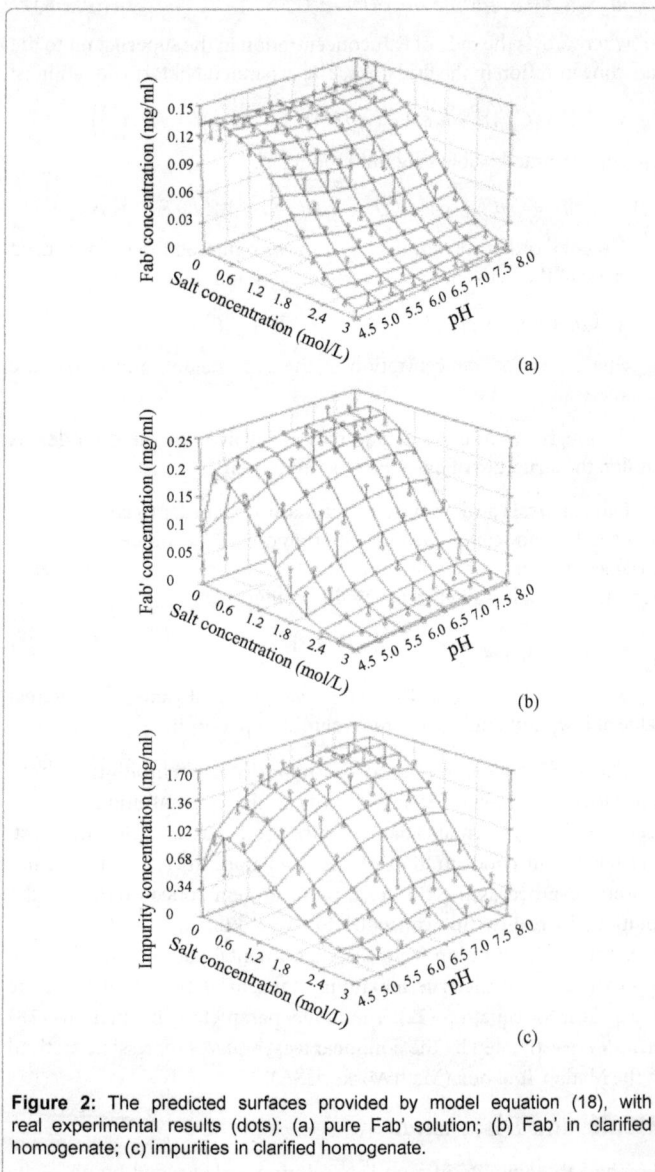

Figure 2: The predicted surfaces provided by model equation (18), with real experimental results (dots): (a) pure Fab' solution; (b) Fab' in clarified homogenate; (c) impurities in clarified homogenate.

	Parameters								R^2	F-test value
	θ	v	σ	δ	α	β	χ	λ		
Pure Fab'	7.97	1.62	0.53	-5.30	1.15	-1.05	0.16	0.03	0.975	624.81
Fab' in clarified homogenate	7.61	-6.60	4.34	-4.27	1.04	-0.75	21.12	0.002	0.972	320.36
Impurities	5.51	-7.88	5.03	-3.43	1.32	-0.41	20.60	0.01	0.945	172.68

Table 1: Parameters, F-test value and R^2 value of developed model using equation (18) for pure Fab' precipitation, Fab' precipitation in clarified homogenate and impurity precipitation.

assumed average properties of a pseudo-single molecule are not constant under all conditions, especially for extreme conditions, e.g. low pH or high salt concentration.

Model modification

It has been shown in Figures 2 that the new model predictions agreed well with the experimental data. However, there are eight parameters in equation (18) and the model exhibits a high level of nonlinearity. A simpler model is more useful for the processes operation and design. It is also beneficial for the parameter estimation since a higher number of parameters have the potential to over-fit the data which could result in less accuracy of the model. The t-test for individual parameter in the model can be used to evaluate and decide if a parameter is necessary. If a parameter fails the t-test, it is either inaccurate or not needed, or both. After carrying out the t-test for each parameter in model (18), the parameters α, β, and χ failed in all three models, i.e., pure Fab', Fab' in clarified homogenate and impurities. Therefore a simplified model is proposed:

$$C_l / C_T = 1/(1+(1/(\delta+C_s)) \exp(\theta C_s + \nu pH + \sigma pH^2 + \lambda)) \quad (24)$$

The simplified models were developed for pure Fab', Fab' in clarified homogenate and impurities in clarified homogenate based on equation (24) by using the experimental data sets again. The parameters are shown in Table 2. The R^2 values were evaluated and were a little lower than previous, but all tests showed that the models had excellent statistical confidence. All parameters passed t-tests with 95% confidence.

Using the values in Table 2, all three models showed similar predictions as in Figure 2. When the salt concentration is low (<0.2 mol/L), the value of the exponential term is small and changes little while the linear term dominates and changes rapidly. It describes the salting-in phenomenon at low salt concentration better. When the salt concentration is high, e.g. in the salting-out range, the value of the exponential term dominates due to the large values of parameter θ, while the effect of the linear term is small. The values of the parameters associated with pH vary relatively little. The parameter of the second order pH term for impurities, σ, is the largest with the fact that impurity concentration was influenced most by the pH in all three materials. The effect of pH around the neutral point is very small due to its small parameter value and its second order structure. However, according to the models there will be large effects describing the protein concentration changes in the experiments at the extreme pH conditions.

The most difficult operation conditions in precipitation to determine are the cutting points either in salting-in at the low salt concentration and salting-out at the high salt concentration ranges. As equation (24) is derived from thermodynamic phase equilibrium, the term $1/(\delta+C_s)$ is strongly related to the salting-in effect, and δ significantly influences the magnitude of the salting-in effect. Term $(1/\delta)\exp(\lambda)$ indicates the potential increase in protein concentration. As shown in Table 2, $(1/\delta)$ values in the models of Fab' in clarified homogenate and impurity are much larger than that of pure Fab', so the salting-in effect is significant

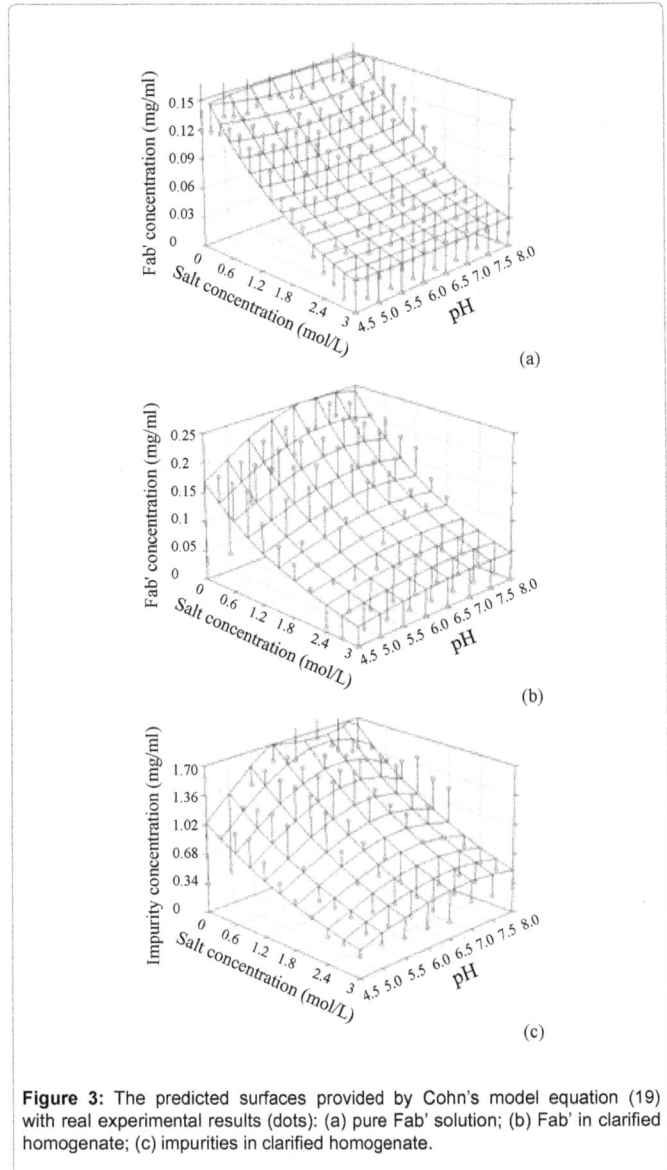

Figure 3: The predicted surfaces provided by Cohn's model equation (19) with real experimental results (dots): (a) pure Fab' solution; (b) Fab' in clarified homogenate; (c) impurities in clarified homogenate.

in these multi-component cases. During salting-out, χ, the coefficient for the interaction between pH and salt, and θ, the coefficient for the salt, play an important role. In the clarified homogenate case, the values of χ in the Fab' model and the impurity model are nearly two hundredfold of that in the pure Fab' model, showing that the impact of multi-components on the salting-out are significant. Together with λ, these three parameters dominate the salting-out effect.

Model validation

Nine independent experiments were carried out and the results used to validate the model. When validating bioprocess models, it is not recommended to use error percentage to evaluate the models because the range of bioprocess data may be very wide, even after scaling, which will introduce mathematical errors. Thus, statistical tests are needed to validate any new model, no matter how good the fitting of the data is in the regression step.

There are however several unusual problems for bioprocess model validation. First of all, the number of samples used for validation is

	Parameters					R^2	F-test value
	θ	ν	σ	δ	λ		
Pure Fab'	8.21	1.49	-1.08	-5.34	0.03	0.973	1037.82
Fab' in clarified homogenate	9.46	-3.16	2.19	-4.60	0.005	0.937	308.28
Impurities	7.02	-5.01	3.70	-3.51	0.02	0.914	195.48

Table 2: Parameters, F-test value and R^2 value of simplified model using equation (24) for pure Fab' precipitation, Fab' precipitation in clarified homogenate and impurities precipitation.

normally small, e.g. nine samples in this case, due to the high cost and long experimental time as well as limited protein solution materials available. Secondly, the distribution of most bioprocess data is normally unknown or the data does not conform to any known distribution, e.g. standard normal distribution. Statistically, normal distribution can only be assumed when the number of samples is large, normally more than 30. Therefore, for a small validation group with an unknown distribution, it is risky to use the paired t-test due to a high probability to fail. Two possible solutions are a Wilcoxon signed-rank test [24] for few samples or to analyse validation samples together with previous regression data by the paired t-test. For a Wilcoxon test, when 2-tailed significance > 0.05, it is regarded as a validation pass. A paired t-test, when significance (the p-value associated with the correlation) > 0.05, can be considered as the null hypothesis, i.e. no difference between the experimental data and model calculated value. Table 3 shows the test results for modified model equation (24) with paired t-test significance > 0.05 and Wilcoxon 2-tailed significance > 0.05 from the Statistical Package for the Social Sciences (SPSS, an IBM Company) calculated for all three materials. It demonstrates that the simplified model passed the validation criteria with strong statistical confidence.

Model comparison

The experimental data sets were also used to estimate the parameters for the four models described by equations (19) to (22). Table 4 presents the R^2 and the F-test values for all four models in different feed stocks. The predicted Fab' and impurity concentration surfaces for the four models are shown in Figures 3 to 6. A large difference between predicted surface and experimental data was observed for all three models based on Cohn's equation in Figures 3a-3c. However as all statistical tests failed with all R^2 values below 0.80, Cohn's equation failed to describe the protein salting-in effect at low salt concentration in multi-component solution as well as salting-out effect due to its linear model structure. Niktari's expansion model and Habib's expansion model can fit quite well for pure Fab' precipitation, as shown in Figure 4a and Figure 5a, with R^2 values of 0.956 and 0.971, F-test values of 402.72 and 400.30 respectively. However, the R^2 for the Fab' precipitation model in multi-components solution and impurity precipitation model were decreased below 0.9 and failed to describe the salting-in effect, as seen in Figure 4b, 4c and Figure 5b, 5c. The R^2 values and F-test values of these two models show poor model fitting and less predicting capability. Besides, both models do not consider the pH effect originally and a second-order polynomial expression for pH expansion may not effectively describe the pH impact.

Theoretically, the polynomial model has the most flexibility to fit

the data. However, the same results were obtained as for the previous two models as shown in Figure 6. The model in pure Fab' precipitation was quite good, with R^2 of 0.95, but the models for Fab' precipitation in multi-components were less good, as R^2 was less than 0.90 for this non-ideal solution. Moreover, at high salt concentration, the predicted concentration surface of the polynomial model was below zero, which also occurred in Habib's model. Therefore, both models were quite misleading as the predicted value below zero had no physical meaning. The negative value can be manually eliminated by constraining the parameters during regression but at the cost of overall accuracy. The polynomial model was not considered adequate for fitting in clarified homogenate precipitation. The higher order polynomial model may be more accurate but will inevitably introduce more parameters.

The comparison studies demonstrate that the structure of the model is crucial. If the structure is not right, the model will not predict the performance well even though there are high number of parameters in the model e.g. 9 parameters in Habib's model.

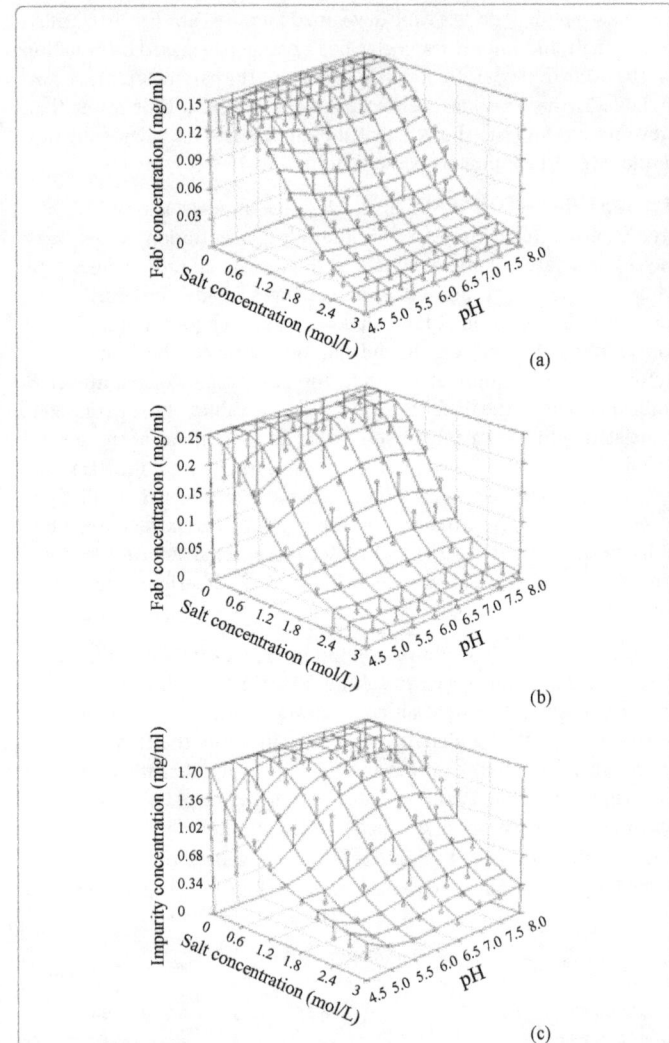

(a)

(b)

(c)

Figure 4: The predicted surfaces provided by Niktari's model equation (20) with real experimental results (dots): (a) pure Fab' solution; (b) Fab' in clarified homogenate; (c) impurities in clarified homogenate.

	Pure Fab'	Fab' in clarified homogenate	Impurities
t-test value	1.201	1.220	-0.198
sig. (2-tailed)	0.232	0.227	0.843
Wilcoxon 2-tailed sig.	0.767	0.086	0.515

Table 3: Results with 9 samples Wilcoxon signed-rank test and all samples paired t-test results for modified model using equation (24).

	Our model		Cohn's equation		Niktari's model		Habib's model		Polynomial model	
	R^2	F-test	R^2	F-test	R^2	F-test	R^2	F-test	R^2	F-test
Pure Fab'	0.973	1037.82	0.795	148.53	0.956	402.72	0.971	400.30	0.950	433.74
Fab' in clarified homogenate	0.937	308.28	0.621	40.88	0.795	46.55	0.864	48.78	0.825	68.90
Impurities	0.914	195.48	0.721	64.50	0.859	73.03	0.877	54.48	0.858	88.00

Table 4: R^2 and F-test values for all four existing models.

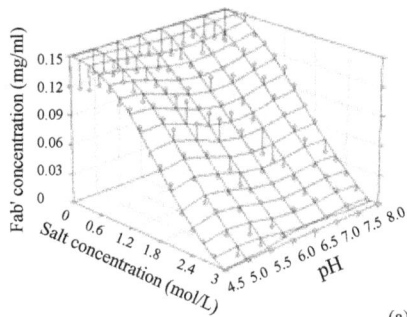

(a)

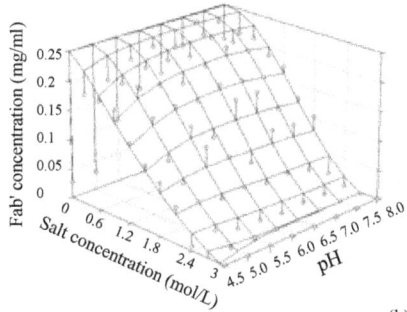

(b)

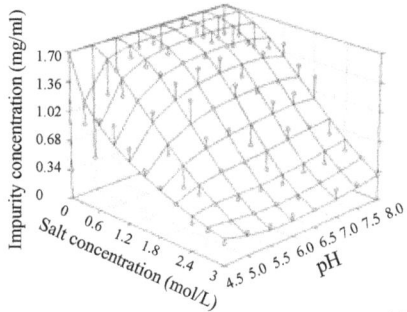

(c)

Figure 5: The predicted surfaces provided by Habib's model equation (21) with real experimental results (dots): (a) pure Fab' solution; (b) Fab' in clarified homogenate; (c) impurities in clarified homogenate.

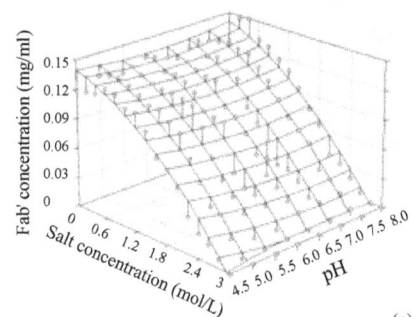

(a)

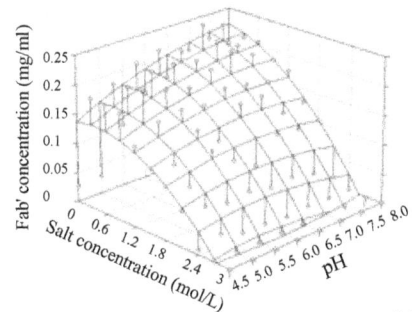

(b)

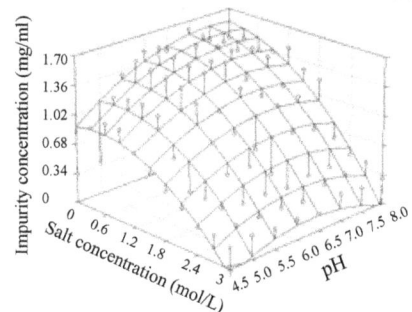

(c)

Figure 6: The predicted surfaces provided by polynomial model equation (22) with real experimental results (dots): (a) pure Fab' solution; (b) Fab' in clarified homogenate; (c) impurities in clarified homogenate.

Conclusions

Phase equilibrium based models have been developed then validated using statistical tests. The model equation (24) can precisely describe the precipitation salting-in and salting-out effects for pure Fab', Fab' in clarified homogenate and impurity with salt concentration and pH change. The model structure, based on a single protein, takes the multi-component factor into consideration and can be applied to a protein in a multi-component solution or a pseudo-single molecule. The estimated parameters in the model passed regression statistical tests proving that the models were accurate with 95% confidence.

Compared to thermodynamic based models, this new model conveniently predicts the precipitation results from operation conditions rather than thermodynamic parameters. The comparison between this new model and four existing empirical models showed that it was superior backed with the statistical tests. As our model structure is derived from fundamental phase equilibrium, it exhibits good prediction even though there are only 4 parameters. Such studies also show the challenge in multi-component protein precipitation

modeling when strong nonlinearity exists. Our model structure is able to reflect the non-linearity of the protein precipitation in both salting-in and salting-out better than existing empirical models.

Acknowledgements

This research was funded by the UK Government Overseas Research Student Scholarship and a University College London Graduate Student Research Scholarship. The provision of E. coli W3110 by UCB and cell paste from the upstream fermentation group, Department of Biochemical Engineering, UCL, was highly appreciated.

References

1. Shahina S, Dalby PA (2011) Thermodynamic parameters for salt-induced reversible protein precipitation from automated microscale experiments. Biotechnol Bioeng 108: 322-332.

2. Cheng YC, Lobo RF, Sandler SI, Lenhoff AM (2006) Kinetics and equilibria of lysozyme precipitation and crystallization in concentrated ammonium sulfate solutions. Biotechnol Bioeng 94: 177-188.

3. Cohn EJ, Strong LE, Hughes WL, Mulford DJ, Ashworth JN, et al. (1946) Preparation and Properties of Serum and Plasma Proteins IV. A System for

the Separation into Fractions of the Protein and Lipoprotein Components of Biological Tissues and Fluids. J Am Chem Soc 68: 459-475.

4. Foster PR, Dunnill P, Lilly MD (1976) The Kinetics of Protein Salting-Out: Precipitation of Yeast Enzymes by Ammonium Sulfate. Biotechnol Bioeng 18: 545-580.

5. Aldington S, Bonnerjea J (2007) Scale-up of monoclonal antibody purification processes. J Chromatogra B 848: 64-78.

6. Shukla AA, Hubbard B, Tressel T, Guhan S, Low D (2007) Downstream processing of monoclonal antibodies-application of platform approaches. J Chromatogr B Analyt Technol Biomed Life Sci 848: 28-39.

7. Gottschalk U (2008) Bioseparation in antibody manufacturing: the good, the bad and the ugly. Biotechnol Prog 24: 496-503.

8. Jiang C, Liu J, Rubacha M, Shukla AA (2009) A mechanistic study of Protein A chromatography resin lifetime. J Chromatogra A 31: 5849-5855.

9. Knevelman C, Davies J, Allen L, Titchener-Hooker NJ (2010) High-throughput screening techniques for rapid PEG-based precipitation of IgG4 mAb from clarified cell culture supernatant. Biotechnol Prog 26: 679-705.

10. Bernaerts K, Van Impe J (2004) Data-driven approaches to the modelling of bioprocesses. T I Meas Control 26: 349-372.

11. Cohn EJ (1925) The Physical Chemistry of the Proteins. Physiol Rev 5: 349-437.

12. Melander W, Horvath C (1977) Salt effects on hydrophobic interactions in precipitation and chromatography of proteins: An interpretation of the lyotropic series. Arch Biochem Biophys 183: 200-215.

13. Ruppert S, Sandler S, Lenhoff AM (2001) Correlation between the osmotic second virial coefficient and the solubility of proteins. Biotechnol Prog 17: 182-187.

14. Winzor DJ, Carrington LE, Harding SE (2001) Analysis of thermodynamic non-ideality in terms of protein solvation. Biophys Chem 93: 231-240.

15. Agena SM, Bogle ID, Pessoa FL (1997) An activity coefficient model for proteins. Biotechnol Bioeng 55: 65-71.

16. Chiew YC, Kuehner D, Blanch HW, Prausnitz JM (1995) Molecular thermodynamics for salt-induce protein precipitation. AIChE J 41: 2150-2159.

17. Kuehner DE, Blanch HW, Prausnitz JM (1996) Salt-induced protein precipitation: Phase equilibria from an equation of state. Fluid Phase Equilibr 116: 140-147.

18. Niktari M, Chard S, Richardson P, Hoare M (1990) The Monitoring and Control of Protein Purification and Recovery Processes. Separations for Biotechnology 2: 622-634.

19. Habib G, Zhou Y, Hoare M (2000) Rapid monitoring for the enhanced definition and control of a selective cell homogenate purification by a batch-flocculation process. Biotechnol Bioeng 70: 131-142.

20. Arakawa T, Timasheff SN (1984) Mechanism of protein salting in and salting out by divalent cation salts: balance between hydration and salt binding. Biochemistry 23: 5912-5923.

21. Shih YC, Prausnitz JM, Blanch HW (1992) Some characteristics of protein precipitation by salts. Biotechnol Bioeng 40: 1155-1164.

22. Kirkwood JG (1943) The theoretical interpretation of the properties of solutions of dipolar ions: Protein, amino acids and peptides as ions and dipolar ions. Cohn EJ, Edsall JT, Reinhold Publishing, New York, USA.

23. Long FA, McDevit WF (1952) Activity Coefficients of Nonelectrolyte Solutes in Aqueous Salt Solutions. Chem Rev 51: 119-169.

24. Wilcoxon F (1945) Individual Comparisons by Ranking Methods. Biometrics Bulletin 1: 80-83.

11

Feasibility of Using Microalgae for Biocement Production through Biocementation

Dessy Ariyanti*, Noer Abyor Handayani and Hadiyanto

Department of Chemical Engineering, Faculty of Engineering, Diponegoro University, Prof Soedarto, SH Kampus Tembalang, Semarang, Indonesia

Abstract

The invention of microorganism's involvement in carbonate precipitation, has lead the exploration of this process in the field of construction engineering. Biocement is a product innovation from developing bioprocess technology called biocementation. Biocement refers to a $CaCO_3$ deposit that formed due to microorganism activity in the system rich of calcium ion. The primary role of microorganism in carbonate precipitation is mainly due to their ability to create an alkaline environment (high pH and DIC increase) through their various physiological activities. Three main groups of microorganism that can induce the carbonate precipitation: (i) photosynthetic microorganism such as cyanobacteria and microalgae; (ii) sulphate reducing bacteria; and (iii) some species of microorganism involved in nitrogen cycle. Microalgae are photosynthetic microorganism and utilize urea using urease or urea amidolyase enzyme, based on that it is possible to use microalgae as media to produce biocement through biocementation. This paper overviews biocement in general, biocementation, type of microorganism and their pathways in inducing carbonate precipitation and the prospect of microalgae to be used in biocement production.

Keywords: Biocement; Biocementation; Microalgae; $CaCO_3$ precipitation

Introduction

Construction engineering consumes a large amount of materials from non-renewable resources, which most of the materials contribute CO_2 emission to the air at their production or application stage. Technology development related to the construction material and their production is necessary, in order to maintain the sustainability and to reduce the production of CO_2 emission. The evidence of microorganism involvement in carbonate precipitation, has lead the development of bioprocess technology in the field of construction material [1,2].

The precipitation of calcium carbonate ($CaCO_3$) may be performed due to microorganism activity and it produces massive limestone or small crystal forms [3]. These deposit of calcium carbonate known as biocement or microbial induced carbonate precipitation (MICP) [3,4]. Biocement has many advantages compared to an ordinary cement, such as: the production process is slightly different with sandstone production, biocement need a much shorter time; it is suitable for in-situ process; raw material of biocement are produced at low temperature, more efficient compared to an ordinary cement which used temperature up to 1500°C in production process; biocement can be used as eco-construction material since it consume less energy and less CO_2 emission in the production process rather than other ordinary cement [3,5].

Recently, research and study of biocement production through biocementation still focused to the nitrogen cycle mechanism using urease enzyme producing bacteria [3-7]. While research using microalgae as media for biocementation still lack in literature, in fact microalgae have a great potency for the objective of biocementation. Overview of biocement, biocementation, type of microorganism, mechanism type and feasibility of microalgae as media for biocement production will briefly described throughout this paper.

Microbial Induced Carbonate Precipitation (MCIP)

Calcium carbonate ($CaCO_3$) precipitation is a common phenomenon found in nature such as marine water, freshwater, and soils [1,6,8]. This precipitation is governed by four key factors: (i) the calcium (Ca^{2+}) concentration, (ii) the concentration of dissolved inorganic carbon (DIC), (iii) the pH (pK2 (CO) = 10.3 at 25°C) and (iv) the availability of nucleation sites [1,9]. Numerous species of microorganism have been detected previously and assumed to be associated with natural carbonate precipitates from diverse environments. The primary role of microorganism in carbonate precipitation is mainly due to their ability to create an alkaline environment (high pH and [DIC] increase) through their various physiological activities [1,6].

There are three main groups of microorganism that can induce the carbonate precipitation: (i) photosynthetic microorganism such as cyanobacteria and microalgae; (ii) sulphate reducing bacteria; and (iii) some species of microorganism involved in nitrogen cycle [1,6,7]. The most common MCIP phenomena appeared in aquatic environments is caused by photosynthetic microorganisms [7,10]. Photosynthetic microorganisms use CO_2 in their metabolic process (equation 1) which is in equilibrium with HCO_3^- and CO_3^{2-} as described in equation 2. Carbon dioxide consumed by photosynthetic microorganisms shift the equilibrium and resulting the increment of pH (equation 3) [7]. When this reaction occurs in the present of calcium ion in the system, calcium carbonate is produced as described at chemical reaction in equation 4 [6].

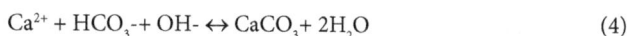

$$CO_2 + H_2O \rightarrow (CH_2O) + O_2 \tag{1}$$

$$2HCO_3^- \leftrightarrow CO_2 + CO_3^{2-} + H_2O \tag{2}$$

$$CO_3^{2-} + H_2O \leftrightarrow HCO_3^- + OH^- \tag{3}$$

$$Ca^{2+} + HCO_3^- + OH^- \leftrightarrow CaCO_3 + 2H_2O \tag{4}$$

*Corresponding author: Dessy Ariyanti, Department of Chemical Engineering, Faculty of Engineering, Diponegoro University, Prof Soedarto, SH Kampus Tembalang, Semarang, Indonesia

The precipitation of calcite ($CaCO_3$) can also be induced by heterotrophic organism. This microorganism produces carbonate or bicarbonate and modified the system so that the carbonate precipitation may occur [1]. Abiotic dissolution of gypsum ($CaSO_4.H_2O$) (equation 5) causes system rich of sulphate and calcium ion. In the presence of organic matter and the absence of oxygen, sulphate reducing bacteria (SRB) can reduce sulphate to H_2S and HCO_3^- as described in equation 6 [1,7]. When the H_2S degasses from the environment, pH of system will increase and the precipitation of calcium carbonate will occur [1].

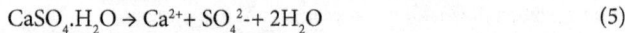

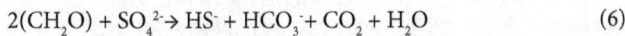

$$CaSO_4.H_2O \rightarrow Ca^{2+} + SO_4^{2-} + 2H_2O \qquad (5)$$

$$2(CH_2O) + SO_4^{2-} \rightarrow HS^- + HCO_3^- + CO_2 + H_2O \qquad (6)$$

Currently, urease enzyme activity in most of microorganism metabolism process has been used as a tool to induce the precipitation of calcium carbonate [11,12]. The hydrolysis of urea by urease enzyme in heterotrophic microorganism will produce carbonate ion and ammonium. This mechanism will result system with higher pH and rich of carbonate ion [12]. One mole of urea hydrolysed intracellularly to one mole ammonia and one mole carbamate (equation 7), which spontaneously hydrolysed to one mole ammonia and one mole carbonic acid (equation 8). Ammonia and carbamate subsequently equilibrate in water to form bicarbonate and 2 moles of ammonium and hydroxide ions as described in equation 9 and 10 [2].

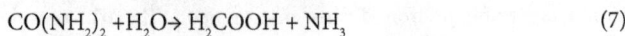

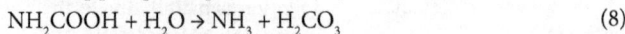

$$CO(NH_2)_2 + H_2O \rightarrow H_2COOH + NH_3 \qquad (7)$$

$$NH_2COOH + H_2O \rightarrow NH_3 + H_2CO_3 \qquad (8)$$

$$2NH_3 + 2H_2O \rightarrow 2NH_4^+ + 2OH^- \qquad (9)$$

$$2OH^- + H_2CO_3 \rightarrow CO_3^{2-} + 2H_2O \qquad (10)$$

Total reaction:

$$CO(NH_2)_2 + 2H_2O \rightarrow 2NH_4^+ + CO_3^{2-} \qquad (11)$$

The presence of calcium ion in the system will lead to the calcium carbonate precipitation once a certain level of supersaturation is reached. The calcium carbonate precipitation mechanism induced by urease enzyme activity illustrated in figure 1.

Calcium ions in the solution are attracted to microorganism cell wall due to the negative charge of the latter. After the addition of urea to the system, microorganism convert urea to dissolved inorganic carbon (DIC) and ammonium (AMM) and released it to the environment (A). The presence of calcium ion cause the supersaturation condition and precipitation of calcium carbonate in microorganism cell wall (B). After a while, the whole cell becomes encapsulated by calcium carbonate precipitate (C). As whole cell encapsulated, nutrient transfer becomes limited and resulting in cell death. Image (D) shows the imprints of microorganism cell involved in carbonate precipitation [6].

Biocementation

Biocementation is a process to produce binding material (biocement) based on microbial induced carbonate precipitation (MICP) mechanism. This process can be applied in many fields such as construction, petroleum, erosion control, and environment. Application in construction field include wall and building coating method, soil strengthening and stabilizing, and sand stabilizing in earthquake prone zone [2].

In application, the precipitation of calcium carbonate (biocement) is combined with other supporting material such as sand. The patented method of producing biocement can be seen in figure 2 [7,4].

Biocementation illustrated in figure 2 uses heterotroph bacteria *Bacillus pasteurii* with urea hydrolysis mechanism. The cementation process occurs in pipe columns filled with commercial sand contained silica. Urea/calcium solution and bacteria solution were mixed immediately and put in the pressurized vessel to be injected to the sand core in pipe column for several time until the sand core fully saturated. Biocementation takes about 24 hours to complete the reaction, after that the biocement were dried in temperature of 60°C [7].

Biocementation were also developed in the process of biological mortar production, crack in concrete remediation and production of bacterial concrete [2,9]. Table 1 shows overview of various construction materials made from biocementation.

In general, mortar refers to "ready to use" binder material contained a binder, and sand or aggregate. Biological mortar consists of three main components such as limestone powder, nutrient and bacterial paste [2]. Biocementation applied in concrete rift remediation and the production of bacterial concrete has been investigated (Santhosh et al. [13]). Specimen of crack in concrete filled with biocement shows the significant increment of strength and stiffness value compared with specimen without biocement [13].

Theoretically, calcium carbonate precipitation occur in nature following several process such as: (i) abiotic chemical precipitation from saturated solution due to evaporation, temperature increase and/or pressure decrease; (ii) production of external and internal skeleton by eukaryotes; (iii) CO_2 pressure derivation under effect of autotrophic processes (photosynthesis, methanogenesis); (iv) fungal mediation;

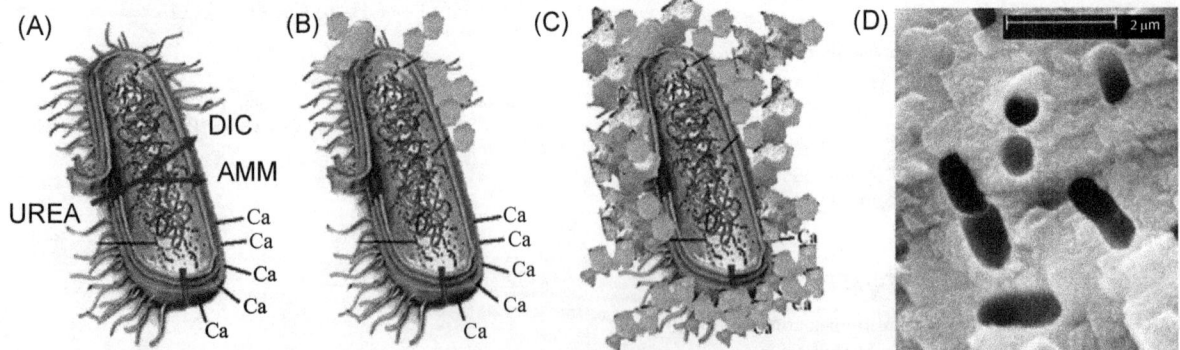

Figure 1: Illustration of calcium carbonate precipitation mechanism induced by urease enzyme activity in microorganism [6].

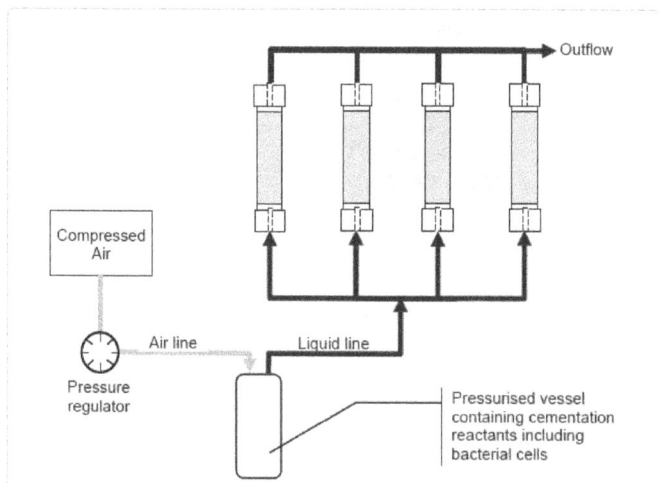

Figure 2: Injection method of cementation liquid (contain calcium/urea solution and bacterial cell) in biocementation [4,7].

Application	Microorganism	Metabolism	Solution	Reference
Biological mortar	*Bacillus cereus*	oxidative deamination of amino acids	Growth media (peptone, extract *yeast*, KNO_3, NaCl) + $CaCl_2.2H_2O$, *Actical, Natamycine*	[2]
Crack in concrete remediation	*Bacillus pasteurii*	Hydrolysis of urea	Nutrient broth, urea, $CaCl_2.2H_2O$, NH_4Cl, $NaHCO_3$	[13]
	Bacillus sphaericus	Hydrolysis of urea	Extract yeast, urea, $CaCl_2.2H_2O$	[9]
Bacterial concrete	*Bacillus pasteurii*	Hydrolysis of urea	*Nutrient broth*, urea, $CaCl_2.2H_2O$, NH_4Cl, $NaHCO_3$	[13]

Table 1: Overview of various construction materials made from biocementation.

(v) heterotrophic bacterial mediation [1]. Most of the mentioned processes above are mediated by microorganism. Both photosynthetic and heterotrophic microorganisms have natural ability to induce the precipitation of calcium carbonate. There are large amount of microorganism in many type of species spreads throughout the world. Table 2 shows several species which is already investigated as media in calcium carbonate precipitation [6].

In biocementation, microorganism that used as media should meet the specific requirement, since the process create a high pH in the environment and involving high concentration of calcium ion. For example, in biocementation based on urea hydrolysis, the process will produce high concentration of ammonium and not all type of microorganism can survive in such condition. Based on that, the selected of microorganism should meet the criteria such as: (i) have a high urease enzyme activity; (ii) ammonium and calcium ion tolerable; (iii) not pathogenic [7].

Feasibility of Using Microalgae in Biocementation

Microalgae are a promising media to be used in biocementation, due to its photosynthetic metabolism. Algae's species like *Spirulina, Arthrospira plantensis* (Cyanophyta), *Chlorella vulgaris* (Chlorophyta), *Dunaliella salina, Haematococcus pluvialis, Muriellopsis sp., Porphyridium cruentum* (Rhodophyta) basically are autotrophic microorganisms that live through photosynthetic process [14-16].

Experiment of nine green algae, a diatom and three cyanobacteria were shown to precipitate $CaCO_3$ in batch culture, where grown in the light in a hard water medium containing 68 mg L^{-1} soluble calcium. The composition of the medium was based on that found in natural marine hard water where precipitation of $CaCO_3$ within algal biofilms occurred. Deposition occurred as a direct result of photosynthesis which caused an increase in the pH of the medium. Once a critical pH had been reached, typically approximately pH 9.0, precipitation began evidenced by a fall in the concentration of soluble calcium in the medium [17]. In other experiment, Synechococcus cyanobacteria, the eukaryotic *Mychonastes sp.*, and *Chlorella sp.*, were found to induce the precipitation of $CaCO_3$ [18]. In all experiments the precipitation process developed in three stages: (1) a pH-drift period, (2) the actual precipitation reaction, and (3) an equilibration phase. The time intervals of the stages as well as the concentration changes found in the work were comparable to the results of other experimental studies on $CaCO_3$ precipitation by algae as shown in table 3 and figure 3 [18].

Several types of microalgae also use urea hydrolysis mechanism to fulfil the needs of nitrogen. For example, *Chorella sp* utilizes urea as a nitrogen source; urea is hydrolysed by urease or urea amidolyase enzyme to produce ammonia and bicarbonate [19]. The activity of urease enzyme also can induce the precipitation of calcium carbonate [11,12].

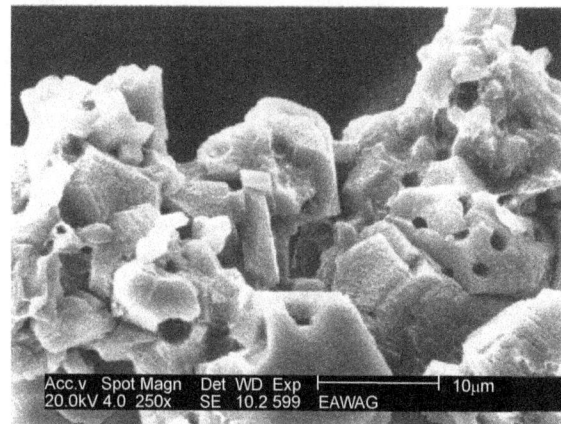

(a)

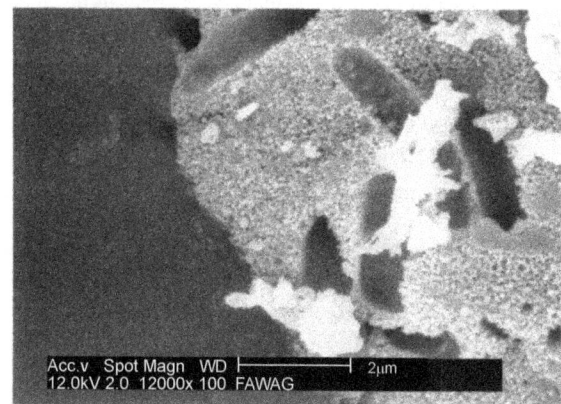

(b)

Figure 3: (a) SEM photograph of carbonate precipitates in presence of eukaryotic picoplankton, holes in the carbonate structure correspond to picoplankton cells and (b) picocyanobacteria [18].

Type of microorganism	System	Chrystal type	Reference
Photosynthetic organism *Synechococcus GL24* *Chlorella*	Meromictic lake Lurcene Lake	Calcite CaCO₃) Calcite (CaCO₃)	[21] [18]
Sulfate reducing bacteria *Isolate SRB LVform6*	Anoxic hypersaline lagoon	Dolomite (Ca(Mg) CO₃)	-
Nitrogen cycle *Bacillus pasteurii* *Bacillus cereus*	Urea degradation in synthetic medium Ammonification and nitrate reduction	Calcite (CaCO₃) Calcite (CaCO₃)	[10] [1]

Table 2: Several species which already investigated as media in calcium carbonate precipitation [6].

Experiment	Cell abundance [10^3cells.ml^{-1}]	Chlorophyll [µg.l^{-1}]	pH drift time [h]	pH at Start of prec.	Length of prec. [h]	% of Ca^{2+} precipitation
Mychonastes sp. (1)	13.2	142	45	9.05	50	41
Mychonastes sp. (2)	22.9	448	18	9.20	30	34
Chlorella sp. (1)	6.85	222	25	9.00	10	26
Chlorella sp. (2)	8.71	379	11	8.95	4	29
Synechococcus (1)	33.4	130	40	8.95	40	13
Synechococcus (2)	94.1	324	30	9.05	8	32

Table 3: Precipitation experiments of CaCO$_3$ induced by several types of algae [18].

There are some advantages of using microalgae as media for biocement production. Microalgae are type of renewable resources that easily cultivated rather than other type of microbe such as bacteria which already proved to be used in biocementation, so that its availability as raw material can be maintained properly. It's easy to grow especially in tropical area, where many non-agricultural landfills can be utilized as a raceway pond for microalgae cultivation. Tropical country also has a good temperature and water with high mineral contained which is very suitable for microalgae cultivation [15]. Another advantage is that the biocement production using microalgae can reduce the CO$_2$ emission, which produced in conventional cement production [5,3].

Based on table 3, the microalga is able to precipitate calcite very effectively within a couple days [18], while using bacteria such as *Sporosarcina pasteurii* is able to precipitate calcite under certain condition within 24 hours. But yet the exact data of experiment and literature still lack for the microalgae carbonate precipitation.

Future Challenge

Biocement is product innovation in material field that can be produce naturally using microorganism such as bacteria and microalgae. Microalgae have a great potential to be developed as media for biocement production through biocementation. Microalgae metabolism activity such as photosynthesize and hydrolysing urea can create the alkaline environment (pH and DIC elevation), so that calcium carbonate precipitation occurs in the presence of calcium ion in the system.

On the other side, microalgae also part of renewable resource that is easily cultivated especially in tropical area, so that its availability as raw material can be maintain properly. Further basic research needs to be done, primary to the theme related to suitable type of microalgae, mechanism used in biocement production through biocementation, the kinetics of process, and also the optimum condition to produce good quality of biocement.

References

1. Castanier S, Levrel GLM, Perthuisot JP (1999) Ca-carbonates Precipitation and Limestone Genesis-The Microbiogeologist Point of View. Sedimentary Geolo 126: 9-23.

2. Muynck WD, Belie ND, Verstraete W (2010) Microbial Carbonate Precipitation in Construction Materials: A Review. Ecolo Eng 36: 118-136.

3. Khanafari A, Khams FN, Sepahy AA (2011) An Investigation of Biocement Production from Hard Water. Middle-East Journal of Scientific Research 7: 964-971.

4. Kucharski ES, Ruwisch RC, Whiffin V, Al-thawadi SM (2008) Microbial Biocementation. US Patent.

5. Jian C, Ivanov V (2009) Biocement - A New Sustainable and Energy Saving Material for Construction and Waste Treatment. Civil Eng Rese 7: 53-54.

6. Hammes F, Verstraete W (2002) Key Roles of pH and Calcium Metabolism in Microbial Carbonate Precipitation. Reviews in Enviro Sci & Biotechnol 1: 3-7.

7. Whiffin Victoria S (2004) Microbial CaCO$_3$ Precipitation for The Production of Biocement, PhD Thesis, School of Biological Science & Biotechnology, Murdoch University.

8. Effendi H (2003) Study of water quality for water resources management and aquatic environment. (3rd edition) Yogyakarta, Kanisius.

9. Belie ND (2010) Microorganisms Versus Stony Materials: A Love-Hate Relationship. Materials and Structure 43: 1191-1202.

10. Mc Connaughey TA, Water HA, Small AM (2000) Community and Environmental Influences on Reef Coral Calcification. Limnology and Oceanography 45: 1667-1671.

11. Mobley HL, Hausinger RP (1989) Microbial Ureases: Significance, Regulation and Molecular Characterization. Microbial Reviews 53: 85-108.

12. Fisher SS, Galinat JK, Bang SS (1999) Microbial Precipitation of CaCO$_3$. Soil Biology and Biochemistry 31: 1563-1571.

13. Santhosh KR, Ramakrishnan V, Sookie SB (2001) Remediation of Concrete using Microorganisms.Materials Journal 98: 3-9.

14. Giordano M, Beardall J, Raven JA (2005) CO$_2$ Concentrating Mechanisms in Algae: Mechanisms, Enviromental Modulation, and Evolution. Annual Review Plant Biol 56: 99-131.

15. Harun R, Singh M, Forde GM, Danquah MK (2010) Bioprocess Engineering of Microalgae to Produce a Variety of Consumer Products. Renewable and Sustainable Energy Reviews 14: 1037-1047.

16. Chen CY, Yeh KL, Aisyah R, Lee DJ, Chang JS (2011) Cultivation, Photobioreactor Design and Harvesting of Microalgae for Biodiesel Production: A Critical Review. Bioresour Technol 102: 71-81.

17. Heath CR, Leadbeater BSC, Callow ME (1995) Effects of inhibitors on calcium carbonate deposition mediated by freshwater algae. J Appl Phycol 7: 367–380.

18. Dittrich M, Kurz P, Wehrli B (2004) The Role of Autotrophic Picocyanobacteria in Calcite Precipitation in An Oligotrophic Lake. Geomicrobiol J 21: 45-53.

19. Perez-Garcia O, Escalante FM, De-Bashan LE, Bashan Y (2011) Heterotrophic Cultures of Microalgae: Metabolism and Potential Products. Water Res 45: 11-36.

20. Bekheet IA, Syrett P (1977) Urea-degrading Enzymes in Algae. Brit Phycol J 12: 137-143.

21. Douglas S, Beveridge TJ (1998) Mineral Formation by Bacteria in Natural Microbial Communities. FEMS Microbiol Ecol 26: 79-88.

12

Degradation of Diesel, a Component of the Explosive ANFO, By Bacteria Selected From an Open Cast Coal Mine in La Guajira, Colombia

Jenny Dussán* and Mónica Numpaque

Centro de Investigaciones Microbiológicas-CIMIC, Departamento de Ciencias Biológicas, Universidad de los Andes, Bogotá, Colombia.

Abstract

Open cast coal mining operations involve the use of the explosive Ammonium Nitrate Fuel Oil (ANFO) for detonation processes. Five bacterial strains belonged to *Pseudomonas sp* and *Pseudomonas stutzeri* were isolated from an open cast coal mine located in La Guajira, Colombia. Degradation of the second component of ANFO, which is diesel, by the five isolates and by a consortium was evaluated. The biodegradation of diesel was determined by gas chromatography with a flame ionization detector. Biodegradation Efficiency (BE) was 96% for the consortium; individual strains had levels between 10-95% of BE. Analysis for the presence and expression of the alkane monooxygenase gene involved in the degradation of diesel was evidenced in two strains showing band size products between 500-600 bp. Results suggest that these bacteria are candidates for diesel bioremediation.

Keywords: *Pseudomonas*; Diesel; Biodegradation; Alkane Monooxygenase

Introduction

Open cast coal mining requires the removal of surface materials and soil to expose sources of coal. The explosive ANFO, composed of 96% ammonium nitrate and 4% diesel fuel, is used for detonation processes in coal mining. Part of the explosive ANFO is deposited in the soil where certain microorganisms are able to survive under these extreme conditions.

Diesel is composed of saturated hydrocarbons such as paraffin, and aromatic hydrocarbons. Some microorganisms are able to use hydrocarbons as an energy source, in fact bacteria have been reported in remediation of diesel contaminated soils, including *Pseudomonas stutzeri* [1] and *Pseudomonas aeruginosa* [2], among others. The ability of bacteria to degrade soil pollutants such as hydrocarbons is useful for bioremediation purposes and their ability to emulsify hydrocarbons has been studied to this respect, making bacteria useful in the removal of these compounds from the environment [3].

Alkane monooxygenase enzyme has been studied in diesel degrading micro organisms, it is responsible for the conversion of alkanes into acetyl-CoA. The model for alkane degradation has been described for *Pseudomonas putida* Gpo1, which has the OCT plasmid [4].

Six bacterial strains were selected from ANFO-contaminated soil, sampled at an open cast coal mine in La Guajira, Colombia. Bacteria were isolated from two locations at the coal mine, known as Patilla and Tabaco. Patilla has been operating for several years and Tabaco belongs to a new area of exploration. Four strains were isolated from the Tabaco pit and two strains from the Patilla pit. The isolates belong to *Pseudomonas sp*, *Pseudomonas stutzeri* and *Arthrobacter sp*. Microorganisms were chosen on the basis of being able to use ANFO as an alternative nitrogen and carbon source, as well as being resistant to its exposure. The purpose of this study was to determine the ability of the six strains selected from the coal mine, to degrade diesel as a carbon energy source as well to determine the presence and expression of the alkane monooxygenase gene (alkB) involved in the degradation of diesel.

Materials and Methods

Bacterial strains and growth conditions

Six bacterial strains were isolated from ANFO-contaminated soil.

Samples were collected from two pits: Patilla (strain: PRIII and PQII) and Tabaco (strains: TRI, TRII, TRIII, TRIV), at a coal mine in La Guajira, northeast Colombia (11°5′ 22″ N, 72°40′ 31″W). Strains were identified as *Pseudomonas sp* (PRIII and TRIV), *Pseudomonas stutzeri* (TRI, TRII and TRIII) and *Arthrobacter sp* (PQ II) according to the V3 to V5 hypervariable region (16S rDNA) PCR method using 325f and 975r primer set, the PCR products were purified and sequenced using the BigDye® Terminator v3.1 Cycle Sequencing Kit. Sequences were assembled using BioEdit v7.0.9.

A selection pressure on solid medium was conducted in which strains were grown under a saturated diesel atmosphere. Strains were incubated at 30°C in a solid Minimal Salts Medium (MSM), with a composition of KH_2PO_4 (0.5 g/L), Na_2SO_4 (2.0 g/L), KNO_3 (2.0 g/L), $CaCl_2$ (0.001 g/L), $FeSO_4$ (0.0004 g/L) and $MgSO_4.7H_2O$ (1.0 g/L). Diesel, a commercial product obtained from the coal mine, was supplied as the sole carbon source. Replica-plating was employed with all strains transferred to fresh media every 8 days for a total of 5 weeks of selection pressure. Strains that survived were kept in 10% glycerol at -80°C for further analysis.

Liquid pressure

The number of cells per colony was established for each strain. Then 10 colonies from each strain were placed in a tight-lidded glass flask with 50 ml of liquid MSM and 5 ml of diesel. Each strain had three replicate flasks. Additionally, an abiotic control was made in triplicate. A consortium was formulated by inoculating equal proportions of all five strains in a flask in the same conditions as the individual tests. The consortium also had three replicate flasks. The 21 flasks were incubated at 30°C for 43 days. Bacterial growth was monitored on days 0, 4, 7, 17

*Corresponding author: Jenny Dussán, Centro de Investigaciones Microbiológicas-CIMIC, Departamento de Ciencias Biológicas, Universidad de los Andes, Bogotá, Colombia

and 43 by serial dilutions of each of the flasks. Data analysis of bacterial growth was conducted using SPSS 18.0 and graphs were drawn in Sigma Plot 10.0.

Chromatography

Diesel degradation on day 0 and day 43 was analyzed by gas chromatography with a flame ionization detector (CG-FID). The gas chromatographic analysis was implemented with a Shimadzu gas chromatograph GC-2014, equipped with a flame ionization detector (FID), a Shimadzu AOC-20i auto-injector, a splitless injection port, and a ZB 5% phenyl-95% dimethylpolysiloxane capillary column (5.30m x 0.25mm x 0.25μm), using GC Data System Solution Release 2.30 software. The injector and detector temperatures were kept at 250ºC and 320ºC respectively. The temperature program was as follows: 50°C (5 min) to 250ºC (7 min) to 7ºC / min, then 10°C/min to 294.3ºC (10 min). Nitrogen was used as the carrier gas (99.995%, Cryogen) at a flow rate of 14 mL/min. The biodegradation efficiency (BE) was calculated according to [5] using the equation: BE (%) = 100–(As×100/Aac) where As=total area of peaks in each sample, Aac=total area of peaks in the appropriate abiotic control, BE (%)=efficiency of biodegradation.

Alkane monoxygenase expression: In order to compare alkane monoxygenase expression four of the strains were chosen, TRI and PRIII that showed the highest BE, TRII that showed the lowest BE and PQ II as a negative control because it was not able to grow under diesel pressure conditions after 17 days of treatment. These four strains were also grown in glass flasks with tight-fitting lids with 50 ml of liquid MSM and 5 ml of glucose 0.05% following the same procedure as the one described in Liquid pressure section.

Design of Primers for the Alkane Monoxygenase Gene (alkb): Primers for alkB were designed by using Primer3 program and checked by BLAST. Alkane monoxygenase gene sequence of *Pseudomonas stutzeri* were obtained from Genbank accession number AAV41375.1. These primers were alkb-f (5′-AACATAACCGTGGCCATCAC-3′) and alkb-r (5′-AACACCACGCTGTACATCCA-3′).

PCR Amplification: Template DNA was obtained from cells grown overnight in Luria Bertani (LB) broth at 30°C/220 rpm. Cells were heat-lysed for 15 min and centrifuged for 2 mins at 13000 rpm, 1.5 μL of supernatant total DNA were placed in a reaction tube with PCR mixes constituted by 1X PCR buffer, 2.5 mM MgCl$_2$, 0.5 U Taq polymerase, 0.2 mM of dNTP's, and 0.3 μM of each primer, in a final volume of 25 μl. PCR amplifications were carried out in an iCycler™ thermal cycler under the conditions: an initial step at 95°C for 4 min, followed by 35 cycles of denaturation at 95°C for 45 s, annealing at 55°C for 45 s for alkB primer pair and elongation at 72°C for 45 s. A final extension cycle was run at 72°C for 10 min. PCR products were analyzed by gel electrophoresis 1% agarose gels. Gels were stained with ethidium bromide and images were captured with a GelDoc imaging system (BioRad).

RNA extraction and RT-PCR amplification: RNA extraction from 7 day samples was made using ZR RNA MiniPrep™ kit (Zymo research) protocol. RNA quality was evaluated by 1% agarose gel electrophoresis and ethidium bromide staining and quantity by 260/280 ratio using Nanodrop technologies. cDNA was obtained using 2 Step PCR Long Range™ Kit (QUIAGEN) and random primers (Invitrogen). alkB gene was amplified from this cDNA following the same conditions described in *PCR amplification*. Products were purified with the Wizard® SV

Gel and PCR Clean-Up System kit and sequenced under BigDye™ terminator cycling conditions. High-quality sequences were compared to GenBank for a similarity analysis using BLAST.

Results and Discussion

Monitoring of bacterial growth

All six strains were able to grow under diesel pressure however strain PQII did not show growth at the 17 day of treatment. The number of cells for each flask was registered on days 0, 4, 7, 17 and 43 these results are shown in figure 1A. On the abiotic control there was no growth at all. Strains showed a decrease in growth in the first days of treatment as a result of the diesel selection pressure and adaptation to the selection media. These strains might be facultative oligotrophs due to the increase in cellular growth in the last days of the incubation period when nutrient-poor conditions exist. In fact, the coal mine with ANFO residues is an oligotrophic environment. The consortium showed the same growth behavior as these strains. These results suggest that strains used diesel as a carbon energy source. An analysis of variance (ANOVA) was performed to verify whether or not mean bacterial growth was equal between strains and days. The ANOVA showed significant differences in bacterial growth between days for all five strains and the consortium (p-value <0.05).

Bacterial growth in strains TRI, TRII, PRIII and PQII was also evaluated by using glucose as the carbon source. The number of cells for each flask was registered in the same days and these results are

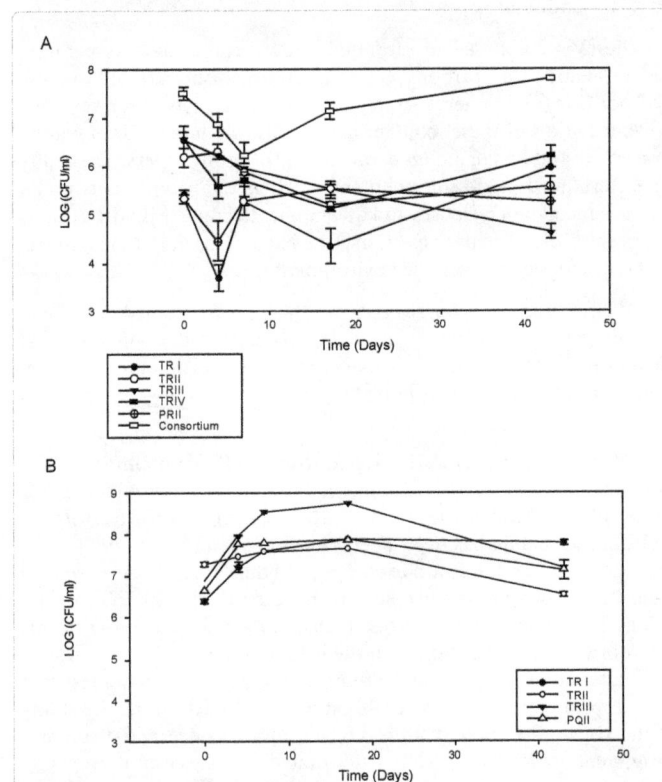

Figure 1: Monitoring of bacterial growth. A) Log values of Colony Forming Units per milliliter (CFU/ml) for all five strains and the consortium. Strains were incubated in liquid minimal salts medium (MSM) supplemented with diesel as the only carbon source at 30°C for 43 days. B) Log values of (CFU/ml) for strains (TRI, TR II, PR III, PQ II) incubated in MSM supplemented with glucose (0.5%) as the only carbon source at 30°C for 43 days. Data are represented as the average of individual triplicate samples. Error bars represent standard error.

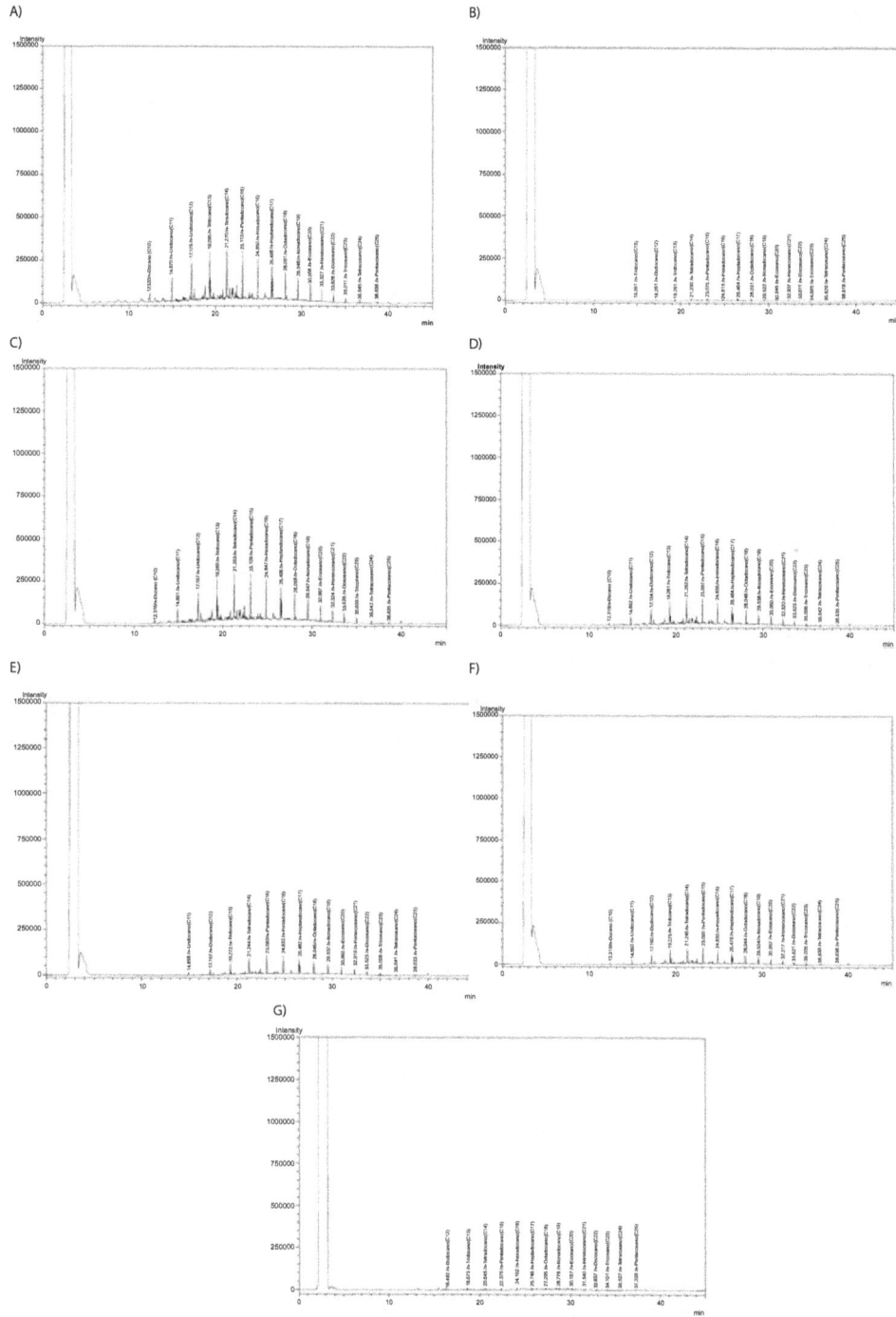

Figure 2: Diesel degradation. Gas Chromatography And Flame Ionization Detector (GC/FID) analysis were made to evaluate the biodegradation of diesel components. GC profiles of all five strains and a consortium after 43 days of incubation in MSM with diesel as the only carbon source. (A) abiotic control, (B) inoculated with strain TRI, (C) inoculated with strain TRII, (D) inoculated with strain TRIII, (E) inoculated with strain TRVI, (F) inoculated with strain PRIII, (G) inoculated with the consortium.

shown in figure 1B. Growth on MMS with addition of glucose was higher than growth on MMS with diesel as a carbon source. This was expected because bacteria can easily use glucose as a carbon energy source. To obtain energy from glucose bacteria employ the glycolysis pathway composed of constitutive enzymes which are always produced by cells independently of the composition of the culture medium.

Chromatography

A GC/FID analysis was performed after 43 days of incubation to investigate the degradation of diesel. The biodegradation of diesel was evidenced by the reduction in the area under the hydrocarbon peaks in the chromatograms when compared to that of the abiotic control suggesting the removal of diesel components. Diesel is composed of hydrocarbons containing 10 to 24 carbon atoms. All strains showed a reduction in the area under the hydrocarbon peaks corresponding to 10 to 25 carbon atoms (C10 to C25) when compared to the abiotic control (Figure 2). Strain PQII was not analyzed by GC/FID as it did not show growth at the 43 day of treatment. The consortium showed the greatest reduction in the total area under the peaks and thus the greatest biodegradation efficiency of 96%, followed by strain TRI, at 95%. PRIII and TRIV also had high BE values of 70% and 69%, respectively, followed by TRIII with a BE of 52%, and TRII with the lowest BE of 10%. The peak corresponding to 10 carbon atoms disappeared entirely in the TRI and TRIV chromatograms. In the consortium chromatogram the peaks corresponding to 10 and 11 carbon atoms disappeared completely. The consortium chromatogram shows a decrease in the area under the hydrocarbon peaks when compared to the DRO standard (Figure 3). Studies report that the mineralization of hydrocarbons is potentiated by the co-existence and cooperation of a group of microorganisms, as is the case in a consortium [5]. Research has shown that the biodegradation rate is high because these bacteria are dealing with simple molecules, such as n-alkanes (C10, C11), however, the rate might be lower for branched alkanes, n-alkyl aromatics, cyclic alkanes and polynuclear aromatics [6]. These results indicate that the Pseudomonas strains employed in this study were able to degrade diesel. In fact, the Pseudomonas genus has been reported as one of the major hydrocarbon degrading groups in previous research [7].

PCR amplification, RNA extraction and RT-PCR (reverse transcriptase) amplification

In diesel biodegradation, the alkane degradation pathway has been reported as one of the principal methods used by microorganisms to degrade diesel [4]. Results suggest that PRIII and TRI strains have the gene encoded in their genomes as shown by the PCR product. Presence of this gene was reflected as a band of the expected size on the electrophoresis. TRI showed a product size of 500 bp and PR III showed two bands with a size product of 600 and 800 bp, the small band was excised and purified. TR II and PQII showed a product of 900 bp this might be a non-specific annealing and extension to the primers (Figure 4). High quality RNA was obtained in concentrations ranging between 500-1180 ng/µl for diesel and between 1635-3163 ng/µl for glucose. The culture was 10^6-10^7 CFU/ml. In the culture with diesel as the carbon source strains TRI and PRIII showed cDNA-PCR products for alkB to be between 500-600 bp. These results are similar to those obtained by the DNA-PCR for alkB. The expression of alkB at seven day of treatment was evidenced in strains TRI and PRIII according to the sequencing of the bands.

Results showed that TRI and PRIII are expressing the alkane monoxygenase (alkB) gene under diesel growth conditions. In fact these two strains showed the higher biodegradation efficiencies. In contrast no expression was found in the culture with glucose as the carbon source. This result indicates that the alkane degradation pathway is induced by the presence of diesel. Bacteria do not synthesize catabolic enzymes unless the substrate for these enzymes is present [8]. The enzymes involved in the alkane degradation pathway are inducible enzymes that are synthesized in response to a particular substrate [9]. As diesel was the only carbon source added to the culture bacteria needed to produce the enzymes involved in the alkane degradation pathway to obtain energy [10]. According to these findings these bacteria should be able to use the alkane degradation pathway to degrade diesel.

By correlating bacterial growth and diesel degradation, it could be concluded that strains that showed a higher cellular growth at the end of the assay, were those with the highest biodegradation efficiency (TRI, TRIV, PRIII and the consortium). In addition, it may be concluded that the presence and expression of the alkB gene in strains TRI and PRIII indicates that these isolates are employing the alkane pathway in a higher metabolic rate; in fact, these two strains have the highest biodegradation efficiency. Results suggest that bacteria employed in this study belonging to Pseudomonas sp and Pseudomonas stutzeri are useful candidates for diesel bioremediation because they are resistant to diesel compounds, which is an important requirement for bioremediation purposes [11].

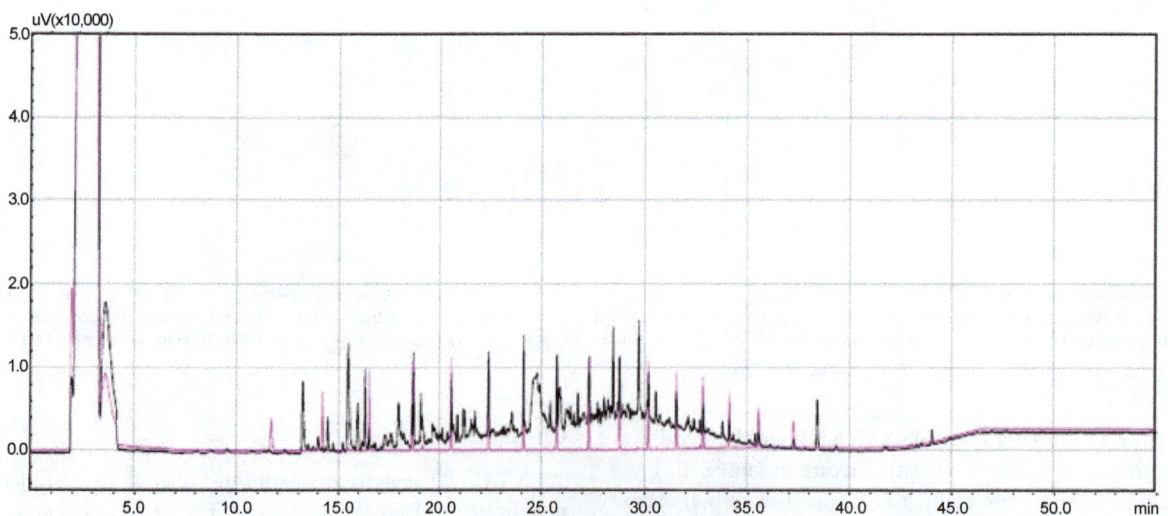

Figure 3: Comparative chromatogram of the consortium after 43 days of incubation in MSM with diesel as the only carbon source and the DRO standard of 25 ppm.

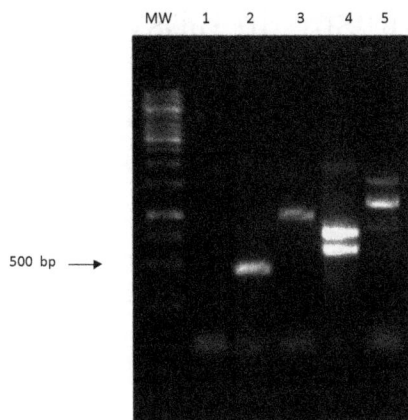

Figure 4: Agarose gel (1%) electrophoresis of PCR amplification of alkB gene. Primer pair alkB, lane MW: 100bp molecular weight marker, lane 1: negative control, lane 2: TRI, lane 3: TRII, lane 4: PRIII and lane 5: PQII.

References

1. Vázquez S, Nogales B, Ruberto L, Hernández E, Christie-Oleza J, et al. (2009) Bacterial Community Dynamics during Bioremediation of Diesel Oil-Contaminated Antarctic Soil. Microb Ecol 57: 598-610.

2. Hong JH, Kim J, Choi OK, Cho KS, Ryu HW (2004) Characterization of a diesel-degrading bacterium, Pseudomonas aeruginosa IU5, isolated from oil-contaminated soil in Korea. World Journal of Microbiology and Biotechnology 21: 381-384.

3. Ganesh A, Lin J (2009) Diesel degradation and biosurfactant production by Gram-positive isolates. African Journal of Biotechnology 8: 5847-5854.

4. Van Beilen JB, Pnanke S, Lucchini S, Franchini A, Rothlisberger et al. (2001) Analysis of Pseudomonas putida alkane-degradation gene clusters and flanking insertion sequences: evolution and regulation of the alk genes. Microbiology 147: 1621-1630.

5. Ghazali FM, Zaliha RN, Rahman RNZA, Salleh AB, Basri M (2004) Biodegradation of hydrocarbons in soil by microbial consortium. International Biodeterioration & Biodegradation 54: 61- 67.

6. Zanaroli G, Di Toro S, Todaro D, Varese GC, Bertolotto A, et al. (2010) Characterization of two diesel fuel degrading microbial consortia enriched from a non acclimated, complex source of microorganisms. Microbial Cell Fact 9: 10.

7. Shukor MY, Hassan NA, Jusoh AZ, Perumal N, Shamaan NA, et al. (2009) Isolation and characterization of a Pseudomonas diesel-degrading strain from Antarctica. J Environ Biol 30: 1-6.

8. Rojo F (2009) Degradation of Alkanes by bacteria. Environ Microbiol 11: 2477–2490.

9. Kato T, Miyanaga A, Kanaya S, Morikawa M (2009) Alkane inducible proteins in Geobacillus thermoleovorans B23. BMC Microbiol 9:60

10. Grundmann O, Behrends A, Rabus R, Amann J, Halder T, et al. (2008) Genes encoding the candidate enzyme for anaerobic activation of n-alkanes in the denitrifying bacterium, strain HxN1. Environ Microbiol 10: 376-385.

11. Kang YS, Park YJ, Jung J, Park W (2009) Inhibitory Effect of Aged Petroleum Hydrocarbons on the Survival of Inoculated Microorganism in a Crude-Oil-Contaminated Site. J Microbiol Biotechnol 19: 1672-1678.

13

Relevant Enzymes,Genes and Regulation Mechanisms in Biosynthesis Pathway of Stilbenes

Di Lu, Wei Zhao, Kuanpeng Zhu and Shu-jin Zhao*

Department of Pharmacy, General Hospital of Guangzhou Military Command, Guangzhou, People's Republic of China

Abstract

Stilbenes are natural phenolic compounds which function as antimicrobial phytoalexins in plants and affect human health as cardioprotective, antibaceteria, antioxidative and antineoplastic agents . In this review, the progresses of study on relevant enzymes, genes and regulation mechanism in biosynthesis pathway of stilbenes are described. Here we introduce a holistic and systematic method of researching relevant enzymes, genes and other regulatory factors in biosynthesis pathway of stilbenes - Systems biology. The application of knowledge of relative enzymes, genes and regulation mechanisms in stilbenes biosynthesis in metabolic engineering which is used as a tool of improving the disease resistance of plants and health caring quality of crops is also discussed.

Keywords: Stilbenes; Relevant enzyme; Regulation mechanism; Biosynthesis pathway; Systems biology

Introduction

Stilbenes,a kind of phytoalexins, are low-molecular-weight defensive substances produced by plants in response to infection after exposure to microorganisms. They are widely distributed in the plant kingdom, being reported from bryophytes and pteridophytes through gymnosperms and angiosperms.Stilbenes display a wide range of biological activitivies, such as antibacterial, antifungal, estrogenic, antitumoral [1], cardioprotective [2] and tyrosinase inhibitory activity [3].There is great interest in their potential health benefits and capacity to improve the disease resistance of plants [4]. Much effort has been directed at the Stilbenes' extraction,structure determination, biological activity over the past decades. In recent years, some success also has been achieved in the metabolic regulation and gene engineering of stilbenes. However, their detailed biosynthesis pathways and metabolic regulation, especially complicated regulation mechanism and expressing of genes and enzymes are unknown. So it is significant to shift focus from previous research priorities to search relevant enzymes, genes, abiotic stress and biotic signals so as to elucidate their detailed biosynthesis pathway and understand metabolic regulation networks.The elucidating of stilbenes' biosynthesis pathway and regulation mechanism is believed to contribute to improve the disease resistance of plants and health caring quality of crops and also provide a opportunity to know more about global regulation networks and coordination between each pathway of secondary metabolism .

The Relevant Enzymes and Regulation in Biosynthesis Pathway of Stilbenes

The relevant enzymes and regulation in phenylpropanoid pathway

The phenylpropanoid pathway is one of the most important plant secondary metabolism pathways and it is involved in the synthesis of a wide variety of important natural products from plants including flavonoids, lignins, coumarins, and stilbenes [5]. Phenylalanine ammonia-lyase (PAL), cinnamic acid 4-hydroxylase (C4H) and 4-coumarate: CoA ligase (4CL) is key enzymes in this pathway [6]. PAL, the first and key enzyme of the phenylpropanoid sequence, is the bridge between primary metabolism and secondary metabolism. PAL catalyzes the formation of trans-cinnamicacid by nonoxidative deamination of L-phenylalanine, which could be the rate-limiting step in the phenylalanine

metabolism pathway.It produces precursors for a variety of secondary metabolites such as flavonoids, lignins, coumarins, and stilbenes. PAL genes are transcriptionally activated after microbial infection or treatment of plant cells with microbial elicitors [7]. The second step in the phenylpropanoid pathway is the hydroxylation of trans cinnamic acid to 4-coumaric acid, which is catalyzed by C4H, a cytochrome P450 monooxygenase [8-9]. C4H is induced by light, elicitors, and wounding [8,10-12]. Its induction often is closely coordinated with PAL induction [13]. The 4-coumaric acid is then activated to its CoA thioester by 4CL. 4-coumaroyl CoA is funneled into branched pathways leading to a wide array of phenolic metabolites, including lignin, flavonoids [5].

Phenylpropanoid biosynthesis comprises reactions through which metabolic channeling may occur. Metabolic channeling offers unique opportunities for enhancing and regulating cellular biochemistry and major advantage of such spatial organization is the transfer of biosynthetic intermediates between catalytic sites without diffusion into the bulk phase of the cell [14]. This phenomenon involves the physical organization of successive pathway enzymes into complexes through which metabolic intermediates are channeled [15].

Studies demonstrate phenylpropanoid pathway and flavonoid metabolism branch are assembled as a linear array of sequential enzymes loosely anchored to the cytoplasmic face of endoplasmic reticulum membranes [7,16-17]. For example, Cytochrome P450 enzymes, such as C4H, Flavanone - 3 - hydroxide transketolase, the ferulic acyl - 5 - hydroxylation enzyme are anchored to the external surface of the endoplasmic reticulum [18-20]. PAL and C4H activities are colocalized on membranes of the endoplasmic reticulum. This organization regulate the partitioning of intermediates among competing pathways and determine the intracellular deposition of end products .PAL, CHS, STS,

*Corresponding author: Shu-jin Zhao, Department of Pharmacy, General Hospital of Guangzhou Military Command, Guangzhou, People's Republic of China

isoflavonids synthase are structure specific enzymes, Flavanone-7-O-methyltransferase, isoflavones-4-O-methyltransferase and isoflavones (isoflavanone) dimethylallyltransferase are modification enzymes in the phenylalanine metabolism pathway.

The relevant enzymes and regulation in biosynthesis pathway of stilbenes

Stilbene phytoalexin is derived from phenylalanine via the general phenylpropanoid pathway [21]. The last step is catalysed by Stilbene synthase (STS) which is the key enzyme of the biosynthesis pathway. STS provides the first committed step by catalyzing the sequential decarboxylative addition of three acetate units from malonyl-CoA to a p-coumaroyl-CoA starter molecule derived from phenylalanine via the general phenylpropanoid pathway (Figure 1). For example,Resveratrol synthase (STS, EC 2.3.1.95) condenses three molecules of malonyl-CoA and one molecule of cumaryl-CoA to form resveratrol.In the same active site,chalcone synthase (CHS) can catalyse the formation of chalcone by p-coumaroyl-CoA and malonyl-CoA via the intramolecular cyclization and aromatization of the resulting linear phenylpropanoid tetraketide [22]. Downstream modification enzymes in branching pathways produce a number of biologically important compounds.

Different kinds of enzymes are involved in the biosynthesis pathway of stilbenes, which include highly species- and substrate- as well as stereospecific enzymes, modification enzymes and others act as regulators. Some well studied enzymes, PAL, 4CL,C4H, Pinosylvin methoxy transferase(PMT) in combination with STS are responsible for the regulation of biosynthesis of stilbenes. There are also other enzymes

which are involved in the in the biosynthesis pathway and metabolic regulation of stilbenes are not definite. However, all the enzymes in the biosynthesis pathway of stilbenes can be classified according to their substrate specificity: stereochemical specific enzymes catalyse the formation of backbone of stilbenes and enzymes catalyse the modification reaction of the products which the first kind of enzymes synthesize. The first kind of enzymes includes PAL, CHS, STS of phenylpropanoid pathway. The stilbene backbone is synthesized from cinnamoyl-CoA and malonyl-CoA by STS. The study on the capacity for building novel and unusual polyketides from alternative substrates of STS has shown that minor modifications can be used to direct the enzyme reaction to form a variety of different and new products [23]. The second kinds of enzymes include hydroxylation enzyme, dehydrogenase, oxidase and glycosyltransferase. The specificity of these enzymes on its substrate group permutation is not high, this may be one of the main reasons why plant tissues can synthesized a series of relevant secondary metabolites [24].

The activity of key enzymes is not the only factor that affects the accumulation of stilbenes in the plants, all the enzymes involved in the formation and regulation of stilbenes as well as the supply of substrates are also associated with the accumulation of end product. The regulation mechanism of metabolic channeling also play an important role in the induction and accumulation of end product in the branch pathway of stilbenes. The activating of metabolic channeling and biosynthesis pathways of homologous synthesis is determined by the expression of the key enzymes, but the accumulation is most determined by the expression of rate-limiting enzymes in plants. Rate-limiting enzymes which usually can be finding at the branch point or the downstream of biosynthesis pathways of secondary metabolites in plants are responsible for the synthesis of precursors of many secondary metabolites [25]. Interactions between transcription factors and coordinate expression of different enzymes in the metabolic channeling have synergistic effects on the accumulation of the stilbenes. Relevant enzymes in metabolic channeling can form multienzyme complexes and coordinated express in different parts of cell. Enhanced coordinate expression of the enzyme complex can lead to a dramatically accumulation of end products. For example, induced coordinate expression of PAL and STS in the biosynthesis pathways of stilbenes can affect the synthesis of the stilbenes [26].

The biosynthesis pathway of the stilbenes in plants is closed in general conditions, which is only activated in response to microbial infections and other environmental inducers. Environmental factors including biotic and abiotic stimuli, carbon-nutrition balance, genotype and ontogenesis usually control and regulate the biosynthesis of secondary metabolites in plants [27-29]. The enzymes involved in the biosynthesis pathway of stilbenes express after the expression of corresponding genes, and the expression level of the enzyme genes is under strict regulation in plant cells due to coordinate control of the biosynthetic genes by transcription factors [30]. Coordinate transcriptional control of biosynthetic genes emerges as a major mechanism dictating the final levels of secondary metabolites in plant cells [31]. This regulation of biosynthesis pathways is achieved by specific transcription factors encoded by genes unlinked to the biosynthetic gene clusters which regulate multiple physiological processes and generally respond to environmental cues such as pH, temperature, nutrition. Transcription factor activity itself is regulated by internal signals, for example plant hormones, or external signals such as microbial elicitors or UV light. Stress hormones, such as ethylene, jasmonic acid, and salicylic acid, induce STS mRNA accumulation in leaves of mature peanut plants .The expression of resveratrol synthase (RS) genes is induced by biotic and abiotic factors in peanut cell cultures [32]. Formation of pi-

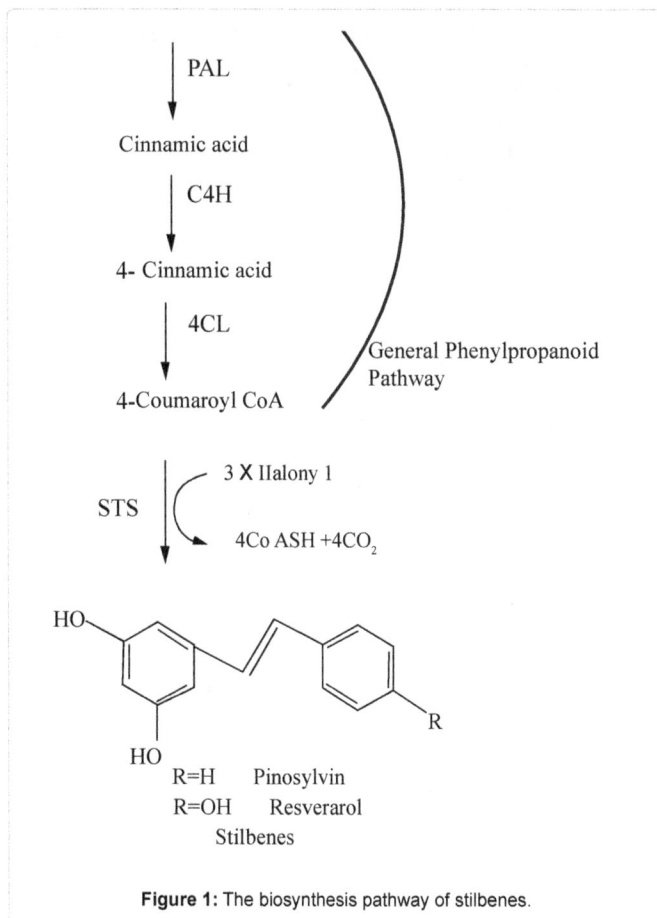

Figure 1: The biosynthesis pathway of stilbenes.

nosylvin (PS) and pinosylvin 3-O-monomethyl ether (PSM), as well as the activities of STS and S-adenosyl-l-methionine (SAM): pinosylvin O-methyltransferase (PMT), were induced strongly in needles of Scots pine seedlings upon ozone treatment, as well as in cell suspension cultures of Scots pine upon fungal elicitation [33]. A modeling method for the induction of resveratrol synthesis by UV irradiation pulses in Napoleon table grapes is proposed. Cantos etc. use the controlled UV irradiation pulses as a simple postharvest treatment to obtain possible "functional" grapes with enhanced health-promoting properties high resveratrol content [34].

The Application of Systems Biology in the Research on Relevant Enzymes and Genes in Biosynthesis Pathway of Stilbenes

Systems biology is a new science which makes us be able to understand biological systems grounded in the molecular level as a consistent framework of knowledge for the first time after the genomics, proteomics etc. were put forward [35]. It is such a rapidly evolving discipline endeavors to study the detailed coordinated workings of entire organisms with the ultimate goal to understand the dynamic networks of regulation and interactions that allows cells and organisms to live in a highly interactive environment [36]. The well studied molecular biology only care about individual gene and protein. However, systems biology is the study of cell signaling and gene regulatory networks and components and functions of the biological system, can also be understood as the study of all components in a biological system (genes, mRNA, protein etc.) and the interactions between these components in a certain circumstances [37]. Systems biology is a powerful tool to comprehensively explore the biological system, the application of its thinking model in secondary metabolites of medicinal plants brings us into a new era of understanding how to connect genes to metabolites by a systems biology approach [38].

Systems biology research methods

The classical molecular biology research is to search for specific genes at the DNA level, and then to study gene functions by gene mutation, gene knockout and other means, it also can be described to study individual genes and proteins by using a variety of means. Genomics, proteomics, transcriptomics etc. are used as single means to research multiple genes or proteins at the same time. But using one of them alone provides only part of information of system without any details of interactions between components of system. Systems biology integrates genomics, proteomics etc. and molecular biology in order to provide complementary datas. It is enabled by recent advances in multidisciplinary scientific disciplines and high-throughput approaches that allow for the parallel large-scale measurement of biomolecules, such as mRNA, proteins and metabolites [39].

Functional genomics with the goal of characterization functions of genes has become an important method of systems biology. It provides comprehensive analysis of gene functions at the genome or system level, which shift focus from research of single gene or protein to multiple genes or proteins using high-throughput experimental methods in combination with mass datas of statistical calculation method. The technology of T-DNA insertion, transposon technology, classic subtraction hybridization, differential screening, cDNA difference analysis, mRNA differential display of gene, and serial analysis of gene expression(SAGE) of systematic analysis, cDNA microarrays and DNA chip are used to analyse information of genome sequence and elucidate the gene functions. Functional genomics is a powerful tool to reveal the

biosynthesis pathways of secondary metabolites, and it will provide a solid theoretical basis for the production of secondary metabolites using metabolic engineering of medicinal plants, as well as the cell or tissue culture combined with metabolic engineering [40].

The methods of transcriptomics include differential display, gene chip, expressed sequence tags (EST) analysis, massively parallel sinature sequencing (MPSS), amplified fragment length polymorphism (AFLP) [41,42]. Two-dimensional electrophoresis, mass spectrometry technology bioinformatics analysis are primary methods of screening and identifying proteomics. Yeast two—hybrid system (Y2H), tandem affinity purification (TAP) can be used to study protein—protein interactions and green fluorescent protein (GFP) as maker to study subcellular localization. Metabonomics is a very important tool to study medicinal plants and promote modernization of traditional Chinese medicine [43,44], including nuclear magnetic resonance (NMR), gas chromatography - mass spectrometry (GC-MS), liquid chromatography-mass spectrometry (LC-MS), combined application of Fourier transformation mass spectrometry (FTMS) and capillary electrophoresis - mass spectrometry (CE-MS)[45].

Technologies of Genomics, proteomics, transcriptomics and metabolomics detect the various molecules and study their functions at DNA, mRNA, protein and metabolite levels.

The application of Systems biology approaches in research on related enzymes and genes in biosynthesis pathway of stilbenes

When we study stilbenes with thinking model of systems biology, various levels of information including the DNA, mRNA, small molecules, proteins and protein interaction networks should to be integrated in order to obtain a series of relevent enzymes, genes or regulatory factors in its biosynthesis pathway. All the information can be used to construct a reasonable model in order to elucidate the biosynthesis pathway, regulation mechanisms of stilbenes [46]. For example, the study of relative enzymes and genes in biosynthetic pathway of tanshinone using thinking model of systems biology.Groups of materials with phenotypic differences are analysed in order to get datas of metabolomics, proteomics, transcriptomics which can be gained with gene chips. Systemic results about genes and enzymes revolved in the biosynthesis pathway of tanshinone were obtained [47-50]. As a result, an unique new branch of two terpene biosynthesis of tanshinone biosynthesis was found.

Our group focused the research on biosynthesis pathway of a kind of stilbenes, stilbene glucoside (2,3,5,4'-tetra-hydroxy-stilbene-2-O-β-D-glucoside)which is the major bioactive principles in Polygonum multiflorum (Figure 2). The application of Systems biology in the research

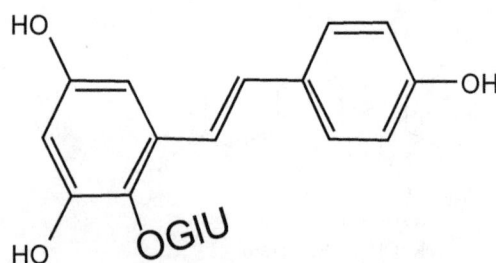

Figure 2: 2,3,5,4'-tetra-hydroxy-stilbene-2-O-β-D-glucoside

on relevant genes and enzymes in biosynthesis pathway of stilbene glucoside can be described as follows.Firstly, the possible relevant genes or enzymes should be identified from Polygonum multiflorum.The clone of a type III polyketide synthase gene(FmPKS) using oligonucleotide primers designed for regions conserved amongst STS(with special attention given to the closely-related species of the Polygonaceae family) in Fallopia multiflora have be conducted by Shujing Sheng [51]. The FmPKS has been identified strongly correlates with the accumulation of stilbene glucoside by using Northern blotting, RNA inference and over expression, suggesting that FmPKS might play an important role in its biosynthesis. Resveratrol (Figure 1), another kind of stilbenes, is similar to stilbene glucoside in structure. Zhongyu Liu has investigated the over expression of a resveratrol synthase gene (PcRS) from Polygonum cuspidatum which is closely related to the Fallopia multiflora in transgenic Arabidopsis .As a result, it causes the accumulation of transpieced with antifungal activity [52]. The foregoing research may be summed up to search for individual possible genes or enzymes involved in biosynthesis pathway of stilbene glucoside and study gene functions using the classical molecular biology.

Secondly, search for the possible relevent genes, enzymes through approaches of functional genomics, transcriptomics, and proteomics. At present, suppression subtractive hybridization (SSH) was performed to search for genomic differences. A subtractive cDNA library was constructed by using cDNA from Polygonum multiforum root tubers with high content of stilbene glucoside as tester and low content as driver for the subtractive hybridization. As a result, 11 clones were obtained as the differentially expressed candidates who play an important role in further validation of genes involved in the biosynthesis pathway of stilbene glucoside in Polygonum multiflorum. Rapid-amplification of cDNA ends (RACE) was carried out to gain the full-length cDNAs. The application of RNA interference and over expression of these genes

will reveal the specific functions of these genes and screen the genes involved in the biosynthesis pathway of stilbenes.

Thirdly, internal components (such as genetic mutation, interference) or external conditions of the Polygonum multiflorum are to be changed. The accumulation of stilbene glucoside, gene expression and enzyme expression in these cases should be detected. Next all the relevant information should be integrated to derivate how they regulate the biosynthesis of stilbene glucoside. The second step and third step ought to be repeated to revise and refine the model through a large number of experimental results, in order to finally gain all the relevent genes, enzymes and regulatory factors in the biosynthesis pathway of stilbene glucoside. These researches are from the thinking model of systems biology at different levels, which has important significance in revealing the biosynthesis pathway and regulation mechanism of stilbene glucoside in Polygonum multiflorum. The discussion and related diagrams give a good idea of the application of systems biology in research on relevent genes and enzymes in biosynthesis pathway of stilbene glucoside (Figure 3).

The Application of the Research on Enzymes and Genes in Biosynthesis Pathway of Stilbenes

The elucidating of stilbenes' biosynthesis pathway and regulation mechanism can accelerate the course of metabolic engineering as a tool for plant disease control and hunman health promotion .The increasing maturity of the plant genetic engineering technology promote the research on the biosynthesis pathway of stilbenes, and the two well studied field are stilbene synthase gene engineering and its transcription factors or regulation engineering.

The genetic engineering research of metabolic key enzymesin biosynthesis pathway of stilbenes.

STS plays an important regulatory role in biosynthesis pathway of

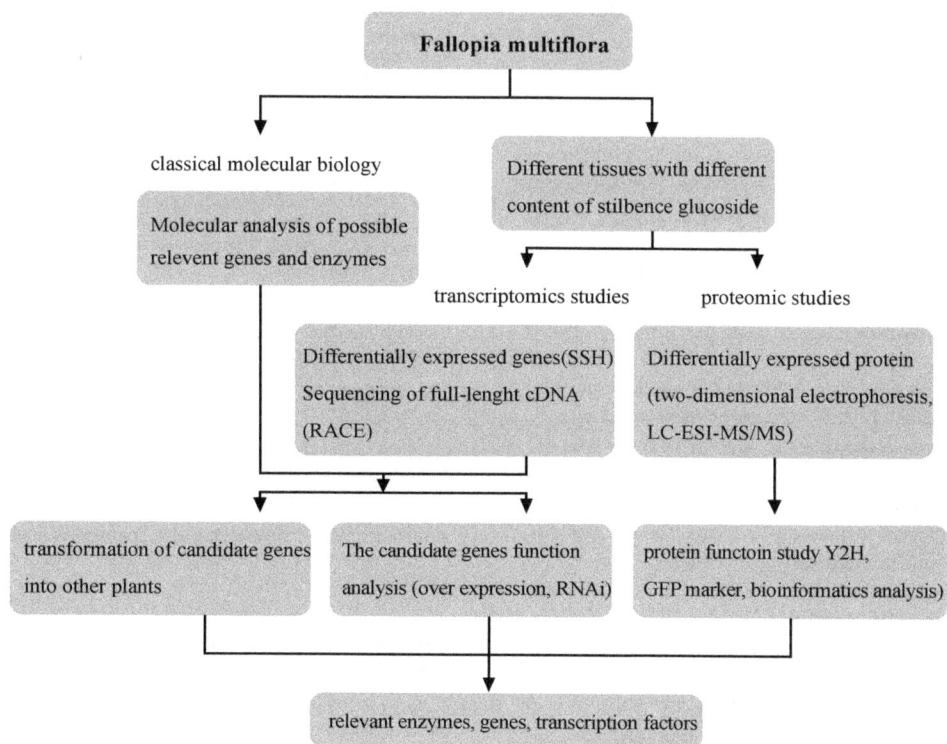

Figure 3: The application of systems biology in research on relevent genes and enzymes in biosynthesis pathway of stilbene glucoside

resveratrol and other stilbene phytoalexin. An important goal of STS gene engineering is the genetic improvement of plants in increasing resistance against diseases. Namely transformate target plant genomes with chimeric strong promoter and STS gene through transgenic technology, which makes transgenic plants, express STS in order to start the stilbene biosynthesis pathway, and change the plant traits and enhance plant defense against external violation [53].

Grape STS gene was introduced into Nicotiana tabacum, which results in transgenic tabacum express the STS and synthesize resveratrol and then dramatically increase its resistance against diseases [54,55]. The transformation of rice, wheat and other crops with grape STS gene also increase the resistance level against disease significantly of transgenic crops [56]. At present, the transgenic plants with STS gene include apple, poplar, papaya, white poplar, rape, wheat, peas, tomatoes, lettuce, hops, Arabidopsis thaliana [57-67]. Through introducing STS gene, most of transgenic plants can exhibit STS activity and synthesize exogenous stilbene phytoalexin.

The gene engineering of regulating gene or transcription factor in biosynthesis pathway of stilbenes

Co transformation of two or several related enzyme genes in downstream of biosynthesis pathway can be used when study stilbenes' biosynthesis involves multiple genes expression. In this way, the new secondary biosynthesis branch can be introduced into the plant then to increase the content of stilbene secondary metabolites in the transgenic plant or synthesize exogenous stilbenes. Coordinated expression of enzyme genes in secondary metabolism channel and the same or similar cisacting elements of regulatory sequences of these genes are regulated by the same transcription factors or regulation genes. Therefore, enhancing the expression of transcription factor genes and regulation genes of these important enzyme genes is a feasible way to achieve enhanced coordinated expression of multiple genes. This requires more understanding about identification and regulation of enzymes gene involved in biosynthesis pathway of stilbenes.

Summary

Now we have only a rudimentary grasp of the basic framework of the main plant secondary metabolic pathways such as alkaloids synthesis pathway, phenylpropanoid biosynthesis pathway and isoprenoid biosynthesis pathway, but lack of research on rate-limiting steps, related isozymes and specific biosynthesis pathways of specific secondary metabolite. Plant secondary metabolism is a complicated dynamic process which is regulated by plant genetic background and growth process, also affected by ecological environment and pathogen infection, insect feeding and stimulation of various elicitors. The induction expression of related genes and enzymes also need be further studied. Because of disease resistance and medicinal health functions of resveratrol, stilbene glucoside and other inducible stilbenes, secondary metabolic engineering can be used to increase the content of exogenous stilbenes in transgenic plants. However, the complexity of enzymes in biosynthesis pathway and regulation of gene expression increase the difficulty in researching metabolism genetic engineering of stilbenes. The thinking model of systems biology can be used to research the biosynthesis pathway of secondary metabolites and understand sequence of events of intermediate products and final products. The enzymes and their genes expression and regulation in each reaction step as well as the interaction of each biosynthesis branch can be definite from the view of metabolic channeling. Then after continuous integration and analysis, the detailed metabolic pathways and regulation mechanism of stilbenes will be finally elucidated. This will be the future research emphasis and direction for people to understand about regulations of secondary metabolic pathways in plants and coordination between the secondary metabolic pathways.

Acknowledgment

We thank Dr. Shu-jing Sheng and Ping Yan for helpful discussions and all those who helped us obtain and identify plants: Prof. Fuwu Xing, South China Botanical Garden; and Prof. Tang Liping, Kun-ming Medical University.

References

1. Delaunois B, Cordelier S, Conreux1 A, Clément C, Jeandet P (2009) Molecular engineering of resveratrol in plants. Plant Biotechnol J 7: 2-12.

2. Lin J K, Tsai SH (1999) Chemoprevention of cancer and cardiovascular disease by resveratrol. Proc Natl Sci Counc Repub China B 23: 99-106.

3. Likhitwitayawuid K, Sritularak B (2001) A new dimeric stilbene with tyrosinase inhibitory activity from Artocarpus gomezianus. J Nat Prod 64: 1457-1459.

4. Shui-Lin HE, Jin-Cui ZHENG, Ming LIN, WANG Yan-Hua (2004) Advances of Biological Function,Regulatory Mechanism of Biosynthesis and Genetic Engineering of in Stilbenes Plant. J Agr Biotechnol 12: 102-108.

5. Zhang HC, Liu JM, Chen HM, Gao CC, Lu HY, et al. (2011) Up-Regulation of Licochalcone A Biosynthesis and Secretion Tween 80 in Hairy Root Cultures of Glycyrrhiza uralensis Fisch. Mol Biotechnol 47: 50-56.

6. Li LI Yue ZHAO, Jun-lan MA (2007) Recent progress on key enzymes:PAL,C4H,4CL of phenylalanine metabolism pathway. China J Bioinfor 4: 187-189.

7. Rasmussen S, Dixon RA (1999) Transgene mediated and elicitor induced perturbation of metabolic channeling at the entry point into the phenylpropanoid pathway. Plant Cell11: 1537-1551.

8. Fahrendorf T , Dixon RA (1993) Stress responses in alfalfa(Medicago sativa L) XVIII Molecular cloning and expression of the elicitor-inducible cinnamic acid 4-hydroxylase cytochrome P450. Arch Biochem Biophys 305: 509-515.

9. Teutsch HG, Hasenfratz MP, Lesot A, Stoltz C, Garnier JM, et al. (1993) Isolation and sequence of a cDNA encoding the Jerusalem artichoke cinnamate 4-hydroxylase,a major plant cytochrome P450 involved in the general phenylpropanoid pathway. ProcNatlAcadSci 90: 4102-4106.

10. Buell CR, Somerville SC (1995) Expression of defense-related and putative signaling genes during tolerant and susceptible interactions of Arabidopsis with Xanthomonas campestris pv.campestris. Mol Plant Microbe Interact 8: 435-443.

11. Batard Y, Schalk M, Pierrel MA, Zimmerlin A, Durst F, et al. (1997) Regulation of the cinnamate 4-hydroxylase (CYP73A1) in Jerusalem artichoke tubers in response to wounding and chemical treatments. Plant Physiol 111: 951-959.

12. Bell-Lelong DA, Cusumano JC, Meyer K (1997) Cinnamate 4-hydroxylase expression in Arabidopsis (Regulation in Response to Development and the Environment). Plant Physiol 113: 729-738.

13. Mizutani M, Ohta D, Sato R (1997) Isolation of cDNA and a genomic clone encoding cinnamate 4-hydroxylase from Arabidopsis and its expression manner in planta. Plant Physiol 113: 755-763.

14. Winkel BSJ (2004) Metabolic channeling in plant. Annual Review of Plant Biology 55: 85-107.

15. Dong XY, Braun EL , Grotewold E (2001) Functional conservation of plant secondary metabolic enzymes revealed by comp lamentation of A rabidopsis flavonoid mutants with maize genes.Plant Physiol 127: 46-57.

16. Burbulis IE, Winkel Shireley B (1999) Interactions among enzymes of the Arabidopsis flavonoid biosynthestic pathway. Proc Natl A cad Sci USA 96: 12929-12934.

17. He XZ, Dixon RA (2001) Genetic manipulation of isoflvone 7-O-methyltransferase enhances biosynthesis of 4'-O-methylated isoflavonoid phytoalexins and disease resistance in alfalfa. Plant Cell 12: 1689-1702.

18. Chapple C (1998) Molecular-genetic analysis of plant cytochrome P450-dependent monooxygenases. Annu Rev Plant Physiol Plant Mol Biol 49: 311-343.

19. Winkel Shirley B (2001) Flavonoids biosynthesis: a colorful model for genetics, biochemistry, cell biology, and biotechnology. Plant Physiol 126: 485-493.

20. Dixon RA (2001) Natural products and plant disease resistance. Nature 411: 843-847.

21. Austin MB, Noel JP (2003)The chalcone synthase su-perfamily of type III polyketide synthases. Nat Prod Rep 20: 79-110.

22. Kati Hanhineva,Harri Kokko,Henri Siljanen(2009) Stilbene synthase gene transfer caused alterations in the phenylpropanoid metabolism of transgenic strawberry (Fragaria x ananassa). J Exp Bot 60: 2093-2106.

23. Morita H, Noguchi H, Schroder J, Abe I (2001) Novel polyketides synthesized with a higher plant stilbene synthase. Eur J Biochem 268: 3759-3766.

24. Rasmussen S, Dixon RA(1999) Transgene-mediated and elicitor-induced perturbation of metabolic channeling at the entry point into the phenylpropanoid pathway. Plant Cell 11: 1537-1551.

25. Shuilin HE, Jingui ZHENG, Xiaofeng WANG (2002) Plant secondary metabolism:function,regulation and gene engineering. Chin J Appl Environ Bio 8: 558-563.

26. Melchior F, Kindl H (1991) Coordinate- and elicitor-dependent expression of stilbene synthase and phenylalanine ammonialyase genes in Vitis cv. Optima. Arch Riochem Biophys 288: 552-557.

27. Kliebenstein DJ (2004) Secondary metabolites and plant/environment interactions: a view through Arabidopsis thaliana tinged glasses. Plant Cell Environ 27: 675-684.

28. Lerdau M, Coley PD (2011)Benefits of the carbon-nutrient balance hypothesis. OIKOS 98: 534-546.

29. Mary Ann Lila (2006) The nature-versus-nurture debate on bioactive phytochemicals: the genome versus terroir. Sci Food Agric 86: 2510-2515.

30. Zhi-lin Yuan, Chuan-chao Dai, Lian-qing Chen (2007) Regulation and accumulation of secondary metabolites in plant-fungus symbiotic system. African Journal of Biotechnology 6: 1266-1271.

31. Debora Vom Endt, Jan W Kijne, Johan Memelink (2002)Transcription factors controlling plant secondary metabolism:what regulates the regulators? Phytochemistry 61: 107-114.

32. Chung IM, Park MR, Rehman S, Yun SJ (2001) Tissue specific and inducible expression of resveratrol synthase gene in peanut Plants. Mol Cells12: 353-359.

33. Chiron H, Drouet A, Claudot A, Eckerskorn C, Trost M, et al. C(2000) Molecular cloning and functional expression of a stress-induced multifunctional O-methyltransferase with pinosylvin methyltransferase activity from Scots pine (Pines sylvestris L). Plant Mol Biol 44: 733-745.

34. Cantos E, Espin JC, Tomas-Barberan FA (2001) Postharvest induction modeling method using U V irradiation pulses for obtaining resveratrol-enriched table grapes: a new functional fruit? J Agric Food Chem 49: 5052-5058.

35. Hiroaki Kitano (2002) Systems biology:Toward system-level understanding of biological systems. Cambridge M A: MIT Press.

36. Martin Latterich (2005) Molecular systems biology at the crossroads: to know less about more, or to know more about less? Proteome Sci 3: 8-11.

37. Hood L, Heath JR, Phelps ME, Lin B (2004) Systems biology and new technologies enable predictive and preventative medicine. Science 305: 640-643.

38. Oksman-Caldentey KM, Inze D, Oresic M (2004) Connecting genes tometabolites by a systems biology approach. PNAS 101: 9949-9950.

39. Aderem A (2005) Systems biology: Its practice and challenges . Cell 121: 511-513.

40. Jingxue WANG,Yi SUN,Peilin XU (2004) Research Progress in Functional Plant Genomics. Biotechnology Bulletin 1: 18-24.

41. Rensink WA, Buell CR (2005) Microarray expression profiling resources for plant genomics. Trends Plant Sci 10: 603-609.

42. Marnik V, Johan DP, Michiel JT van Eijk (2007) AFLP-based tran-scriptprofiling (cDNA-AFLP) forgenome-wide expression analysis. NatProtoc 2: 1399-1413.

43. Jia W, Liu P, Jiang J, Chen MJ, Zhao LP, et al. (2006) Application of metabonomics in complicated theory system research of traditional Chinese medicine. China Journal of Chinese Materia Medica 31: 621-625.

44. Lianwen QI, Ping LI,Jing ZHAO (2006) Metabonomics and Modernization of TMC Research. World Science and Technology 8: 79-87.

45. Sumner LW, MendesP, Dixon RA (2003) Plantmetabolomics:Large-scale phytochemistry in the functional genomics era. Phytochemistry 62: 817-836.

46. Huang L, Gao W, Zhou J, Wang R (2012) System s biology applications to explore secondarymetabolites in medicinal plants. China Journal of Chinese Materia Medica 35: 8-12.

47. Cui GH, Huang LQ, Tang XJ, Qiu DY, Wang XY, et al. (2007) Functional genomics studies of Salvia miltiorrhiza1.Establish cDNA microarray of S. miltiorrhiza. China Journal of Chinese Materia Medica 32: 1137-1142.

48. Wei GAO, Guang-hong CUI, Jian-qiang KONG (2008) Optimizing expression and purification of recombinant Salvia miltiorrhiza copalyl diphosphate synthase protein in E.coli and preparation of rabbit antiserum against SmCPS. Acta Pharmaceutica Sinica 43: 766-772.

49. Wang XY, Cui GH, Huang LQ, Gao W, Yuan Y (2008) A full length cDNA of 4-(cytidine 5'-diphospho)-2-C-methyl-D-erythritol kinase cloning and analysis of introduced gene expressing in sava miltiorrhiza. Acta Pharmaceutica Sinica 43: 1251-1257.

50. Gao W, Hillwig ML, Huang L, Cui G, Wang X, et al. (2009) A functional genomics approach to tanshinone biosynthesis provides stereo-chemical insights. OrgLett 11: 5170-5173.

51. Shu-Jing SHENG, Zhong-Yu LIU, Wei ZHAO (2010) Molecular analysis of a type III polyketide synthase gene in Fallopia multiflora. Section Cellular and Molecular Biology 65: 939-946.

52. Liu Z, Zhuang C, Sheng S, Shao L, Zhao W, et al. (2011) Overexpression of a resveratrol synthase gene (PcRS) from Polygonum cuspidatum in transgenic Arabidopsis causes the accumulation of trans-piceid with antifungal activity. Plant cell rep 30: 2027-2036.

53. Raiber S, Schroder G, Schroder J (1995) Molecular and enzymatic characterization of two stilbene synthases from Eastern white pine (Pinus strobus) A single Arg/His difference determines the activity and the pH dependence of the enzymes. FEBS Lett 361: 229-302.

54. Hain R, Bleseler B, KindlH, Schroder G, Stocker R (1990) Expression of a stilbene synthase gene in Nicotiana tabacumresults in synthesis of the phytoalexin resveratrol. Plant Mol Biol 15: 325-335.

55. Hain R, Reif HJ, Krause E, Langebartels R, Kindl H, et al. (1993) Disease resistance results from foreign phytoalexin expression in a novel plant. Nature 361: 153-156.

56. Tian WZ, Ding L, Cao SY, Shun-Hong D, Song-Qing YE, et al. (1998) Rice transformation with a phytoalexin gene and bioassay of the transgenic plants. Bot Sci Acta 40: 803-808.

57. Serazetdinova L,Oldach KH,Lörz H (2005) Expression of transgenic stilbene synthases in wheat causes the accumulation of unknown stilbene derivatives with antifungal activity. J Plant Physiol 162: 985-1002.

58. Szankowski I, Briviba K, Fleschhut J, Schönherr J, Jacobsen HJ, et al. (2003) Transformation of apple (Malus domestica Borkh.) with the stilbene synthase gene from grapevine (Vitis vinifera L.)and a PGIP gene from kiwi(Actinidia deliciosa). Plant Cell Rep 22: 141-150.

59. Seppänen SK, SyrjäläL, von Weissenberg K, Teeri TH, Paajanen L (2004) Antifungal activity of stilbenes in in vitro bioassays and in transgenic Populus expressing a gene encoding pinosylvin synthase. Plant Cell Rep 22: 584 -593.

60. Zhu YJ,Agbabani R,Jackson MC, Tang CS, Moore PH (2004) Expression of the grapevine stilbene synthase gene VST1 in papaya provides increased resistance against diseases caused by Phytophthora palmivora. Planta 220: 241-250.

61. Giorcelli A,Sparvoli F,Mattivi F, Tava A, Balestrazzi A, et al. (2004) Expression of the stilbene synthase (StSy) gene from grapevine in transgenic white poplar results in high accumulation of the antioxidant resveratrol glucosides. Transgenic Res 13: 203-214.

62. Hüsken A, Baumert A, Milkowski C, Becker HC, Strack D, et al. (2005) Resveratrol gluco-Side (Piceid) synthesis in seeds of transgenic oilseed rape (Brassica napus L).Theor Appl Genet 111: 1553-1562.

63. Morelli R, Das S, Bertelli A, Bollini R, Lo Scalzo R, et al. (2006) The introduction of the stilbene synthase gene enhances the natural antiradical activity of Lycopersicon esculentum mill. Mol Cell Biochem 282: 65-73.

64. Richter A, Jacobsen HJ, de Kathen A, de Lorenzo G, Briviba K, et al. (2006) Transgenic peas (Pisum sativum)expressing polygalacturonase inhibiting pro-

tein from raspberry (Rubus idaeus.) and stilbene synthase from grape (Vitis vinifera). Plant Cell Rep 25: 1166-1173.

65. Liu S, Hu Y, Wang X, Zhong J, Lin Z (2006) High content of resveratrol in lettuce transformed with a stilbene synthase gene of Parthenocissus henryana. Agric Food Chem 54: 8082-8825.

66. Schwekendiek A, Spring O, Heyerick A, Pickel B, Pitsch NT (2007) Constitutive expression of a grapevine stilbene synthase gene in transgenic hop (Humulus lupulus L.) yields resveratrol and its derivatives in substantial quantities. Agric Food Chem 55: 7002-7009.

67. Lo C, Le Blanc JC, Yu CK, Sze KH, Ng DC, et al. (2007) Detection, characterization, and quantification of resveratrol glycosides in transgenic Arabidopsis over—expressing a sorghum stilbene synthase gene by liquid chromatography / tandem mass spectrometry. Rapid Commun Mass Spectrom 21: 4101-4108.

14

Rapid Conversion of Chicken Feather to Feather Meal Using Dimeric Keratinase from *Bacillus licheniformis* ER-15

Ekta Tiwary and Rani Gupta*

Department of Microbiology, University of Delhi, South Campus, New Delhi, India-110021

Abstract

Dimeric keratinase from *Bacillus licheniformis* ER-15 completely degraded 25g boiled native chicken feather to feather meal within 8h at pH 8, 50°C and 150rpm. Feather degradation was a linear function of enzyme concentration and 2.5g chicken feather was degraded in presence of 1200U keratinase. Process for feather meal production comprised soaking of 25g feather in 250ml water followed by boiling for 10min-20min before enzyme addition. Feather meal thus produced was dried at 80°C and ground to obtain feather meal powder. Feather meal contained 14% nitrogen, 44% carbon with all essential amino acids and showed 73% *in-vitro* digestibility.

Keywords: *Bacillus licheniformis*; Feather Degradation; Feather Meal, *in-vitro* Digestibility, Keratinase

Introduction

Feather is protein rich waste product of poultry processing industries which are being generated in billion of tons every year [1-4]. These feathers are generally land filled or burnt which cause environmental problems [4]. Feather are also degraded to feather meal which is used as animal feed, organic fertilizers, feed supplements because it is made up of >90% protein and rich in hydrophobic amino acids and important amino acids like cystine, arginine, threonine [5,6]. Most popular method of feather meal production is by hydrothermal process where feather are cooked under high pressure at high temperature. However, hydrothermal treatment, results in destruction of essential amino acids like methionine, lysine, tyrosine, tryptophan and has poor digestibility and low nutritional value [7,8]. In this respect, microbial degradation of feather into feather meal has gained importance and new microbes are being looked upon for efficient degradation of feather. Feather are degraded during fermentation process where consortium of thermophilic/mesophilic bacterial cultures such as *Bacillus, Streptomyces, Vibrio, Chryseobacterium* strains are used [4,6,9]. During fermentation not more than 0.5-2% w/v can be used and also essential amino acids are utilized by micro-organism which decreases the nutritional value of feather meal. To combat this, focus of the research is changing towards developing of enzymatic methods of feather degradation using special class of proteases, the keratinases. Till date not more than 10% feather degradation is reported in the presence of keratinases however a novel dimeric keratinase from *Bacillus licheniformis* ER-15 was observed to degrade feather completely into feather meal [10]. Here various process parameters for enzymatic degradation of feather to feather meal have been standardized. Amino acid analysis and *in-vitro* digestibility of the feather meal is also compared with existing reports.

Material and Methods

Chemicals used for buffer and medium were obtained from Sisco Ranbaxy Laboratory (SRL, India). Soy flour and chicken feather were collected from local market only.

Analytical methods

Keratinase assay: Keratinase activity was measured using 20mg feather, 1ml of properly diluted enzyme and 4ml, pH 10 buffer (50mM Glycine-NaOH buffer). Reaction mixture was incubated at 60°C for 1h and stopped with 4ml of 5% w/v trichloro acetic acid (TCA) followed by incubation at room temperature for 30min and centrifugation at 8000rpm for 10min. Absorbance of supernatant was measured at A280. Similarly, control reaction was set up with 1ml of 5% w/v TCA Enzyme unit was defined as amount of enzyme required to release protein equivalent to absorbance of 0.01 from feather keratin under standard assay condition [2].

Protein determination: Protein in supernatant was measured at A280 and 1 absorbance was considered as 1mg/ml protein using bovine serum albumin as standard.

Production and downstream processing of keratinase from *Bacillus licheniformis* ER-15

Enzyme production: Keratinase was produced in soyflour feather medium in 60h as reported earlier [10]. A loopful bacterial culture was inoculated in 50ml nutrient broth in 250ml Ehrlenmeyer flask, grown at 37°C, 200rpm for 16h and used as seed culture. This seed culture was again grown in nutrient broth for 16h in similar conditions and used as inoculum for production medium. Four hundred ml of production medium in 2L flask (0.4%w/v soyflour, 3%w/v glucose, 0.3w/v KH2PO4, 0.9%w/v K2HPO4, 0.5%w/v feather) was inoculated with 4% v/v inoculum and incubated at 250rpm, 37°C for 60h.

Sedimentation and microfiltration: After 60h of production, fermentation broth was collected in a beaker and kept at room temperature in static condition for 24h. Most of the bacterial cells were settled along with feather meal at the bottom of beaker, supernatant was decanted and micro filtered through 0.2µ filters (MDI, India) using vacuum pump. This micro filtered supernatant was concentrated and used as enzyme.

Enzyme concentration and shelf life: Micro filtered supernatant was concentrated using 85% saturation of ammonium sulphate. Precipitated enzyme was collected after centrifugation, dissolved in pH 7

*__Corresponding author:__ Rani Gupta, Department of Microbiology, University of Delhi, South Campus, New Delhi, India-110021

phosphate buffer and stored at 4°C. concentrated enzyme was used for feather meal production after required dilution. Enzyme was checked for shelf life for a period of 1 year at room temperature.

Feather processing and feather meal production

Procurement of feather and feather processing: Chicken feather were procured from the local market. Feathers were washed with detergent and detergent was removed by several washing with tap water followed by distilled water. Washed feather were dried at 80°C for 6h and were used for subsequent experimentation.

Standard protocol for feather degradation: Feather (2.5g) was autoclaved at 15psi for 15min in a 250ml flask containing 25ml, 25mM pH 8 phosphate buffers. Volume was made upto 50ml with properly diluted enzyme. Flasks were kept at 150rpm, 50°C for 12h or till specified time. After degradation, feather meal was filtered through 2mm sieve and residual feather were dried at 80°C till constant weight. Percent degradation was calculated on the basis of dry weight. Experiments were set up in triplicate and repeated twice. Data is presented as mean (± SD).

Process Parameters

Effect of enzyme concentration

Effect of enzyme concentration on feather degradation was checked using 150-1500U keratinase on 2.5g feather in 50ml volume, for a period of 12h and percent degradation was studied using dry weight method.

Effect of temperature and time

Feather degradation was studied as a function of time for the period of 2h-12h or till complete degradation at 37°C and 50°C using 2.5g from in 50ml volume and 1200U enzyme under standard conditions. Percent feather degradation, protein release and residual keratinase activity was determined after every 2h. Structural changes in feather were also analyzed by scanning electron microscopy (SEM). Feather were washed with 50mM phosphate buffer and dried at room temperature for scanning studies. Feather were coated with gold particle and observed using scanning electron microscopy (LEO 435VP SEM, Carl Zeiss NTS, GmbH, Germany) at Department of Anatomy, All India Institute of Medical Sciences, New Delhi, India, Department of Science and Technology supported service centre.

Scale-up

Keratinase degradation was further scaled up for degradation of 5, 10, 15 20 and 25g feather in a 2L flask at pH 8 and 50°C under optimized condition (5%w/v feather, 150rpm and 1200U enzyme/2.5g feather).

Pre-soaking and boiling of feather

Pre-soaking and boiling method was substituted for autoclaving. Twenty five gram feather was soaked in 250ml water for 2h and boiled for 10-20 min with intermittent mixing instead of autoclaving in standard protocol. It was cooled to room temperature and 12000U enzyme along with remaining 50% moisture i.e. 250ml pH 8 buffer was added and mixed properly. It was kept at 50°C and 150rpm till complete degradation of feather. Feather meal was dried at 80°C and was ground to form homogenous powder.

Amino acid and CHN analysis of feather meal

One g of feather meal powder was hydrolyzed with 10ml of 6N HCl

at 60°C for 12h, filtered. Filtrate was concentrated by speed vac and analyzed for amino acid profiling by HPLC (Agilent 1100 HP-HPLC) after derivatization with orthopthalaldehyde. Sample was run into the Aminex column using mobile phase (A (20mM sodium acetate + 0.018% triethylamine) and B (20% of 100mM sodium acetate + 40% methanol + 40% acetonitrile) with the flow rate of 0.5ml/ min at 40°C and detected at 338nm using VW detector. Amino acids were quantified using HPLC standards at Shanker Nethrayala, Chennai, India.

Carbon, nitrogen, sulfur and hydrogen contents of feather meal were analyzed using 1g feather meal powder at USIC facility, Delhi University, North campus, Delhi, India Using CHNS analyzer (Elementar, Vario El, Germany).

In-vitro digestibility of feather meal

For *in-vitro* digestibility, 1g feather meal was resuspended in 10ml, 2N HCl, 2mg/ml pepsin was added and mixture was incubated 37°C for 2h. Further, pH of the mixture was adjusted to pH 8 by adding 2M sodium bicarbonate and 2mg/ml trypsin was added and incubated at 37°C for 16h [11]. After pepsin and trypsin treatment peptide release was measured at 280nm. Percent digestibility was calculated by total protein released/ total protein of 1g feather.

Results and Discussion

Keratinase from *Bacillus licheniformis* ER-15 was produced in soy flour feather medium for 60h in a 2L flask as reported earlier [10]. Fermentation broth was kept at room temperature for 24h which allowed the settling of >90% bacterial biomass onto feather meal. Feather are made up hydrophobic amino acids which may have facilitated settling of microbial cells on the degraded feather specially the *Bacillus* sp. which is known to produce biosurfactant making the cell surface hydrophobic [1,12]. Sedimentation step can easily substitute centrifugation in downstream processing of fermentation broth of biosurfactant producing microbes.

The keratinase was concentrated with 85% ammonium sulphate saturation which resulted in > 80% enzyme recovery. Concentrated enzyme was stored at room temperature with shelf life of upto a year with almost no loss in activity (data not shown).

Standardization of feather degradation

Keratinase from *B. licheniformis* ER-15 exhibited maximum activity at pH 11 and 70°C [10]. Although, alkaline pH and high temperature would facilitate rapid feather degradation by reducing disulfide bonds [2] but is not often recommended for the direct use of feather meal in feed due to loss of some essential amino acids [3,13]. Therefore, feather degradation was studied at pH 8 and 50°C where present keratinase exhibited >60% activity [10].

Effect of enzyme concentration

Degradation of feather was observed to be a linear function of enzyme concentration with >600U (Figure 1). No visible degradation was observed upto 600U even after prolonged treatment. By increasing enzyme concentration from 600 to 900U, 60% feather degradation was achieved which increased to >90% with dissolution of shaft as concentration was increased to 1200-1500U/2.5g feather. Feather degradation using keratinase have been reported in presence of reducing agents like hypochlorite, dithiothretol, glutathione or in presence of live cells which provides reducing environment [1,14]. Majority of the feather meal production involves fermentation using keratinolytic microbes and subsequent fermentation broth was regarded as feather meal

[2,6,11,15]. During fermentation, feather degradation is supposed to be achieved by co-operative action of protease and cell redox [16-18]. In this context, present process is better than the existing ones since no fermentation is required for feather meal production.

Effect temperature on feather degradation

Feather degradation was studied at 37°C and 50°C and >90% degradation was observed after 24h at 37°C and 8h at 50°C. Further dissolution of shaft was observed only at 50°C. complete degradation at 50°C may be a result of faster breakdown of disulfide bonds at higher temperature which may have resulted in dissolution of shaft [1]. Further, the present enzyme has optimal activity at 70°C with >90% activity at 50°C and 67% at 37°C [10]. Thus, the faster degradation at 50°C may be result of both enzyme concentration and temperature.

Protein release and structural changes during feather degradation

Protein release and structural changes in feather was studied at 50°C (Figure 2). From figure 2, it can be observed that maximum protein was released in the first 2h with only 20% degradation (Figure 2A) and not much visible changes in feather (Figure 2B). First visual observation of feather degradation was made with shedding of barbules after 4h of incubation accompanied by additional 20% loss in weight. However, no substantial release of protein in supernatant was observed as was after first 2h. Similar trend in protein release was observed on further incubation upto 8h where complete degradation of feather was obtained. This suggests that, smaller peptides are released mostly during first 2h of degradation and later on most of the protein remains in the feather meal. This is the first report where keratinase alone could degrade feather completely within 8h. Hydrothermal hydrolysis of degradation also required longer time (16h) and high temperature (120°C) for feather degradation [1]. Further microbial degradation of feather generally requires more than 24h which may extend upto 5 week with the microbe used [2].

During feather degradation enzyme stability was also studied and 80% enzyme activity was recovered at 50°C after 8h which is higher than earlier report where half life was 5h at 50°C. This observation suggests that thermo-stability of present keratinase was enhanced in presence of substrate. This is in confirmation with commonly observed phenomenon that substrate protects enzyme against thermal destabilization [19].

Scale up

Feather degradation was scaled upto 5-25g feather in 2L flask in standardized condition and >90% degradation was achieved within 12h in >10g feather. Thus, complete feather degradation of feather in large volume was obtained in same condition and process was successfully scaled up.

Presoaking and boiling method

Since, for bulk degradation, 5% w/v feather involves large volume and would lead to cumbersome downstream process therefore pre-soaking and boiling was attempted with 25g feather. Feather was soaked in 250ml water for 2h and approximately 50% absorption of water. This pre-soaked feather was boiled for 10-20min and cooled till room temperature. Further 250ml enzyme (12000U) in pH 8 buffer was added to the boiled feather and >90% feather degradation was achieved at 50°C within 12h. This process formed a thick meal which was dried, ground and stored directly (Figure 3).

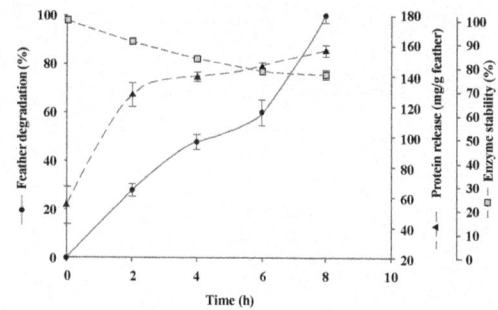

A.

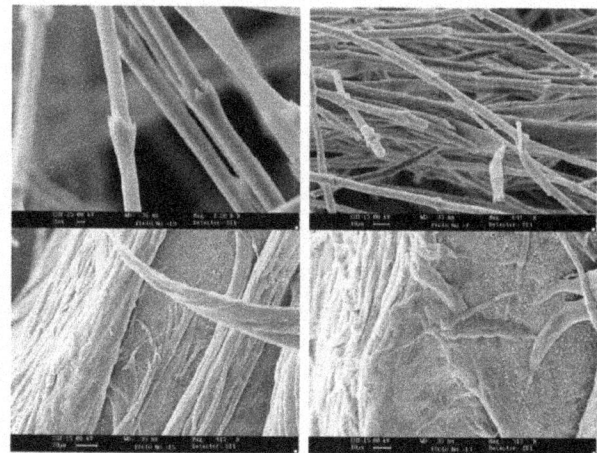

B.

Figure 2: Time kinetics (A) and structural changes (B) of feather during degradation.
A. Feather degradation was performed at pH 8, 50°C and 150rpm till complete degradation.
B. Structure changes of feather at 0h (a), 2h (b), 4h (c) and 6h (d) after enzyme addition.

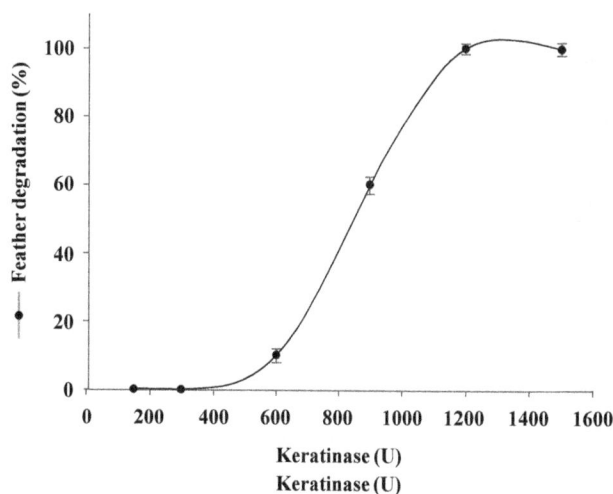

Figure 1: Feather degradation as a function of enzyme
Feather degradation was performed with 250mg feather in pH 8 buffer at 50°C, 150 rpm for 12h. Percent degradation was calculated by measuring residual feather, dried at 80°C for 12h.

Figure 3: Various step of conversion of feather to feather meal. A. 25g feather before pre-soaking and boiling, B. Feather after pre-soaking and boiling, C. Feather meal after 12h with 12,000U enzyme treatment, D. Dried feather meal.

Amino acid	mg/g of feather (present work)	Steam hydrolyzed mg/g feather (Eggum, 1970)	acid hydrolyzed mg/g feather (Eggum, 1970)	mg amino acid/g CP (present work)	mg amino acid/g CP (Grazziotin et al., 2006)
Aspartic acid	51	-	-	55.85	57.8
Glutamic acid	56.1	-	-	61.45	92.2
Serine	73.8	-	-	80.85	108.3
Histidine	65.9	7.2	6.3	72.15	9.3
Glycine	51.05	-	-	55.9	59.6
Threonine	50.9	4.84	4.87	55.75	36.6
Alanine	46.3	-	-	50.7	54.2
Arginine	32.85	2.08	2.3	35.95	84.3
Tyrosine	37.85	2.8	3.11	41.45	32.9
Valine	33.15	7.25	7.73	36.3	85.6
Methionine	31.6	.72	0.76	34.6	17.0
Phenyl alanine	69.95	4.61	4.85	76.6	54.2
Isoleucine	51.85	4.82	5.55	56.8	62.8
Leucine	17.15	8.25	8.27	18.75	66.9
Lysine	14.85	2.08	2.23	16.25	24.1

Table 1: Amino acid composition of feather meal and comparison with reported feather meals.

Quality of feather meal

Quality of feather meal was checked by CHN analysis as well as amino acids profiling. The feather meal contained 14%w/w nitrogen, 44%w/w carbon, 3.2%w/w sulfur and 1.4%w/w hydrogen which suggests that feather meal is a protein rich meal with 87% protein by weight. Amino acid composition of feather meal is presented in Table 1 and compared with steam cooked, acid hydrolyzed, culture supernatant hydrolyzed (CSH) feather meal [11,20]. Enzymatic hydrolysis of feather meal was observed to be rich in essential amino acids in comparison to steam cooked/ acid hydrolyzed feather meal. Amino acid content of present feather meal was comparable to CSH produced by keratinolytic bacterium *Vibrio* kr6 [11] except for few quantitative differences. The present feather meal had higher content of essential amino acids histidine, phenylalanine, methionine and threonine while CSH revealed high content of glutamate, serine, arginine and leucine. These differences may be due to the different processes used for feather degradation. Since the CSH was a result of crude supernatant after feather degradation during fermentation by *Vibrio* kr 6 and present process was a cell free enzymatic degradation.

In-vitro digestibility of feather meal was determined by pepsin fol-lowed by trypsin treatment. Feather meal was found to be digestible by pepsin and trypsin by releasing 670 mg protein/g feather meal after 18h of digestion. *In- vitro* digestibility results showed that digestibility of the present feather meal is 0.734 i.e. 73.4% which is better than commercial feather meal (0.578) or milled feather (0.096) and comparable to whole cell hydrolysate (WCH) (0.834) feather meal produced by fermentation [11]. This suggests that present feather meal can be used as feed for chickens, cattle and fish as reported earlier for feather meal produced by fermentation [21].

Conclusion

The present process of bioconversion of feather into feather meal is completely an enzymatic process. To the best of our knowledge, this is the only process where no additional redox has been provided. The process is not only simple and time saving but at the same time economically viable as it does not require any bioreactor for feather degradation. Thus, bulk feather can be easily recycled into feather meal using keratinase from *Bacillus licheniformis* ER-15 within 12h.

Acknowledgments

Authors thank to Delhi University for Dean Research grant (R&D/2010/1311) and DU-DST PURSE grant for financial assistance and Ekta Tiwary thanks Council of Scientific and Industrial Research (CSIR), New Delhi for Senior Research Fellowship grant (9145(1080)/2011-EMRI). Department of Anatomy, All India Institute of Medical Sciences, New Delhi, India is also acknowledged for providing SEM facility.

References

1. Onifade AA, Al-Sane NA, Al-Musallam AA, Al-Zarban S (1998) A review: Potentials for biotechnological applications of keratin-degrading microorganisms and their enzymes for nutritional improvement of feathers and other keratins as livestock feed resources. Bioresour Technol 66: 1–11.

2. Gupta R, Ramnani P (2006) Microbial keratinases and their prospective applications: an overview. Appl Microbiol Biotechnol 70: 21–33.

3. Brandelli A (2008) Bacterial keratinases: Useful enzymes for bioprocessing agroindustrial wastes and beyond. Food Bioprocess Technol 1: 105-116.

4. Vasileva-Tonkova E, Gousterova A, Neshev G (2009) Ecologically safe method for improved feather wastes biodegradation. Int Biodeterior Biodegrad 63: 1008–1012.

5. Coward-Kelly G, Agbogbo FK, Holtzapple MT (2006) Lime treatment of keratinous materials for the generation of highly digestible animal feed: 2. animal hair. Bioresour Technol. 97: 1344–1352.

6. Brandelli A, Daroit D J, Riffel A (2010) Biochemical features of microbial keratinases and their production and applications. Appl Microbiol Biotechnol 85: 1735-1750.

7. Papadopolous MC, El-Boushy AR, Roodbeen AE, Ketelaars EH (1986) Effects of processing time and moisture content on amino acids composition and nitrogen characteristics of feather meal. Animal Feed Sci Technol 14: 279-290.

8. Wang X, Parson CM (1997) Effect of processing systems on protein quality of feather meal and hog hair meals. Poultry Sci 76: 491-496.

9. Zaghloul TI, Embaby AM, Elmahdy AR (2011) Biodegradation of chicken feathers waste directed by Bacillus subtilis recombinant cells: Scaling up in a laboratory scale fermentor. Bioresour Technol 102: 2387-2393.

10. Tiwary E, Gupta R (2010) Medium optimization for a novel 58 kDa dimeric keratinase from Bacillus licheniformis ER-15: Biochemical characterization and application in feather degradation and dehairing of hides. Bioresour Technol 101: 6103–6110.

11. Grazziotin A, Pimentel FA, de Jong EV, Brandelli A (2006) Nutritional improvement of feather protein by treatment with microbial keratinase. Animal Feed Sci Technol 126: 135-144.

12. Czaczyk K, Bialas W, Myszka K (2008) Cell surface hydrophobicity of Bacillus spp. as a function of nutrient supply and lipopeptides biosynthesis and its role in adhesion. Pol J Microbiol 57: 313-319.

13. Hood CM, Healy MG (1994) Bioconversion of waste keratins: wool and feathers. Resources Conversion Recycling 11: 179-188.

14. Karthikeyan R, Balaji S, Sehgal PK (2007) Industrial applications of keratins—a review. J Sci Ind Res 66: 710–715.

15. Williams CM, Lee CG, Garlich JD, Shih JCH (1991) Evaluation of a bacterial feather fermentation product, feather lysate, as a feed protein. Poult Sci 70: 85-90.

16. Bockle B, Muller R (1997) Reduction of disulfide bonds by Streptomyces pactum during growth on chicken feathers. Appl Environ Microbiol 63: 790-792.

17. Yamamura S, Morita Y, Hasan Q, Yokoyama K, Tamiya E (2002) Keratin degradation: a cooperative action of two enzymes from Stenotrophomonas sp. Biochem Biophys Res Commun 294: 1138–1143.

18. Ramnani P, Gupta R (2007) Keratinases vis-à-vis conventional proteases and feather degradation. World J Microbiol Biotechnol 23: 1537–1540.

19. Sharpe DJ, Wong LJ (1990) Effect of substrates on the thermal stability of nuclear histone acetyltransferase. Biochimie 72: 323-326.

20. Eggum BO (1970) Evaluation of protein quality of feather meal under different treatments. Acta Agricultura Scandinavic 20: 230-234.

21. Bertsch A, Coello N (2005) A biotechnological process for treatment and recycling poultry feathers as a feed ingredients. Bioresour Technol 96: 1703-1708.

15

Total Microbial Populations in Air-Conditioned Spaces of a Scientific Museum: Precautions Related to Biodeterioration of Scientific Collections

Antonio Carlos Augusto da Costa*, Lucia Alves da Silva Lino and Ozana Hannesch

Museum of Astronomy and Related Sciences, Department of Documentation and Archives, 586 Rua General Bruce, S. Christopher, Rio de Janeiro, Brazil

Abstract

Air-conditioned areas of a museum were monitored for the presence of total microbes in the air. The results were evaluated based on a Brazilian resolution that regulates accepted contamination levels in air-conditioned spaces, as well as based on the parameters from the World Health Organization. The results indicated low levels of bacterial and fungal populations in four distinct spaces, with total counts smaller than 50CFU m^{-3}. These results, compared to the monitoring performed in the outside area of the museum indicated a very low internal to external count ratio, the highest one around 0.131, a value far beyond the acceptable limit of 1.5, as predicted by the Brazilian legislation. Even though those values clearly indicate low levels of contamination for human comfort, in the spaces monitored, the marked presence of fungi from the genera *Cladosporium*, *Aspergillus* and *Penicillium* deserve particularly attention due to their possible cellulolytic activity. The spaces are permanently controlled for their temperature and relative humidity levels, to be used as a permanent repository for scientific and historical documents in Brazil, and the presence of these potential cellulose-degrading microbes can markedly jeopardize the effective occupation of the areas due to their biodeterioration effects.

Keywords: *Cladosporium*; *Aspergillus*; *Penicillium*; Museum; Cellulolytic activity; CFU counting

Introduction

Microbiological monitoring of air-conditioned environments has become a common practice throughout the world, particularly in areas designed to keep important paper collections, such as museums and archives. It is known that microorganisms of various types are present in the environment, often associated with particulate matter resulting from the lack of preventive and corrective maintenance services of air-conditioners [1,2].

Pasanema et al. [3] investigated fungal growth and the maintenance of its viability in building materials under controlled humidity. The materials were subjected to various environmental conditions with varying absorption of water and relative humidities. After appropriate treatment, the authors observed the proliferation of fungi and actinomycetes, after two weeks of incubation. The results showed that when water was absorbed by capillary action, fungal growth was faster in early wood-based materials under 20% (w/v) humidity. Condensation under varying humidity and temperature was responsible for the the rapid growth of different fungal populations, particularly under high humidities. It is noteworthy that the fungal species were particularly tolerant to fluctuations in temperature and humidity, with very little effect viability.

Hyvarinena et al. [4] studied the diversity of fungi and actinomycetes in spaces designed with building materials of different natures, amenable to turn into microenvironments for their proliferation. In particular, cellulosic materials and ceramics, paints and plastics, in a clear state of decomposition, were evaluated for the presence of microbial populations. The authors found approximately 100 CFU g^{-1} (Colony Forming Units per gram of material), with the largest microbial populations associated with the presence of ligno-cellulosic and paper-based materials. The authors also observed that bacterial populations were lower than fungal populations, in contrast to the great fungal diversity, particularly of the genus *Penicillium*, as well as a great number of yeasts. In paper-based materials the main fungal genera found were *Cladosporium* and *Stachybotrys*; in glues and paints the most prevalent genus was *Acremonium* and the species *Aspergillus versicolor*. Accord-

ing to the authors, the main contribution of that research was to show the association between microbial growth and its occurrence in building materials of different nature. In particular, the authors highlighted a certain degree of specificity between the type of material and the predominant fungal genera.

Nielsena et al. [5] studied the influence of relative humidity and temperature on the growth and metabolism of selected fungal species in various types of building materials. The authors evaluated the microbial metabolic diversity, after incubation of several samples of building materials based on wood, starch and composite materials, at temperatures varying from 5 to 25oC, under 69 to 95% relative humidity, during seven months. The authors observed a high diversity of species present on the materials, with a prevalence of the genera *Penicillium*, *Aspergillus* and *Eurotium*, all of them mycotoxins producers.

Giannantonio et al. [6] observed the presence of incrustations on concrete surface, under controlled laboratory conditions, due to the direct action of the fungal genera *Alternaria*, *Cladosporium*, *Epicoccum*, *Fusarium*, *Mucor*, *Penicillium*, *Pestalotiopsis* and *Trichoderma* on the concrete.

Hoang et al. [7] evaluated the susceptibility of "green" building materials to biodeterioration by *Aspergillus niger*, an indoor reference fungus. The detection of spores and the presence of external compounds acting as nutrients contributed to the growth of *Aspergillus niger* on

*Corresponding author: Antonio Carlos Augusto da Costa ,State University of Rio de Janeiro,Institute of Chemistry, Department of Biochemical Process Technology,Rua São Francisco Xavier 524, Rio de Janeiro, Brazil, CEP: 20550-013

walls and ceilings gypsum. The authors found a strong correlation between the content of the mixture and organic materials by observing the time for coating of 50% of surface area by fungi. The results suggested that the presence of organic matter in a given material appears to be an important factor for the diagnosis of fungal susceptibility to a subsequent possible biodeterioration. Not only the materials are responsible for the spread of fungal spores and bacteria, but also the climatic conditions that regulate the environment, internally or externally.

Aira et al. [8] found some fungi in the architectural complex of the Cathedral of Santiago de Compostela in Spain, observing the presence of 35 different genera, mainly *Alternaria, Aspergillus, Cladosporium* and *Penicillium*. Interestingly, the authors did not find differences between populations inside and outside the Cathedral, with a maximum occurrence between spring and summer. The amount of fungi found was relatively small at various points of the central nave of the Cathedral, while in Corticela Chapel this number reached 6,500CFU m^{-3}. There was also a higher incidence of microorganisms around 13:00 h (around 400 CFU m^{-3}), where the flow of visitors reaches a peak.

Mesquita et al. [9] used advanced techniques of molecular biology to elucidate the fungal morphology and to evaluate the infection of historical documents. The researchers identified a wide diversity of fungi, on parchment, laid and wood pulp paper. Authors identified fourteen genera of fungi, the most frequent *Cladosporium, Penicillium* and *Aspergillus*, and less abundantly the presence of the genera *Alternaria, Botrytis, Chaetomium, Chromelosporium, Epicoccum, Phlebiopsys* and *Toxicocladosporium*.

Abe [10] found fungal contamination of materials stored in an art museum, which was monitored according to a biological index related to climatic parameters, giving an indication of the environmental capacity to maintain and proliferate fungal cells. To determine this index, fungal spores were encapsulated, followed by the observation of the germination of spores, and measurements of the extent of fungal hyphae. The authors identified a predominant occurrence of Aspergillus and Eurotium penicillioides. A number of other microbial populations is reported in the literature, specific to intrinsic characteristics of materials where populations grow, as well as with environmental factors that regulate proliferation.

All these studies suggest that the microorganisms that colonize building materials or are commonly found in the air are the same microbes that create microenvironments that colonize surfaces, including those consisting of lignocellulosic materials, such as documentary collections.

Thus, it is important to predict, for practical purposes, the effects of archives transfers between areas under different climate conditions. Given the wide distribution of microbial species in environments subject to temperature and relative humidity control, the objective of this study was to quantify fungi and bacteria in four selected rooms from the Museum of Astronomy and Related Sciences, for purposes of qualification of the area and also to evaluate the effect of transferring collections between rooms under distinct climate conditions. The final evaluation will be based on Brazilian Anvisa Resolution 176/2000, a resolution that presents reference standards for indoor air quality in climate-controlled spaces.

Materials and Method

Monitored areas

The Museum of Astronomy and Related Sciences (MAST) is locat-

ed in the Imperial Quarter of São Cristóvão, Rio de Janeiro City, and is responsible for the safety and custody of scientific archives from Brazilian researchers and institutions. Four selected rooms from the Coordination of Documentation and Archives (CDA) from the MAST, located in the basement of the building, in September 2010, as well as internal offices and external areas were selected for the monitoring of microbial growth, as represented in Figure 1.

Sample collection was performed every meter of each space, according to the Anvisa Resolution 176/2000. At each sampling point two open Petri dishes were placed, each one containing the appropriate culture medium, under the air conditioning system on. This resolution intends to reach the "Technical guidance on standards benchmarks indoor air quality in climate-controlled environments for public use, as regards the setting of maximum recommended values for biological, chemical and physical parameters of indoor air, the identification of pollution sources of biological, chemical and physical analytical methods and recommendations for control". Only microbiological parameters were investigated, since Resolution 176/2000 also recommends monitoring the chemical environment. The number of samples collected in each area is presented in Table 1.

Control samples (EXT) are those from environments where it is expected a high incidence of bacterial and fungal populations (outdoors, negative control). Samples from indoor environments (LAB, ESC, HIG, ENC, positive controls), are from areas where it is expected a low microbial contamination due to the characteristics of the area. All the samples were collected in duplicate.

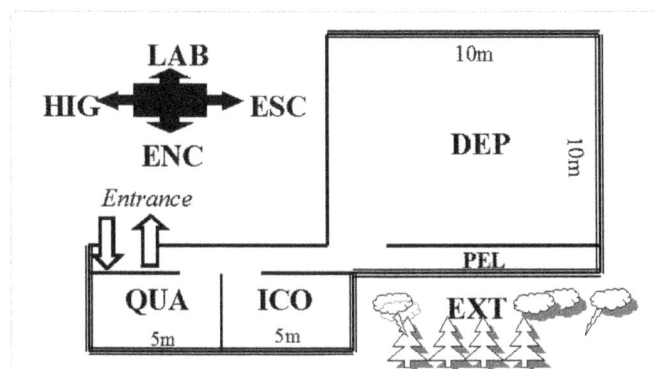

Figure 1: Schematic representation of the monitored areas of CDA/MAST, located in the basement of the building for the Conservation of Collections. Legend: Downstairs: QUA (Quarantine Room), ICO (Iconography Room), PEL (Film Storage Room); DEP (Repository of Collections Room). Upstairs: LAB (Research Laboratories); ESC (Offices); HIG (Hygienisation Laboratory); ENC (Book Binding Room). Outside the building: EXT (External Area).

Environment	Number of samples
QUA	20
ICO	20
PEL	20
DEP	40
LAB	05
HIG	05
ENC	05
ESC	05
EXT	10

Table 1: Total samples collected in the monitored areas of the CDA/MAST for microbiological monitoring.

Reference standards

The reference standards adopted according to Anvisa Resolution 176/2000 were: The maximum recommended value for microbiological contamination should be less than, or equal to, 750 CFU m^{-3} microbes, for an I/E ratio smaller than or equal to 1.5, where I is the amount of microbes in the indoor environment and E is the amount of microbes in the outdoor environment. When the I/E ratio is greater than or equal to 1.5, it is necessary to diagnose the sources for corrective actions.

Culture media for fungi and bacteria

The culture media used were PCA (Plate Count Agar, Merck, Darmstadt, Germany) for the quantification of total bacteria and PDA (Potato Dextrose Agar, Darmstadt, Germany) for quantification of total fungi. The media were dissolved in distilled water, autoclaved at 121°C for 20 minutes, distributed in sterile Petri dishes and solidified after cooling.

Sampling

Two Petri dishes were distributed to each square meter of each room monitored, as well as in the areas that represent positive and negative controls. For each point indicated one set of two Petri dishes were placed for the quantification of total heterotrophic bacteria and one set for the quantification of total fungi. The Petri dishes were distributed at the sampling points, opened at approximately 3m below the output of central air-conditioners, for two hours. Direct impact of the air over the medium contained in Petri dishes was responsible for the dissemination of the microbial cells. The samples were incubated (Incubator New Ethics, Model D-411, São Paulo, Brazil) at temperatures adjusted to grow bacteria (35°C) and fungi (25°C), for two and seven days, respectively. Colonies were counted using a colony counter coupled to a magnifying glass and digital record of the data (Phoenix, Model CP600, São Paulo, Brazil). For each space average values of quantification were reported.

Results and Discussion

Figure 2 presents the results of the quantification of fungi in the reported areas. It can be observed that, except for the external environment (EXT) the remaining areas showed a low average concentration of fungi, ranging from 4.4 to 16.8 CFU m-3 in the areas QUA, ICO, PEL and DEP. Values ranged from 0.4 to 2.2 CFU/m^3 in the administrative and laboratory areas (LAB, HIG, ENC and ESC). These values are extremely low compared with values obtained in the external environment (greater than 300CFU m^{-3}). This indicates that the values of the relationships I/E is low, as shown in Table 2.

* For purposes of calculation, the reference value for the external environment (EXT) was considered 300 UFC/m^3, as recommended by most microbiological reviews.

The results indicate that the results of the I/E ratio were extremely low, indicating that the selected areas are in perfect condition for the movement of staff, according to the Anvisa Resolution 176/2000. Moreover, in all cases, the concentration of microorganisms was smaller than 750 CFU m^{-3}. This indicates that both criteria from the Resolution were observed, corroborating that the rooms from the CDA/MAST are suitable with respectto indoor air quality and the setting of maximum recommended for biological contamination. From microscopical observation and colony characteristics it can be concluded that colonies from the genera *Cladosporium*, *Aspergillus* and *Penicillium* predominate. These observations are not conclusive and must be confirmed by DNA tests or complementary genotypic assessments. It is important to mention that those genera are the most typical fungi found in these environments.

Jaffal et al. [11] using a mechanical air samples for the enumation of fungal CFU in residential environments, found five groups of fungi, ainly members of the genus *Aspergillus*. The authors concluded that although their high numbers, the fungal cells presented little effect on human health.

As well, Shelton et al. [12] examined 12,026 fungal air samples (9,619 indoor samples and 2,407 outdoor samples) from 1,717 buildings located across the United States; these samples were collected during indoor air quality investigations performed from 1996 to 1998. The most common culturable airborne fungi, both indoors and outdoors and in all seasons and regions, were *Cladosporium*, *Penicillium*, nonsporulating fungi, and *Aspergillus*. *Stachybotrys chartarum* was identified in the indoor air in 6% of the buildings studied and in the outdoor air of 1% of the buildings studied.

Room	I/E Ratio*
QUA	0.056
ICO	0.015
PEL	0.040
DEP	0.021
LAB	0.001
HIG	0.005
ENC	0.006
ESC	0.007

Table 2: I/E ratio obtained for fungal contamination in selected areas of the CDA/MAST.

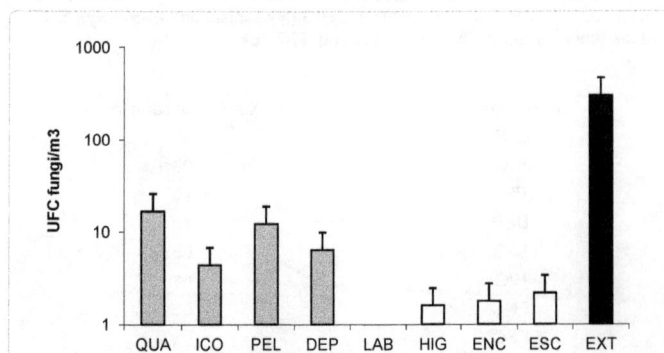

Figure 2: Quantification of fungal colonies in the selected areas of the CDA/MAST. Bars represent average standard deviations.

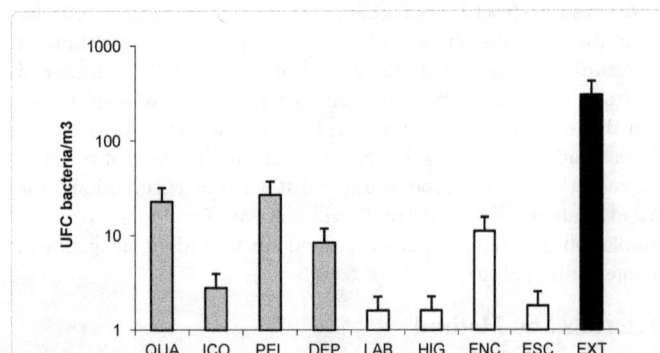

Figure 3: Quantification of bacterial colonies in the selected areas of the CDA/MAST. Bars represent average standard deviations.

Similarly, Figure 3 presents the results of the quantification of bacterial populations in the same environments. Quantification of bacterial populations presented similar values in comparison to fungal populations, with small differences between populations inside the same environment. Thus, the relationships I/E remained practically the same, also in accordance to Anvisa Resolution 176/2000. This means that the environments monitored presented bacterial populations in equivalent quantities as fungal populations, although fungi have a remarkable role in the degradation of cellulose through their cellulolytic enzymes.

Unlike fungi, phenotypic analysis of the grown bacterial populations did not identify the most representative groups. One conclusion that can be drawn about the presence of bacterial populations in the same proportion as observed for fungal populations, is due to the highest specificity of the PCA culture medium, and the small size of the colonies, typically observed in bacterial growth in this culture medium [13].

Anvisa Resolution 176/2000 does not specify between bacterial and fungal populations to characterize the microbial population quantified. Indeed, for purposes of environmental monitoring of air conditioned environments, there is increasing concern about fungal populations due to their natural ability to produce spores, which can be easily disseminated by ducts and pipes of air-conditioning systems. However, the ability to produce spores is not a unique feature of fungi, this property being also characteristic of some bacterial strains. Nevertheless, this property is much less common in bacterial than in fungal structures, supporting the ongoing concern about the monitoring of fungi, often neglecting the occurrence of bacteria. Because the Anvisa Resolution 176/200 does not state clear boundaries for that matter, we proceeded to quantify the overall fungal and bacterial populations, as depicted in Figure 4.

It is noteworthy from Figure 4, even considering the sum of bacterial and fungal populations, that all environments are in accordance to the Anvisa Resolution 176/2000, for microbial populations (bacteria and fungi) below the recommended lower limit of 750 CFU m^{-3}. Moreover, the I/E ratio is still far below the limit value of 1.5, confirming the suitability of the environment for the movement of staff. Table 3 presents the I/E ratio for the total microbial populations quantified.

These results show that the relationships I/E are in a range of extremely low values, confirming that all environments are suitable for the human comfort, as required by Anvisa Resolution. It is noteworthy that the largest I/E ratio was equal to 0.131, less than 10% of the value that would characterize the environment as an environment corrective actions are needed.

However, it must be stressed that these particular environments

Room	I/E Ratio*
QUA	0.131
ICO	0.024
PEL	0.129
DEP	0.049
LAB	0.006
HIG	0.010
ENC	0.042
ESC	0.013

Table 3: I/E ratio for total microbial contamination in selected areas of the CDA/MAST

were monitored for handling and storage of papers and archives of historical nature, subject to deterioration from chemical, physical or microbiological attack. The microbiological attack can result from excessive proliferation of bacterial and fungal species according to the malfunctioning of air conditioners and humidity controllers.

* For purposes of calculation, the reference value for the external environment (EXT) was considered 300 UFC/m^3, as recommended by most microbiological reviews.

Particulate matter suspended in the indoor and outdoor atmosphere are potential carriers of microorganisms, being responsible for microbiological contamination is observed in most artificially cooled environments. Brickus et al. [14] observed concentrations of particulate matter between 42.7 and 91.7 mg m^{-3} for indoor air and between 39.9 and 151.0 mg m-3 for the outside air in office environments on several floors of a commercial building. The authors use the same I/E ratio to evaluate the air quality. It can be observed from the data discussed above that, probably in many environments, the I/E ratio is above the value of 1.5. In addition to this high value of I/E ratio Brazilian CONAMA established a maximum value of 80 mg m^{-3} for particulate matter concentration in the air. One must consider that in 1997 the Anvisa Resolution 176/2000 not yet published, this being an adaptation of the current limits in the 90's. The following year, Dantas [15] conducted a thorough review of the weather conditions as agents of chemical, physical and biological integrity of air-conditioning systems through its transport system, with corrective solutions when identifying risk associated with each of the factors described.

Kulcsar Neto and Siqueira [16] suggested reference standards for microbiological quality in interiors, both in terms of quality and quantity. The authors state that it is not acceptable the presence of the following pathogenic species or toxigenic fungus: *Histoplasma capsulatum, Cryptococcus neoformans, Paracoccidioides brasiliensis, Aspergillus fumigatus, Aspergillus parasiticus, Aspergillus flavus, Stachybotrys atra* and *Fusarium moniliforme*.

When bacteria are considered, it is not allowed in indoor environments the presence of Legionella *pneumophila* (present in the condensation water air-conditioning and not associated with particulate matter) and Gram-negative pseudomonads. The authors consider the possibility of naturally occurring Gram-positive bacteria of the genera *Micrococcus, Streptococcus* and *Staphylococcus*. Although there is no consensus in the literature, some authors propose an acceptable concentration of fungal spores indoors up to 1,000 CFU m^{-3}, although there are obvious signs of impairment of health, with the occurrence of allergic alveolitis associated with fungal concentrations below this value. With regard to bacterial populations there are different reference standards, as this proposed by Toth [17]: for bacteria that inhabit the human tract without causing damage or disease, the author suggests the limit of 200 CFU m^{-3}.

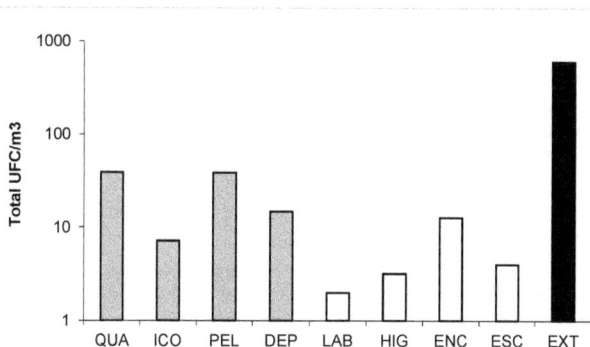

Figure 4: Quantification of total colonies in the selected areas of the CDA/MAST. Bars represent average standard deviations.

Hood [18] suggests a limit of 500 CFU/m³ for the presence of Gram-negative, attempting for the need for maintenance of air-conditioning systems.

There are also the limits recommended by the World Health Organization, with different criteria, all more restrictive than those previously mentioned, which will be only briefly mentioned: (a) to 50 CFU m-3 should proceed to a prompt investigation, if the fungus is unique, (b) Up to 150 CFU m⁻³ if the presence of more than one species, (c) Up to 500 CFU m⁻³ is the predominant species is *Cladosporium* sp. or other fungi commonly found in the environment. There are also other parameters that consider the occurrence of other species such as *Alternaria* sp., which will not be described here. With respect to bacteria the authors mention the lack of knowledge and monitoring for artificially air conditioned environments. Anyway, even considering the parameters suggested by WHO, the present results are in accordance to this rule.

Bortoletto [19] reports the presence of fungal contamination in a large library due problems related to control of the heating and air-conditioning system and air humidity, which gave rise to outbreaks of occurrence of fungi, whose removal was being done in a manner concomitant with the repair of the cooling system. After the outbreaks structural interventions, corrective and preventive actions have been suggested, which included fumigation of environments for inactivation of fungal structures, complete cleaning of the ducts by mechanical cleaning, followed by a new fumigation after washing. Books in the collection found in the monitored environment presented fungal cells of *Aspergillus*, *Penicillium*, *Cladosporium* and *Trichoderm*, all present in the environments selected from the library, with populations of about 800 fungal and bacterial populations CFU m⁻³ of 500 CFU m⁻³, respectively. Burge et al. [20] collected volumetric culture plate air samples on 14 occasions over the 18-month period immediately following a building occupancy. On each sampling occasion, the authors collected duplicate samples from three sites on three floors of the building, and an outdoor sample. Fungal concentrations indoors were consistently below those outdoors, and no sample clearly indicated fungal contamination in the building, although visible growth appeared in the ventilation system during the course of the study.

Regardless of the resolution considered, all the monitored environments in CDA/MAST presented acceptable microbiological concentrations, both for fungal, bacterial or total populations. This supports the hypothesis that human comfort is achieved, with respect to the presence of microorganisms in the air-conditioned areas monitored. Regarding the transfer of documentary collections of MAST or its maintenance in these areas, it is needed a complementary identification of bacterial and fungal species found, only to confirm if these species are those with a more cellulose degrading ability, despite their very low concentration. Both from the standpoint of human comfort, as well as the point of view of a possible cellulolytic activity on the documents a deeper investigation, involving their biochemical identification and genotypic confirmation that would render more conclusive results about archives biodeterioration.

Conclusions

In none of the room of CDA/MAST microbiological contamination was superior to that obtained in the external environment. This indicates that the ventilation system and air conditioning and/or the presence of humidity, do not favor the concentration of microbial species in these environments. Regarding the adequacy of the microbiological parameters according to Anvisa Resolution 176/2000 [21], all environ-

ments can be considered suitable for movement of staff, since the relationship between microbial populations in the internal and external areas is not higher than 1.5. Other Resolutions based on the same parameters, including the World Health Organization [22], indicated that there was no incidence of fungal or bacterial microbial populations in numbers that justify the need for decisions in relation to the operation of ventilation systems or relative humidity control. Environments being monitored are intended to keep scientific and historical archives, and this deserves a more thorough investigation regarding the identification of microbial species isolated in order to identify the possible occurrence of species that have the cellulolytic enzyme complex, which may jeopardize the long-term permanence of historical documents, by direct biochemical action or through a possible actions of secreted metabolites.

Acknowledgements

Authors thank the Ministry of Science and Technology and CNPq for support to conduct this work.

Ethical Standards

The authors declare that these experiments comply with the current laws of Brazil.

References

1. Abe K (2010) Assessment of the environmental conditions in a museum store-house by use of a fungal index. Intnl Biodet Biodegr 64: 32-40.

2. ABNT Associação Brasileira de Normas Técnicas, NBR 6401 Central installations of air-conditioners for human comfort – Basic project parameters (In Portuguese), 1980.

3. Aira MJ, Jato V, Stchigel AM, Rodrigues-Rajo FJ, Piontelli E (2007) Aeromycological study in the Cathedral of Santiago de Compostela (Spain). Intnl Biodet Biodegr 60: 231–237.

4. ANSI/ASHARAE, American Society of Heating, Refrigerating and Air-Conditioning Engineers, 62. Standard-Ventilation for Acceptable Indoor Air Quality, 1990.

5. ANVISA, Agência Nacional da Vigilância Sanitária, Resolução n.º 176 de 24/10/2000.

6. Bergey (2005) Bergey's Manual of Systematic Bacteriology 2nd Edition. Volume 2, The Proteobacteria, Garrity, GM Brenner, DJ Krieg, NR. e Staley, J. T. (eds.).

7. Bortoletto ME (1998) Fungal contaminations in indoor environments: the case of Manguinhos Library (In Portuguese). Rev Brasindoor 2:1-7.

8. Brickus LSR, Cardoso JN, Aquino Neto FR (1997) Determination of suspended particulate materials in indoor and outdoor áreas of a commercial building in Rio de Janeiro City (In Portuguese). Rev Brasindoor 2:12-22.

9. Burge HA, Pierson DL, Groves TO, Strawn KF, Mishra SK (1999) Dynamics of airborne fungal populations in a large office building. Curr Microbiol 40:10-16.

10. Dantas EHM (1998) Air conditioner: villain or ally? A critical review. Rev Brasindoor 2:4-14.

11. Giannantonio DJ, Kurth JC, Kurtis KE, Sobecky PA (2009) Effects of concrete properties and nutrients on fungal colonization and fouling. Intnl Biodet Biodegr 63:252–259.

12. Hoang CP, Kinney KA, Corsi RL, Szaniszlo PJ (2010) Resistance of green building materials to fungal growth. Intnl Biodet Biodegr 64:104-113.

13. Hood MA (1990) Gram-negative bacteria as aerosols. In: Biological contaminants in indoor environments. Edited by PR Morey, JC Feeley, JA Otten. American Society for Testing and Materials, Philadelphia, Pennsylvania, USA.

14. Hyvarinena A, Meklina T, Vepsalinena A, Nevalainen A (2002) Fungi and actinobacteria in moisture-damaged building materials-concentrations and diversity. Intnl Biodet Biodegr 49:27–37.

15. Kulcsar-Neto F, Siqueira LFG (1999) Reference standars for the analysis of indoor air microbiological quality. References for health quality in Brazil (In Portuguese). Rev Brasindoor 2:4-21.

16. Nielsena KF, Holma G, Uttrupa LP, Nielsen PA (2004) Mould growth on building materials under low water activities. Influence of humidity and temperature on fungal growth and secondary metabolism. Intnl Biodet Biodegr 54: 325–336.

17. Jaffal AA, Banat LM, El Mogheth AA, Nsanze H, Bener A, Ameen AS (1997) Residential indoor airborne microbial populations in the United Arab Emirates. Enviornment international 23: 529-533

18. Mesquita N, Portugal NMA, Videira S, Eccheverria SR, Bandeira AML, et al. (2009) Fungal diversity in ancient documents. A case study on the Archive of the University of Coimbra. Intnl Biodet Biodegr 63: 626-629.

19. Pasanena AL, Kasanena JP, Rautialla S, Ikaheimoa M, Rantamaki J, et al. (2000) Fungal growth and survival in building materials under fluctuating moisture and temperature conditions. Intnl Biodet Biodegr 46: 117-127.

20. Shelton BG, Kirkland KH, Flanders WD, Morris GK (2002) Appl Environ Microbiol 2002 68:1743-1753.

21. Toth C (1992) Microbials in the overall context of indoor air quality investigation. Proceedings of the First Annual IAQ. Conference and Exposition 255-259.

22. World Health Organization. Indoor air quality: biological contaminants; Copenhagen, Denmark, 1983 (European Series nº 31).

16

Enhancement of Enzymatic Process by Electric Potential Application

Nadia Abdi[1], Lila Bensaadallah[1], Nadjib Drouiche*[1,2], Hocine Grib[1], Hakim Lounici[1], Andre Pauss[3] and Nabil Mameri[3]

[1]Ecole Nationale Polytechnique d'Alger, B.P. 182-16200, El Harrach, Alger, Algeria
[2]University of Technology of Compiegne, Departement Genie chimique, B.P. 20.509, 60205 Compiègne cedex, France
[3]Centre de Recherche en Technologie des Semi-conducteurs pour l'energetique (CRTSE), 2, Bd Frantz Fanon BP140, Alger-7 Merveilles, 16000, Algeria

Abstract

The purpose of this study is to investigate a new bio-electrochemical technique based on the utilisation of electric potential to enhance the enzymatic reaction. The efficiency of bio-electrochemical reactor has been achieved by studying the production of reducing sugar by enzymatic hydrolysis of olive mill. The results indicate that the application of a continuous electric potential of about 50 mV allowed a significant increase of the saccharification efficiency by about 25% (compared to an enzymatic process without electric potential). For an electric potential higher than 60 mV, the saccharification efficiency decreased, suggesting that the enzyme, a biological substance, could be damaged at high electric potential. It has been shown that the kinetics of the bio-catalyzed reactions could be controlled by an applied electric potential.

Keywords: Biotechnology; Bioprocess; Biochemical engineering; Electrochemistry; Bio-electrochemical reactor; Enzymatic reactor efficiency

Introduction

The use of an electric field or an electric potential to improve the performance of the processes has been applied through several techniques such as ultrafiltration and electrosorption [1-3]. The electric field has also been used to increase the output capacity of the olive oil and fruits juice extraction [4]. The use of an electric field has not been studied yet in an enzymatic process. Recently, a new analysis technique using a field-effect bio-detector based on the application of a gating voltage to immobilized enzymes on the working electrode of the detector was presented [5,6]. Also, it was reported that the enzyme biospecificity was preserved in the presence of the applied field.

The main purpose of this work was the determination of the ability of the electric potential to increase the enzymatic process efficiency. The effect of the electric field on the enzymatic reactor efficiency was studied using the Trichoderma reesei enzyme and olive mill substrate as a biocatalytic system. This work was aimed to be achieved by adopting optimal conditions obtained from a previously batch mode experiment [7], i.e. pH 5, T=50°C and Enzyme to Substrate ratio of 0.1 g enzymes/g olive mill.

Material and Methods

The batch bio-electrochemical reactor equipped with electrodes is presented in Figure 1. The electric potential has been applied by integrating two electrodes within the enzymatic batch reactor by means of a generator (TACUSSEL-France). The vessel volume of the reactor is of about 0.5 dm3. The enzyme (Trichoderma reesei) and the solid substrate (olive mill) have been submitted to continuous electric potential.

The commercial enzyme Trichoderma reesei (Sigma, France) has been used during the experiments. Its activity is 11.2 units mg-1 (one unit will liberate 1 μmole of glucose from cellulose in 1 hour at pH 5, T=37°C, for an incubation time of 2 hours).

Enzyme solutions have been prepared with an acetic acid/sodium acetate buffer solution (0.05 mol/L, pH 5).

The substrate is solid olive mill residue (SOMR), collected from an olive oil plant (Tadmait Kabylia region: Algeria) and transported to the laboratory at T=4°C.

Although the olive mill residue was already chopped in the oil manufacturing process, it was submitted to a fine size reduction step (granular size of 460 μm) and finally dried at T=70°C. For all experiments, the olive mill residue was finally pre-treated with NaOH to remove the most of the lignin content, as reported in a previous investigation [7].

The main characteristics of this lignocellulosic material are given in Table 1. Abdi et al. [7] detailed the analytical methods [7]. It is important to note that the chemical composition of the crude SOMR indicates that carbohydrates (cellulose and hemicellulose) are the main components confirming that this material would be an interesting substrate.

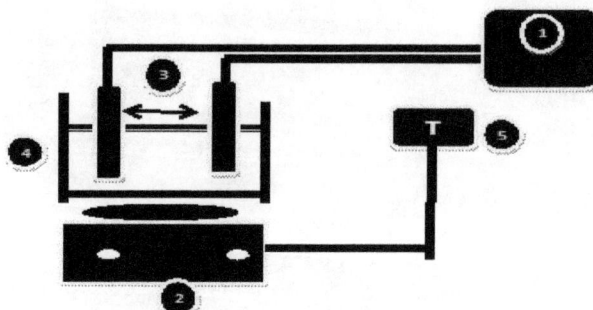

Figure 1: Batch bio-electrochemical reactor.

*Corresponding author: Dr. Nadjib Drouiche, University of Technology of Compiegne, Departement Genie chimique, B.P. 20.509, 60205 Compiègne cedex, France, Tel: 213 21 279880 extn 192

Parameter	Crude SOMR (%)	Treated SOMR (%)
Dried matter	88.23	85.12
Organic matter	92.11	96.11
Fat matter	4.02	1.88
Total ashes	4.55	10.45
Total nitrogen matter	7.33	6.26
Crude cellulose	42.54	40.30
Hemicellulose	19.43	7.23
Lignin	21.53	0.45
Potassium	0.11	0.19
Sodium	0.20	10.16
Calcium	0.39	--

[a]Concentrations expressed in percentage of dry matter

Table 1: Chemical composition of the solid olive mill residue (SOMR) before and after pre-treatment with NaOH[a].

Results and Interpretation

Feasibility of the bio-electrochemical reactor

The feasibility of this process has been examined first, by operating a fixed electric potential of 100 mV. To evaluate the electric potential influence on the enzymatic reactor performance, the reducing sugars concentrations were determined: Pre-treated SOMR without electric potential, pre-treated SOMR with E=100 mV and a no pre-treated SOMR with E=100 mV respectively.

The results are presented in Figures 2a and 2b. It was observed that the SOMR must be pre-treated by NaOH to be efficiently hydrolysed by the T. reesei's enzyme. Indeed, after the first hour of hydrolyse, it was noticed that the pre-treated SOMR (with or without application of electric field) revealed better results than the untreated SOMR. After 24 hours the saccharification yield reached about 35% for untreated SOMR, while 70 and 80% were respectively obtained for pre-treated SOMR without and with electric potential application. With a fixed electric potential of 100 mV a slight increase of reducing sugar (compared to an enzymatic process without electric potential of the production) by enzymatic hydrolysis of olive mill, was obtained. Indeed, during the first five hours, the electric potential did not provide significant benefit for the saccharification of SOMR, suggesting that a polarization of the latter onto the electrode must be obtained before an improvement of the enzyme efficiency.

Influence of the electric potential on the efficiency of the bio-electrochemical reactor

The influence of the electric field on the enzymatic saccharification of the olive mill was realised through batch process, under optimal conditions obtained from an earlier study [7], with applications of electric potential varying from 10 to 200 mV

The reducing sugars concentration, produced according to the time, is presented in Figures 3a and 3b. The results obtained for 0.04 V and 0.05 V are practically in the same order and close to the best saccharification efficiency value. These results indicate that the enzyme was most improved at electric potentials ranging from 0.04 to 0.05 V.

The lower results obtained by the enzymatic process for potentials higher than 60 mV could be explained by the fact that the enzyme might be deposited under strong electric field on the electrodes and then destroying the enzyme activity. Similar results were previously reported [8] for phenol degrading bacteria in the presence of high electric current. A bioelectrokinetic test suggested that bacterial inactivation might occur by interaction with the surfaces of the electrodes, resulting in cell wall or membrane degradation through oxidation or reduction. The cell viability may be also affected, at high applied potential or induced field, by irreversible permeabilization of the cell membrane which results in direct oxidation or reduction of the cellular constituents [8,9].

In contrast, a lower electric potential (less than 60 mV) preserved the enzyme's activity and its biospecificity. Furthermore, the results obtained indicate that the application of an optimal electric potential close to 50 mV allows the increase of the saccharification efficiency by a significant percentage of about 25% (as compared to results obtained without application of an electric field [7]). The maximal efficiency values obtained for each electric potential after 24 hours of hydrolyses are plotted in Figure 4. The difference between enzymatic process without and with an application of electric potential of about 50 mV was relatively very high and then allowed to deduce the efficiency of the electric potential to improve the enzymatic reactor performances.

These results may be explained by the fact that during the catalytic cycle, enzyme alternates between two conformational states, E1 and E2. The use of an electric field allows driving oscillation or fluctuation of enzyme conformation between the E1 and the E2 states. Tsong TY [10] reported that the electric field-induced conformational oscillation or fluctuation leads to uphill pumping of the cation, K+ into and Na+ out of a cell, by the enzyme without consumption of ATP. Biochemical specificity of the catalysis was also preserved in their study and data indicated that Na, K-ATPase can harvest energy from the applied electric field to perform chemical work.

In the present investigation, it was demonstrated that the kinetics of the bio-catalyzed reactions can be controlled by an applied potential.

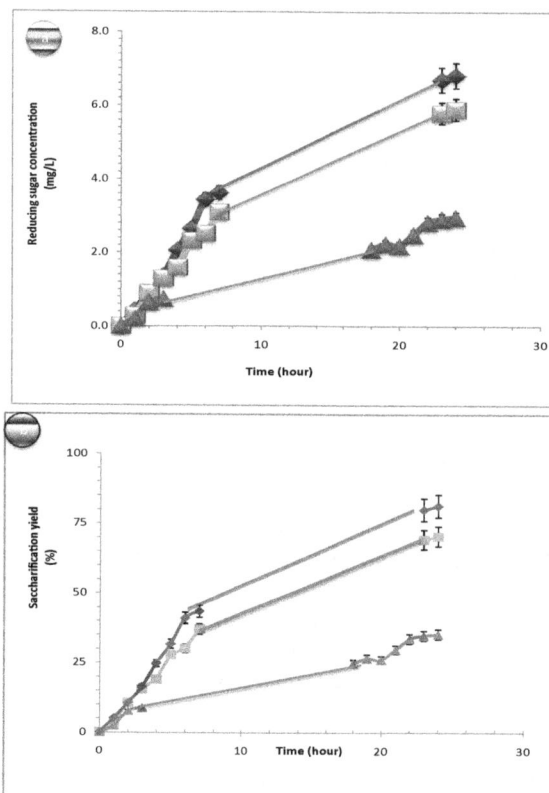

Figure 2: Graph for (a) reducing sugar concentration vs. time. (b) Saccharification yield vs. time.

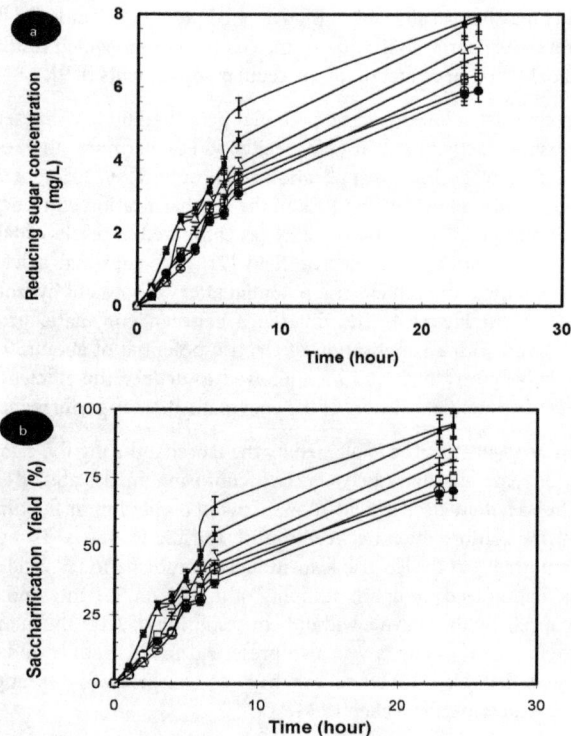

Figure 3: Graph for (a) reducing sugar concentration vs. time. (b) Saccharification yield vs. time.

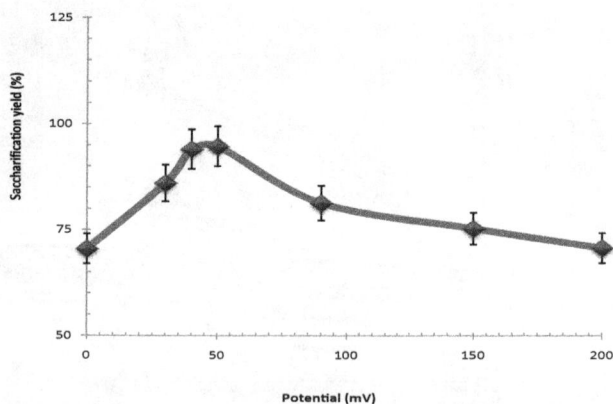

Figure 4: Graph for saccharification yield vs. potential.

An optimal value was determined indicating that oscillation or fluctuation of enzyme conformation between the E1 and the E2 states need a precise energy necessity given by applied potential close to 50 mV. These interesting results obtained after enzymatic experiments confirm the notion of the specific range of the field strength (a so called window) which may be utilized for bacterial manipulation [11].

Conclusion

The ability of the electric potential to increase enzymatic reactor performance was demonstrated in the current work. The kinetics of the bio-catalyzed reactions could be controlled by an applied potential which was an accomplishment in this study. Furthermore, the enzyme's

biospecificity was preserved in the presence of electric potential less than 60 mV.

In addition, the effect of the electric potential on the enzymatic activity was highlighted. An improvement of about 25% the olive mill saccharification performance was achieved under an electric potential close to 50 mV which confirms the notion of the specific range of the field strength. The relationship between the increase of the enzyme performance and the changes of the enzyme cell surface and shape, determined by scanning electron microscope analysis, could be stated to understand the bio-catalyzed reactions under electric potential carrying out additional studies.

References

1. Parmar NR, Majumder SK (2013) Microbubble generation and microbubble-aided transport process intensification—A state-of-the-art report. Chem Eng Process 64: 79-97.

2. Cheikh H, Grib N, Drouiche N, Abdi H, Lounici N, et al. Water denitrification by a hybrid process combining a new bioreactor and conventional electrodialysis. Chem Eng Process 63: 1-6.

3. Lounici H, Addour L, Belhocine D, Elmidaoui A, Bariou B, et al. (2001) Novel technique to regenerate activated alumina bed saturated by fluoride ions. Chem Eng J 81: 153-160.

4. Vorobiev E, Lebovka N, Jemai A, Bouzrara H, Bazhal B (2005) Pulsed Electric Field Assisted Extraction of Juice from Food Plants. In: Barbosa-Canovas GV, Tapia MS, Cano MP (eds.) Novel Food Processing Technologies. Marcel Dekker/CRC Press, Boca Raton, FL, USA, 105-130.

5. Choi Y, Yau ST (2009) Field-Effect Enzymatic Amplifying Detector with Pico-Molar Detection Limit. Anal Chem 81: 7123-7126.

6. Choi Y, Yau ST (2011) Ultrasensitive biosensing on the zepto-molar level. Biosens Bioelectron 26: 3386-3390.

7. Abdi N, Halet F, Belhocine D, Lounici H, Grib H, et al. (2000) Enzymatic treatment of solid residue of olive mill in a batch reactor. Biochem Eng J 6: 177-183.

8. Luo Q, Wang H, Zhang X, Qian Y (2005) Effect of direct electric current on the cell surface properties of phenol-degrading bacteria. Appl Environ Microbiol 71: 423-427.

9. Dreesa KP, Abbaszadegan M, Maiera RM (2003) Comparative electrochemical inactivation of bacteria and bacteriophage. Wat Res 37: 2291-2300.

10. Tsong TY (2002) Na, K-ATPase as Brownian Motor: Electric Field–induced conformational fluctuation leads to uphill pumping of cation in the absence of ATP. J Biol Phys 28: 309-325.

11. Alshawabkeh AN, Maillacheruvu K (2002) Electrochemical and biogeochemical interactions under DC electric fields. Physicochemical groundwater remediation 73-90.

17

Aqueous Two-Phase Extraction Advances for Bioseparation

Arafat M Goja[1,2], Hong Yang[1,3,4,5]*, Min Cui[6,7], Charles Li[8]

[1]College of Food Science and Technology, Huazhong Agricultural University, Wuhan, Hubei 430070, China
[2]Department of Food Science and Technology, Faculty of Agriculture and Natural Resources, Bakht Alruda University, ED Dueim, 1311, Sudan
[3]Key Laboratory of Environment Correlative Dietology, Huazhong Agricultural University, Ministry of Education, Wuhan, Hubei 430070, China
[4]National R&D Branch Center for Conventional Freshwater Fish Processing (Wuhan), Wuhan, Hubei 430070, China
[5]Aquatic Product Engineering and Technology Research Center of Hubei Province, Wuhan, Hubei 430070, China
[6]State Key Laboratory of Agricultural Microbiology, Huazhong Agricultural University, Wuhan, Hubei 430070, China
[7]Laboratory of Animal Virology, College of Veterinary Medicine, Huazhong Agricultural University, Wuhan, Hubei 430070, China
[8]College of Food Science, Fujian Agricultural and Forestry University, Fuzhou, China

Abstract

Aqueous two-phase extraction (ATPE), unique liquid-liquid extraction, involves a transfer of solute from one aqueous phase to another. ATPE includes polymer–polymer type and polymer–salt type systems for the recovery of proteins. The protein must be recovered in a highly purified form in order to improve its quality, decrease energy consumption, reduce waste and minimize costs. To acquire the high value and achieve good control over processes, the reliable, multi-component products are required especially those with the ability to investigate complex processing conditions. The current reviewing paper discusses the most recent progresses for the recovery of biomolecules by using the ATPE, covering the mechanism, which controls the phase formation and the behavior of solute partitioning in aqueous two-phase systems (ATPS) processes. The review discusses also the increasing application for the recovery of high-value bioproducts, the recent development of alternative low cost ATPS and disadvantages attributed to ATPS.

Keywords: Aqueous Two-Phase Extraction; Protein; Bioseparation

Introduction

Aqueous two-phase extraction (ATPE) system is composed of either two different immiscible mixing polymers, or one polymer with salt, which are water-soluble in a certain concentration [1,2]. It has been well known as a useful technique for separation and purification of biomolecules, such as proteins [3-14] and antibodies [15-17]. Partitioning of biomolecules in ATPE systems is affected by many factors, including molecular weight/size of polymer and concentration of polymer. Also, the ionic strength of salt and the addition of salts, such as NaCl, improve the hydrophobic resolution of the system. Furthermore, the partitioning of biomolecules can also be influenced by the degree of pH and affinity of the macromolecule for the phase-forming polymer [1,18-21].

The conventional liquid–liquid extraction using organic-aqueous phase systems was previously established. However, due to the poor solubility and possible denaturation of the protein in organic solvents, the technique leads to limit their application in partitioning of many bimolecular products [22]. Now, the application of liquid–liquid extraction based on aqueous two-phase systems (ATPS) has been increased. The recoveries of high-value biomolecules were achieved from various plants using different applications, such as papain from Papaya fruit [23], α- and β-amylases from Zea mays malt [9], glutenin flour from special wheat [24] and recombinant protein from alfalfa [25]. In addition, the high-value bioproducts also are obtained from different fruits, such as bromelain (EC 3.4.22.33) from pineapple [26], serine protease from Mango [12], invertase from tomato [11] and papain from Papaya fruit [23]. In recent years, the high-quality biomolecules have been recovered from various sources, such as theanine from waste liquid of tea [27] and flavonoids from pigeon pea roots [28]. There are also other bioproducts, such as recombinant human serum albumin from Pichia pastoris broths [29], luciferase from fireflies [30] and immunoglobulin G [17,31].

The ATPE has some advantages in the downstream processing of biomolecules, for instance, the system characterized with high-water content (70–80%, w/w) and low interfacial tension between conjugated phases. This, in fact, provides a secure separation and purification technique for biomolecules [1,32,33]. In addition, it decreases energy consumption, reduces waste and minimizes costs due to a few steps of unit operation, which require low-energy input and easy to scale up. The ATPE also increases biomolecules recovered in a highly purified form. Moreover, it can be used in combination of other separation methods such as liquid chromatography [1,34], packed column [17], magnetic particle adsorption [35].

The ATPE in downstream processing of biomolecules in the bench-scale prototype has been successful with potentially commercial application. However, the scale-up of the ATPS of some biological products were not achieved [15]. Most of those methods have some limitations such as low capacity, several steps and fewer chemicals and proteolytic stability, which may lead to the contamination of the final product [36,16]. Moreover, there are maybe some difficulties during the large scale-up of the ATPS in the industry, especially polyethylene glycol (PEG)/salt system causing corrosion of equipment and precipitation of target product. Thus, the development of alternative methods is very important to achieve the desired biomolecules in high quality and purity. The objective of this paper reviews recent progresses for the recovery of biological products by using the ATPE and the factors affecting partitioning in ATPS processes. The paper also reviews

*Corresponding author: Hong Yang, College of Food Science and Technology, Huazhong Agricultural University, Wuhan, Hubei 430070, China

the modern development of the alternative low cost ATPS, and some disadvantages associated with the ATPS.

Separation and Purification of Bioproducts by the Atpe System

The ATPS is considered a powerful and versatile technique, having low-cost and good efficiency downstream process, A high-water content (>70 % (w/w) water) generates low interfacial tension, non-flammable, slightly toxic and safe to the environment. This system has high selectivity and recovery yield of biomolecules [37,38]. It has been widely used in the field of biotechnology for separation and purification of various biological products, such as proteins, amino acids, enzymes, cells, antibodies and other bioproducts [1,7,39,40]. Also the ATPS method has received a special application in the area of non-biotechnology, including the recovery of glycosaminoglycans from tannery wastewater [41], crocins from saffron stigmas [42], papaverin from pericarpium papaveris [43].

Protein is one of the most important bio-molecules in the living organism, which is responsible for many reactions and functionalities, such as metabolism, bioprocess, signal transduction, cellular and extracellular structures [44]. A protein in a purified form would be very useful in the applications (e.g. food, chemical and pharmaceuticals). The challenging problem is that the downstream processing mostly accounts for 50 – 80% of the total production costs of proteins [45]. Conventional methods for separation and purification of proteins usually are expensive, time-consuming and difficult to scale up [44,46]. The good news is that the ATPE, an achievable alternative method, has been recognized as an economical and effective technique for recovery and purification of proteins with a variety of advantages, such as simple, fast, low-cost and easily scaled-up. These make it possible strategy for purification of a desired protein in large-scale [35,47]. Partitioning of proteins in the ATPS depends on many factors, namely, hydrophobicity, molecular size, weight and conformation, net electrical charge and environmental conditions [48-50]. A protein purified by this method will be considered to be very important bimolecular product for the bioprocess as it can be used in commercial scale at low cost with relative reliability and accessibility. Optimizing conditions of some selected examples of bioproducts (proteins, enzymes and other products) are summarized in Tables 1-3, respectively.

Mechanism Controlling Phase Formation and the Behavior Solute Partitioning in the Atps Processes

Partitioning of biomolecules in the ATPS is decided by main electrostatic, hydrophobic and steric hindrance interactions that are very important to the ATPS composition. The ATPS made up by the polymer and the nonionic surfactant results in hydrophobic interaction, [8]. Although electrostatic interactions and salting-out effects during protein extraction in ionic liquid-based aqueous two-phase extraction are important for the transfer of the proteins, the thermodynamics of hydrophobic interactions plays the most important role as a main driving force [44].

Generally, the partitioning of biological products is a result of Van der Waals and ionic interactions of biomolecules with the surrounding phase [37]. However, it is still not well understood the responsible mechanism for the partitioning of biomolecules in the ATPS, which is very important for developing the reliable technique for the industrial application. Therefore, the factors influencing the partitioning of biomolecules are a useful way to study the behavior of solute partitioning in the ATPS process.

Factors Influencing Partitioning of Biomolecules in the Atps

Impact of polyethylene glycol (PEG) characteristics

The PEG characteristics, including weight, size and concentration, are very important factors in the properties of the phase-forming system [51]. The influences of these factors on the partitioning of biomolecules have been reported previously [11,33,41,52].

Molecular weight and size: The partitioning of biomolecules depends on the molecular weight of polymers and the other components constituting the phase. Molecular weight has a strong effect on the partitioning behavior of biomolecules [11,12,53]. Higher molecular weight of PEG has less coefficient factor and then lower polymer concentration needed for high separation [38]. The low molecular weight of PEG has a hydrophilic end group with shorter polymer chains that reduces the hydrophobicity [41], while better partitioning can be achieved due to the low interfacial tension of low molecular weight. An increase in the PEG molecular mass reduces free volume by increasing the chain length of the PEG polymer [11,25,30], resulting in partitioning of the biomolecules to the bottom phase [4]. The increase in polymer weight causes the reduction of free volume of the top phase, so the partition of biomolecules in the salt-rich bottom phase decreases the partitioning coefficient [54].

Another study indicated the influence of different molecular weights of PEG (4,000, 6,000, and 8,000) on partitioning of myoglobin and

Bioproduct	Source	System and Composition (%w/w)	K_p	Yield (%)	TLL (%w/w)	pH	Reference
Protein	Cheese whey	PEG6000/potassium phosphate (11.7/10%)	0.9	83.4	23.9	7.0	[67]
Glutenin	Wheat flour	PEG1500/Li$_2$SO$_4$ (14.0/13.23%)	5.90	84.9	35.51	-	[24]
Recombinant protein	Alfalfa	PEG8000/phosphate (16.1/10.0%)	0.1	88	35.7	7.0	[25]
(α-la), (β-lg) and (Gmp)	Milk whey	PEG2000/phosphate (13.0/13.0%)	0.48, 0.01, 0.92	81.1, 97.3, 97.8	-	6.7	[6]
OVA	Chicken egg	PEG4000/poly(acrylic acid) (PAA) + 1 M NaCl	5.5	87.4	54.7	8.0	[37]
Protein	Zea mays malt	PEG 6000/CaCl$_2$	4.2	-	-	7.0	[39]
Bromelain	Pineapple Peel	PEG3000/MgSO$_4$ (15/ 20%)	2.93	108.45	-	9.0	[71]
(α-AI)	Wheat flour	PEG2000/(FBP) trisodiumsalt (11.7/19%)	-	79	-	7.0	[69]
Protein	Corn	PEG1450/Na$_2$SO$_4$-8.5% NaCl	-	93	-	7.0	[5]

Kp = partition coefficient of protein, TLL= Tie line length, (-) = the value was not given, α-la= α-lactalbumin, β-lg= β-lactoglobulin, Gmp= glycomacropeptide, α-AI= α-amylase inhibitor, FBP= fructose-1,6-bisphosphate, OVA= Ovalbumin

Table 1: Optimizing partitioning of proteins in the various ATPS recovered from foods.

Bioproduct	Source	System and Composition (%w/w)	Kp	Yield%	TLL%	pH	Reference
Serine Protease	Mango Peel	PEG8000/ phosphate (-/ 4.5%)	84.2	97.3	17.2	7.5	[12]
Inverts	Tomato	PEG 3000/ Na$_2$SO$_4$ (15/12%) +5% KCl	1.1	90	-	4.5	[11]
Serine Protease	Kesinai (*Streblus asper*)	PEG6000/rich- MgSO$_4$ (16/ 15%)	-	96.7	-	7.0	[59]
Papain	Papaya	PEG6000/ (NH$_4$)$_2$SO$_4$ (8/ 15%)	-	89.9	-	5.0	[23]
Phytase	*Aspergillus niger*	PEG6000+8000/ Citrate (10.5/ 20.5%)	0.96	96.0	-	5.6	[13]
Protease	Tuna (*Thunnus alalunga*)	PEG2000/MgSO$_4$(15/15%)	0.86	89.1	-	-	Nalinanon et al.
Soybean peroxidase	Soybean	PEG4000–IDA–Cu^{2+}/Na$_2$SO$_4$ (4/ 10%)	0.05	64	-	-	[72]
G6PDH	Sigma (USA)	PEG)/phosphate buffer(17.5/ 15%)	351	97.7	-	7.5	[55]
β-gala.; β –gluc.	Barley	PEG1500/ (NH$_4$)$_2$SO$_4$ (14/ 13%)	2.7; 2.8	98.26; 92.58	19.65	6.9; 6.5	[53]
Lipase (E.C. 3.1.1.3)	*Burkholderia pseudomallei*	2-propanol/phosphate (ATPS) (16/ 4.5%)	13.5	99	-	9.0	[40]
PPL	Sigma (USA)	PEG1500/potassium phosphate (17/13%)	12.7	94.7	-	7.0	[14]
PheDH	*Bacillus sphaericus*	PEG-6000/(NH$_4$)$_2$SO$_4$ (8.5/17.5%)	0.027	94.42	39.89	8.0	[56]
Invertase	Baker's yeast	PEG3000/MgSO$_4$(15/123%)+5%MnCl$_2$	-	98	-	5.5	[54]
Plant-esterase	Wheat flour	PEG1000/NaH$_2$PO$_4$(27.0/13.0%) and PEG1000/NaH$_2$PO$_4$/(NH$_4$)$_2$SO$_4$ (27.0/13.0/6.0%)	-	83.16	-	5.0	[62]

Kp = partition coefficient of protein, TLL= Tie line length, IDA= Iminodiacetic acid, G6PDH= Glucose-6-phosphate dehydrogenase, pk= purification factors, β-gala =β-galactosidase, β –glu =β –glucosidase, PheDH= phenylalanine dehydrogenase, PPL= porcine pancreatic lipase, (-) = the value was not given

Table 2: Optimizing partitioning of different enzymes recovered in the various ATPS.

Bioproduct	Source	System and composition (%w/w)	Kp	Yield (%)	TLL (%)	pH	Reference
IgG	Chinese Hamster Ovary (CHO)	PEG3350/phosphate-rich phase (cont, ATPE)	-	85	-	6.0	[98]
Lectin	*Canavalia grandiflora Benth*	PEG400/sodium citrate (20/20%)	8.67	104	-	6.0	Porto et al.
Lutein	Green microalga (*Chlorella protothecoides*)	PEG 8000/phosphate (22.9/10.3%)	-	81.0	49.4	7.0	Cisneros et al.
Luciferase	Fireflies (*Photinus pyralis*)	PEG1500 rich (NH$_4$)$_2$SO$_4$(4/20.5%)	-	13.69 fold in pk	-	-	[30]
Crocins	Saffron stigmas (*Crocus sativus*)	Ethanol/ potassium phosphate (19.8/16.5%) +0.1 M NaCl	-	>75	25	-	[42]
Glycosaminoglycans	Tannery wastewater	PEG4000/(PAA)	-	91.50	54.7	8.0	[41]

Kp = partition coefficient of protein, TLL= Tie line length, (-) = the value was not given, IgG=Human immunoglobulin G

Table 3: Optimizing partitioning of some selected products in the various ATPS recovered from different sources.

ovalbumin, the partitioning of both proteins is higher if the molecular mass of PEG is lower [37]. In PEG (4,000)–phenylacetic acid (PAA) system, the percentage yields of extracted myoglobin at 20oC and pH 8.0 in 1 M NaCl increases from 75.2% to 95.2% with the increase of tie line length (TLL). This is due to the increase of hydrophobicity and partitioning coefficient of the ATPS. It also affects the partitioning of proteins present in the phase system, whereas in the case of ovalbumin, the increase is from 67% to 87.4%. Different molecular weights of PEG (1,000, 2,000, 3,000, 4,000, 6,000 and 8,000) have been used to purify invertase enzyme from tomato, while the partitioning of invertase in PEG–Na2SO4 system is strongly dependents on the molecular weight of the PEG [11]. Nearly all invertases partitioned into the top phase with PEG-3,000, while most contaminating proteins were partitioned into the lower phase in the ATPS containing the other PEGs (1,000, 2,000, 4,000, 6,000 and 8,000). A wide range of molecular weight PEG (400, 1000, 1,500, 4,000, 6,000 and 8,000) was screened for differential partitioning of α-galactosidase and β-glucosidase from the barley [53]. For low molecular weight PEG, both the enzymes partitioned to top phase whereas for high molecular weight PEG, both the enzymes partitioned to bottom phase. PEG 1,500 had better partitioning of both of enzymes. The highest partitioning (97.22%) of luciferase enzyme from fireflies (*Photinus pyralis*) was obtained in 1,500PEG/(NH4)$_2$SO$_4$ system [30]. In the smaller molecular mass PEG (300, 400 and 600), there was a tendency for glucose-6-phosphate dehydrogenase (G6PDH) to stay in the top phase (PEG phase) [55]. In the higher value of PEG (1,000 and 1,500), the larger amount of G6PDH stayed in the bottom phase. However, the low molecular mass PEG is also unsuitable for adequate partitioning, due to the decrease of the exclusion effect [56]. It allows the polymer to attract all the proteins to the upper phase. Thus, the choice of the most suitable intermediate molecular mass of PEG is very important for increasing the extraction efficiency of the ATPE system [54,57,58].

PEG concentration: Several studies demonstrated the influence of different PEG concentrations (7–21%, w/w) on protein partition coefficient (K_p) and enzyme partition coefficient (K_E) from various sources [53,54,59]. Their results showed the significant effect of the PEG concentration on the partitioning of biomolecules in the ATPS. The highest concentration in PEG lowered K_p and K_E. The intermediate concentration of PEG/salt is more applicable for good separation and purification. When serine protease was extracted from Kesinai (*Streblus asper*) leaves, the K_E of the enzyme decreased significantly at a low concentration of molecular weight (8% PEG4,000) [59]. On the other hand, the high PEG concentration and high molecular weight gave negative effect on the partition coefficient of the enzyme. In partitioning of a yeast invertase at different PEG-3,000 concentrations (7.5–20%, w/w) together with 20% (w/w) $MgSO_4$ at pH 5.0, partitioning of invertase in the ATPS is affected significantly by the PEG concentration [54]. The PEG at concentration of 15% (w/w) resulted in the highest value (3.2-fold) of purification factors (PF) with the yield of 134%. At concentration of 20% (w/w), purification factors and yield decreased dramatically to 2.1-fold and 46%, respectively. Similarly, high purification factors and yield were obtained when invertase was partitioned in a 15% (w/w) PEG/12.5% (w/w) Na_2SO_4 at pH 5.0 [11].

Impact of salt concentration

The impact of salt concentration has been widely studied. Increases in salt concentration result in an increase in partition coefficients of bioproducts in upper phase or interface due to salting out [11,53,60]. In general, proteins with the negative charge tend to partition to the top phase in PEG/salt systems while those with the positive charge usually go to the bottom phase [54,61,62].

Varied salt concentrations have been used in separation of proteins (cytochrome c, lysozyme, trypsin, bovine serum albumin and myoglobin) by the ATPE system combined with high-performance liquid chromatography (HPLC) [34]. When $(NH4)_2SO_4$ concentration increased from 10% to 18%, the top phase volume decreased and the bottom phase volume increased. All proteins retained in the lower phase except lysozyme, which was partitioned in the two phases. Moreover, 23% (w/w) $MgSO_4$ caused the best partition behavior of a yeast invertase at different concentrations (15–25%, w/w) in the PEG/$MgSO_4$ aqueous [54]. Besides the salt type, the distribution of invertase is mainly controlled by concentration of the salt. The partition coefficient (K_E) of luciferase increased rapidly as compared to that of total protein (K_p) with an increase in salt concentration from 12% to 16% [30]. The optimum condition for cephalexin antibody separation was at a concentration of 20% for both PEG and Na_2SO_4 [60]. Similar finding was reported in the case of penicillin G and PAA extraction (phenylacetic acid) in the PEG/Na_2SO_4 ATPS at pH 8.0 [60]. The optimized salt concentration for purification of lipase enzyme was 18% potassium phosphate with 20% PEG at the level of 8,000 g/mol [63].

Impact of the type of salts

The selection of salts for the ATPS depends on their ability to promote hydrophobic interactions between biomolecules [64]. The PEG/phosphate system is widely used for recovery of bioproducts [65]. Other salts having similar properties to phosphate, such as sulphate and citrate, have been also used. The changes of the environmental phase system, due to use of different salts, lead to change the behaviors of partitioned protein [62]. Anions are the most effective in partitioning ($SO_4^{-2} >$ $HPO_4^{-2} >$ acetate) than cations ($NH_4^+ > K > Na^+ > Mg^{2+} > Ca^{2+}$) [66]. Recently, use of salts, like citrate (biodegradable) and ammonium carbonate (volatile) are favored because of their high selectivity, less

pollution, biocompatibility and easy to scale-up [38,50]. The partition coefficient values of methionine in systems containing Na_3PO_4 are greater than values in systems containing NaH_2PO_4 or Na_2HPO_4 because of more abilities of Na_3PO_4 to enhance hydrophobic interactions between particles [7]. The PEG 6,000/potassium phosphate system of 23.9% (w/w) TLL gave the best partitioning results with the highest recovery of proteins from cheese whey than the PEG/ammonium sulphate and PEG/potassium dihydrogen phosphate systems [67]. In fact, a significant aggregation tendency of proteins was observed in PEG/ammonium sulphate and PEG/potassium dihydrogen phosphate systems, whose pH was close to isoelectric pH of whey proteins. The PEG/ammonium sulphate could be the suitable purification phase for differential partitioning of β-galactosidase and β-glucosidase rather than other types of salts, such as sodium sulphate, sodium phosphate, potassium phosphate and sodium citrate. This PEG/ammonium sulphate was found to be the best in terms of activity recovery and differential partitioning of both the enzymes [53], while potassium salts have much better effects on partition of lipase than that sodium salts and ammonium salts. This is due to the partition coefficients increased according to the following order: $K^+ > Na^+ > NH_4^+$ [68].

Impact of pH

Partitioning of proteins and enzymes to the phases in the ATPE system depends on their isoelectric points (pI) [63]. The pH of the system, however, affects the charge of target protein and ion composition as well as introduces differential partitioning into the two phases [56]. Accordingly, the initial pH of the system must be above the pI of target bio-molecules [65]. A pH value above 7 is suitable for the PEG/phosphate system and a pH below 6.5 is compatible with the PEG/sulphate system. Most of the biomolecules, especially proteins and enzymes, are stable at neutral pH that is favorable condition to conduct the ATP partitioning. At pH 7, 79% α-amylase inhibitor was recovered with 3.2 purification factor in 11.7% (w/w) PEG-2000 and 19% (w/w) fructose-1,6-bisphosphate trisodium salt [69]. When the pH rose from 5.8 to 8.0, the K_E, yield and recovery of phenylalanine dehydrogenase (pI 5.3) increased, and the optimal values could be obtained at pH 8.0 [56]. However, an increase in pH of the ATPS (PEG-4000/K_2HPO_4, 12/13%) from 7.0 to 9.0 reduced the partition coefficient of lipase from 7.94 to 4.45 and activity recovered from 81.1 to 70.6% [70]. Enzyme stability slightly reduced in the acidic area, but it was dramatically lost at pH above 9.0 [71]. Generally, the efficacy of the pH can be either by changing the charge of the solute or by altering the ratio of the charged species presents [37].

Tie line length (TLL)

Tie line length can affect biomolecule partitioning by hydrophobicity and interfacial tension between phases of the ATPS. The ATPS becomes more hydrophobic with increasing TLL due to reduction of water availability [8]. An increase in the TLL causes an increase in the protein partition coefficient that, in turns, increases the yield of proteins in the top phase due to reduction of the bottom phase [12]. Increasing TLL in the PEG-salt system makes salting-out more effectively, leading to shift of proteins to the top PEG-rich phase [72]. If protein solubility in the PEG phase is insufficient, protein will precipitate at the interface. Solubility and salting-out limits depend on the properties of individual proteins. Therefore, a different response will be expected when a mixture of protein is handled [39]. It was reported that glutenin partitions from wheat flour, an increase in TLL caused an increase in protein transfer to the upper phase in systems formed by PEG-1,500/sulphate salts (lithium or sodium) [24]. However, in terms of systems that composed of PEG-4,000/sulphate salts (lithium or sodium), an increase in TLL caused an increase in protein transfer to the lower phase.

Influence of the addition of NaCl

In general, addition of neutral salts, such as NaCl, to the ATPS results in an increase in the hydrophobic difference due to generation of an electrical potential difference between two phases [5,12,73]. An increase in the hydrophobicity will decrease the amount of bound water, which keeps the final composition of the systems constantly [15]. Furthermore, it increases the ionic strength and enhances the migration of low molecular weight compounds towards the polymer-phase, especially in PEG<4,000 [65]. However, the addition of high concentration of neutral salts may cause denaturation of proteins existing in the system, thus low concentration range from 0.0 to 1.0 M is preferred. High yield (97.3%) of serine protease from mango peel was obtained with addition of 4.5% of NaCl to the PEG/dextran ATPS [12]. The addition of different concentrations of salt (0.0 - 1.0 M) increased the partition coefficient of myoglobin and ovalbumin from 4.20 to 15.77 and 2.82 to 5.51, respectively, in the PEG-4,000/PAA system [37]. Similar results were observed with bovine serum albumin (BSA) in the ATPS [74]. In 6% (w/v) of NaCl, the purification factor of lipase enzyme increased significantly from 59.93 to 141.65 fold. However, further addition of NaCl decreased the K_E of lipase [63].

Recent Process in Atpe, Applications Economic Costs

The ATPE is a powerful method commonly used for separation and purification of biomolecules, such as proteins, enzymes and antibodies. It is composed either polymer (PEG)–polymer (dextran) or polymer (PEG)–salts. However, the polymer-polymer interaction dominates due to the low solubility of amphiphilic proteins in the PEG-salt, which has a high tendency to aggregate in aqueous solution that may damage fragile proteins [37]. The environmental problems were raised from elevated salt concentration in waste disposal [4,75]. Large chemical consumption during the phase-forming [37] will lead to additional cost for the phase recycling in the system. However, the ATPE based on polymer (PEG)-polymer (dextran) is very expensive because of the high cost of some forming phase polymers, such as dextran [37] and ethylene oxide–propylene oxide copolymers [65]. This, in fact, limits the implementation of this system at the large scale.

In recent years, some progress in the ATPE technique has been reported, including some modification introduced to the ATPE system to recover biomolecules in high value, good quality and low cost, providing basic materials for bio-product processes and increasing the application of the non-biotechnology. The alternative polymers used as a substitute for dextran, are generally safe, low-cost and compatible with the system, such as PEG/nonionic surfactant polymers (Triton X-100 and Tween 80) [8] and ionic liquid-based ATPE [42,44]. The dextran polymer also substituted by alcohol/salt ATPE system [42,76], microfluidic aqueous PEG/detergent ATPE system [77] and acid polymer PEG-poly (acrylic acid) system [37,41]. These alternatives reduce the cost and make the process simpler. On the other hand, many studies have demonstrated that the PEG-salts have some advantages such as simple, fast, easy recyclable and low-cost and viscosity [11,38,65,78]. Thus, they are also attractive for commercial applications as a rapid and continuous protein separation [11]. Affinity ligands are introduced into the ATPE system as a free ligand that appears to make the process easier [79-81]. The ATPS in a microfluidic platform was designed and tested for mAbs extraction. This system indicated the potential to be an effective tool to accelerate Bioprocessing design and optimization [82].

Currently, the ATPE based either on a PEG/salt system or a polymer/polymer system has been rarely used in a large scale due to the high cost of the polymers and the difficulty in isolating the extracted biomolecules from the viscous polymer phase by back extraction [29]. To overcome this limitation, there are studies on the ATPE system composed of alcohols or hydrophilic organic solvents and salts. These studies are characterized with low cost of extraction, easy recovery of solvent by evaporation and simple scale-up in the recovery of many products from different sources [42,76,83-85]. For example, K_2HPO_4/ethanol is used for partitioning of proteins, such as lysozyme, chymotrypsinogen solution, BSA and partially purified DNA polymerase from Thermus aquaticus [86]. Ethanol/K_2HPO_4 ATPS was combined with hydrophilic interaction chromatography for the isolation and purification of recombinant human serum albumin (rHSA) [29].

In the last few years, some studies focused on the development of excellent polymers forming ATPE for higher recovering. A lot of novel copolymers forming ATPS are synthesized, such as one pH-response copolymer P_{ADB} [87], a light-response polymer PNNC [88], light-response reversible polymer (P_{NBAC}) and copolymers (P_{NDBC}) [89]. Also other copolymers have been designed, including pH-response polymer (P_{ADB}) and one light-response polymer (P_{NBC}) [90], pH-thermo pH-response polymer (P_{ADB}) and one thermo-response polymer (PNB) [91], pH-response random copolymers (P_{ADB} and P_{ADBA}) [92], and copolymer membrane of poly(acrylonitrile-acrylamide-styrene) [93].

In addition, the reverse micelle solvent system is another ATPE that can be used as an alternative technique for protein separation due to its simplicity and feasibility of large-scale sample loading [94]. The reverse micelle solvent system is a very attractive system for protein separation [95,96]. Lastly, there is an application of a novel continuous ATPE prototype for the recovery of biomolecules [97]. It was used for recovery of protein and α-amylase from soybean, to extract the low-abundant protein from complex mixtures. A continuous ATPE process incorporating three various steps (extraction, back-extraction, and washing) has been introduced with a pump mixer-settler battery. The ATPS process recovered 99% purity of IgG from a CHO cell supernatant and 100% of IgG from a PER.C6 [98]. This new process indicates the ability to successfully recover and purify different antibodies. It could overcome some of the limitations encountered using the typical chromatographic processes, besides inherent advantages of scalability, process integration, capability of continuous operation, and economic feasibility.

Generally, the continuous operation increases partition coefficients with higher recovery efficiencies. The processing time is reduced at least three folds, compared to the batch ATPS. Furthermore, it achieves higher enzyme partitions coefficient (K_p>4) and a top phase enzyme recovery (81%) with the purification factor 40-fold than a batch system. It is suggested that the continuous ATPE model can be used in an industry field for the recovery of bioproducts.

Disadvantages Attributes to the Atps

The development and application of the ATPE have some drawbacks due to lack of information about the exact mechanism of partitioning and unpredictable [56]. There are two restrictions to limit the wide application of the ATPE [99]. Firstly, it is difficult to predict exactly the behavior of target proteins in the ATPE system. Secondly, monitoring the characteristics of proteins is the basic requirement for assessment of bioprocesses, which are affected frequently by the presence of high concentrations of polymers or salts. Considerable time is needed to build first recovery process of the experiment, while big budget is needed for installation and limited output of the purification units [69]. Experimental design is needed to determine the optimal ATPS system for partitioning of desired products [65]. Compared with a novel alcohol/salt ATPE, polymer/polymer, polymer/salt and surfactant ATPE systems have some disadvantage such as the high cost and viscosity as well as slow segregation. There are difficulties in isolating the extracted molecules from the polymer phase or micellar phase by re-extraction and environmental pollution resulting from the recycling of phase-forming polymers [68,76].

Conclusion

The ATPE is the suitable technique for separation and purification of bioproducts in biological and biotechnological fields, especially the PEG/salts system. However, the implementation of the PEG/salt in the large scale may cause environmental problems due to the great amount of chemicals (salts and polymers) needed to phase forming and the high cost resulting from effectively recycling. To overcome it, some modifications are introduced into the system to purify good quality of biomolecules with low-cost and safely to the environment. This modification includes synthesized polymers (P_{NBAC}, P_{ADBA}), copolymers (P_{NDBC}, P_{ABC}), reverse micelle solvent ATPS system and interaction of HPLC with the system to increase the extraction rate. Furthermore, the application of a continuous ATPE system is an alternative technique for the recovery of proteins in large scale. This, in fact, provides the ability to recover proteins in higher efficiencies, increases partition coefficients and shortens processing time. Therefore, the continuous separation ATPE system can be used in pharmaceuticals, foods and chemical industries for purification of high-value bioproducts.

Although the ATPE has been extensively used for recovering of biomolecules, the poor understanding of mechanism that governs phase formation and solute partitioning in the ATPE hinders its application in some cases. Therefore, using of the ATPS with some modifications, such as affinity ligands, a substitute of polymers or using copolymers and a combination of the ATPS with other methods (e.g. chromatography), are more effective due to the high efficiency of separation, purification and increasing recovery yields of bioproducts.

Acknowledgement

The authors would like to thank the financial support from the Fundamental Research Funds for the Central Universities (Project 2013PY096), the Ministry of Scientific and Technology China (Grant No. 2012BAD28B06) and National Natural Science Foundation, China (Grant No. 31172294).

References

1. Alberttson PA (1986) Partition of Cell Particles and Macromolecules. Separation and Purification of biomolecules, cell organelles, membranes and cells in aqueous polymer two-phase systems and their use in biochemical analysis and biotechnology. Wiley-Interscience, New York.

2. Zaslavsky BY (1994) Aqueous Two-phase Partitioning, Physical Chemistry and Bioanalytical Applications. Marcel Dekker Inc, New York.

3. Truust H, G Johansson (1998) Fractionation of Wheat Gliadins by Counter-Current Distribution using an Organic Two-Phase System. J Chromatogr B 711: 245-254.

4. Hatti-Kaul R (2000) Aqueous Two-Phase Systems: Methods and Protocols. Humana Press 11.

5. Gu Z, CE Glatz (2007) Aqueous Two-Phase Extraction for Protein Recovery from Corn Extracts. J Chromatogr B 845: 38-50.

6. Alcantara LAP, Minim LA, Minim VPR, Bonomo RCF, da Silv LHM, et al. (2011) Application of the Response Surface Methodology for Optimization of Whey Protein Partitioning in PEG/Phosphate Aqueous Two-Phase System. J Chromatogr B 879: 1881-1885.

7. Salabat A, Sadeghi R, Moghadam ST, Jamehbozorg B (2011) Partitioning of

8. Liu Y, Wu Z, Zhang Y, Yuan H (2012) Partitioning of Biomolecules in Aqueous Two-Phase Systems of Polyethylene Glycol and Nonionic Surfactant. Biochem Eng J 69: 93-99.

9. Biazus JPM, Santana JCC, Souza RR, Jordao E, Tambourgi EB (2007) Continuous Extraction of A-And B-Amylases from Zea Mays Malt in A PEG4000/Cacl2 ATPS. J Chromatogr B 858: 227-233.

10. Ling Y, Nie HL, Su SN, Branford-White C, Zhu L (2010) Optimization of Affinity Partitioning Conditions of Papain in Aqueous Two-Phase System using Response Surface Methodology. Sep Purif Technol 73: 343-348.

11. Yucekan I, S Onal (2011) Partitioning of Invertase from Tomato in Poly (Ethylene Glycol)/Sodium Sulfate Aqueous Two-Phase Systems. Process Biochem 46: 226-232.

12. Mehrnoush A, Mustafa S, Sarker MI, Yazid AMM (2012) Optimization of Serine Protease Purification from Mango (Mangifera Indica cv. Chokanan) Peel in Polyethylene Glycol/Dextran Aqueous Two Phase System. Int J Mol Sci 13: 3636-3649.

13. Bhavsar K, V Ravi Kumar, J Khire (2012) Downstream Processing of Extracellular Phytase from Aspergillus Niger: Chromatography Process Vs. Aqueous Two Phase Extraction for its Simultaneous Partitioning and Purification. Process Biochem 47: 1066-1072.

14. Zhou Y, Hu C, Wang N, Zhang W, Yu X (2013) Purification of Porcine Pancreatic Lipase by Aqueous Two-Phase Systems of Polyethylene Glycol and Potassium Phosphate. J Chromatogr B 926: 77- 82.

15. Rosa PAJ, Azevedo AM, Sommerfeld S, Mutter A, Aires-Barros MR, et al. (2009) Application of Aqueous Two-Phase Systems to Antibody Purification: a Multi Stage Approach. J Biotechnol 139: 306-313.

16. Rosa PAJ, Azevedo AM, Sommerfeld S, Bäcker W, Aires-Barros MR (2011) Aqueous Two-Phase Extraction as a Platform in the Biomanufacturing Industry: Economical and Environmental Sustainability. Biotechnol Adv 29: 559-567.

17. Rosa PAJ, Azevedo AM, Sommerfeld S, Mutter M, Bäcker W, et al. (2012) Continuous Aqueous Two-Phase Extraction of Human Antibodies using a Packed Column. J Chromatogr B 880: 148-156.

18. Brooks DE, Fisher D (1985) Partitioning in aqueous two-phase systems. Academic Press.

19. Brooks E, Sharp KA, Fischer D (1985) Partitioning in Aqueous Two-Phase Systems: Theory, Methods, Uses and Applications to Biotechnology. Academic Press.

20. Asenjo JA, BA Andrews (2012) Aqueous Two-Phase Systems for Protein Separation: Phase Separation and Applications. J Chromatogr A 1238: 1-10.

21. Rosa PAJ, Ferreira IF, Azevedo AM, Aires-Barros MR (2010) Aqueous Two-Phase Systems: A Viable Platform in the Manufacturing of Biopharmaceuticals. J Chromatogr A 1217: 2296-2305.

22. Raghavarao KSMS, Rastogi NK, Gowthaman MK, Karanth NG (1995) Aqueous Two-Phase Extraction for Downstream Processing of Enzymes/Proteins. Adv Appl Microbol 41: 97-171.

23. Nitsawang S, R Hatti-Kaul, P Kanasawud (2006) Purification of Papain from Carica Papaya Latex: Aqueous Two-Phase Extraction versus Two-Step Salt Precipitation. Enzyme Microb Tech 39: 1103-1107.

24. do Nascimento ISB, Coimbra JSR, Martins JP, Silva LHM, Bonomo RCF, et al. (2010) Partitioning of Glutenin Flour of Special Wheat using Aqueous Two-Phase Systems. J Cereal Sci 52: 270-274.

25. Ibarra-Herrera CC, O Aguilar, M Rito-Palomares (2011) Application of an Aqueous Two-Phase Systems Strategy for the Potential Recovery of a Recombinant Protein from Alfalfa (Medicago Sativa). Sep Purif Technol 77: 94-98.

26. Yin L, Sun CK, Han X, Xu L, Xu Y, et al. (2011) Preparative Purification of Bromelain (EC 3.4. 22.33) from Pineapple Fruit by High-Speed Counter-Current Chromatography using a Reverse-Micelle Solvent System. Food Chem 129: 925-932.

27. Junwei Z, Yan W, Qijun P (2013) Extraction of Theanine from Waste Liquid of Tea Polyphenol Production in Aqueous Two-Phase Systems with Cationic and Anionic Surfactants. Chinese J Chem Eng 21: 31-36.

28. Zhang D, Zu Y, Fu Y, Wang W, Zhang L, et al. (2012) Aqueous Two-Phase

l-methionine in aqueous two-phase systems containing poly (propylene glycol) and sodium phosphate salts. J Chem Thermody 43: 1525-1529.

Extraction and Enrichment of Two Main Flavonoids from Pigeon Pea Roots and the Antioxidant Activity. Sep Purif Technol 102: 26-33.

29. Dong Y, Zhang F, Wang Z, Du L, Hao A, et al. (2012) Extraction and Purification of Recombinant Human Serum Albumin from Pichia Pastoris Broths using Aqueous Two-Phase System Combined with Hydrophobic Interaction Chromatography. J Chromatogr A 1245: 143-149.

30. Priyanka BS, Rastogi NK, Raghavarao KSMS, Thakur MS (2012) Downstream Processing of Luciferase from Fireflies (Photinus Pyralis) using Aqueous Two-Phase Extraction. Process Biochem 47: 1358-1363.

31. Wu Q, Lin DQ, Yao SJ (2013) Evaluation of Poly (Ethylene Glycol)/ Hydroxypropyl Starch Aqueous Two-Phase System for Immunoglobulin G Extraction. J Chromatogr B 928: 106-112.

32. Antov MG, DM Peričin, MG Dašić (2006) Aqueous Two-Phase Partitioning of Xylanase Produced by Solid-State Cultivation of Polyporus Squamosus. Process Biochem 41: 232-235.

33. Su CK, Chiang BH (2006) Partitioning and Purification of Lysozyme from Chicken Egg White using Aqueous Two-Phase System. Process Biochem 41: 257-263.

34. Zhao XY, Qu F, Dong M, Chen F, Luo AQ, et al. (2012) Separation of Proteins by Aqueous Two-Phase Extraction System Combined with Liquid Chromatography. Chinese J Anal Chem 40: 38-42.

35. Gai Q, Qu F, Zhang T, Zhang Y (2011) Integration of carboxyl modified magnetic particles and aqueous two-phase extraction for selective separation of proteins. Talanta 85: 304-309.

36. Srinivas N, A Narayan, K Raghavarao (2002) Mass Transfer in a Spray Column during Two-Phase Extraction of Horseradish Peroxidase. Process Biochem 38: 387-391.

37. Saravanan S, Rao JR, Nair BU, Ramasami T (2008) Aqueous Two-Phase Poly (Ethylene Glycol)–Poly (Acrylic Acid) System for Protein Partitioning: Influence of Molecular Weight, Ph and Temperature. Process Biochem 43: 905-911.

38. Raja S, Murty VR, Thivaharan V, Rajasekar V, Ramesh V (2011) Aqueous Two Phase Systems for the Recovery of Biomolecules–A Review. Sci Tech 1: 7-16.

39. Ferreira GB, Evangelista AF, Junio JBS, de Souza RR, Santana JCC, et al. (2007) Partitioning Optimization of Proteins from Zea Mays Malt in ATPS PEG 6000/Cacl2. Braz Arch Biol Techn 50: 557-564.

40. Ooi CW, Tey BT, Hii SL, Kamal SMM, Lan JCW, et al. (2009) Purification of Lipase Derived from Burkholderia Pseudomallei with Alcohol/Salt-Based Aqueous Two-Phase Systems. Process Biochem 44: 1083-1087.

41. Rao JR, Nair BU (2011) Novel Approach Towards Recovery of Glycosaminoglycans from Tannery Wastewater. Bioresource Technol 102: 872-878.

42. Montalvo-Hernández B, Rito-Palomares M, Benavides J (2012) Recovery of Crocins from Saffron Stigmas (Crocus Sativus) in Aqueous Two-Phase Systems. J Chromatogr A 1236: 7-15.

43. Cao Q, Li S, He C, Li F, Liu F (2007) Extraction and Determination of Papaverin in Pericarpium Papaveris Using Aqueous Two-Phase System of Poly (Ethylene Glycol) –(NH4)2SO4 Coupled with High-Performance Liquid Chromatography. Analytica Chimica Acta 590: 187-194.

44. Pei Y, Wang J, Wu K, Xuan X, Lu X (2009) Ionic Liquid-Based Aqueous Two-Phase Extraction of Selected Proteins. Sep Purif Technol 64: 288-295.

45. Pietruszka N, Galaev IY, Kumar A, Brzozowski ZK, Mattiasson B (2000) New Polymers Forming Aqueous Two-Phase Polymer Systems. Biotechnol Progr 16: 408-415.

46. Boeris V, Spelzini D, Farruggia B, Pico G (2009) Aqueous Two-Phase Extraction and Polyelectrolyte Precipitation Combination: A Simple and Economically Technologies for Pepsin Isolation from Bovine Abomasum Homogenate. Process Biochem 44: 1260-1264.

47. Hustedt H, Kroner KH, Papamichale N (1988) Continuous Cross-Current Aqueous Two Phase Extraction of Enzymes from Biomass. Process Biochem 23: 129-137.

48. Schmidt AS, Ventom AM, Asenjo JA (1994) Partitioning and Purification of A-Amylase in Aqueous Two-Phase Systems. Enzyme Microb Tech 16: 131-142.

49. Lin Q, Mie HL, You TW, Shan JY, Zhu Q (2003) Modeling the Protein Partitioning in Aqueous Polymer Two-Phase Systems: Influence of Polymer Concentration and Molecular Weight. Chem Eng Sci 58: 2963-2972.

50. Marini A, Imelio N, Picó G, Romanini D, Farruggia B (2011) Isolation of a Aspergillus Niger Lipase from a Solid Culture Medium with Aqueous Two-Phase Systems. J Chromatogr B 879: 2135-2141.

51. Mattiasson B, Kaul R (1986) Use of Aqueous Two-Phase Systems for Recovery and Purification in Biotechnology. Separation, Recovery, and Purification and Biotechnology 7: 78-92.

52. Ashipala OK, He Q (2008) Optimization of Fibrinolytic Enzyme Production by Bacillus Subtilis DC-2 in Aqueous Two-Phase System (Poly-Ethylene Glycol 4000 And Sodium Sulfate). Bioresource Technol 99: 4112-4119.

53. Hemavathi AB, Raghavarao KSMS (2011) Differential Partitioning of B-Galactosidase and B-Glucosidase using Aqueous Two Phase Extraction. Process Biochem 46: 649-655.

54. Karkas T, Önal S (2012) Characteristics of Invertase Partitioned in Poly (Ethylene Glycol)/Magnesium Sulfate Aqueous Two-Phase System. Biochem Eng J 60: 142-150.

55. Ribeiro MZ, Silva DP, Vitolo M, Roberto IC, Pessoa-Jr A (2007) Partial Purification of Glucose-6-Phosphate Dehydrogenase by Aqueous Two-Phase Poly (Ethyleneglycol)/Phosphate Systems. Braz J Microb 38: 78-83.

56. Mohamadi H, Omidinia E, Dinarvand R (2007) Evaluation of Recombinant Phenylalanine Dehydrogenase Behavior in Aqueous Two-Phase Partitioning. Process Biochem 42: 1296-1301.

57. Pico G, Romanini D, Nerli B, Farruggia B (2006) Polyethyleneglycol Molecular Mass and Polydispersivity Effect on Protein Partitioning in Aqueous Two-Phase Systems. J Chromatogr B 830: 286-292.

58. Yue H, Yuan Q, Wang W (2007) Purification of Phenylalanine Ammonialyase in PEG 1000/ Na2SO4 Aqueous Two-Phase System by a Two-Step Extraction. Biochem Eng J 37: 231-237.

59. Mehrnoush A, Mustafa S, Yazid AMM (2011) 'Heat-Treatment Aqueous Two Phase System' for Purification of Serine Protease from Kesinai (Streblus Asper) Leaves. Molecules 16: 10202-10213.

60. Bora MM, Borthakur S, Rao PC, Dutta NN (2005) Aqueous Two-Phase Partitioning of Cephalosporin Antibiotics: Effect of Solute Chemical Nature. Sep Purif Technol 45: 153-156.

61. Tomatani EJ, Vitolo M (2007) Production of High-Fructose Syrup using Immobilized Invertase in a Membrane Reactor. J Food Eng 80: 662-667.

62. Yang L, Huo D, Hou C, He K, Lv F, et al. (2010) Purification of Plant-Esterase in PEG1000/Nah2po4 Aqueous Two-Phase System by a Two-Step Extraction. Process Biochem 45: 1664-1671.

63. Barbosa JMP, Souza RL, Fricks AT, Zanin GM, Soares CMF, et al. (2011) Purification of Lipase Produced by a New Source of Bacillus in Submerged Fermentation using an Aqueous Two-Phase System. J Chromatogr B 879: 3853-3858.

64. Franco TT, Andrews AT, Asenjo JA (1996) Conservative Chemical Modification of Proteins to Study the Effects of a Single Protein Property on Partitioning in Aqueous Two-Phase Systems. Biotechnol Bioeng 49: 290-299.

65. Benavides J, Rito-Palomares M (2008) Practical Experiences from the Development of Aqueous Two-Phase Processes for the Recovery of High Value Biological Products. J Chem Technol Biot 83: 133-142.

66. Roe S (2000) Protein Purification Techniques: A Practical Approach. Oxford University Press.

67. Anandharamakrishnan C, Raghavendra SN, Barhate RS, Hanumesh U, Raghavarao KSMS (2005) Aqueous Two-Phase Extraction for Recovery of Proteins from Cheese Whey. Food Bioprod Process 83: 191-197.

68. Li X, Wan J, Cao X (2010) Preliminary Application of Light-Ph Sensitive Recycling Aqueous Two-Phase Systems to Purification of Lipase. Process Biochem 45: 598-601.

69. Chen X, Xu G, Li X, Li Z, Ying H (2008) Purification of an A-Amylase Inhibitor in a Polyethylene Glycol/Fructose-1, 6-Bisphosphate Trisodium Salt Aqueous Two-Phase System. Process Biochem 43: 765-768.

70. Zhang Y, Liu J (2010) Purification and InSitu Immobilization of Lipase From of a Mutant of Trichosporon Laibacchiiusing Aqueous Two-Phase Systems. J Chromatogr B 878: 909-912.

71. Ketnawa S, Sai Ut S, Theppakorn T, Chaiwut P, Rawdkuen S (2009) Partitioning of Bromelain from Pineapple Peel (Nang Lae Cultv.) by Aqueous Two Phase System. Asian J Food Agro-Indus 2: 457-468.

72. da Silva ME, Franco TT (2000) Purification of Soybean Peroxidase (Glycine Max) by Metal Affinity Partitioning in Aqueous Two-Phase Systems. J Chromatogr B 743: 287-294.

73. Hachem F, Andrews BA, Asenjo JA (1996) Hydrophobic Partitioning of Proteins in Aqueous Two-Phase Systems. Enzyme Microb Tech 19: 507-517.

74. Tubio G, Nerli B, Pico G (2004) Relationship between the Protein Surface Hydrophobicity and its Partitioning Behaviour in Aqueous Two-Phase Systems of Polyethyleneglycol–Dextran. J Chromatogr B 799: 293-301.

75. Diamond AD, Hsu JT (1990) Protein Partitioning in PEG/Dextran Aqueous Two-Phase Systems. AIChE J 36: 1017-1024.

76. Tan ZJ, Li FF, Xu XL (2013) Extraction and Purification of Anthraquinones Derivatives from Aloe Vera L. using Alcohol/Salt Aqueous Two-Phase System. Bioprocess Biosys Eng 36: 1105-1113.

77. Hu R, Feng X, Chen P, Fu M, Chen H, et al., (2011) Rapid, Highly Efficient Extraction and Purification of Membrane Proteins using a Microfluidic Continuous-Flow Based Aqueous Two-Phase System. J Chromatogr A 1218: 171-177.

78. Rito-Palomares M (2004) Practical Application of Aqueous Two-Phase Partition to Process Development for the Recovery of Biological Products. J Chromatogr B 807: 3-11.

79. Giuliano KA (1991) Aqueous Two-Phase Protein Partitioning using Textile Dyes as Affinity Ligands. Anal Biochem 197: 333-339.

80. Teotia S, Lata R, Gupta MN (2001) Free Polymeric Bioligands in Aqueous Two-Phase Affinity Extractions of Microbial Xylanases and Pullulanase. Protein Expres Purif 22: 484-488.

81. Xu Y, Vitolo M, Albuquerque CN, Pessoa JrA (2002) Affinity Partitioning of Glucose-6-Phosphate Dehydrogenase and Hexokinase in Aqueous Two-Phase Systems with Free Triazine Dye Ligands. J Chromatogr B 780: 53-60.

82. Silva DF, Azevedo AM, Fernandes P, Chu V, Conde JP, et al., (2012) Design of a Microfluidic Platform for Monoclonal Antibody Extraction using an Aqueous Two-Phase System. J Chromatogr A 1249: 1-7.

83. Zhi W, Deng Q (2006) Purification of Salvianolic Acid B from the Crude Extract of Salvia Miltiorrhiza with Hydrophilic Organic/Salt-Containing Aqueous Two-Phase System by Counter-Current Chromatography. J Chromatogr A 1116: 149-152.

84. Wang H, Dong Y, Xiu ZL (2008) Microwave-Assisted Aqueous Two-Phase Extraction of Piceid, Resveratrol and Emodin from Polygonum Cuspidatum by Ethanol/Ammonium Sulphate Systems. Biotechnol Lett 30: 2079-2084.

85. Jiang B, Li ZG, Dai JY, Zhang DJ, Xiu ZL (2009) Aqueous Two-Phase Extraction of 2, 3-Butanediol from Fermentation Broths using an Ethanol/Phosphate System. Process Biochem 44: 112-117.

86. Louwrier A (1998) Model Phase Separations of Proteins using Aqueous/Ethanol Components. Biotechnol Tech 12: 363-365.

87. Wang W, Wan J, Ning B, Xia J, Cao X (2008) Preparation of a Novel Light-Sensitive Copolymer and its Application in Recycling Aqueous Two-Phase Systems. J Chromatogr A 1205: 171-176.

88. Kong FQ, Cao XJ, Xia JA, Hour B (2007) Synthesis and Application of a Light-Sensitive Polymer Forming Aqueous Two-Phase Systems. J Ind Eng Chem 13: 424-428.

89. Chen JP, Miao S, Wan JF, Xia JA, Cao XJ (2010) Synthesis and Application of Two Light-Sensitive Copolymers Forming Recyclable Aqueous Two-Phase Systems. Process Biochem 45: 1928-1936.

90. Ning B, Wan JF, Cao XC (2009) Preparation and Recycling of Aqueous Two-Phase Systems with Ph-Sensitive Amphiphilic Terpolymer PADB. Biotechnol Progr 25: 820-824.

91. Miao S, Chen JP, Cao XJ (2010) Preparation of a Novel Thermo-Sensitive Copolymer Forming Recyclable Aqueous Two-Phase Systems and its Application in Bioconversion of Penicillin G. Sep Purif Technol 75: 156-164.

92. Yan B, Cao X (2012) Preparation of Aqueous Two-Phase Systems Composed of Two Ph-Response Polymers and Liquid–Liquid Extraction of Demeclocycline. J Chromatogr A 1245: 39-45.

93. Xing J, Li F (2009) Separation and Purification of Aloe Polysaccharides by a Combination of Membrane Ultrafiltration and Aqueous Two-Phase Extraction. Appl Biochem Biotech 158: 11-19.

94. Dong XY, Meng Y, Feng XD, Sun Y (2010) A Metal-Chelate Affinity Reverse Micellar System for Protein Extraction. Biotechnol Progr 26: 150-158.

95. Osakai T, Shinohara A (2008) Electrochemical Aspects of the Reverse Micelle Extraction of Proteins. Anal Sci 24: 901-906.

96. Fileti AMF, Fischer GA, Tambourgi EB (2010) Neural Modeling of Bromelain Extraction by Reversed Micelles. Braz Arch Biol Technol 53: 455-463.

97. Vazquez-Villegas P, Espitia-Saloma E, Rito-Palomares M, Aguilar O (2013) Low-Abundant Protein Extraction from Complex Protein Sample using a Novel Continuous Aqueous Two-Phase Systems Device. J Separ Sci 36: 391-399.

98. Rosa PAJ, Azevedo AM, Sommerfeld S, Mutter M, Bäcker W, et al., (2012) Continuous Purification of Antibodies from Cell Culture Supernatant with Aqueous Two-Phase Systems: From Concept to Process. Biotechnol J 8: 352-362.

99. González-González M, Mayolo-Deloisa K, Rito-Palomares M, Winkler R (2011) Colorimetric Protein Quantification in Aqueous Two-Phase Systems. Process Biochem 46: 413-417.

18

A *Bacillus* Strain Able to Hydrolyze Alpha- and Beta-Keratin

Soltana Fellahi[1,2*], Taha I Zaghloul[3], Elisabeth Feuk-Lagerstedt[2] and Mohammad J Taherzadeh[2]

[1]*Laboratory of Microbiology and Plant Biology, Department of Biotechnology Faculty of Sciences of Nature and Life, Mostaganem University, Mostaganem, Algeria*
[2]*Swedish Centre for Resource Recovery, University of Borås, Borås, Sweden*
[3]*Institute of Graduate Studies and Research, University of Alexandria, Alexandria, Egypt*

Abstract

The ability to hydrolyze keratin, a rigid and strongly cross-linked protein in the waste of poultry feather and sheep wool, has made keratinase production by microorganisms highly important to the biotechnological industry. A protein-degrading bacterium (C_4) was isolated from compost. Based on morphology and biochemical tests, along with 16S rRNA sequencing, the isolated C_4 was tentatively identified as *Bacillus sp.* C_4 (2008). The proteolytic activity of the *Bacillus sp.* C_4 strain was broadly specific; it degraded keratinous and non-keratinous proteins to different degrees. Pea pods as substrate generated the highest protease production, followed by soybean meal and sheep wool. Notwithstanding, using wool keratin as a sole source of carbon and nitrogen yielded the highest level of soluble proteins. Furthermore, the C_4 bacterium grew well, and produced a significant level of keratinase when using wool and feather as substrates. Supplementing the medium with yeast extract and peptone shortened the time required for feather degradation, but delayed the onset of the wool keratin hydrolysis with two days. The predominant amino acids released in feather hydrolysate were tyrosine, phenylalanine, and histidine. In contrast, the wool lysate was rich in aspartic acid, methionine, tyrosine, phenylalanine, histidine, and lysine. Results established that utilizing the C_4 strain for keratin degradation in waste management holds considerable potential.

Keywords: Proteolytic enzymes; *Bacillus*; Keratinase; Chicken feather; Sheep wool

Introduction

Every year, large amounts of keratin containing wastes are generated from poultry, leather, and meat processing industries. The annual global feather waste from the poultry processing industry alone reaches 8.5 million tons. At present, the poultry feathers are dumped, buried, used for land filling, or incinerated, resulting in problems in terms of storage, handling, emission control, and ash disposal [1].

Poultry feathers consist to 90% of keratin, which is rich in hydrophobic amino acids, but also contains important amino acids like cysteine, arginine, and threonine [2]. The feather waste can be processed to feather meal, hydrothermal processes or chemical treatments being the most popular methods. These processes result however in the destruction of essential amino acids, yielding a product with poor digestibility and low nutritional value [3,4]. Bioconversion of feather represents an alternative method for improving the nutritional value of feather waste [5].

Also sheep production and wool processing generate a significant amount of keratin waste, and as the world market for wool has dropped dramatically, this has brought about a huge amount of wool waste that cannot be processed [6]. Wool contains a different kind of keratin (α-keratin) than feather, due to the high concentration of cysteine crosslinking in the exocuticle of the wool fiber, and it is consequently not as rigid as feather keratin (β-keratin) [7]. Both types of keratin are intensively cross-linked with disulfide bridges, hampering their degradation by common proteases such as trypsin, pepsin, and papain [8].

In nature, keratin waste is continuously and efficiently decomposed by a large number of bacteria and fungi that produce special proteolytic enzymes, "keratinases" [9]. These enzymes are proteases, and they target insoluble keratin substrates [9,10]. They show high affinity toward hard-to-degrade proteins and have broad substrate specificity in comparison to conventional proteases. They are generally alkaline and thermostable by nature [9,11], and have consequently gained importance not only within various conventional biotechnological sectors, e.g. detergents,

feed, and fertilizers [9,12], but are also applied for environmental cleanup of feather keratin, converting it into feather meal for multiple purposes [13,14]. These proteases are also considered for the leather industry, as they are superior for enzymatic dehairing, and may also be applied in pharmaceutical preparations and in the cosmetic industry as ungual enhancers [15], etc. [16,17]. A recent finding, disclosing that keratinases cause enzymatic breakdown of prion protein PrP^{SC}, opens the door for novel relevant applications of a broad range of keratinases [18].

A large number of microorganisms have been reported to produce keratinases [9-11]. So far, the vast majority of the identified keratinase producing organisms appear to be able to hydrolyze only β-keratin in chicken feather [19]. Few organisms, e.g. *Bacillus subtilis* and *Strenotrophomonas*, are known to be able to hydrolyze both α- and β-keratin. The most likely explanation for this is that β-keratin is more susceptible to the enzyme, as it contains less cysteine residues, and thus also fewer disulfide bonds than α-keratin.

In spite of the broad range of applications for keratinases, only a few commercial preparations of these enzymes are available on the market. Research needs to be focused on screening keratinases in order to succeed in extending the range of their applications, ultimately bridging the gap between demand and availability.

Owing to the complexity and variability of substrates, specific keratinases are required for each application. Advances in keratinase

*Corresponding author: Soltana Fellahi, Swedish Centre for Resource Recovery, University of Boras, Boras, Sweden

research should therefore be directed toward searching for novel keratinases with broader substrate specificity and a higher catalytic efficiency. Hence, we seized different aquatic and terrestrial bacteria in an attempt to obtain a bacterial strain that produced a keratinase with broad substrate specificity and with good catalytic efficiency.

Materials and Methods

Media

Bacterial isolates were cultivated on Peptone Yeast extract (PY) medium [20] (Bactopeptone 10 g/l, Difco yeast extract 5 g/l, NaCl 5 g/l, pH 7.0), PA (PY supplemented with 15 g/l agar), milk agar (yeast extract 0.5 g/l, skimmed milk 10 g/l, agar 15 g/l, pH 7.0), basal medium II [5] (NH_4Cl 0.5 g/l, NaCl 0.5 g/l, K_2HPO_4 0.5 g/l, KH_2PO_4 0.4 g/l, $MgCl_2·6H_2O$ 0.1 g/l, yeast extract 0.1 g/l, pH 7.0), and modified basal medium II (basal medium II supplemented with yeast extract 1.5 g/l and peptone 1g/l). Basal medium II and modified basal medium II (the latter supplemented with 10 g/l chicken feather, sheep wool, soybean meal, or pea pods) were used for protease production.

Isolation, screening, and identification of the most potent protease-producing microorganism

Egyptian aquatic and terrestrial samples were used for the experiments. Thirty-one bacterial isolates were obtained from the material, twenty from the Red Sea, two from Wadi Natrun (EL HAMRA spring), one from olive oil waste, one from compost, five from Sohag soil, and two from Aswan soil.

Protease-producing bacterial isolates were obtained by using the spread plate method on PA and milk agar. The ratio (X/Y) was inferred to indicate protease activity, (X) being the diameter of the bacterial colony and (Y) the diameter of the clearing zone.

The bacterial isolate displaying the highest protease production, based on the ratio (X/Y), was identified by means of morphological, physiological, and biochemical characteristics, and by conducting 16S rRNA gene sequencing (Ribotyping). The ensuing data were compared with the standard descriptions in Bergey's Manual of Determinative Bacteriology [21].

DNA isolation was carried out in accordance with a method previously described by Zaghloul et al. [22], and the 16S rRNA gene was amplified by PCR in a thermocycler (Progene, England). The forward and reverse primers were:

Fw: 5'AGAGTTTGATCMTGGCTCAG-3`

Rv: 5'TACGGYTACCCTGTTACGACTT-3`

The primers were designed based on the conserved zones within the rRNA operon in *E. coli*. The PCR reaction was executed with 30 pmole of each primer, 10 µM dNTPs, and 2 units of *Taq* polymerase enzyme, producing a final volume of 50 µl. The process comprised an initial denaturation at 95°C for 5 min, followed by 30 cycles, each comprising 94°C for 1 min, 55°C for 1 min, and 72°C for 1.5 min, followed by a final extended period at 72°C for 10 min. The PCR product was purified by using the spin column (The Wizard SV Gel & PCR Clean-Up System, Promega, Madison, USA).

The nucleotide sequences of the 16S rDNA gene were determined by Eurofins Genomics, Germany. Similarity to other bacteria was investigated by using the nucleotide search engine "Blastn", accessible at NCBI (http://blast.ncbi.nlm.nih.gov/Blast.cgi). The16S rDNA gene sequence was subsequently submitted to the NCBI GenBank. A

dendrogram was generated by the neighbor-joining method, using the BioEdit software (http://bioedit.software.informer.com/).

Measurement of enzyme activity

Proteolytic activity was measured as described by Cliffe and Law [23], using Hide Powder Azure (HPA, Sigma) as substrate. The amount of enzyme causing a change in absorbance of 0.1 (in comparison with a control at 595 nm) after 30 minutes at 37°C equals 1 unit. Keratinolytic activity of the bacterial cells, as indicated by free amino ($-NH_2$) groups being released during biodegradation of feather and wool, was determined by using ninhydrin [24]. Soluble proteins were determined by the method of Bradford [25].

Ability to hydrolyze different protein-based substrates

The isolate showing the highest protease production in the screening process was further tested for its ability to hydrolyze the following protein substrates: Chicken feathers, sheep wool, pea pods, and soybean. Chicken feathers (collected from medium sized white hens), sheep wool, and pea pods were washed with tap water, followed by distilled water, and then dried overnight at 60°C. Chicken feathers, sheep wool, and pea pods were chopped into smaller particles. The dried pea husks were finely powdered [26].

For enzyme production 500 ml Erlenmeyer flasks containing 50 ml basal medium II supplemented with 10 g/l of the following: pea pods, soybean, chicken feathers, or sheep wool were used. In addition, 500 ml Erlenmeyer flasks containing 50 ml of modified basal medium II were used for the keratin substrates (Chicken feather and sheep wool). The batch cultures were inoculated with 3.16×10^9 CFU/ml and maintained at 37°C under shaking conditions (160 rpm; New Brunswick Scientific, USA) for three days. Protein content, Colony-Forming Units (CFU/ml) as well as proteolytic activity (for the four substrates in basal medium II) and keratinolytic activity (for the substrates in the modified basal medium II) were determined daily.

Analysis of the released amino acids

The amino acids released from feather and wool keratin into the modified basal medium II upon the action of bacterial isolate were analyzed (Beckman 119 CL AAA, Palo Alto, USA) [27], and the concentration of each amino acid (with the exception of tryptophan) was calculated on day 0 and day 3.

Results and Discussion

Selection of the most potent protease-producer

In accordance with the ratio X/Y (diameter of bacterial colony/ diameter of the clearing zone) resulting from the hydrolysis on milk agar, three bacterial isolates (NS_1, NS_2, and C_4) were deemed to be high protease producers. In spite of using several liquid media for cultivating the bacterial isolates NS_1 and NS_2, neither growth nor proteolytic activity were successful. Being the most potent producer of extracellular protease, the isolate C_4 was thus selected for further experimental studies. Our results concur with bacteria being the most dominant group of alkaline protease producers, with the genus *Bacillus* as the most prominent one [28-30].

Identification of the isolate C_4

Electron microscopic examination disclosed that the selected isolate was a spore-forming, Gram-positive bacterium. Cells were shaped as short rods, and the endospores were oval and located centrally in a slightly swollen sporangium (Figure 1). Morphological and

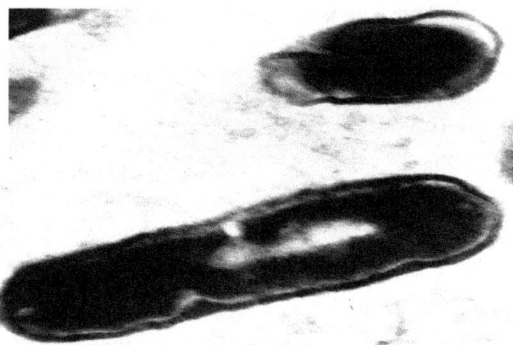

Figure 1: Transmission Electron Microscope (TEM) micrograph of the protease-producing isolate of the selected C₄ strain, showing a longitudinal section of a vegetative cell at division stage, and a spore. (40,000 X).

Character	Reaction
Morphological characteristics	
Gram stain	+
Cell shape	Short rods
Spore	+ (Central)
Salinity tolerance:	
5 g/l	+
30 g/l	+
70 g/l	+
Growth at:	
55°C	+
60°C	-
Hydrolysis of:	
Casein	+
Gelatin	+
Starch	-
Biochemical characteristics:	
Voges-Proskauer (VP) test	-
Nitrate reductase	+
Methyl red (MR)	-
Catalase	+
Gas from glucose	-
Egg yolk lecithinase	-
Formation of indole	-
Acid from:	
Glucose	+
Arabinose	-
Xylose	-
Mannitol	+
Maltose	-
Lactose	-
Sucrose	+
Motility	Motile
O-F test	Fermentative
Gelatin liquefaction	+
Urease	-
Citrate utilization	+

Table 1: Some morphological, physiological, and biochemical characteristics of the most potent protease-producing isolate.

biochemical characteristics of the bacterial isolate, summarized in table 1, suggested that the bacterial isolate C₄ belongs to family Bacillaceae, and is a member of the genus *Bacillus*. The genus was confirmed by a phylogenetic analysis of the 16S rDNA gene. The 16S rDNA sequence of the isolate C₄ were highly similar to the sequences of a group comprising several *Bacillus* strains (Figure 2), e.g. agreeing to 96% with *Bacillus aerophilus sp.* nov. and *Bacillus stratosphericus sp.* nov. Morphological, biochemical, and physiological characteristics, along with the results from the comparative sequence analysis of the 16S rDNA (RNA) gene of the isolate C₄ with that of other 16S rDNA sequences (available in the GenBank database), indicate that our bacterial isolate should be categorized as a *Bacillus* species. We have hence tentatively named our bacterial isolate "*Bacillus sp.* C₄ (2008)", submitting its16S rDNA gene sequence to the NCBI GenBank (accession number FJ214667).

Ability of *Bacillus sp.* C₄ to hydrolyze protein substrates

Proteolytic activity and release of soluble proteins proceeded throughout the cultivation of *Bacillus sp.* C₄ in basal medium II (Figure 3), pea pods displaying the highest protease activity of the substrates, and peaking on day 2 (10.36 U/Log CFU). The soybean meal substrate showed the second highest activity, and peaked on day 1, but declined only after 2 days (6.10 U/Log CFU). Proteolytic activity showed a linear increase in sheep wool, with no sign of decline on day 3 (5.94 U/Log CFU). The least successful substrate was chicken feather, showing a small increase that declined after day 2. The release of soluble proteins did not follow the proteolytic activity curves for the four substrates; the release from soybean, for instance, dropped abruptly after one day. It did however gradually recover as the cultivation period continued, and the level of soluble proteins from soybean eventually approximated the level emitted on day 3 by chicken feather, which reached 0.208 mg/ml. The level of soluble proteins from pea pods did not change much during incubation, the maximum level reaching 0.337 mg/ml at the end of the experiment (three days). Sheep wool appeared to be the most effective substrate, with soluble protein levels reaching 0.595 mg/ml on day 3 after a rather steep rise (Figure 4).

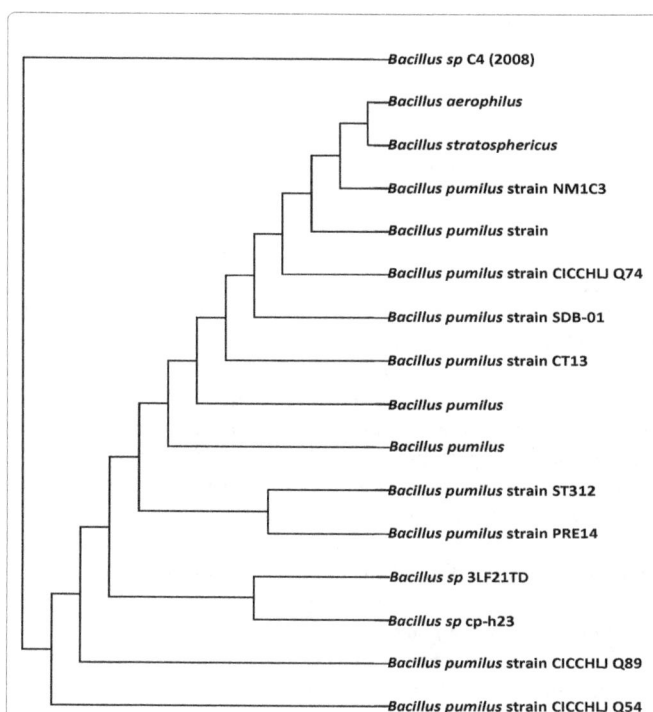

Figure 2: Phylogenetic relation of the 16S rDNA sequence of the *Bacillus sp.* C₄ (2008) strain to the 16S rDNA sequence of the 15 bacteria displaying the highest similarity.

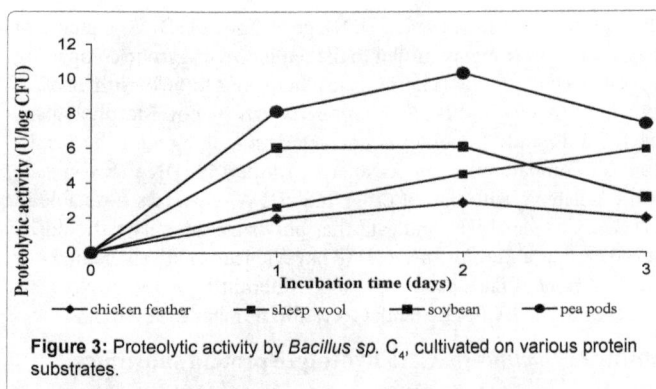

Figure 3: Proteolytic activity by *Bacillus sp.* C$_4$, cultivated on various protein substrates.

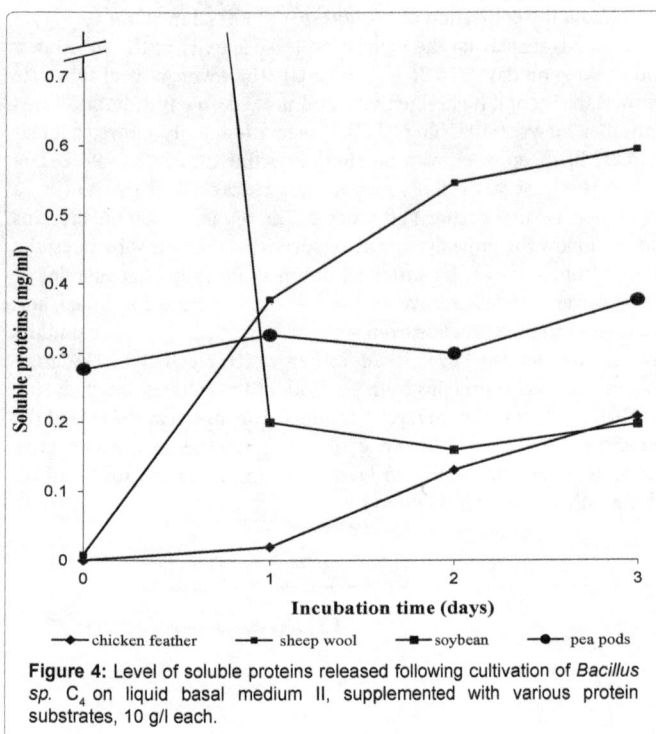

Figure 4: Level of soluble proteins released following cultivation of *Bacillus sp.* C$_4$ on liquid basal medium II, supplemented with various protein substrates, 10 g/l each.

Several studies report on the use of agro-industrial residues for the production of proteases, such as the following combinations: nug meal and *Bacillus sp.* AR 009 [31], pigeon pea and *Bacillus sp.* JB-99 [26], wheat bran and *Rhizopus oryzae* [32], green gram husk and *Bacillus sp.* [33], soybean meal and *Bacillus sp.* L21 [34]. The proven ability of bacteria to grow and to produce sizable levels of protease from various wastes and agro-industrial substrates offers tremendous potential for developing biotechnological methods for hydrolysis of such by-products. An aspect of great significance is that these natural materials are cheap and readily available substrates, and might thus be of interest for industrial-level production of protease. In the production of industrial enzymes, the growth substrate accounts for up to 30-40% of the production cost [35]. Utilizing natural residues, particularly in countries where they are generated in abundance, might hence considerably reduce the cost of enzyme production [36]. Furthermore, using agro-industrial substrates or environmental wastes might greatly bring down pollution problems [37].

The 3-day monitoring of *Bacillus sp.* C$_4$ hydrolyzing keratin-rich materials, such as chicken feather and sheep wool (10 g/l each), manifested bacterial growth and protease production, using both substrates as sole source of carbon, energy, nitrogen, and sulfur. The modified basal medium generated a higher level of proteolytic activity than the unmodified basal medium. Furthermore, when supplementing the medium with yeast extract and peptone, the proteolytic activity of feather and sheep wool displayed a 4.3- and 2.7-fold increase, respectively (Figure 5a), which conformed with previously reported results [38,39].

Growing *Bacillus sp.* C$_4$ on feather keratin in modified basal medium enhanced the keratinolytic activity 1.53 times on day 2, while in the basal medium, a similar enhancement occurred already on day 1 (Figure 5b). However, while the activity in the modified basal medium continued to rise on day 3, it showed a dramatic drop in the unmodified medium. In contrast, keratinolytic activity of sheep wool in modified basal medium did not rise until day 3, and was not prominent, while in the unmodified basal medium, it increased steeply already on day 1, although figures fluctuated after that. Moreover, as illustrated in Figure 6, yeast extract and peptone supplementation shortened the time required for feather degradation from three days to two. The time required for wool decomposition was however not shortened (data not shown). Our results are in good agreement with previous research [40,41].

The degradation of feather and wool keratin by *Bacillus sp.* C$_4$ was accompanied by a release of soluble proteins in the culture medium. The protein release from feather degradation varied between the two media, but showed consistency in the wool incubation. Soluble proteins, liberated at the end of the experiment, exceeded 0.2 and 0.5 mg/ml in basal medium II containing feather and wool, respectively (Figure 5c), the latter producing as much as 0.475 mg/ml already on day 1, but in the modified basal medium. Our results disclose that *Bacillus sp.* C$_4$ favored yeast extract and peptone as nutrients, and thus wool keratin hydrolysis started only on day 2 in the supplemented modified medium. Interestingly, this strain of bacterium appears to develop considerable amounts of soluble proteins in comparison with other feather-degrading microorganisms. For instance, *Bacillus subtilis* DB100 (p5.2) released a maximum of 0.2 mg/ml soluble proteins after four days of incubation on 10 g/l feather [39], *Streptomyces fradiae* yielded a maximum of 0.12 mg/ml soluble proteins after 5 days [40,42], while *Trichoderma atroviride* strain F6 and a *Streptomyces* strain produced somewhat more, maximum values reaching 0.323 and 0.423 mg/ml after 8 and 3 days, respectively, when cultivated in 10 g/l feathers [43,44]. Growing a recombinant strain of keratinase-producing *Bacillus subtilis* on sheep wool-based distilled water medium yielded however as much as 1 mg/ml soluble protein after 3 days [45].

Analysis of amino acids released during feather and wool bioconversion

The amino acid profile in feather and wool hydrolysates is presented in table 2. Feather hydrolysate was rich in tyrosine, phenylalanine, and histidine residues, while aspartic acid, methionine, tyrosine, phenylalanine, histidine, and lysine were the predominant amino acids released in wool hydrolysate. Aspartic acid was the most abundant amino acid liberated in wool hydrolysate, but the nutritionally essential amino acids, such as methionine, tyrosine, histidine, and lysine were (as mentioned above) also present in this lysate. The levels of the other amino acids were rather modest in both feather and wool lysate, suggesting the possibility of these amino acids being consumed by *Bacillus sp.* C$_4$ during the fermentation process.

The two amino acid profiles mapped in the present study were compared with those obtained by the action of various keratinases

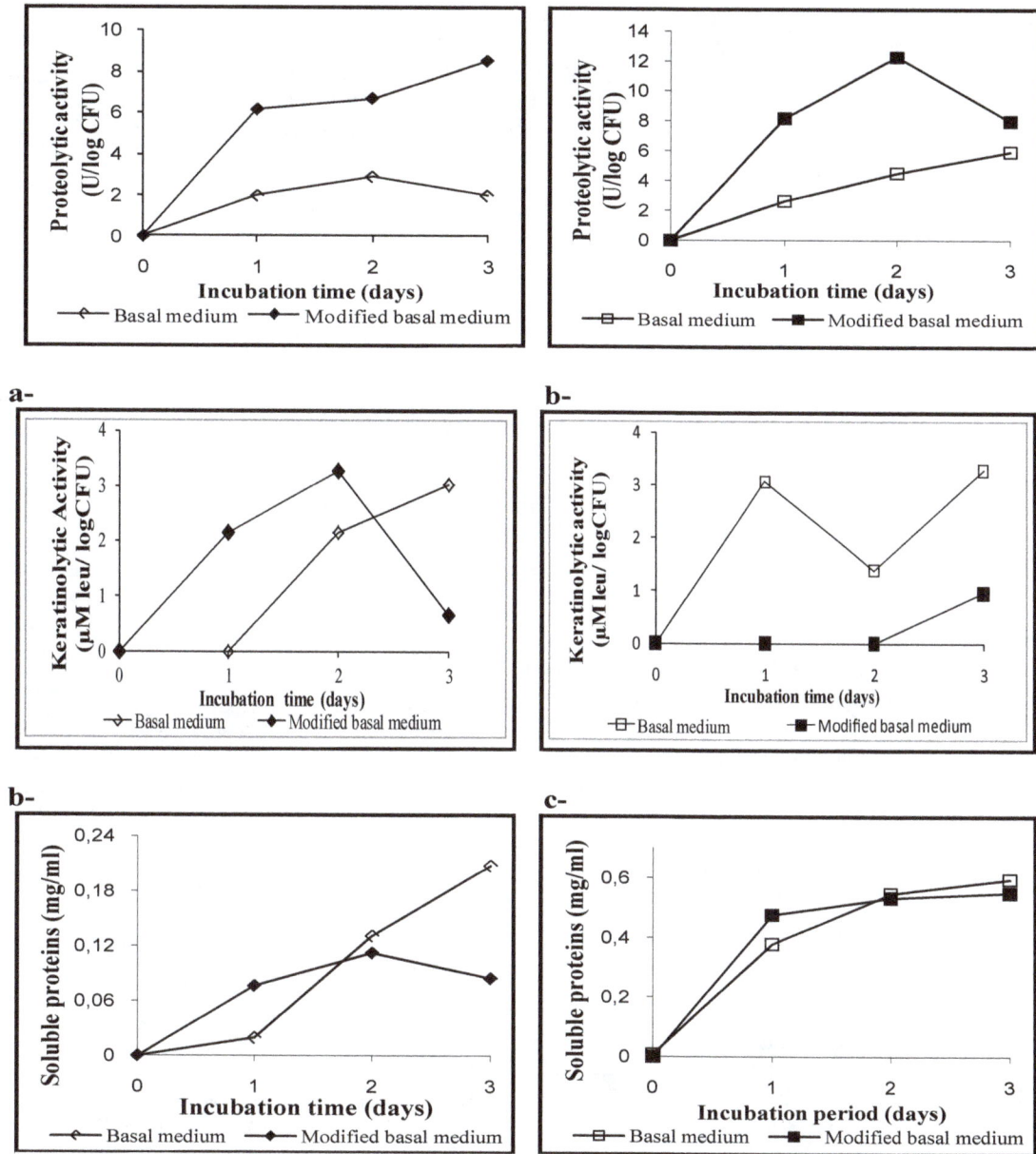

Figure 5: Proteolytic (**a**) and keratinolytic (**b**) activity as a result of cultivating *Bacillus sp.* C$_4$ on basal medium II (empty marks) and on modified basal medium II (filled marks), supplemented with chicken feather (diamond shape) and sheep wool (square shape), 10 g/l each. (**c**) Illustrates the levels of soluble proteins produced under the same conditions.

Figure 6: Physical appearance of chicken feather at three different stages: (**A**) before incubation, (**B**) on day 3 after the action of a *Bacillus sp.* C$_4$ strain on basal medium II, and (**C**) after 2 days of incubation on modified basal medium II.

Free Amino acids	Feather-keratin			Wool-keratin		
	Day 0	Day 3	Day 3/Day 0	Day 0	Day 3	Day 3/Day 0
Aspartic acid	170.65	0.75	0.004	1.00	67.86	67.86
Threonine	101.10	36.77	0.007	105.88	8.24	0.08
Serine	75.27	27.37	0.36	26.28	14.01	0.53
Glutamic acid	288.05	8.81	0.03	32.95	18.24	0.55
Proline	29.29	18.45	0.63	8.86	7.38	0.83
Glycine	116.53	1.99	0.01	20.09	9.42	0.46
Alanine	169.44	4.03	0.02	29.93	12.41	0.41
Cysteine	8.09	3.23	0.40	13.09	3.08	0.23
Valine	103.64	13.84	0.13	18.93	3.32	0.17
Methionine	79.05	11.03	0.14	26.87	97.61	3.63
Isoleucine	79.44	7.03	0.09	11.92	10.69	0.89
Leucine	137.12	8.23	0.06	26.00	2.57	0.09
Tyrosine	34.12	249.09	7.30	34.80	204.73	5.88
Phenylalanine	78.25	684.07	8.74	40.57	305.05	7.51
Histidine	19.38	62.88	3.24	30.53	86.79	2.84
Lysine	56.85	32.49	0.57	10.83	17.26	1.59
Arginine	86.45	5.40	0.06	27.99	8.65	0.31
Ammonia	118.06	214.07	1.81	167.25	184.17	1.10

Table 2: Levels of released amino acids and ammonia in cell-free supernatants of *Bacillus sp.* C_4 on day 0 and day 3 of cultivation in modified basal liquid medium II. Values are given in mg/l.

derived from other bacteria, and it appears that the amino acid profile depends on the bacterial strain. For instance, the amino acid composition of feather lysate generated by the action of *Fervidobacterium islandicum* AW-1 was rich in alanine, proline, serine, and cysteine [46], while serine, leucine, and glutamate residues dominated among the amino acids released in feather hydrolysate of *Vibrio sp.* strain kr2 [47]. In contrast, the action of *B. cereus* protease on wool keratin generated an amino acid profile rich in glutamate, serine, leucine, proline, arginine, aspartic acid, and threonine residues [48], while phenylalanine, tyrosine, and lysine were the main amino acids produced in wool lysate of a recombinant strain of *Bacillus subtilis* [45].

Conclusion

We have in the present study accomplished the isolation of a strain of *Bacillus sp.*, capable of hydrolyzing both α- and β-keratin in three days. The strain proved suitable for degradation of avian feathers and sheep wool, and its keratin hydrolysates hold a strong potential for future applications in biotechnological processes.

Acknowledgement

This research was supported by the Ministry of Higher Education and Scientific Research, Algeria.

References

1. Agrahari S, Wadhwa N (2010) Degradation of chicken feather a poultry waste product by keratinolytic bacteria isolated from dumping site at Ghazipur poultry processing plant. Int J Poult Sci 9: 482-489.

2. Tiwary E, Gupta R (2012) Rapid conversion of chicken feather to feather meal using dimeric keratinase from *Bacillus licheniformis* ER-15. J Bioprocess Biotech 2: 123.

3. Wang X, Parsons CM (1997) Effect of processing systems on protein quality of feather meals and hog hair meals. Poult Sci 76: 491-496.

4. Habbeche A, Saoudi B, Jaouadi B, Haberra S, Kerouaz B, et al. (2014) Purification and biochemical characterization of a detergent-stable keratinase from a newly thermophilic actinomycete *Actinomadura keratinilytica* strain Cpt29 isolated from poultry compost. J Biosci Bioeng 117: 413-421.

5. Williams CM, Richter CS, Mackenzie JM, Shih JC (1990) Isolation, identification, and characterization of a feather-degrading bacterium. Appl Environ Microbiol 56: 1509-1515.

6. Zheljazkov VD (2005) Assessment of wool waste and hair waste as soil amendment and nutrient source. J Environ Qual 34: 2310-2317.

7. Liu B, Zhang J, Li B, Liao X, Du G, et al. (2013) Expression and characterization of extreme alkaline, oxidation-resistant keratinase from *Bacillus licheniformis* in recombinant *Bacillus subtilis* WB600 expression system and its application in wool fiber processing. World J Microbiol Biotechnol 29: 825-832.

8. Papadopoulos MC, El Boushy AR, Ketelaars EH (1985) Effect of different processing conditions on amino acid digestibility of feather meal determined by chicken assay. Poultry Sci 64: 1729-1741.

9. Gupta R, Ramnani P (2006) Microbial keratinases and their prospective applications: an overview. Appl Microbiol Biotechnol 70: 21-33.

10. Onifade AA, Al-Sane NA, Al-Musallam AA, Al-Zarban S (1998) A review: Potentials for biotechnological applications of keratin-degrading microorganisms and their enzymes for nutritional improvement of feathers and other keratins as livestock feed resources. Bioresour Technol 66: 1-11.

11. Brandelli A, Daroit DJ, Riffel A (2010) Biochemical features of microbial keratinases and their production and applications. Appl Microbiol Biotechnol 85: 1735-1750.

12. Brandelli A (2008) Bacterial keratinases: Useful enzymes for bioprocessing agroindustrial wastes and beyond. Food Bioprocess Tech 1: 105-116.

13. Thanikaivelan P, Rao JR, Nair BU, Ramasami T (2004) Progress and recent trends in biotechnological methods for leather processing. Trends Biotechnol 22: 181-188.

14. Karthikeyan R, Balaji S, Sehgal PK (2007) Industrial applications of keratins-a review. J Sci Ind Res 66: 710-715.

15. Villa AL, Aragão MR, Dos Santos EP, Mazotto AM, Zingali RB, et al. (2013) Feather keratin hydrolysates obtained from microbial keratinases: effect on hair fiber. BMC Biotechnol 13: 15.

16. Bertsch A, Coello N (2005) A biotechnological process for treatment and recycling poultry feathers as a feed ingredient. Bioresour Technol 96: 1703-1708.

17. Mohorcic M, Torkar A, Friedrich J, Kristl J, Murdan S (2007) An investigation into keratinolytic enzymes to enhance ungual drug delivery. Int J Pharm 332: 196-201.

18. Sharma R, Gupta R (2010) Extracellular expression of keratinase Ker P from *Pseudomonas aeruginosa* in *E. coli*. Biotechnol Lett 32: 1863-1868.

19. Gupta R, Sharma R, Beg QK (2013) Revisiting microbial keratinases: next generation proteases for sustainable biotechnology. Crit Rev Biotechnol 33: 216-228.

20. Bernhard K, Schrempf H, Goebel W (1978) Bacteriocin and antibiotic resistance plasmids in *Bacillus cereus* and *Bacillus subtilis*. J Bacteriol 133: 897-903.

21. Preer JR, Preer LB (1989) Endosymbionts of protozoa. Bergey's manual of systematic bacteriology. Holt JG (editor). Volume 1. The Williams & Wilkins. Baltimore, London, pages 795-813.

22. Zaghloul TI, Kawamura F, Doi RH (1985) Translational coupling in *Bacillus subtilis* of a heterologous *Bacillus subtilis-Escherichia coli* gene fusion. J Bacteriol 164: 550-555.

23. Cliffe AJ, Law BA (1982) A new method for the detection of microbial proteolytic enzymes in milk. J Dairy Res 49: 209-219.

24. Pearce K, Karahalios DA, Friedman M (1988) Ninhydrin assay for proteolysis in ripening cheese. J Food Sci 53: 432-435.

25. Bradford MM (1976) A rapid and sensitive method for the quantitation of microgram quantities of protein utilizing the principle of protein-dye binding. Anal Biochem 72: 248-254.

26. Johnvesly B, Manjunath BR, Naik GR (2002) Pigeon pea waste as a novel, inexpensive, substrate for production of a thermostable alkaline protease from thermoalkalophilic *Bacillus sp.* JB-99. Bioresour Technol 82: 61-64.

27. Moore S, Spackman DH, Stein WH (1958) Automatic recording apparatus for use in chromatography of amino acids. Fed Proc 30: 1190-1206.

28. Kumar CG, Takagi H (1999) Microbial alkaline proteases: from a bioindustrial viewpoint. Biotechnol Adv 17: 561-594.

29. Verma T, Baiswar V (2013) Isolation and Characterization of Extracellular Thermoalkaline Protease Producing Bacillus cereus isolated from Tannery Effluent. Int J Eng Sci 2: 23-29.

30. Tekin N, Cihan AÇ, Takaç ZS, Tüzün CY, Tunç K, et al. (2012) Alkaline protease production of *Bacillus cohnii* APT5. Turk J Biol 36: 430-440.

31. Gessesse A (1997) The use of nug meal as a low-cost substrate for the production of alkaline protease by the alkaliphilic *Bacillus sp.* AR-009 and some properties of the enzyme. Bioresour Technol 62: 59-61.

32. Aikat K, Bhattacharyya BC (2000) Protease extraction in solid state fermentation of wheat bran by a local strain of *Rhizopus oryzae* and growth studies by the soft gel technique. Process Biochem 35: 907-914.

33. Prakasham RS, Rao ChS, Sarma PN (2006) Green gram husk--an inexpensive substrate for alkaline protease production by *Bacillus sp.* in solid-state fermentation. Bioresour Technol 97: 1449-1454.

34. Tari C, Genckal H, Tokatli F (2006) Optimization of a growth medium using a statistical approach for the production of an alkaline protease from a newly isolated *Bacillus sp.* L21. Process Biochem 41: 659-665.

35. Ramnani P, Kumar SS, Gupta R (2005) Concomitant production and downstream processing of alkaline protease and biosurfactant from *Bacillus licheniformis* RG1: Bioformulation as detergent additive. Process Biochem 40: 3352-3359.

36. De Azeredo LA, De Lima MB, Coelho RR, Freire DM (2006) Thermophilic protease production by *Streptomyces sp.* 594 in submerged and solid-state fermentations using feather meal. J Appl Microbiol 100: 641-647.

37. Wang SL, Hsu WT, Liang TW, Yen YH, Wang CL (2008) Purification and characterization of three novel keratinolytic metalloproteases produced by *Chryseobacterium indologenes* TKU014 in a shrimp shell powder medium. Bioresour Technol 99: 5679-5686.

38. Gessesse A, Hatti-Kaul R, Gashe BA, Mattiasson B (2003) Novel alkaline proteases from alkaliphilic bacteria grown on chicken feather. Enzyme Microb Technol 32: 519-524.

39. Ouled Haddar H, Zaghloul TI, Saeed HM (2009) Biodegradation of native feather keratin by *Bacillus subtilis* recombinant strains. Biodegradation 20: 687-694.

40. Grazziotin A, Pimentel FA, Sangali S, de Jong EV, Brandelli A (2007) Production of feather protein hydrolysate by keratinolytic bacterium *Vibrio sp.* kr2. Bioresour Technol 98: 3172-3175.

41. Ignatova Z, Gousterova A, Spassov G, Nedkov P (1999) Isolation and partial characterisation of extracellular keratinase from a wool degrading thermophilic actinomycete strain *Thermoactinomyces candidus*. Can J Microbiol 45: 217-222.

42. Hood CM, Healy MG (1994) Bioconversion of waste keratins: wool and feathers. Resour Conserv Recy 11: 179-188.

43. Cao L, Tan H, Liu Y, Xue X, Zhou S (2008) Characterization of a new keratinolytic *Trichoderma atroviride* strain F6 that completely degrades native chicken feather. Lett Appl Microbiol 46: 389-394.

44. Mohamedin A (1999) Isolation, identification and some cultural conditions of a protease-producing thermophilic *Streptomyces* strain grown on chicken feather as a substrate. Int Biodeterior Biodegrad 43: 13-21.

45. Zaghloul TI, Embaby AM, Elmahdy AR (2011) Key determinants affecting sheep wool biodegradation directed by a keratinase-producing *Bacillus subtilis* recombinant strain. Biodegradation 22: 111-128.

46. Nam GW, Lee DW, Lee HS, Lee NJ, Kim BC, et al. (2002) Native-feather degradation by *Fervidobacterium islandicum* AW-1, a newly isolated keratinase-producing thermophilic anaerobe. Arch Microbiol 178: 538-547.

47. Grazziotin A, Pimentel FA, De Jong EV, Brandelli A (2006) Nutritional improvement of feather protein by treatment with microbial keratinase. Anim Feed Sci Technol 126: 135-144.

48. Sousa F, Jus S, Erbel A, Kokol V, Cavaco-Paulo A, et al. (2007) A novel metalloprotease from *Bacillus cereus* for protein fibre processing. Enzyme Microb Technol 40: 1772-1781.

19

A Comparative Study of Aerobic Granule and Activated Sludge Based Dynamic Membrane Reactors for Wastewater Treatment

Tianyin Huang*, Yongxin Gui, Wei Wu, Kai Feng and Feng Liu

School of Environmental Science and Engineering, Suzhou University of Science and Technology, Suzhou, Jiangsu, 215011, P.R. China

Abstract

Dynamic membrane bioreactor (DMBR), as a new development of membrane bioreactor (MBR) technology, has attracted increasing attention recently. However, the effluent quality is usually affected by an incomplete rejection of sludge flocs and other suspended solids (SS) by the meshes. Herein, a novel aerobic granule dynamic membrane bioreactor (AGDMBR) is proposed. A comparison of the AGDMBR and DMBR systems was made in this study. According to the results, AGDMBR not only had higher NH4+ and turbidity removal, but also exhibited better anti-fouling capability than DMBR, implying a high potential of this technology for low-cost and efficient wastewater treatment.

Keywords: Aerobic granule (AG); Activated sludge (AS); Dynamic membrane bioreactor (DMBR); Membrane fouling; Wastewater treatment

Introduction

Membrane bioreactor (MBR), as a compact and high-efficient biological wastewater treatment technology, has seen rapid development and widespread application in the past few decades. However, high cost and membrane fouling are still two major challenges of this technology in practical application [1,2]. This has intrigued intensive studies to lower the membrane cost and improve the anti-fouling ability of membranes [3]. Among these, one important research direction is to substitute the conventional microfiltration (MF) or ultrafiltration (UF) membranes with low-cost macro-pore materials (usually above 10 μm), such as nylon mesh, stainless steel mesh and nonwoven fabrics [4-6]. These materials are used as a support media to facilitate the built-up of a layer of biocake, which features numerous micropores and micro-channel structures and can thus serve as the real filter for sludge- water separation [7,8]. Since the thickness and composition of this biocake layer may vary dynamically with the proceeding of filtration, such reactors are also referred to as dynamic membrane bioreactor (DMBR) [9,10]. This biocake can be readily in-situ removed and rapidly rebuilt on the surface of macro-pore materials, thus membrane fouling can be easily controlled [8]. Moreover, the low material cost and gravity-driven filtration mode further add up to its economic benefits over conventional MBR [11]. However, one common disadvantage of most DMBRs is a relatively high SS in the effluent, attributed to an unstable biocake layer that cannot reject all the SS. This is especially true at the initial stage of biocake formation when the size of pores is relatively large. In that case, the fine floc sludge and other particles may directly penetrate through the thin biocake layer, leading to poor effluent quality [12]. The effluent SS would decline with the built-up of a dense biocake. But when the biocake becomes too thick and dense, the filtration resistance would increase rapidly, leading to "membrane" fouling [8]. Thus, a key to the success of this technology is to maintain an appropriate biocake layer.

In view of the fact that biocake properties are closely associated with the sludge characteristics [13,14], it is thus reasonable to expect that better filtration performance can be achieved by a proper manipulation of the sludge characteristics. This creates a possibility of integrating aerobic granules (AG) into DMBR operation. AG have also attracted increasing interest recently for wastewater treatment application, attributed to its many superior properties, such as higher settleability, better treatment efficiency and easier separation, over activated sludge (AS) flocs [15,16]. On the one hand, AG, with larger size than AS, would have less chance to directly penetrate through the biocake and mesh, thus ensuring better effluent quality. On the other, a more porous biocake layer would form attributed to the more compact and strong structure of AG than AS, thus further lowering the filtration resistance and extending the stable filtration time.

Therefore, this study aims to validate the above two assumptions and compare the performances of AGDMBR and conventional DMBR for wastewater treatment. Apart from sludge characteristics, the filtration and treatment performances of this process are also significantly affected by the filtration flux. Thus, the filtration and wastewater treatment performances of these two systems under different filtration flux were evaluated.

Materials and Methods

Reactor setup and operation

The AGDMBR was consisted of a 4L column plexiglass reactor and a dynamic membrane module. A schematic diagram of the reactor and membrane module configurations is shown in Figure 1. The total effective membrane surface was 0.02 m2, and the average pore size of nylon mesh was 70 μm. Wastewater was continuously fed into the reactor at a constant rate of 0.6 L/h, except when the impact of different filtration fluxes was investigated. Effluent was discharged semi-continuously under the control of a solenoid valve at cycles of 10-min on/ 10-min off. Aeration was provided at the bottom of the

*Corresponding author: Tianyin Huang, School of Environmental Science and Engineering, Suzhou University of Science and Technology, Suzhou, Jiangsu, 215011, P.R. China

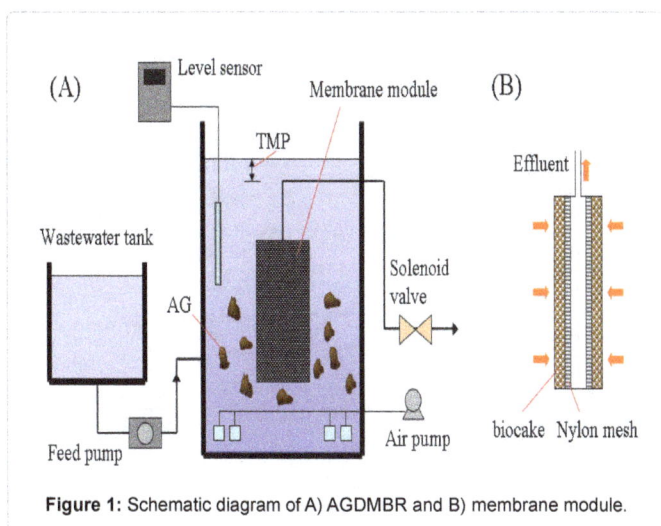

Figure 1: Schematic diagram of A) AGDMBR and B) membrane module.

reactor by an air pump. For comparison, an identical reactor but under normal DMBR mode (continuous effluent discharge and seeded with AS) was used as the control.

Gravity-driven filtration was adopted in both reactors. The transmembrane pressure (TMP) can be reflected by the water head drop across the membrane. Once the TMP increase exceeded 4 cm in water head, the membrane module was taken out to physically remove the biocake through water flushing, and then installed back for a next cycle of filtration.

AG and AS, both from other bench-scale reactors, were seeded into the AGDMBR and DMBR respectively. Synthetic wastewater, with similar composition as reported by Kimura et al. [17] except for a chemical oxygen demand (COD) concentration of 400 mg/L, was used for the experiment. During the operation, the mixed liquor suspended solids (MLVSS) concentration in both reactors were maintained at about 4 g/L, and the aeration rate were 0.1 m3/h. All the reactors were operated at ambient temperature of 25°C.

Analysis

The water levels of the reactors were real-time monitored by a level sensor (LD187, Leide Electronic Technology, China). The MLVSS, sludge volume index (SVI), COD and ammonia (NH4+) concentrations were measured following the Standard Methods (APHA-AWWA-WEF, 1998). The effluent turbidity was measured using a turbidimeter (WGZ-20, XinRui Instrument Co., China). The EPS content was analyzed using the method described in previous studies [18]. In addition, a piece of fouled nylon mesh cut from the membrane module, after removal of the surface biocake, was observed using scanning electron microscopy (SEM, XL-30 ESEM, FEI Co., USA). The procedures of pretreatment and SEM analysis were the same as described in Li et al. [8].

Results and Discussion

Characteristics of AS and AG

When treatment performances of the reactors became stable, the AG and AS sample were taken from the reactors and characterized. Compared with the fine flocs of AS, AG exhibited larger size, more compact structure, better settleability (reflected by a lower SVI) and higher content of extracellular polymer substances (EPS) (Figure 2 and Table 1).

Wastewater treatment performances

The wastewater treatment performances of the two reactors in terms of COD, NH4+ and turbidity removal were evaluated. As shown in Figure 3A, AGDMBR showed comparable high COD removal rates with DMBR (up to 97%) throughout the experimental period. However, a slightly higher NH4+ removal was achieved in the AGDMBR (95% in average) than the DMBR (89% in average), possibly attributed to a better retention of the slow-growing nitrifying bacteria in AG than AS (Figure 3B). More significant difference was observed in the effluent turbidity of the two reactors (Figure 3C), indicating a rejection of more solid particles in the AGDMBR.

Anti-fouling capabilities

Figure 4 illustrates the filtration performances of the two reactors during the 28d operation. Once the TMP exceeded 4 cm, the membrane was taken out and flushed to remove the biocake layer. During this period, the membrane was flushed for three times for the DMBR, but only once for the AGDMBR, indicating a better anti-fouling capability of the AG system than AS. Although AG had a higher EPS content than AS, this did not cause severe membrane fouling because the EPS is not a major pollutant in such DMBR system during short-period operation [8].

Furthermore, the impact of filtration flux on the fouling behaviors of the two reactors was investigated. To facilitate the comparison, both systems were operated on a continuous basis with a filtration flux of 60 L/ (m2.h). It is shown in Figure 5 that AG system showed a lower fouling rate than AS system, which is consistent with the results in Figure 4.

This better anti-fouling capability of AG might be attributed to the formation of a more porous biocake layer than AS. While the AS biocake tends to be compacted when it gets thick, the AG can remain a good porous structure attributed to its higher mechanical strength. After the 18h of filtration, the membranes were taken from both reactors to remove the surface biocake layer before SEM observation. As shown Figure 6, there was no pore blocking in the AG system attributed to a higher AG size than the mesh pore size. In contrast, a

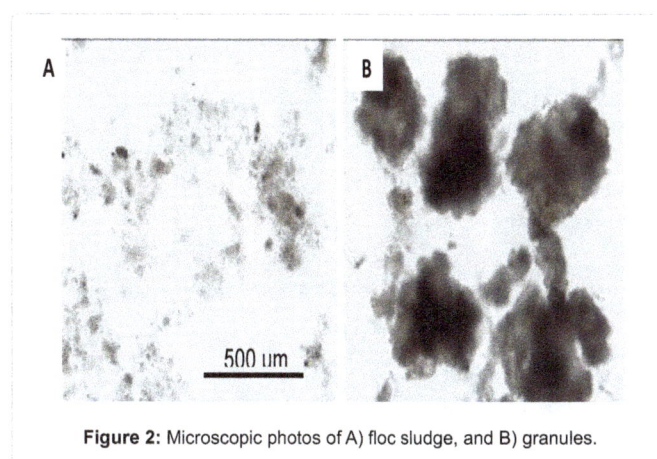

Figure 2: Microscopic photos of A) floc sludge, and B) granules.

Properties	AS	AG
SVI (mL/g)	112 ± 7	41 ± 4
Size (mm)	<0.1	0.4~1.1
EPS content (mg/g-MLVSS)	135.0 ± 10.7	94.8 ± 6.5

Table 1: Characteristics of AS and AG in DMBRs.

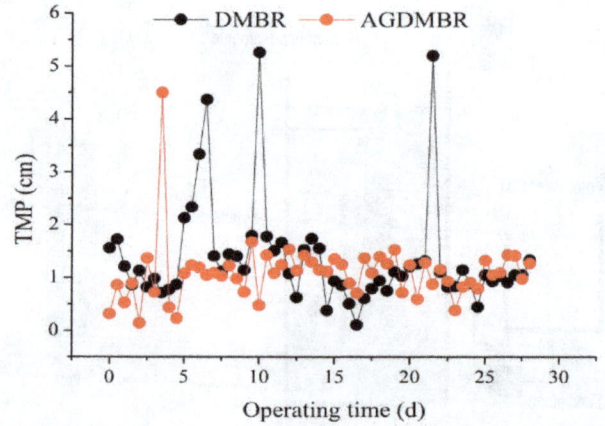

Figure 4: TMP variations during filtration operation of DMBR and AGDMBR.

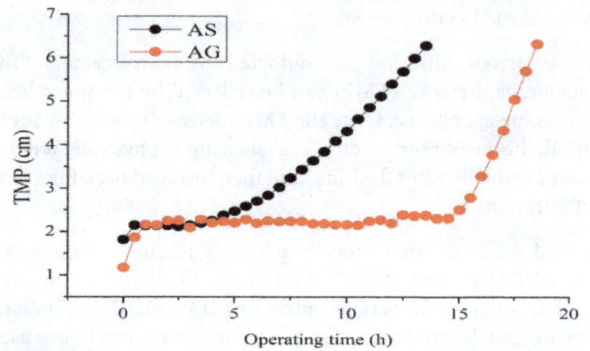

Figure 5: TMP profiles for continuous process with filtration flux of 60 L/ (m².h).

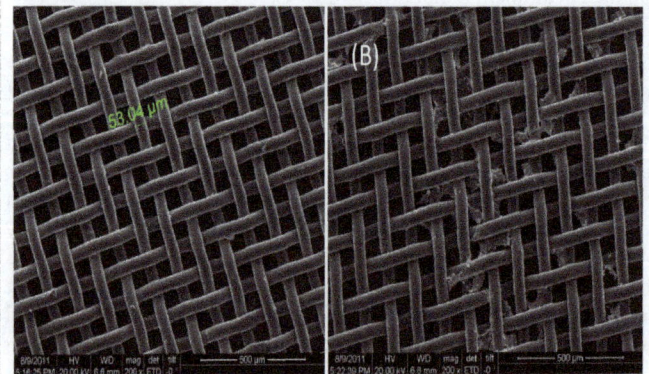

Figure 6: SEM images of fouled nylon mesh, after cleaning, from: A) AG system; B) AS system.

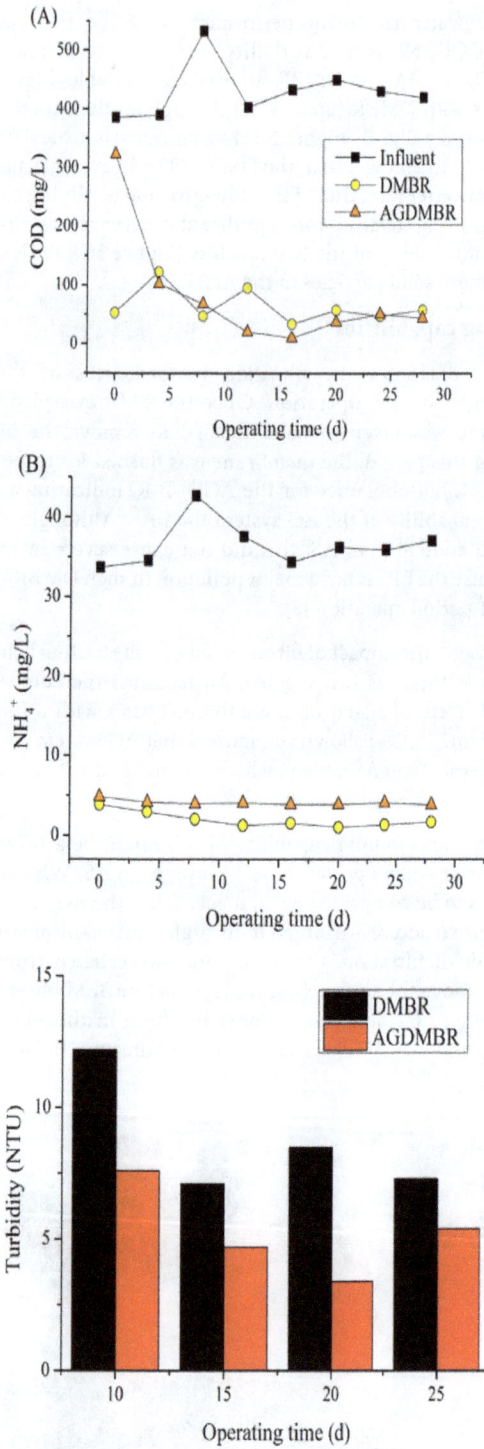

Figure 3: Treatment performance of DMBR and AGDMBR: A) COD removal; B) NH4+ removal; C) effluent turbidity.

considerable amount of AS was found in the pores, leading to higher filtration resistance.

Conclusion

In this study, a novel AGDMBR system was introduced for wastewater treatment, and its performance was compared with an AS-based DMBR. The AGDMBR featured excellent COD removal and higher NH4+ and turbidity removal than the DMBR. In addition, it also exhibited better anti-fouling capacity than DMBR attributed to the formation of a more porous biocake layer, implying a high potential of this ADGMBR process for low-cost and efficient wastewater treatment.

Acknowledgements

This study was supported financially by the Taihu lake in Jiangsu Province of the China Management Research, Grant No. TH2012201.

References

1. Meng F, Chae SR, Drews A, Kraume M, Shin HS, et al. (2009) Recent advances in membrane bioreactors (MBRs): membrane fouling and membrane material. Water Res 43: 1489-1512.

2. Amy G (2008) Fundamental understanding of organic matter fouling of membranes. Desalination 231: 44-51.

3. Phattaranawik J, Leiknes T (2011) Feasibility study of moving-fiber biofilm membrane bioreactor for wastewater treatment: process control. Water Res 45: 2227-2234.

4. Kiso Y, Jung YJ, Ichinari T, Park M, Kitao T, et al. (2000) Wastewater treatment performance of a filtration bio-reactor equipped with a mesh as a filter material. Water Res 34: 4143-4150.

5. Ren X, Shon HK, Jang N, Lee YG, Bae M, et al. (2010) Novel membrane bioreactor (MBR) coupled with a nonwoven fabric filter for household wastewater treatment. Water Res 44: 751-760.

6. Chang WK, Hu AYJ, Horng RY, Tzou WY (2007) Membrane bioreactor with nonwoven fabrics as solid-liquid separation media for wastewater treatment. Desalination 202: 122-128.

7. Liu H, Yang C, Pu W, Zhang J (2009) Formation mechanism and structure of dynamic membrane in the dynamic membrane bioreactor. Chem Eng J 148: 290-295.

8. Li WW, Sheng GP, Wang YK, Liu XW, Xu J, et al. (2011) Filtration behaviors and biocake formation mechanism of mesh filters used in membrane bioreactors. Sep Purif Technol 81: 472-479.

9. Fan B, Huang X (2002) Characteristics of a self-forming dynamic membrane coupled with a bioreactor for municipal wastewater treatment. Environ Sci Technol 36: 5245-5251.

10. Zhou XH, Shi HC, Cai Q, He M, Wu YX (2008) Function of self-forming dynamic membrane and biokinetic parameters' determination by microelectrode. Water Res 42: 2369-2376.

11. Poostchi AA, Mehrnia MR, Rezvani F, Sarrafzadeh MH (2012) Low-cost monofilament mesh filter used in membrane bioreactor process: Filtration characteristics and resistance analysis. Desalination 286: 429-435.

12. Fuchs W, Resch C, Kernstock M, Mayer M, Schoeberl P, et al. (2005) Influence of operational conditions on the performance of a mesh filter activated sludge process. Water Res 39: 803-810.

13. Fan F, Zhou H, Husain H (2006) Identification of wastewater sludge characteristics to predict critical flux for membrane bioreactor processes. Water Res 40: 205-212.

14. Farizoglu B, Keskinler B (2006) Sludge characteristics and effect of crossflow membrane filtration on membrane fouling in a jet loop membrane bioreactor (JLMBR). J Membrane Sci 279: 578-587.

15. Adav SS, Lee DJ, Show KY, Tay JH (2008) Aerobic granular sludge: recent advances. Biotechnol Adv 26: 411-423.

16. Liu XW, Sheng GP, Yu HQ (2009) Physicochemical characteristics of microbial granules. Biotechnol Adv 27: 1061-1070.

17. Kimura K, Naruse T, Watanabe Y (2009) Changes in characteristics of soluble microbial products in membrane bioreactors associated with different solid retention times: Relation to membrane fouling. Water Res 43: 1033-1039.

18. Sheng GP, Yu HQ (2006) Chemical-equilibrium-based model for describing the strength of sludge: taking hydrogen-producing sludge as an example. Environ Sci Technol 40: 1280-1285.

20

Extraction of Proteins from Mackerel Fish Processing Waste Using Alcalase Enzyme

Ramakrishnan VV, Ghaly AE*, Brooks MS, Budge SM

Department of Process Engineering and Applied Science, Dalhousie University, Halifax, Nova Scotia, Canada

Abstract

Fish proteins are found in the flesh, head, frames, fin, tail, skin and guts of the fish in varying quantities. Unutilized fish and fish processing waste can be used to produce fish proteins which contain amino acids and many bioactive peptides. After removing the fish flesh during the fish processing operation, all other parts are considered wastes which are not properly utilized. The aim of this study was to evaluate the enzymatic extraction of protein from mackerel fish processing waste. Enzymatic extraction of proteins was carried out using alcalase enzyme at three concentrations (0.5, 1 or 2%) and four hydrolysis times (1, 2, 3 and 4 h). The fish protein hydrolysate was dried using a spray dryer to obtain protein powder. The highest protein yield (76.30% from whole fish and 74.53% from the frame) was obtained using 2.0% enzyme concentration after 4 h of hydrolysis. The results showed that increasing the enzyme concentration from 0.5 to 2% (400%) increased the protein yield by 3.13- 43.52% depending upon the fish part and reaction time used. Increasing the enzyme concentration by 4 fold for a small increase in protein yield may appear unjustified. Therefore, the enzyme concentration of 0.5% should be used for the protein extraction unless the enzyme is recycled or an immobilized reactor is used in order to reduce the cost associated with the enzyme. Also, increasing the hydrolysis time from 1 to 4 h (400%) increased the protein yield by 16.45 - 50.82% depending upon the fish part and enzyme concentration used. Increasing the hydrolysis time by 4 fold for a small increase in protein yield will increase the capital and operating costs of protein production. A shorter hydrolysis time will allow more throughput and/or reduce the volume of the reactor thereby reducing the cost of protein extraction. Therefore, a 1 h reaction time for protein extraction is recommended. The results showed that the combined fish waste can be used for protein extraction without any segregation.

Keywords: Fish; Fish processing; Fish waste; Protein; Oil; Alcalase; Spray drying; Zipper mechanism

Introduction

Canada is currently ranked 19th in marine fisheries and 26th in aquaculture production in the world. Fish production is one of the major industries in Canada with total landings of 876,277 tones having a value of $2,165,608 in the year 2011. Aquaculture production contributed 163,036 tones which is about 15.6% of the total fish production [1]. Statistics Canada [2] reported that in the year 2011/2012, 55,587 metric tons of fish waste was produced in Canada.

However, there are several environmental issues related to the fishing industry. The Government of Canada has granted permission to the provincial fisheries to dump fish waste into the ocean. This included 459 permits to the Atlantic fisheries to dump 732,770 tones, 40 permits to the Quebec fisheries to dump 22,520 tones and 1 permit to the Pacific fisheries and Yukon to dump 2,800 tons of fish waste in the sea during the period of April 2000 to March 2011 [3]. In addition, the aquaculture industry generates significant amounts of fish wastes that pollute surface and ground water. These organic wastes include nitrates and phosphates and can stimulate algal bloom resulting in oxygen depletion and destruction of aquatic life [4]. However, the underutilized and spoiled whole fish and fish processing wastes can be utilized for the production of value added products such as proteins, amino acids, biodiesel, glycerol, omega-3-fatty acids, meal, silage, lactic acid, ethanol and methanol.

Fish protein concentrates (FPC) are usually prepared from whole fish or fish waste by removing most of the water and fat contents from minced fish by chemical processes. The process is quite complicated and is carried out with solvents such as ethanol or propanol at 75°C. This process yields FPC having 80% protein content. FPC is made commercially in several countries including Canada, USA, UK, Norway, Sweden, Morocco, South Africa, Chile and Peru. However, the main problems with this process are: (a) the fishy flavor of the product, (b) some of the flavor is lost during this process, (c) the presence of fat which causes rancidity, (d) oxidized fat is undesirable in the final product and (e) complete removal of solvent is difficult [5].

Enzymatic hydrolysis of underutilized and spoiled fish was previously carried out to extract proteins for use as animal feed or fertilizer [6-8]. However, fish processing waste can also be converted into proteins for human consumption [9-12] and oil for bio-energy [13-16]. The work in this study involves a laboratory scale enzymatic hydrolysis of whole mackerel fish and fish waste (frames, head, fin, tail, skin and gut) for the production of protein and further hydrolysis to amino acids.

Objectives

The main aim of this study was to evaluate the enzymatic extraction of protein from whole mackerel fish and fish parts (whole fish, head, fin, tail, skin and gut, and frames). The specific objectives were: (a) to study the effectiveness at various concentrations of the enzyme alcalase (0.5, 1 and 2%) for extracting protein from fish and fish waste and (b) to study the effect of the hydrolysis time (1, 2, 3 and 4 h) on the extraction yield of protein.

*Corresponding author: Ghaly AE, Professor, Department of Process Engineering and Applied Science, Faculty of Engineering, Dalhousie University, Halifax, Nova Scotia, Canada

Materials and Methods

Fish samples

Whole frozen mackerel fish was obtained from Clearwater, Bedford, Nova Scotia, Canada. The mackerel fish waste was obtained from Sea Crest Fisheries, Nova Scotia, Canada. The fish and fish waste were collected in sealed plastic bags and transported to the Biological Engineering Laboratory and stored in a freezer at -20°C.

Chemicals and enzymes

The enzyme alcalase (≥2.4 U/g) used in this study was obtained from Sigma (Catalogue No. P4860, Sigma-Aldrich, Oakville, Ontario, Canada). The chemicals used in the study included: potassium phosphate monobasic, potassium phosphate dibasic, concentrated sulfuric acid, concentrated hydrochloric acid, Bovine serum albumin, copper sulfate, sodium carbonate, sodium tartrate, 2N Folin Ciocalteu's Phenol Reagent, trichloroacetic acid and acetone. All the chemicals were obtained from Sigma-Aldrich, Oakville, Ontario, Canada. The reagents used in this study included: 1N hydrochloric acid, 6N hydrochloric acid, 1N sodium hydroxide, 20% trichloroacetic acid, 1M potassium phosphate monobasic and 1M potassium phosphate dibasic. 1M pH 8 phosphate buffer was prepared by adding 94.7 ml of 1M potassium phosphate mono-basic and 5.3 ml of 1M potassium phosphate dibasic with 100 ml of distilled water.

Experimental procedure

The enzymatic extraction of protein was carried out according to the procedure described in (Figure 1).The whole mackerel fish was minced in a homogenizer (Model No.4532s/s, Hobart Manufacturing Co. Ltd, Ontario, Canada) without adding water. The minced fish sample (50 g) was first placed in a 500 ml glass bottle and heated in a water bath (Precision 280 Series, Thermo Scientific, Marietta, Ohio, USA) at 90°C for 10 min before the extraction to deactivate the endogenous enzymes. Then, 50 ml of 1M potassium phosphate buffer (pH 7.5) was added to the fish in the ratio of 1:1 (fish : buffer) and mixed with the mixture using a magnetic stirrer (Corning Magnetic Stirrer PC 210, Thermo Scientific, Marietta, Ohio, USA). The total volume was found to be 100 ml. The pH of the mixture was measured using a pH meter (Orion 5 Star pH meter, Thermo Scientific, Billerica, Massachusetts, USA) and adjusted to 7.5 with 1N NaOH. The glass bottle was then placed in a water bath shaker (Precision 2870 Series, Thermo Scientific, Marietta, Ohio, USA) operating at 140 rpm and 55°C and kept for 30 min. The temperature was measured using a thermometer (Fisher Scientific, Montreal, Quebec, Canada). The enzymatic hydrolysis was started by adding 0.5% (by weight of raw material) alcalase. After hydrolyzing the protein for 1h, the mixture was taken and placed in another water bath (Precision 280 Series, Thermo Scientific, Marietta, Ohio, USA) operating at 90°C for 5min to inactivate the enzymes.

The mixture was then allowed to cool and centrifuged (Sorvall RT1 Centrifuge, Thermo Scientific, Marietta, Ohio, USA) at 4100 rpm for 40 min. Four layers were formed in the centrifuge tubes: an upper oil layer, a light-lipid layer, a soluble clear protein layer and a bottom sludge layer containing the remaining fish tissues (Figure 2). The upper oil layer was carefully removed using a pipette. The soluble protein layer and the lipid layer were removed by tilting the centrifuge tubes and pouring the material into a filtration funnel without disturbing the sludge. The remaining sludge layer was discarded. The protein hydrolysate was obtained by filtering out the lipid layer from the liquid protein using Fisher P5 filter paper (Catalogue No. S47571C, Fisher Scientific, Montreal, Quebec, Canada). The protein hydrolysate was first analyzed using Lowry method. Then, the filtered protein was spray dried using

a spray dryer (Mini Pulvis Spray GS-310, Yamato Scientific America, Santa Clara, California, USA) at an inlet temperature of 130°C, outlet temperature of 90°C and flow rate of 1-2 ml/min. The spray dried protein was stored at -20°C until used for yield determination.

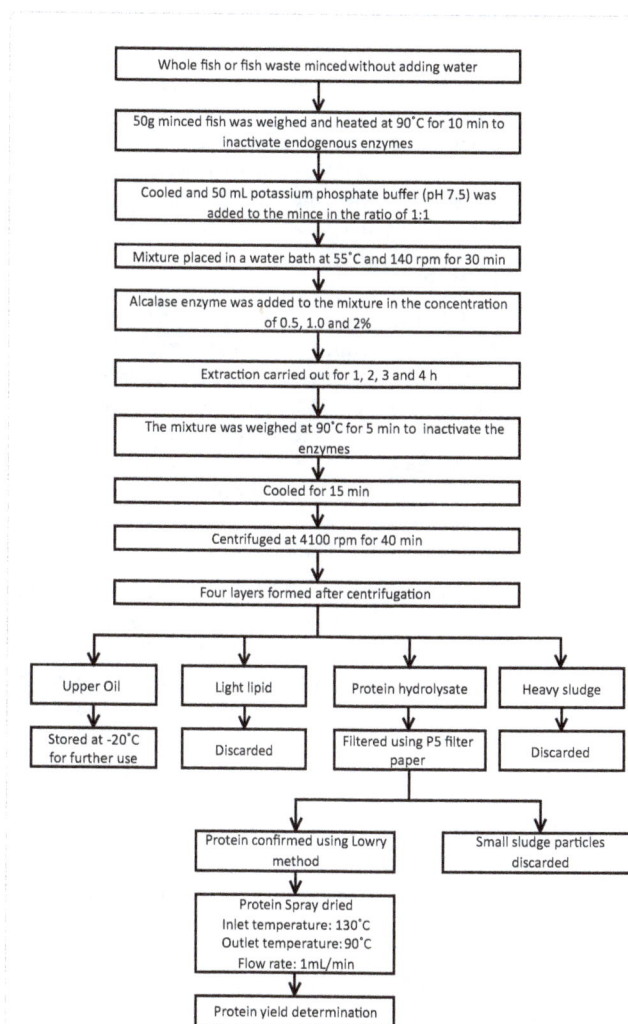

Figure 1: Enzymatic extraction of protein from whole fish.

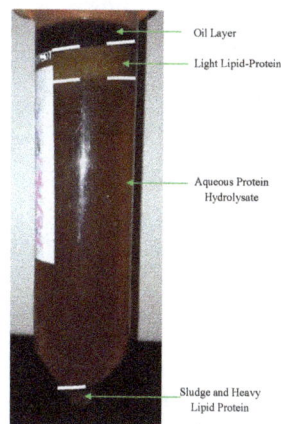

Figure 2: Four layers when recovering soluble fish protein hydrolysate.

The enzymatic hydrolysis of proteins from the various parts of mackerel fish (head, frames and fins, tails, gut and skin together) was carried out using the same procedure. The same procedure was also repeated with all reaction times and enzyme concentrations.

Chemical analyses

The moisture content was analyzed by the oven drying method, the fat content was analyzed using the chloroform/methanol method of lipid extraction, the protein content was analyzed using the Kjeldhal method, the ash content was analyzed using the standard ASTM E1755-01 and the carbohydrate content was analyzed using the standard ASTM D5896-96 in the fish samples. These analyses were performed by Nova West Laboratory Ltd., Saulnierville, Nova Scotia, Canada.

Determination of protein

The protein concentration was determined by the Lowry method according to the procedure described by Gerhardt et al. [17]. In this procedure, the proteins were first pretreated with copper ions in an alkali solution. The aromatic amino acids in the treated sample reduced the phosphomolybdic-phosphotungstic acid present in the Folin reagent. Since the endpoint of the reaction has a blue color, the amount of protein in the sample could be estimated by reading the absorbance using a spectrophotometer (Genesys 10 S UV-VIS spectrophometer, Thermo Scientific, Ohio, USA) at 750 nm.

Solution A was prepared by mixing 2.8598 g NaOH and 14.3084 g Na_2CO_3 in 500 ml distilled water in a volumetric flask. Solution B was made by adding 1.4232 g $CuSO_4.5H_2O$ to 100 ml distilled water in a volumetric flask. Solution C was prepared by adding 2.85299 g sodium tartarate to 100 ml water in a volumetric flask. Lowry solution was prepared fresh daily by combining solutions A + B + C together in the ratio 100:1:1. Folin Reagent was prepared fresh by adding 5 ml of 2N Folin Ciocalteu's Phenol Reagent to 6 ml distilled water. Bovine serum albumin (BSA) was used as the standard protein solution.

Several concentrations were prepared for the standard curve. 0.05 g of BSA was added to a 500 ml volumetric flask containing distilled water. The final concentration of the BSA solution was 100 mg/l. Dilutions ranging from 0 to 100 mg/L were prepared. The absorbance values of the BSA standards were plotted against the BSA concentrations as shown in (Figure 3).

The percentage of protein recovery from the fish was defined as the ratio of protein yield obtained during the extraction process by spray drying to the amount of protein estimated Kjeldhal protein multiplied by 100 and was calculated as follows:

$$Percent\ Protein\ Recovery\,(\%) = \frac{Recovered\ protein\,(\%)}{Kjeldhal\,protein\,(\%)} \times 100 \qquad (1)$$

The total protein yield from the fish was defined as the concentration of protein in the raw material multiplied by the average weight of part and was calculated as follows:

$$Total\ Protein\ Yield\,(g) = \frac{Recovered\ protein\,(g)}{Weight\ of\ raw\ material\,(g)} \times Average\ weight\ of\ fish\ part\,(g)\,(2)$$

Statistical analyses

Statistical analyses were performed on protein and oil results using Minitab Statistics Software (Version 16.2.2, Minitab Inc., Canada). Both analysis of variance (ANOVA) and Tukey's grouping were carried out.

Results

Weight distribution and nutritional composition

The average weight of a whole fish was 487.11 g. The weight distribution of the different parts of the fish is shown in (Table 1). The flesh, head, frame, fins and tails, skin and gut make up 286.91g (58.90%), 75.87g (15.58%), 37.12g (7.62%), 5.71g (1.17%), 34.74g (7.13%), and 36.69g (7.53%), respectively. About 3.26% of fish tissue was lost during the cutting of fish and preparing the samples.

The nutritional composition (moisture, protein fat, carbohydrate and ash contents) of whole fish and fish parts are shown in (Table 2).The average protein, fat, carbohydrate and ash contents were 15.57, 16.52, 0.65 and 1.68% for the whole fish, 12.30, 17.16, 1.17 and 3.74% for the head, 14.16, 10.43, 0.31 and 3.48% for the frame and 12.18, 20.84, 0.00 and 1.36% for the fins, tails, skin and gut, respectively. The whole fish had the highest protein content (15.57%) while the fins, tails, skin and gut had the highest fat content (20.84%) and the head had the highest carbohydrate content (1.17%).

Protein yield

Fish protein was extracted from the whole fish (WF), head (H), frames (F) and fin, tail, skin and gut (FTSG). The extraction was carried out using three enzyme concentrations (0.5, 1.0, or 2.0%) and four reaction times (1, 2, 3 and 4 h). After extraction, the fish protein hydrolysate was obtained by centrifugation and then spray dried to obtain dried protein powder. The protein powder was weighed and the amount of protein recovered (yield) from each sample was calculated. The results are shown in Table 3.

Analysis of variance (ANOVA) and Tukey's grouping were performed on the protein yield data as shown in Tables 4 and 5. The effects of fish parts, enzyme concentration and hydrolysis time were significant at the 0.001 level. There were significant interactions among the various parameters at the 0.001 level. The results obtained from the Tukey's grouping indicated that head and frame were not significantly different from each other but were significantly different from the whole fish (WF) and fin, tail, skin and guts (FTSG) at the 0.05 level. The highest protein yield (61.02%) was obtained from the frame (F). The three, enzyme concentrations (0.5, 1.0 and 2.0%) were significantly different from each other at the 0.05 level. The highest protein yield (65.30%) was achieved at the 2% enzyme concentration.

Effect of enzyme concentration on protein yield

The effect of enzyme concentrations (0.5, 1 and 2%) at different hydrolysis times (1, 2, 3 and 4 h) on the protein yield from different parts are shown in Figure 4. The results indicated that increasing enzyme concentration from 0.5 to 2% increased the protein yield for all fish parts.

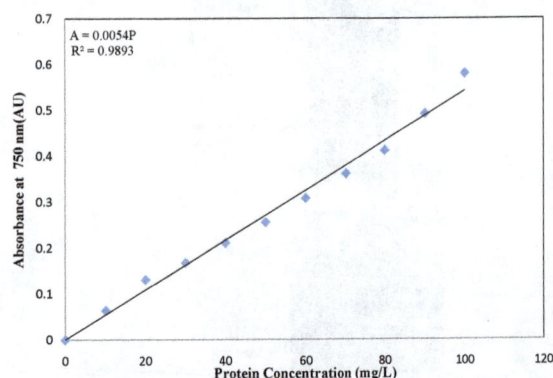

Figure 3: Standard curve for protein concentration.

Sample	Whole (g)	Flesh (g)	Waste (g)	Head (g)	Frames (g)	Fins & Tails (g)	Skin (g)	Gut (g)	Loss (g)
1	470.03	256.41	213.62	82.23	41.87	7.62	31.51	38.53	11.86
2	403.18	213.20	189.98	66.34	19.48	3.67	41.96	32.73	25.80
3	514.74	302.63	212.11	79.41	38.71	9.73	32.43	32.18	19.65
4	492.58	301.65	190.93	81.70	26.58	5.32	38.08	39.25	0.00
5	366.48	187.79	178.69	63.57	36.66	4.77	30.73	31.17	11.79
6	580.41	340.69	239.72	85.73	53.06	6.07	30.30	53.12	11.44
7	438.25	246.08	192.17	65.76	30.79	3.42	38.26	32.3	21.64
8	557.25	343.86	213.39	80.94	34.05	6.23	30.95	32.3	28.92
9	529.35	313.98	215.37	73.57	52.96	4.79	38.17	37.85	8.03
10	518.85	302.85	216.00	79.49	38.96	5.5	35.01	37.52	19.52
Average	487.11 ± 67.89	286.91 ± 52.79	206.20 ± 17.93	75.87 ± 7.97	37.31 ± 10.56	5.71 ± 1.87	34.74 ± 4.12	36.69 ± 7.53	15.87 ± 8.79
Percentage	100.00	57.67	42.33	15.58	7.66	1.17	7.13	7.53	3.26

Table 1: Weight distribution of mackerel fish parts.

Sample	Moisture (%)	Protein (%)	Fat (%)	Carbohydrate (%)	Ash (%)
Head	65.63	12.30	17.16	1.17	3.74
Frames	71.62	14.16	10.43	0.31	3.48
Fins, Tails, Skin and Gut	65.62	12.18	20.84	0.00	1.36
Whole Fish	65.58	15.57	16.52	0.65	1.68

Table 2: Nutritional composition of mackerel fish and fish waste.

Enzyme Concentration (%)	Hydrolysis Time (h)	Protein Yield							
		Whole Fish		Head		Frame		FTSG	
		(g)	(%)	(g)	(%)	(g)	(%)	(g)	(%)
0.5	1	3.53 ± 0.15	7.06	3.28 ± 0.03	6.56	3.52 ± 0.11	7.04	2.48 ± 0.06	4.96
	2	3.86 ± 0.12	7.72	3.55 ± 0.02	7.10	3.97 ± 0.01	7.94	2.75 ± 0.04	5.50
	3	4.53 ± 0.11	9.06	3.75 ± 0.04	7.50	4.19 ± 0.02	8.38	2.95 ± 0.03	5.90
	4	5.21 ± 0.01	10.42	3.82 ± 0.02	7.64	4.39 ± 0.03	8.78	3.28 ± 0.02	6.56
1	1	3.67 ± 0.03	7.34	3.22 ± 0.02	6.44	3.81 ± 0.04	7.62	2.96 ± 0.02	5.92
	2	4.17 ± 0.10	8.34	3.53 ± 0.02	7.06	4.25 ± 0.02	8.50	3.37 ± 0.05	6.74
	3	5.14 ± 0.12	10.28	3.75 ± 0.02	7.50	4.46 ± 0.02	8.92	3.58 ± 0.04	7.16
	4	5.54 ± 0.01	11.08	3.92 ± 0.03	7.84	4.68 ± 0.02	9.36	3.78 ± 0.02	7.56
2	1	4.24 ± 0.11	8.48	3.39 ± 0.02	6.78	3.63 ± 0.02	7.26	3.45 ± 0.04	6.90
	2	4.86 ± 0.05	9.72	4.00 ± 0.09	8.00	4.71 ± 0.03	9.42	3.95 ± 0.03	7.90
	3	5.44 ± 0.02	10.88	4.22 ± 0.02	8.44	4.97 ± 0.05	9.94	4.14 ± 0.02	8.28
	4	5.94 ± 0.02	11.88	4.36 ± 0.03	8.72	5.28 ± 0.02	10.56	4.26 ± 0.02	8.52

Sample size= 50 g
Whole fish = 7.78% (15.57%)
Head =6.15% (12.30%)
Frame = 7.08% (14.16%) and
Fin, tail, skin and gut (FTSG) = 6.09% (12.18%)

Table 3: Protein yield.

For the 0.5% enzyme concentration and 1 h hydrolysis time, the protein yield was 45.34, 53.39, 49.67 and 40.78% for the whole fish (WF), head (H), frame (F) and fin, tail, skin and guts (FTSG), respectively. When the enzyme concentration was increased from 0.5 to 1%, the protein yield increased from 45.34 to 47.18% (4.06%), from 40.78 to 48.60% (19.19%) and from 49.67 to 53.86% (8.43%) for the whole fish (WF), fin, tail, skin and guts (FTSG) and frame (F), respectively and decreased from 53.39 to 52.41% (1.83%) for head (H). However, when the enzyme concentration was further increased from 1 to 2%, the protein yield decreased from 53.86 to 51.22% (4.89%) for frame (F), and increased from 47.18 to 54.46% (15.42%), from 48.60 to 56.70% (16.66%), from 52.41 to 55.18% (3.29%), from whole fish (WF), fin, tail, skin and guts (FTSG) and head (H), respectively. Similar trends were observed with other hydrolysis times (2, 3 and 4 h). The decrease in the yield for the head (H) is because the low enzyme concentration of 0.5% enzyme concentration and the small hydrolysis time of 1 h were not enough to release all the protein into the system.

Effect of hydrolysis time on protein yield

The effect of hydrolysis time (1, 2, 3 and 4 h) on the protein yield at different enzyme concentrations (0.5, 1 and 2%) is shown in Figure 5. The results indicated that increasing hydrolysis time from 1 to 4 h increased the protein yield from the different parts of fish. There was no protein yield observed at zero time from any of the fish parts. However, there was a significant increase in the protein yield from all fish parts during the first hour.

For the 0.5% enzyme concentration, an increase in the hydrolysis times from 1 to 4 h, increased the protein yield from 45.34 to 66.92% (47.59%), from 40.78 to 53.80% (31.94%), from 53.39 to 62.17% (16.44%) and from 49.67 to 61.96% (24.73%) for the whole fish (WF), fin, tail, skin and gut (FTSG), head (H) and frame (F) increased, respectively. Similar trends were observed at the other enzyme concentrations. Increases of 50.82, 27.59, 21.71 and 22.64% and 40.09, 23.36, 28.39 and 45.50% with the 1% and 2% enzyme concentrations were observed for

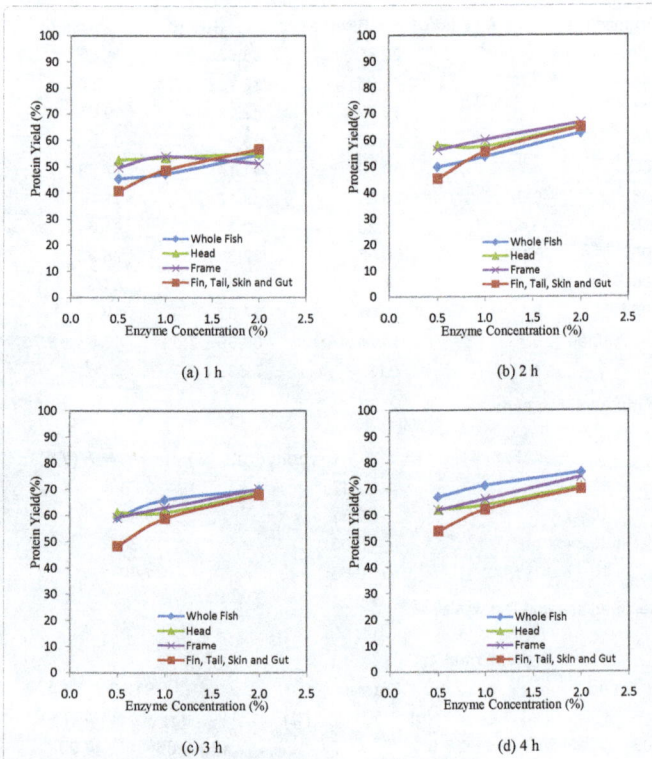

Figure 4: Effect of enzyme concentrations on protein yield from different fish parts at different reaction times.

Source	DF	SS	MS	F	P
Total	143	9836.37			
Model					
Parts	3	587.33	195.78	395.34	0.001
EC	2	2888.51	1444.26	2916.43	0.001
HT	3	5083.29	1694.43	3421.61	0.001
Parts*EC	6	486.24	81.04	163.65	0.001
Parts*HT	9	486.34	54.04	109.12	0.001
EC*HT	6	128.11	21.35	43.12	0.001
Parts*EC*HT	18	129.00	7.17	14.47	0.001
Error	96	47.54	0.50		

DF = Degree of freedom
SS = Sum of square
MS =Mean of square
EC = Enzyme concentration
HT = Hydrolysis time
R^2= 99.52
CV = 13.94%

Table 4: Analysis of variance for protein yield.

whole fish (WF), fin, tail, skin and gut (FTSG), head (H) and frame (F), respectively.

Total protein recovery

The total protein recovered from various fish parts is shown in (Table 6). The results indicated the highest amount of protein was recovered from whole fish (WF), followed by the head (H) and fin, tail, skin and gut (FTSG) and the lowest amount of protein was recovered from the frame (F).

Discussion

The selection of enzyme plays an important role in the extraction

of proteins from fish and fish waste. Several researchers [9,10,18-20] reported that alcalase was the best enzyme for the extraction of proteins from fish and fish waste. Kristinsson and Rasco [9] stated that alcalase is prominently used in the hydrolysis of proteins from fish due to its high degree of hydrolysis in a relatively short time.

In this study, the enzymatic hydrolysis of the whole fish and fish waste (head, frame, fin, tail, skin and gut) was carried out with the enzyme alcalase to obtain protein for further use as a substrate for production of amino acids. The extraction process of protein was carried out at 55°C which is the optimum temperature for alcalase as reported by several authors [12,19-22]. The effects of enzyme concentration (0.5, 1 or 2%) and hydrolysis time (1, 2, 3 and 4 h) were investigated in order to determine the protein yield from whole fish and various fish parts. The highest protein yield from whole fish (WF), head (H), frame (F)

Factors		Level	N	Mean Yield (%)	Tukey Grouping
Parts		Whole fish	36	60.17	A
		Head	36	60.73	B
		Frame	36	61.02	B
		FTSG	36	56.03	C
Enzyme Concentration (%)		0.5	48	54.40	A
		1.0	48	58.76	B
		2.0	48	65.30	C
Hydrolysis time (h)		1	36	50.73	A
		2	36	57.81	B
		3	36	62.78	C
		4	36	66.63	D

Groups with the same letter are not significantly different from each other at the 0.05 level.
FTSG = Fin, tail, skin and gut

Table 5: Tukey's grouping on protein yield.

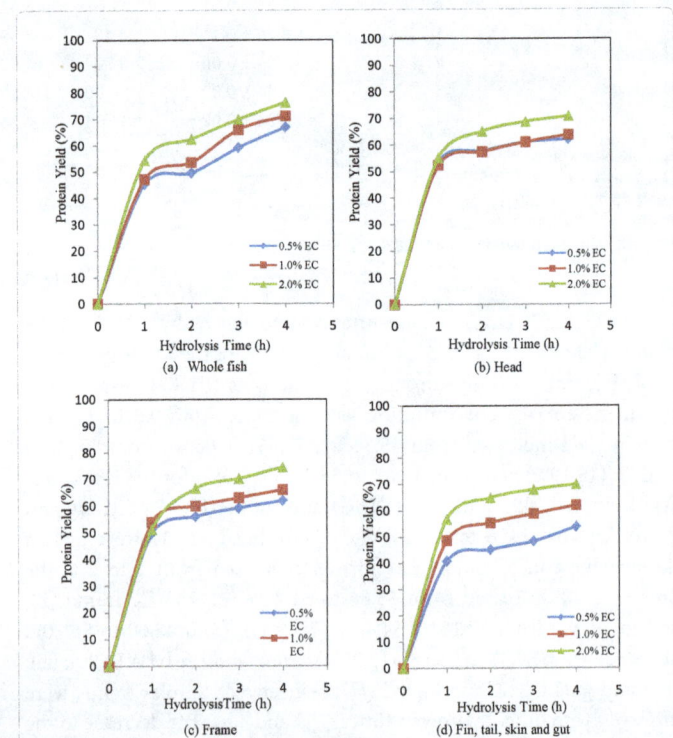

Figure 5: Effect of hydrolysis time on protein yield from different fish parts at different enzyme concentrations.

Enzyme Concentration (%)	Hydrolysis Time (h)	Whole Fish		Head		Frame		FTSG	
		Protein Yield (g)	Total Protein (g)	Protein Yield (g)	Total Protein (g)	Protein Yield (g)	Total Protein (g)	Protein Yield (g)	Total Protein (g)
0.5	1	3.53 ± 0.15	34.39	3.28 ± 0.03	4.98	3.52 ± 0.11	2.61	2.48 ± 0.06	3.83
	2	3.86 ± 0.12	37.60	3.55 ± 0.02	5.39	3.97 ± 0.01	2.95	2.75 ± 0.04	4.24
	3	4.53 ± 0.11	44.13	3.75 ± 0.04	5.69	4.19 ± 0.02	3.11	2.95 ± 0.03	4.55
	4	5.21 ± 0.01	50.76	3.82 ± 0.02	5.80	4.39 ± 0.03	3.26	3.28 ± 0.02	5.06
1.0	1	3.67 ± 0.03	35.75	3.22 ± 0.02	4.89	3.81 ± 0.04	2.83	2.96 ± 0.02	4.57
	2	4.17 ± 0.10	40.62	3.53 ± 0.02	5.36	4.25 ± 0.02	3.16	3.37 ± 0.05	5.20
	3	5.14 ± 0.12	50.07	3.75 ± 0.02	5.69	4.46 ± 0.02	3.31	3.58 ± 0.04	5.52
	4	5.54 ± 0.01	53.97	3.92 ± 0.03	5.95	4.68 ± 0.02	3.47	3.78 ± 0.02	5.83
2.0	1	4.24 ± 0.11	41.31	3.39 ± 0.02	5.14	3.63 ± 0.02	2.69	3.45 ± 0.04	5.32
	2	4.86 ± 0.05	47.35	4.00 ± 0.09	6.07	4.71 ± 0.03	3.50	3.95 ± 0.03	6.09
	3	5.44 ± 0.02	53.00	4.22 ± 0.02	6.40	4.97 ± 0.05	3.69	4.14 ± 0.02	6.39
	4	5.94 ± 0.02	57.87	4.36 ± 0.03	6.62	5.28 ± 0.02	3.92	4.26 ± 0.02	6.57

Whole Fish = 487.11 g;
Head = 75.87 g;
Frame = 37.12 g
Fin Tail Skin and Gut (FTSG) = 77.14 g

Table 6: Total protein recovered from fish parts.

and fin, tail, skin and gut (FTSG) after 4 h of hydrolysis using a 2% enzyme concentration was 76.30, 70.84, 74.53 and 69.95%, respectively. The enzyme concentration and the reaction time had significant effects on the protein yield.

Enzyme concentration

In this study, the enzyme concentration was kept below 2%. The results showed that when the enzyme concentration was increased from 0.5 to 1%, while using 1 h hydrolysis time, the increases in protein yield were 4.06, 1.82, 8.43 and 19.19%. When the enzyme concentration was further increased from 1 to 2% using the same hydrolysis time, additional increases in protein yields of 15.42, 5.27, 4.89 and 16.16% were obtained from the whole fish (WF), head (H), frame (F) and fin, tail, skin and gut (FTSG), respectively. Similar trends were seen with other hydrolysis times for all fish parts. The increase in enzyme concentration (from 0.5 to 2%) increased the protein yield because more enzyme molecules became associated with fish particles thus releasing more protein molecules into the system [9,19]. Benjakul and Morrissey [23] obtained similar results and indicated that the increase in the alcalase enzyme concentration increased the overall proteolysis rate and the solubilization of protein. Tello et al., Marquez and Vazquez and Moreno and Cuadrado [24-26] reported similar results from kinetic studies involving the hydrolysis of whey-proteins, bovine hemoglobin and vegetable proteins, respectively.

Shahidi et al. [19] reported that during the initial phase of hydrolysis, bulk soluble proteins were released and no increase in the release of soluble hydrolysates was seen when additional enzyme was added to the system during the stationary phase of the hydrolysis. This may be due to the product inhibition present during the hydrolysis or due to total cleavage of all the susceptible peptide bonds. The authors also suggested that removal of products during the hydrolysis can improve the rate and the recovery of proteins.

Gildberg [6] reported that an increase in the enzyme concentration increased the rate of reaction. The author suggested that the fish tissues are very complex substrates and contain a large amount of proteinase inhibitors which makes it difficult to explain the hydrolysis process. Kristinsson and Rasco [9] stated that kinetics of the fish protein hydrolysis process is complicated due to the presence of various types of peptide bonds present and their specificity for the attack by enzymes

during the process. Diniz and Martin [21] reported that once all the substrates present in the system are attached to the active sites of enzyme, there will be free enzymes which may inhibit the hydrolysis process and may even hydrolyze it. Therefore, increasing the enzyme concentration above 4% is not recommended and would not be cost effective.

The results obtained from the present study showed that increasing the enzyme concentration by 400% (from 0.5 to 2%) increased the protein yield by of 3.13- 43.52% depending upon the fish part and reaction time used as shown in (Table 7). Increasing the enzyme concentration for a small increase in protein yield seems unjustified. Therefore, the concentration of 0.5% should be used for the protein extraction unless the enzyme is reused or an immobilized reactor is used in order to reduce the cost associated with the enzyme.

Hydrolysis time

Generally, increasing the hydrolysis time from 0.5 to 2% increased protein yield. All the hydrolytic curves obtained at different enzyme concentrations (0.5, 1 and 2%) and different times (1, 2, 3 and 4 h) tend to have an initial rapid phase during the first 1 h followed by a phase of slow increase after the core proteins have been hydrolyzed (amount of substrate in the system is decreased). Even though the protein yield slowly increased after the first hour till the end of the fourth hour, the rate of increases in protein yield started to decrease for all enzyme concentrations and all fish parts. The decrease in the rate was due to the decrease in the substrate concentration.

Shahidi et al., Guerard et al. and Gbogouri et al. [19,22,10] reported on hydrolysis studies in which hydrolysis time ranging from 1 to 5 h were used and observed similar trends. Guerard et al. [20] reported that increasing the time above 5.5 h did not cause any insignificant increase in the protein yield and stopped the hydrolysis process after 4 h.

Liaset et al. [27] reported that during the enzymatic hydrolysis of cod using pepsin, alcalase and Neutrase, the reaction mechanism followed two first order kinetics processes, in which the first process involves an initial fast reaction in which loosely bound polypeptide chains were cleaved to form insoluble protein particles and in the second process the compact proteins were digested. The author also suggested that this mechanism of slow reaction at the end may be due to decrease in enzyme activity, substrate saturation or product inhibition.

Parts	Time (h)	Increase in Protein Yield (%)
Whole fish	1	20.11
	2	25.91
	3	18.00
	4	14.01
Head	1	3.35
	2	12.57
	3	12.62
	4	13.95
Frame	1	3.13
	2	18.72
	3	18.62
	4	20.29
Fins, Tail, Skin and Gut	1	39.04
	2	43.52
	3	40.50
	4	30.01

Table 7: The increase in protein yield as a result of increase in enzyme concentration from 0.5 to 2%.

Parts	Enzyme concentration (%)	Increase in Protein Yield (%)
Whole Fish	0.5	47.59
	1.0	50.82
	2.0	40.09
Head	0.5	16.45
	1.0	21.72
	2.0	28.39
Frame	0.5	22.64
	1.0	24.74
	2.0	45.50
Fins, Tail, Skin and Gut	0.5	31.95
	1.0	27.59
	2.0	23.36

Table 8: The increase in protein yield as a result of increase in reaction time from 1 to 4 h.

Parameters	Optimum Level
Enzyme	alcalase
Enzyme concentration	0.5 %
Reaction Time	1 h
pH	7.5
Temperature	55°C

Table 9: Effective parameters for protein extraction from fish and fish waste.

Gbogouri et al. [10] observed same phenomenon when using alcalase to extract proteins from salmon byproducts. Guerard et al. [22] compared umamizyme with alcalase during hydrolysis of tuna waste and observed a decrease of the hydrolysis rates suggesting enzyme deactivation or enzyme inhibition.

Guerard et al. [22] reported that protein proteolysis can happen sequentially, releasing one peptide at a time (one by one mechanism) or by forming intermediates that are further hydrolyzed as the time progresses (zipper mechanism). In this study, during the hydrolysis of fish, there was rapid burst phase of reaction followed by a steady phase till the end of the experiment indicating that the enzymatic reaction followed the zipper mechanism. Guerard et al. [20] reported that in the presence of sufficient substrate, the initial rate of reaction increased linearly when the enzyme concentration was increased up to 3%, after

which the reaction followed the zipper mechanism as it depended upon the reaction time during the extraction process.

Choisnard et al. [28] reported that during pepsin hydrolysis of protein, one molecule of pepsin cleaved the protein substrate once and then dissociated from it following one-by-one mechanism. In another case, they found one pepsin molecule would cleave the protein substrate multiple times without dissociating from it, thereby generating intermediate peptide products following a zipper mechanism. They concluded that the pepsin hydrolysis can be between two phases of reaction and it was not clear if the rate of decay of the starting substrate followed first order kinetics for both phases.

The results obtained from this study showed that increasing the hydrolysis time (from 1 to 4 h) (400%) increased the protein yield by 16.45 - 50.82% depending upon the fish part and enzyme concentration used, as shown in (Table 8). Increasing the hydrolysis time 4-folds for a small increase in protein yield will increase the capital and operating costs of production. A shorter reaction time will allow more throughput and/or reduce the volume of the reactor thereby reducing the cost of protein extraction. Therefore, a 1 h hydrolysis time for protein extraction is recommended.

Fish Parts

The results from weight composition indicated that the average weight of whole fish was 487.11 g and fish part was between 37.12 - 77.14 g. The chemical analyses results indicated that the highest amount of protein was present in whole fish (15.57%) and the protein content in the fish parts ranged from 12.18 - 14.16%. After hydrolysis and centrifugation, four layers were observed: upper oil layer, light-lipid layer, soluble clear protein layer and bottom sludge layer. Similar observations were reported by Spinelli et al. and Gildberg [29,30]. The protein hydrolysate was spray dried to obtain protein powder to determine the protein yield as described. The amount of protein recovered depended upon the amount of protein present in the raw material and the hydrolysis conditions (enzyme concentration and hydrolysis time). The protein and oil recovery from whole fish (WF), fin, tail, skin and gut (FTSG), head (H) and frame (F) increased with increases in the enzyme concentration (from 0.5% to 2.0%) and the reaction time (from 1 to 4 h). The results indicated that the highest protein yield was obtained from whole fish (WF). The amount of protein obtained from fish waste (frames (F), head (H) and fin, tail, skin and gut (FTSG)) was lower than the whole fish (WF). However, fish wastes (together) can be utilized for the extraction of proteins without any segregation of fish waste parts during fish processing. The recommended extraction parameters for the protein and oil extraction from mackerel fish waste are shown in (Table 9).

Response surface methodology is a collection of statistical and mathematical techniques used for developing, improving and optimizing processes. Diniz and Martin [21] used response surface methodology to describe the optimum conditions for the protein extraction from dog fish (*Squalus acanthias*) using alcalase. The results indicated that a pH range of 8-8.5 and a temperature range of 50-60°C and enzyme concentration of 3.6% were optimal for alcalase-assisted hydrolysis. Gbogouri et al. [10] hydrolyzed salmon and reported a protein recovery of 71.0% at a 5.5% enzyme concentration, a pH of 8 and a temperature of 58°C after 2 h, which is less than the protein yield (76.30%) achieved at 2% enzyme concentration in this study. This agrees with the findings of other researchers [20,21]. Guerard et al. [20] extracted proteins from yellowfin tuna (*Thunnus albacares*) using alcalase (0.2-3% w/w) at a pH of 8.0 and a temperature of 55°C for 6 h and obtained the highest protein recovery at 3% enzyme concentration. However, increasing the enzyme concentration beyond 1% only slightly increased the protein yield.

Beaulieu et al. [31] extracted proteins from Atlantic mackerel (*Scomber scombrus*) using protamex under optimum conditions (0.001% enzyme concentration, a pH of 8 and a temperature of 40°C for 120 min) and achieved 77.8% recovery, which is similar to the protein recovery (76.30%) achieved from whole fish in this study.

Shahidi et al. [19] extracted proteins from capelin (*Mallotus villosus*) fish using four different enzymes including alcalase (1.05% (w/w)), Neutrase (1.05% (w/w)), papain (0.14% (w/w)) and endogenous enzymes. The results indicated that the capelin treated with alcalase gave a superior protein recovery of 70.6% after 120 min of hydrolysis compared to those of 51.6, 57.1 and 22.9% from Neutrase, papain and endogenous enzymes, respectively and is less than the protein yield (76.30%) achieved at 2% enzyme concentration but it is similar to the protein yield (71.16%) achieved at 1% enzyme concentration from whole fish in this study.

Vieira et al. [8] reported a protein recovery yield of 61.9, 44.9 and 70.1% using papain, pepsin and fungal protease at 0.5% from Brazilian lobster heads after 5 h of hydrolysis. Bhaskar et al. [32] extracted proteins from visceral waste of catla (*Catla catla*) using alcalase under optimal condition (enzyme concentration of 1.5%, a pH of 8.5, a temperature of 50°C for 135 min) and achieved a protein yield of 71.54%. Bhaskar and Mahendrakar [11] extracted proteins from visceral waste of catla (*Catla catla*) using multifect-neural under optimal condition (enzyme concentration of 1.25%, temperature of 55°C for 165 min) and achieved a protein yield of 70.54%. Holanda and Netto [33] recovered protein from shrimp processing waste using alcalase and pancreatin (enzyme concentration of 3%, a temperature of 60°C for 30 min) and achieved a protein yield of 59.50 and 50.55%, respectively. Alcalase also recovered 18% more proteins than pancreatin. The protein yields from these studies are similar to the protein yield achieved at 2% enzyme concentration in this study.

Ovissipour et al. [12] reported that the time and temperature plays an important role in determining the protein yield using alcalase from Persian sturgeon. The results indicated that when the time was increased from 30 to 205 min, the protein recovery increased by 9.91% and the yield was 38.38% at 35°C. When the temperature increased from 35 to 55°C, the protein recovery increased by 61.43% and the protein recovery was 61.96% at 205 min, indicating the highest protein yield was obtained at 55°C and 205 min which is lesser than the protein recovery (76.30%) achieved at 2% enzyme concentration from whole fish in this study.

The maximum protein recovery was obtained at 4 h and 55°C for whole fish (WF), fin, tail, skin and guts (FTSG), head (H), and frames (F) are similar to those reported by Guerard et al. [20] and Kristinsson and Rasco (2000) [9]. The highest protein yield (76.30%) from whole fish (WF) at 2% enzyme concentration after 4 h of hydrolysis was superior to those reported by Gbogouri et al., Ovissipour et al. and Shahidi et al. [10,12,19]. The protein yield at 2% enzyme concentration and 4 h hydrolysis from fin, tail, skin and guts (FTSG), head (H) and frame (F) were similar to those reported by Vieira et al., Bhaskar et al., Bhaskar and Mahendrakar and superior to those report by Holanda and Netto [8,11,32,33].

Conclusions

The effects of the alcalase enzyme concentration (0.5, 1 and 2%)

and time (1, 2, 3 and 4 h) on the extraction of proteins from the whole fish (WF) and fish waste (head (H), frame (F), fin, tail, skin and gut (FTSG)) were studied. The protein yield increased with increases in enzyme concentration (from 0.5 to 2%) for the whole fish and fish waste parts because the enzyme molecules become associated with the fish particles, thereby releasing more protein molecules into the system. The results indicated that increasing the enzyme concentration by 400% (from 0.5 to 2%) increased the protein yield by 3.13- 43.52% depending upon the fish part and reaction time used. Increasing the enzyme concentration for a small increase in protein yield seems unjustified. Therefore, the concentration of 0.5% should be used for the protein extraction unless the enzyme is reused or an immobilized reactor is used in order to reduce the cost associated with the enzyme.

The protein yield increased with increases in reaction time (1, 2, 3 and 4 h) for the whole fish and fish waste parts. The protein yield increased rapidly in the first 1.5 h and then increased slowly until the hydrolysis was stopped at 4 h. The highest protein yield was obtained at 4 h from the whole fish and fish waste. The results showed that increasing the hydrolysis time from 1 to 4 h (400%) increased the protein yield by 16.45 - 50.82% depending upon the fish part and enzyme concentration used. Increasing the hydrolysis time 4 folds for a small increase in protein yield will increase the capital and operating costs of production. A shorter reaction time will allow more throughput and/or reduce the volume of the reactor thereby reducing the cost of protein extraction. Therefore, a 1 h hydrolysis time for protein extraction is recommended.

The results indicated that the highest protein yield was obtained from whole fish (WF). The amount of protein obtained from fish waste (frames (F), head (H) and fin, tail, skin and gut (FTSG)) was lower than the whole fish (WF). The total fish wastes (all parts together) can be utilized for the extraction of proteins without any segregation of fish waste parts during fish processing.

Acknowledgement

The research was supported by the Natural Sciences and Engineering Research Council (NSERC) of Canada.

References

1. FOC (2011) Fisheries and Oceans Canada. Statistics for commercial landings and aquaculture.

2. Statistics Canada (2013).

3. Statistics Canada (2012) Human Activity and the Environment: Waste Management in Canada. Statistics Canada, Catalogue Number: 16-201-X.

4. Statistics Canada (2009) Human Activity and the Environment: Annual Statistics. Statistics Canada, Catalogue Number: 16-201-X.

5. Windsor ML (2001) Fish protein concentrate. Food and Agricultural Organization of United Nations, Torry Research Station, Torry Advisory Note No. 39.

6. Gildberg A (1994) Enzymic processing of marine raw materials. Process Biochem 28: 1-15.

7. Himonides TA, Taylor KDA, Morris AJ (2011) Enzymatic Hydrolysis of Fish Frames using Pilot Plant Scale Systems. FNS 2: 586-593.

8. Vieira GHF, Martin MA, Saker-Sampaiao S, Omar S, Goncalves RCF (1995) Studies on the enzymatic hydrolysis of Brazilian lobster (Panulirus spp) processing wastes. J Sci Food Agr 69: 61-65.

9. Kristinsson HG, Rasco BA (2000) Fish protein hydrolysates: production, biochemical, and functional properties. Crit Rev Food Sci Nutr 40: 43-81.

10. Gbogouri GA, Linder M, Fanni J, Parmentier M (2004) Influence of Hydrolysis Degree on the Functional Properties of Salmon Byproducts Hydrolysates. J Food Sci 69: C615-C622.

11. Bhaskar N, Mahendrakar NS (2008) Protein hydrolysate from visceral waste

proteins of Catla (Catla catla): optimization of hydrolysis conditions for a commercial neutral protease. Bioresour Technol 99: 4105-4111.

12. Ovissipour M, Abedian A, Motamedzadegan A, Rasco B, Safari R, et al. (2009) The effect of enzymatic hydrolysis time and temperature on the properties of protein hydrolysates from Persian sturgeon (Acipenser persicus) viscera. Food Chem 115: 238-242.

13. Fukuda H, Kondo A, Noda H (2001) Biodiesel fuel production by transesterification of oils. J Biosci Bioeng 92: 405-416.

14. Shimada Y, Watanabe Y, Sugihara A, Tominaga Y (2002) Enzymatic alcoholysis for biodiesel fuel production and application of the reaction to oil processing. J Mol Catal B-Enzym 17: 133-142.

15. Gog A, Roman M, Tosa M, Paizs C, Irimie FD (2012) Biodiesel production using enzymatic transesterification- Current state and perspectives. Renew Energ 39: 10-16.

16. Yang Z, Galsius M, Xu X (2012) Enzymatic Transesterification of Ethyl Ferulate with Fish Oil and Reaction Optimization by Response Surface Methodology. Food Technol Biotech 50: 88-97.

17. Gerhardt P, Murray RGE, Wood WA, Krieg NR (1994) Methods for general and molecular bacteriology, ASM Washington DC, 518.

18. Hoyle NT, Merritt JH (1994) Quality of Fish Protein Hydrolysates from Herring (Clupea harengus). J Food Sci 59: 76-79.

19. Shahidi F, Han XQ, Synowiecki J (1995) Production and characteristics of protein hydrolysates from capelin (Mallotus villosus). Food Chem 53: 285-293.

20. Guerard F, Dufosse L, De La Broise D, Binet A (2001) Enzymatic hydrolysis of proteins from yellowfin tuna (Thunnus albacores) wastes using Alcalase. J Mol Catal B-Enzym 11: 1051-1059.

21. Diniz FM, Martin AM (1996) Use of response surface methodology to describe the combined effects of pH, temperature and E/S ratio on the hydrolysis of dogfish (Squalus acanthias) muscle. Int J Food Sci Tech 31: 419-426.

22. Guerard F, Guimas L, Binet A (2002) Production of tuna waste hydrolysates by a commercial neutral protease preparation. J Mol Catal B-Enzym 19-20: 489-498.

23. Benjakul S, Morrissey MT (1997) Protein Hydrolysates from Pacific Whiting Solid Wastes. J Agric Food Chem 45: 3423-3430.

24. Gonzàlez-Tello P, Camacho F, Jurado E, Paez MP, Guadix EM (1994) Enzymatic hydrolysis of whey proteins: I. Kinetic Models. Biotechnol Bioeng 44: 523-528.

25. Marquez MC, Vazquez MA (1999) Modeling of enzymatic protein hydrolysis. Process Biochem 35: 111-117.

26. Moreno MCM, Cuadrado VF (1993) Enzymic hydrolysis of vegetable proteins: mechanism and kinetics. Process Biochem 28: 481-490.

27. Liaset B, Lied E, Espe M (2000) Enzymatic hydrolysis of by-products from the fish-filleting industry; chemical characterization and nutritional evaluation. J Sci Food Agr 80: 581-589.

28. Choisnard L, Froidevaux R, Nedjar-Arroume N, Lignot B, Vercaigne-Marko D, et al. (2002) Kinetic study of the appearance of an anti-bacterial peptide in the course of bovine haemoglobin peptic hydrolysis. Biotechnol Appl Biochem 36: 187-194.

29. Spinelli J, Dassow JA (1982) Fish Proteins: their modification and potential uses in the food industry. In: Chemistry and Biochemistry of Marine Food Products. AVI publishing Company, Westport, CT, 13-25.

30. Gildberg A (1992) Recovery of proteinases and protein hydrolysates from fish viscera. Bioresource Technol 39: 271-276.

31. Beaulieu L, Thibodeau J, Bryl P, Carbonneau ME (2009) Proteolytic processing of Atlantic mackerel (Scomber scombrus) and biochemical characterisation of hydrolysates. Int J Food Sci Tech 44: 1609-1618.

32. Bhaskar N, Benila T, Radha C, Lalitha RG (2008) Optimization of enzymatic hydrolysis of visceral waste proteins of Catla (Catla catla) for preparing protein hydrolysate using a commercial protease. Bioresour Technol 99: 335-343.

33. Holanda HD, Netto FM (2006) Recovery of Components from Shrimp (Xiphopenaeus kroyeri) Processing Waste by Enzymatic Hydrolysis. J Food Sci 71: C298-C303.

21

Effect of Different Processing Parameters on Quality Factors and Image Texture Features of Bread

Mahdi Karimi[1], Milad Fathi[1,2]*, Zahra Sheykholeslam[1], Bahareh Sahraiyan[3] and Fariba Naghipoor[3]

[1]*Khorasan Razavi Agricultural and Natural Resources Research Center, Iran*
[2]*Department of Food Science and Technology, Isfahan University of Technology (IUT), Isfahan, 84156-83111, Iran*
[3]*Department of Food Science and Technology, Ferdowsi University of Mashhad (FUM), Mashhad, Iran*

Abstract

In this research, image texture analysis as a nondestructive and rapid method was applied for estimation of mechanical texture property of bread. Bread samples were formulated using different emulsifiers (i.e. SSL, DATEM and E471) in three concentrations (0.2, 0.4 and 0.6%) at three proofing times (25, 35 and 45min) and sensory properties, specific volume, color components (i.e. L^*, a^* and b^*), porosity, hardness and four image texture features (i.e. contrast, correlation, energy and homogeneity) were determined. The results indicated that emulsifier treated samples showed better ($p < 0.05$) sensory perceptions in compression to control samples. However, the proofing time did not show significant effect. Application of E471 significantly increased the specific volume of bread. Emulsifier treated samples had higher lightness values. Application of higher concentration and longer proofing time led to higher porosity of bread. Emulsifiers, their concentrations and proofing times had positive significant ($p < 0.05$) effects on softness of bread. On the other side of this research, contrast, correlation, energy and homogeneity were calculated from Grey Level Co-Occurrence Matrix (GLCM). They showed splendid correlations with hardness of bread (0.958, 0.973, 0.966 and 0.91, respectively). Multiple Linear Regression (MLR) was conducted between hardness and four image texture features and the mathematical equation could predict hardness with high correlation of coefficient of 0.994. These results strongly suggest that image texture analysis can be applied as a nondestructive and rapid method for estimation of hardness of bread.

Keywords: Bread; Image Texture Analysis; Mechanical Texture Property; Emulsifier; Proofing Time

Introduction

Bread is one of the oldest and most popular diets all over the world. Consumer demand is toward consumption of high quality fresh bread. However, this product has a short shelf life. Therefore, application of some additives (e.g. emulsifiers) and process modifications are necessary to overcome this limitation and improve other quality parameters of bread such as sensory and rheological features. The main mechanism by which the emulsifiers retard the firming or retrogradating of crumb is based on their capability to form inclusion complexes with amylose part of starch during the baking process [1-2]. Several studies have been carried out to show the potential application of emulsifiers in the bakery products [3-5]. On the other hand, proofing time is also one of the important processing parameters affecting final quality of bread due to production of gas bubbles [6].

Mechanical texture features are recognized as one of the most important quality aspects affecting sensory perception and shelf life of bread. However, determination of these physical properties using physical instruments are both time consuming and destructive. Image textures are important image features and have been recently applied in food sector for quality evaluation as a nondestructive, objective and rapid method [7-11]. Published data revealed that image texture features can be used for determination of mechanical properties of food materials [12-13]. Image texture is defined as the spatial organization of intensity variations of pixels in gray level image, which corresponded to both brightness value and pixel locations [14]. Image texture features are usually classified into four categories namely, statistical, structural, model-based and transform-based textures [15]. In the food systems, statistical texture is the most commonly used method for quality evaluations. This method includes Grey Level Co-Occurrence Matrix (GLCM), grey level pixel-run length matrix, and neighboring grey level dependence matrix [9]. The former that has been proposed by Haralick et al. [16], is the widely applied statistical texture analysis method, in which texture features such as entropy, homogeneity, correlation and contrast are extracted by some statistical approaches from the co-occurrence matrix of gray scale image histogram. GLCM has been used for classification of cereal grain and dockage [17], and apple [18].

There is not published data in the literature on mechanical and image texture properties of bread. Therefore, the objectives of the present work were to investigate the effect of different emulsifiers, concentrations and proofing times on bread quality factors and to study the efficiency of image texture analysis for prediction of mechanical texture property.

Materials and Methods

Materials

Commercial wheat flour (Golmakan Co., Khorasan Razavi, Iran), dry active yeast (Razavi Co., Khorasan, Iran) and emulsifiers (Sodium Stearoyl Lactylate (SSL), Diacetyl Tartaric Acid Ester of Monoglyceride (DATEM) and Mono-and diglycerides of fatty acids (E471)) (Pars Behbod Asia Co, Tehran, Iran) were bought from authoritative company. Other ingredients were purchased from local supermarket.

*Corresponding author: Milad Fathi, Department of Food Science and Technology, Isfahan University of Technology (IUT), Isfahan, 84156-83111, Iran

Methods

The bread formula consisted of flour (100 parts); water (50 parts), active dried yeast (1 part); salt (1 part); sugar (1 part); shortening (1 part) and emulsifier (0.2, 0.4 and 0.6% of applied flour). The baking procedure was followed based on typical methods [19]. Three different proofing times (25, 35 and 45 min) were considered at 45° C. Baking was carried out using a laboratory air impingement oven (Minicombo rotor oven, Zucchelli, Italy) at 240° C for 10 min. After cooling, bread samples were packed in polyethylene bags of 20 μm thickness, stored at room temperature until future analysis.

Chemical analysis: Moisture (44–16 A), ash (08–07), dry gluten (38-12.02) and falling number (56–81) were determined according to AACC-approved methods [20]. Flour protein was assessed based on Kjeldahl using a Kjeltec auto protein tester (model 1030, Tecator Co., Hoeganaes, Sweden).

Specific volume: Specific volume was determined an hour after baking based on rapeseed displacement method [21-22].

Mechanical texture analysis: The hardness of bread samples was measured with a Texture Analyzer (CNS Farnell, Hertfordshire, UK) equipped with a 50 N load cell. A cylindrical probe with a diameter of 20 mm and height of 23mm with the speed of 30mm/min was used for cubic samples with dimensions of 25×25×25. The mechanical texture analysis of bread samples were performed after one day of production at room temperature based on the method of AACC [23].

Sensory analysis: The sensory analysis was conducted with a group of 10 semi-trained panelists, applying a hedonic scale of 5 point. Panelists were asked to assess the bread's quality factors including crumb color, crust color, taste, aroma, staling and total acceptance, to rate samples from 0 to 5 (0 unacceptable, 5 very acceptable).

Image analysis: For each treatment, three samples (crust and crumb) were scanned with desktop flatbed scanner (HP, Scanjet G3010; at Optical Resolution of 4800 dpi×9600 dpi) and the images were saved as BMP format. To study the effect of processing parameters on color components of bread, the RGB color space images were converted to $L*a*b$ space [24]. For determination of the bread porosity using image analysis, the color images were first grayscaled and then thresholded using isodata algorithm. The porosity was measured from the ratio of white to the total numbers of pixels.

Grey level co-occurrence matrix and image texture analysis: The first procedure for extracting image textural features was presented by Haralick et al. [16]. Each textural property is computed from a set of GLCM probability distribution matrices for a given image. The GLCM shows the probability that a pixel of a particular grey level occurs at a specified direction and distance from its neighboring pixels. Gray level co-occurrence matrix is represented by $P_{d,\theta(i,j)}$ where counts the neighboring pair pixels with gray values i and j at the distance of d and the direction of θ.

In this study, four image texture features namely, contrast, correlation, entropy and homogeneity were calculated based on equations 1-4 [12]. Contrast measures the local variation in an image (ranging from 0 to [size (GLCM, 1)-1]²) and a high contrast value indicates a high degree of local variation. Correlation is an indicator of linear dependency of intensity values in an image (ranging from -1 to 1). For an image with large areas of similar intensities, a high value of correlation is measured. Energy (angular second moment) returns the sum of squared elements in the GLCM (ranging from 0 to 1) and

homogeneity indicates the uniformity within an image (ranging from 0 to 1).

$$Contrast = \sum_{i=0}^{N-1}\sum_{j=0}^{N-1}\left(i-j\right)^2 P_{d,\theta}\left(i,j\right) \tag{1}$$

$$Correlation = \frac{\left[\sum_{i=0}^{N-1}\sum_{j=0}^{N-1}(ij)P_{d,\theta}\left(i,j\right)\right]-\mu_x\mu_y}{\sigma_x\sigma_y} \tag{2}$$

$$Energy = \sum_{i=0}^{N-1}\sum_{j=0}^{N-1}P_{d,\theta}\left(i,j\right)^2 \tag{3}$$

$$Homogeneity = \sum_{i=0}^{N-1}\sum_{j=0}^{N-1}\frac{P_{d,\theta}\left(i,j\right)}{1+\left|i-j\right|} \tag{4}$$

Property	value
Moisture (g/100g, w.b.)	13.8 ± 0.45
Protein (g/100g)	10.8 ± 0.60
Ash (g/100g)	0.86 ± 0.08
Dry gluten (g/100g)	8.7 ± 0.43
Falling number (s)	423 ± 1.83

Table 1: Quality characteristic of wheat flour and dough.

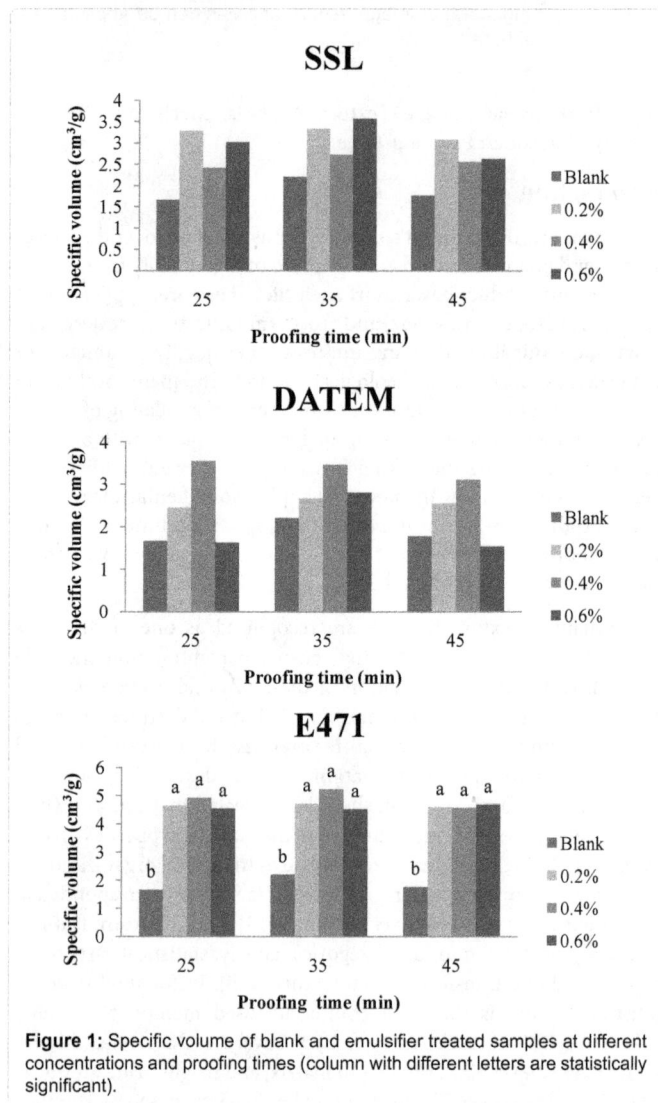

Figure 1: Specific volume of blank and emulsifier treated samples at different concentrations and proofing times (column with different letters are statistically significant).

Emulsifier	Concentration (%)	Proofing time (min)	Sensory properties					
			Total acceptance	Crumb color	Taste	Crust color	Aroma	Staling
Blank	0	25	2.7 ± 0.67c	2.5 ± 0.84c	2.2 ± 0.66c	2.7 ± 0.67b	2.1 ± 0.73c	1.8 ± 0.63b
	0	35	3 ± 0.66c	2.7 ± 0.67c	2.5 ± 0.66c	2.5 ± 0.66b	2.4 ± 0.63c	2.1 ± 0.73b
	0	45	2.8 ± 0.63c	2.7 ± 0.67c	2.5 ± 0.66c	2.4 ± 0.62b	2.4 ± 0.63c	2.2 ± 0.66b
E471	0.2	25	3.6 ± 0.51b	3.4 ± 0.96b	3.4 ± 0.84b	3.5 ± 0.52a	2.9 ± 0.73b	2.8 ± 0.63ab
	0.2	35	3.9 ± 0.56b	3.4 ± 0.73b	3.6 ± 0.51b	3.2 ± 1.13a	3 ± 0.66b	3.1 ± 0.73ab
	0.2	45	3.6 ± 0.82b	3.4 ± 0.84b	3.7 ± 0.82b	3.2 ± 0.63a	3 ± 0.63b	3.2 ± 0.81ab
	0.4	25	3.6 ± 0.51b	3.4 ± 0.82b	3.7 ± 0.82b	3.3 ± 0.82a	3 ± 0.66b	2.9 ± 0.73ab
	0.4	35	4 ± 0.47ab	3.7 ± 0.82ab	3.8 ± 0.42ab	3.3 ± 0.82a	3 ± 0.66b	3.2 ± 0.66ab
	0.4	45	3.5 ± 0.52b	3.6 ± 0.61ab	3.6 ± 0.84b	3.3 ± 0.94a	3.1 ± 0.73b	3.5 ± 0.84a
	0.6	25	3.5 ± 0.52b	3.3 ± 0.64b	2.9 ± 0.74bc	3.2 ± 0.91a	3 ± 0.66b	3.2 ± 0.91ab
	0.6	35	3.9 ± 0.74b	3.7 ± 0.82ab	3.5 ± 0.70b	3.3 ± 0.78a	3.1 ± 0.73b	3.1 ± 0.83ab
	0.6	45	3.9 ± 0.73b	3.5 ± 0.70ab	3.3 ± 0.95b	3.3 ± 0.82a	3 ± 0.66b	3.1 ± 0.74ab
DATEM	0.2	25	3.2 ± 0.91bc	3.2 ± 0.91b	3.2 ± 0.91b	3.5 ± 0.70a	2.7 ± 0.67bc	2.8 ± 0.63ab
	0.2	35	3.7 ± 0.94b	3.1 ± 0.73b	3.6 ± 0.51b	3.2 ± 0.91a	2.7 ± 0.67bc	3.4 ± 0.95ab
	0.2	45	3.4 ± 0.84b	3 ± 0.84b	3.3 ± 0.82b	3 ± 0.84a	2.6 ± 0.69bc	3.1 ± 0.73ab
	0.4	25	3.4 ± 0.84b	3.2 ± 0.75b	3.5 ± 0.52b	3.1 ± 0.73a	2.8 ± 0.63b	2.9 ± 0.73ab
	0.4	35	3.8 ± 0.78b	3.5 ± 0.70ab	3.6 ± 0.69b	3.1 ± 0.87a	2.9 ± 0.87b	3.2 ± 0.91ab
	0.4	45	3.3 ± 0.94b	3.4 ± 0.84b	3.4 ± 0.84b	3 ± 0.66a	2.9 ± 0.73b	3 ± 0.66ab
	0.6	25	3.1 ± 0.73bc	3 ± 0.47b	2.6 ± 0.84c	2.9 ± 0.73a	2.6 ± 0.69bc	2.9 ± 0.73ab
	0.6	35	3.7 ± 0.94b	3.4 ± 0.86b	3.2 ± 0.91b	3 ± 0.66a	2.8 ± 0.63b	3.1 ± 0.73ab
	0.6	45	3.3 ± 0.94b	3.2 ± 0.91b	3 ± 0.66b	3 ± 0.0a	2.7 ± 0.67bc	3.1 ± 0.56ab
SSL	0.2	25	3.7 ± 0.82b	3.5 ± 0.52ab	3.6 ± 0.84b	3.6 ± 0.51a	3.1 ± 0.87b	3.4 ± 0.96ab
	0.2	35	4.1 ± 0.87ab	3.8 ± 0.63a	3.9 ± 0.56ab	3.4 ± 0.64a	3.1 ± 0.73b	3.4 ± 0.69ab
	0.2	45	3.7 ± 0.67b	3.7 ± 0.67ab	3.5 ± 0.52b	3.4 ± 0.84a	3.1 ± 0.73b	3.2 ± 0.78ab
	0.4	25	3.7 ± 0.67b	3.6 ± 0.51ab	3.6 ± 0.96b	3.7 ± 0.67a	3.5 ± 0.52a	3.2 ± 0.91ab
	0.4	35	4.3 ± 0.48a	4 ± 0.47a	4.3 ± 0.48a	3.7 ± 0.82a	3.8 ± 0.91a	3.5 ± 0.70a
	0.4	45	3.9 ± 0.87b	3.9 ± 0.87a	4.1 ± 0.56a	3.4 ± 0.84a	3.7 ± 0.82a	3.3 ± 0.82ab
	0.6	25	3.4 ± 0.84b	3.5 ± 0.70ab	3.5 ± 0.70b	3.6 ± 0.96a	3.4 ± 0.69b	3.2 ± 0.91ab
	0.6	35	3.6 ± 0.51b	3.8 ± 0.78a	4.2 ± 0.63a	3.7 ± 0.94a	3.7 ± 0.82a	3.6 ± 0.84a
	0.6	45	3.1 ± 0.56bc	3.7 ± 0.82ab	4.1 ± 0.56a	3.3 ± 0.82a	3.5 ± 0.52b	3.4 ± 0.84ab

Values in each column with different letters are statistically significant ($p < 0.05$).

Table 2: Sensory properties of blank and emulsifier treated bread.

Emulsifier	Concentration (%)	Proofing time (min)	L*	a*NS	b*NS	Porosity (%)
Blank	0	25	54.45d	11.98	39.2	20.7h
	0	35	56.82dc	15.28	46.26	23.5gh
	0	45	59.64bc	12.45	40.75	25.1g
E471	0.2	25	69.76a	6.94	41.49	31.0f
	0.2	35	65.23ab	9.86	44.56	31.3f
	0.2	45	67.35a	9.85	46.23	32.2f
	0.4	25	62.65b	10.46	41.7	32.5f
	0.4	35	61.86b	10.92	41.1	33.5ef
	0.4	45	60.54b	12.47	40.73	33.8ef
	0.6	25	55.19dc	14.4	39.78	33.5ef
	0.6	35	62.19b	13.13	62.19	33.7ef
	0.6	45	56.21dc	14.42	14.3	35.4de
DATEM	0.2	25	53.68d	10.45	45.82	31.3f
	0.2	35	62.15b	14.46	44.76	35.9e
	0.2	45	61.12b	13.83	42.65	37.4d
	0.4	25	65.72ab	13.59	48.9	34.4e
	0.4	35	65.95ab	12.05	65.95	36.8de
	0.4	45	62.19b	13.13	62.19	37.1d
	0.6	25	65.4ab	11.1	43.86	38.3d
	0.6	35	64.55ab	14.4	47.76	38.8d
	0.6	45	62.41b	14.65	46.94	38.3d
SSL	0.2	25	67.35a	9.85	46.23	40.0c
	0.2	35	56.21c	14.42	44.3	40.3c
	0.2	45	63.01b	13.53	45.95	42.2b
	0.4	25	59.5bc	15.7	51.04	40.5c
	0.4	35	62.73b	9.72	44.35	41.5b
	0.4	45	61.27b	10.75	43.95	42.1b
	0.6	25	60.73b	13.74	43.33	41.5b
	0.6	35	69.9a	9.99	42.79	42.7b
	0.6	45	63.21b	9.12	42.92	43.4a

Values in each column with different letters are statistically significant ($p < 0.05$); NS, not statistically significant.

Table 3: Image color properties and porosity values extracted from bread image.

where μ_x, μ_y and σ_x, σ_y are the mean and standard deviation of the sums of rows and columns in the matrix, respectively, and N is the dimension of square matrix of GLCM. In this study, the four mentioned textural features were computed using the mean of the four values of different orientations (0°, 45°, 90° and 135°) at d=1 applying a program developed in MATLAB 7.0.

Statistical analysis: Analysis of variance (ANOVA) was performed using a computerized statistical program called "MSTAT" version C, and determination of significant differences of means was carried out by "Duncan" test at 95% confidence level applying the above software program. Regression equations and correlation coefficients (R) between the mechanical and image texture features were obtained using Minitab software, version 14. Each experiment was conducted applying factorial design at least in three replications.

Results and Discussions

In this study, 27 different bread samples were formulated using three emulsifiers in three concentrations at three proofing times. The physicochemical compositions of wheat and dough are tabulated in Table 1. These characteristics in the range of typical values of medium strong flour.

The results indicated that application of E471 in bread formulation

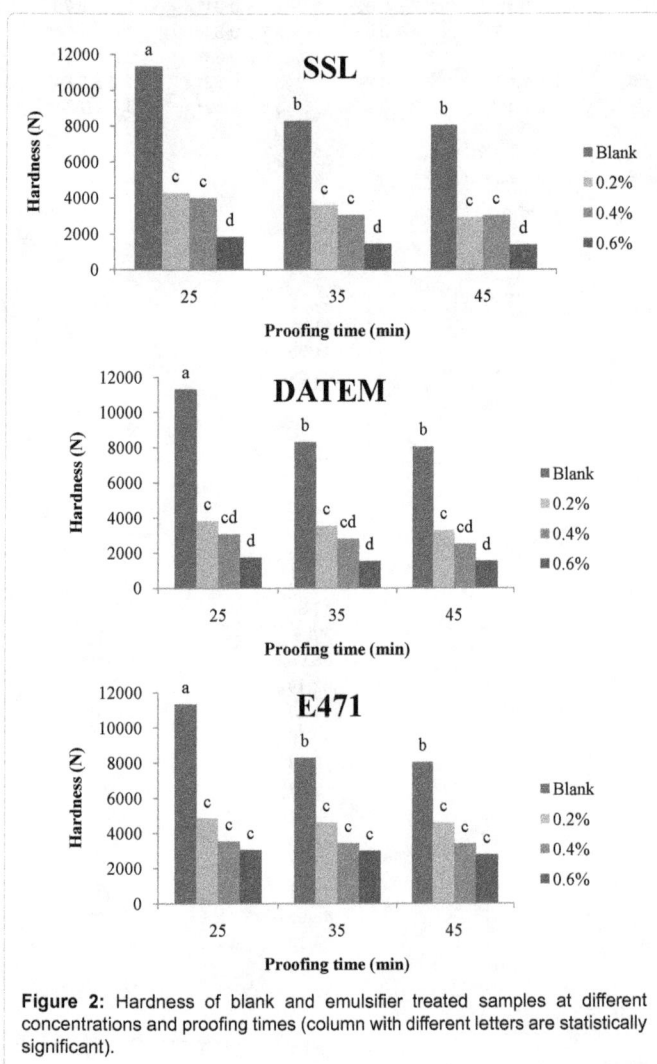

Figure 2: Hardness of blank and emulsifier treated samples at different concentrations and proofing times (column with different letters are statistically significant).

led to a significant increase (p<0.05) in specific volume (Figure 1). It may be attributed to ability of emulsifier to permit dough expansion during baking process. However, the effects of proofing time, SSL and DATEM were not statistically significant. Ribotta et al. [25] reported that emulsifier treated bread had the higher specific volume in comparison to control samples.

Table 2 indicates the sensory attributes of blank and emulsifier treated samples which were evaluated a day after baking. As one can see, emulsifier addition has significant influences on all sensory properties (test, aroma, crumb color, crust color and staling and total acceptance). Taste and aroma are two important quality characteristics in food acceptance, which are difficult to determine by instrumental measurements. Taste and aroma of emulsifier treated breads gained higher scores in comparison to blank samples. The crumb and crust colors were enhanced by using emulsifiers. The staling results indicated that the applied emulsifiers could delay, about two times, the retrogradation process. Among different emulsifiers, SSL showed better sensory attributes. In most cases, the proofing time did not show significant effect on sensory properties. The calculated averages of all quality scores of treated samples were higher than control breads; therefore, sensory analysis allows concluding that addition of emulsifiers improves sensory properties of fresh bread.

The average values of L^*, a^*, b^* and porosity extracted from bread images tabulated in table 3. The results indicated that the effects of treatments were not statistically significant on color components of a^* and b^*. However, lightness (L^*) of bread increased by emulsifier addition, which could be attributed to the positive effect of emulsifiers to enhance lightness of bakery products. The porosity of bread enhanced by emulsifier addition and increasing emulsifier concentration and proofing time. It may be the consequence of higher incorporation of air into dough with increasing proofing time and high ability of emulsifier-amylose complex to retain the gas.

The effect of process variables on hardness of bread was depicted in Figure 2. All applied emulsifiers showed significant effect (p<0.05) on hardness. By increasing emulsifier concentration and proofing time the hardness of bread diminished. However, the effect of SSL pronounced more strongly, which caused about five times decrease in hardness of bread. It could be due to interaction of amylose and emulsifiers which led to retardation of retrogradation and decrease of hardness [1]. On the other hand, increasing proofing time caused an increase in CO_2 production, raising porosity and consequently decreasing of hardness.

Aforementioned statements indicate that determination of mechanical texture properties of food products is both destructive and time-consuming. Four image texture features (i.e. contrast, correlation, energy and homogeneity) of bread were calculated from GLCM in four orientations at the distance of one pixel and their average values for different emulsifiers, concentrations and proofing times were shown in Table 4. The contrast decreased in results of emulsifier addition due to diminish of local variation of pixels. The softer the texture the lower the contrast, which is due to lower pixel value difference between two neighbors [14]. On the other hand, the increasing trends of energy, correlation and homogeneity values revealed improvement of uniformity and smoothness of the images due to decrease of coalescence and increase of softness of bread texture. Similar trends were also reported by Qiao et al [12] for image textural properties of nugget. In this research an effort has been made to apply image texture analysis as a nondestructive and rapid method to predict mechanical properties of bread. To evaluate the capability of image texture features

Emulsifier	Concentration (%)	Proofing time (min)	Image texture feature			
			Contrast	Correlation	Energy	Homogeneity
Blank	0	25	0.360175[a]	0.7651[e]	0.17215[e]	0.746075[f]
	0	35	0.289825[b]	0.8242[d]	0.249075[d]	0.763025[e]
	0	45	0.288575[b]	0.825275[d]	0.24584[d]	0.7769[d]
SSL	0.2	25	0.210925[c]	0.884175[c]	0.29993[c]	0.904575[c]
	0.2	35	0.202225[c]	0.890925[bc]	0.31075[c]	0.908325[c]
	0.2	45	0.20025[c]	0.89385[bc]	0.325225[bc]	0.91649[b]
	0.4	25	0.20265[c]	0.886175[c]	0.293775[c]	0.905[c]
	0.4	35	0.200225[c]	0.893475[bc]	0.33335[b]	0.915525[bc]
	0.4	45	0.200108[c]	0.89305[bc]	0.320025[bc]	0.915075[bc]
	0.6	25	0.15575[d]	0.9001[b]	0.331125[b]	0.9235[b]
	0.6	35	0.1491[de]	0.91865[a]	0.339985[a]	0.929825[a]
	0.6	45	0.145375[e]	0.91995[a]	0.342225[a]	0.931958[a]
DATEM	0.2	25	0.207325[c]	0.887[c]	0.298275[c]	0.904[c]
	0.2	35	0.2023[c]	0.889025[c]	0.310925[c]	0.9094[bc]
	0.2	45	0.2017[c]	0.892625[c]	0.31173[c]	0.909933[bc]
	0.4	25	0.200725[c]	0.893075[bc]	0.314675[c]	0.9151[bc]
	0.4	35	0.200305[c]	0.8948[bc]	0.32585[bc]	0.915775[bc]
	0.4	45	0.199225[cd]	0.8997[b]	0.32995[b]	0.91609[bc]
	0.6	25	0.1669[d]	0.900025[b]	0.33245[a]	0.9267[b]
	0.6	35	0.149975[de]	0.91446[a]	0.33685[a]	0.92905[a]
	0.6	45	0.14903[de]	0.914925[a]	0.33608[a]	0.929625[a]
E471	0.2	25	0.22021[bc]	0.877546[c]	0.29486[c]	0.89393[c]
	0.2	35	0.21593[c]	0.87678[c]	0.296767[c]	0.897543[c]
	0.2	45	0.214[c]	0.877767[c]	0.297878[c]	0.899768[c]
	0.4	25	0.20222[c]	0.889787[bc]	0.311233[c]	0.9094[c]
	0.4	35	0.20121[c]	0.889967[bc]	0.311876[c]	0.909756[c]
	0.4	45	0.201021[c]	0.890004[bc]	0.311212[c]	0.9098[bc]
	0.6	25	0.20107[c]	0.89306[bc]	0.32[bc]	0.9154[bc]
	0.6	35	0.20012[c]	0.89452[bc]	0.32012[bc]	0.9162[bc]
	0.6	45	0.200001[c]	0.89676[bc]	0.32434[bc]	0.9168[bc]

Values in each column with different letters are statistically significant (p<0.05).

Table 4: Image texture features (contrast, correlation, energy and homogeneity) of blank and treated bread.

for estimation of hardness of bread, linear regressions were conducted between contrast, correlation, energy and homogeneity and hardness (Figure 3). The mathematical linear equations between image texture features and hardness and their correlation coefficients (Figure 3) indicated that all four image texture features showed excellent correlations with hardness (0.958, 0.973, 0.966 and 0.91 for contrast, correlation, energy and homogeneity, respectively). The Multiple Linear Regressions (MLR) between image texture features and hardness of bread (Eq. 5) showed the capability of image properties to predict a mechanical feature of bread (R^2=0.994).

Hardness = 9654 + 21208 [Contrast] + 9317 [Correlation – 31645] [Energy] – 9816 [Homogeneity] R^2 = 0.994   (5)

The ANOVA results of MLR are shown in Table 5, indicate that the coefficients of regression equation are statistically significant. The above results reveal that image texture features can be strongly suggested as a nondestructive and rapid method for quality control of mechanical properties of bread.

Conclusion

The main aim of this work was to apply image texture analysis for prediction of a mechanical texture property of bread and the sub-aim was to investigate the effect of emulsifier and proofing time on some sensory and physical properties of bread. The following conclusions have been conducted for this research:

i. E471 significantly increased (p<0.05) specific volume of bread. Whereas, DATEM, SSL, emulsifier concentration and proofing time did not had significant effect.

ii. All sensory properties including crumb color, crust color, test, aroma and staling and total acceptance positively and significantly affected by emulsifier addition. The effect of SSL pronounced more strongly. However, proofing time did not show significant effect on almost all sensory features.

iii. Lightness and porosity were calculated from crust and crumb images. Lightness increased by emulsifier addition and porosity also improved by increasing emulsifier concentration and proofing time.

iv. Hardness of bread decreased (up to five times) by increasing emulsifier concentration and proofing time.

v. Contrast, correlation, energy and homogeneity were calculated from GLCM. They showed high correlations with hardness of bread (0.958, 0.973, 0.966 and 0.91, respectively). Multiple Linear Regression (MLR) between hardness and four image texture features could predict hardness with high correlation of coefficient of 0.994.

The results of current research strongly suggest that applied emulsifiers could improve quality factors of bread and image texture

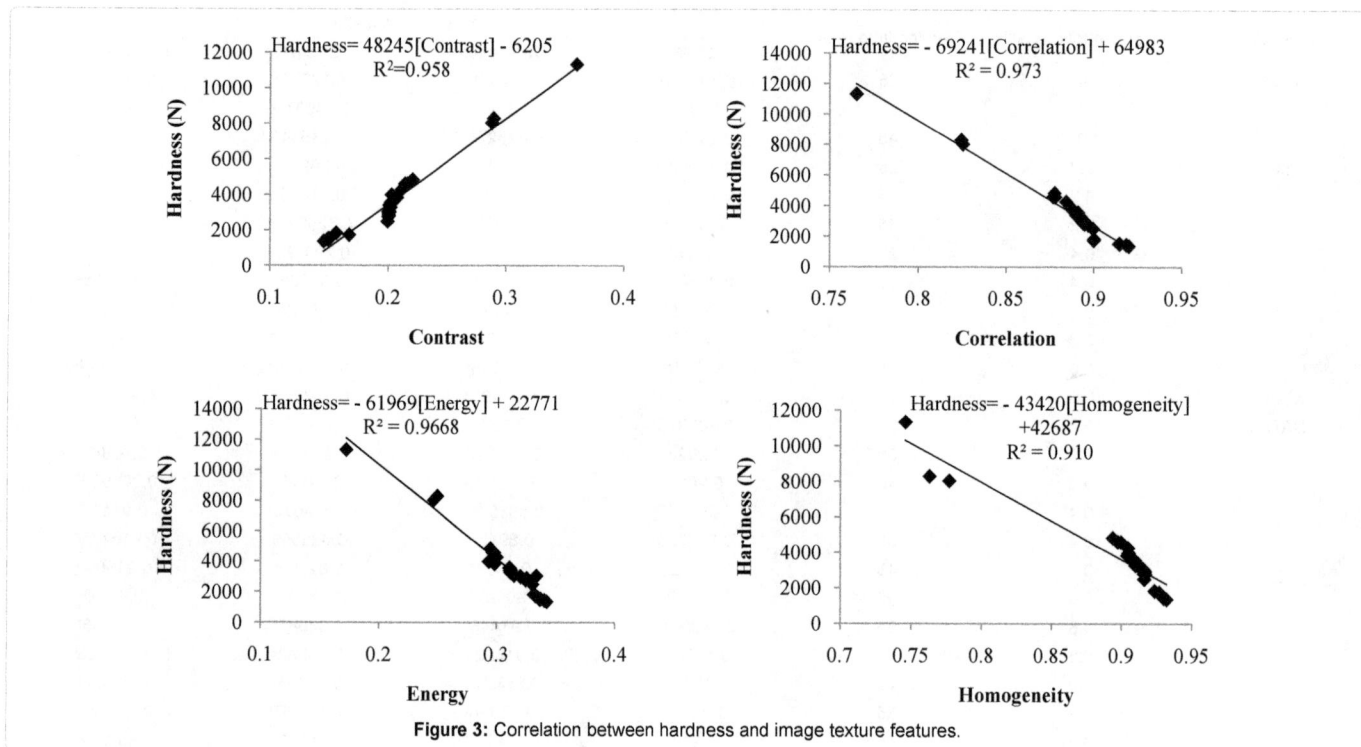

Figure 3: Correlation between hardness and image texture features.

Source	DF	SS	MS	F	P
Regression	4	133933597	33483399	632.35	0.00
Residual error	25	1323768	52951		
Total	29	135257365			

Table 5: The ANOVA results of multiple linear regressions between hardness and four image texture features.

analysis can be applied as a nondestructive rapid method for estimation of hardness of bread.

References

1. Azizi MH, Rao GV (2005) Effect of storage of surfactant gels on the bread making quality of wheat flour. Food Chem 89: 133-138.

2. Krog N, Jensen BN (1970) Interaction of monoglycerides in different physical states with amylose and their anti-firming effects in bread. Int J Food Sci Tech 5: 77-87.

3. Koocheki A, Mortazavi SA, Mahalati MN, Karimi M (2009) Effect of emulsifiers and fungal α-amylase on rheological characteristics of wheat dough and quality of flat bread. J Food Process Eng 32: 187-205.

4. Nunes MHB, Moore MM, Ryan LAM, Arendt EK (2009) Impact of emulsifiers on the quality and rheological properties of gluten-free breads and batters. Eur Food Res Technol 228: 633-642.

5. Moayedallaie S, Mirzaei M, Paterson J (2010) Bread improvers: Comparison of a range of lipases with a traditional emulsifier. Food Chem 122: 495-499.

6. Chin NL, Tan LH, Yusof YA, Rahman RA (2009) Relationship between aeration and rheology of breads. J Texture Stud 40: 727-738.

7. Dan H, Azuma T, Kohyama K (2007) Characterization of spatiotemporal stress distribution during food fracture by image texture analysis methods. J Food Eng 81: 429-436.

8. Borah S, Hines EL, Bhuyan M (2007) Wavelet transform based image texture analysis for size estimation applied to the sorting of tea granules. J Food Eng 79: 629-639.

9. Zheng C, Sun DW, Zheng L (2006) Recent applications of image texture for evaluation of food qualities-a review. Trends Food Sci Tech 17: 113-128.

10. Gonzales-Barron U, Butler F (2008) Discrimination of crumb grain visual appearance of organic and non-organic bread loaves by image texture analysis. J Food Eng 84: 480-488.

11. Fathi M, Mohebbi M, Razavi SMA (2009) Application of image texture analysis for evaluation of Osmotically Dehydrated Kiwifruit Qualities. 5th International Symposium on Food Rheology and Structure-Isfrs.

12. Qiao J, Wang N, Ngadi MO, Kazemi S (2007) Predicting mechanical properties of fried chicken nuggets using image processing and neural network techniques. J Food Eng 79: 1065-1070.

13. Thybo AK, Szczypinski PM, Karlsson AH, Dønstrup S, Stodkilde-Jorgensen HS, et al. (2004) Prediction of sensory texture quality attributes of cooked potatoes by NMR-imaging (MRI) of raw potatoes in combination with different image analysis methods. J Food Eng 61: 91-100.

14. Pietikanen MK (2000) Texture analysis in machin vision. World Scientific, London.

15. Bharati MH, Liu JJ, MacGregor JF (2004) Image texture analysis: methods and comparisons. Chemometr Intell Lab 72: 57-71.

16. Haralick RM, Shanmugam K, Dinstein I (1973) Textural features for image classification. IEEE T Syst Man Cyb SMC3: 610-621.

17. Paliwal J, Visen NS, Jayas DS, White NDG (2003) Cereal Grain and Dockage Identification using Machine Vision. Biosyst Eng 85: 51-57.

18. Kavdir I, Guyer DE (2002) Apple sorting using artificial neural networks and spectral imaging. T ASAE 45: 1995-2005.

19. Pourfarzad A, Khodaparast MHH, Karimi M, Mortazavi SA, Davoodi MG, et al. (2011) Effect of polyols on shelf-life and quality of flat bread fortified with soy flour. J Food Process Eng 34: 1435-1448.

20. AACC (2000) Approved Methods of the AACC. Methods 44–16 A, 08–07, 30–10, 38–11, 56–81. American Association of Cereal Chemists, St. Paul, MN.

21. Sabanis D, Tzia C, Papadakis S (2008) Effect of Different Raisin Juice Preparations on Selected Properties of Gluten-Free Bread. Food Bioprocess Tech 1: 374-383.

22. Barcenas ME, Rosell CM (2006) Different approaches for improving the quality and extending the shelf life of the partially baked bread: low temperatures and HPMC addition. J Food Eng 72: 92-99.

23. AACC (2004) Approved Methods of the AACC. Methods 10-05, 74-09. American Association of Cereal Chemists, St. Paul, MN.

24. Fathi M, Mohebbi M, Razavi SMA (2011) Application of Image Analysis and Artificial Neural Network to Predict Mass Transfer Kinetics and Color Changes of Osmotically Dehydrated Kiwifruit. Food Bioprocess Tech 4: 1357-1366.

25. Ribotta PD, Pérez GT, Añón MC, León AE (2010) Optimization of Additive Combination for Improved Soy–Wheat Bread Quality. Food Bioprocess Tech 3: 395-405.

22

Effects of Culture Conditions on a Micropatterned Co-culture of Rat Hepatocytes with 3T3 cells

Kohji Nakazawa*, Yukako Shinmura, Ami Higuchi and Yusuke Sakai

Department of Life and Environment Engineering, The University of Kitakyushu 1-1 Hibikino, Wakamatsu-ku, Kitakyushu, Fukuoka 808-0135, Japan

Abstract

We investigated the effect of culture conditions on the micropatterned co-culture of rat hepatocytes with 3T3 cells. A micropatterned chip was prepared using polydimethylsiloxane (PDMS) microstencil such that the chip contained 724 hepatocyte islands, each 500 μm in diameter, in a triangular arrangement with 800-μm pitch, in which hepatocytes were co-cultured with 3T3 cells. The hepatocytes in micropatterned co-culture exhibited hepatocellular morphology, and the micropatterned configuration of hepatocyte islands was maintained for several weeks of culture by supporting the heterotypic interface between the hepatocytes and 3T3 cells. The albumin secretion activity of hepatocytes was highest in the micropatterned co-culture but decreased in the random co-culture, micropatterned mono-culture (hepatocytes only), and random mono-culture (hepatocytes only) in that order. Furthermore, earlier formation of co-culture promoted higher functional activity of hepatocytes as compared to later formation, and hepatocyte functions were induced with an increasing the density of inoculated 3T3 cells. These results suggest that the formation of a heterotypic interaction at an early stage is important for maintaining high levels of hepatocyte functions. The findings of this study will provide information useful for designing co-culture conditions for liver tissue engineering and pharmacological and toxicological studies.

Keywords: Micropatterned culture; Co-culture, Microstencil; Rat hepatocytes; 3T3 cells; Albumin secretion

Introduction

The liver plays many essential roles in maintaining normal physiology. Therefore, primary hepatocytes have been used for various applications such as liver tissue engineering and pharmacological, toxicological, and fundamental cell biology studies. For the success of such applications, hepatocytes have to express liver-specific functions at a high level and maintain these functions over a long term. Expression of liver-specific functions in the hepatocytes is closely related to the *in vitro* cell configuration and culture environments. Various approaches have been adopted to preserve hepatocyte functions *in vitro*. These include optimization of medium components [1–3], studies of extracellular matrices [4–6], construction of spheroid culture [7–9], and adoption of co-culture techniques. Co-culturing hepatocytes with non-parenchymal cells or fibroblasts is one approach that is useful for regulating culture environments because it is known that hepatocytes can maintain cell viability and liver-specific functions such as albumin synthesis, urea production, and detoxification, including cytochrome P450 activities, in long-term culture [10–15].

Although many previous studies on co-culture have used random culture in which cell distribution is heterogeneous, micropatterned co-cultures have been established in recent studies [16–25]. Micropatterned cultures have the following advantages over random cultures: they can control cellular microenvironments by regulating cell arrangement on a micro-scale. Furthermore, micropatterned co-cultures of hepatocytes with non-parenchymal cells or fibroblasts have provided beneficial information regarding heterotypic cell–cell communication, demonstrating that the expression of hepatocyte functions is stabilized by the increase of heterotypic cell–cell contacts between the hepatocytes and other cell types [19–22,25].

Among all techniques for micropatterned co-culture, a microstencil method is a simple yet effective technique: the cell pattern is easily formed on the culture plate by peeling off a microstencil, which is a thin membrane with orderly through-microholes [23–25]. Furthermore, this method can also be used to evaluate the relationship between co-culture conditions and functional behaviors of cells. Determining such a relationship may require establishing the optimum conditions for micropatterned co-culture.

In this study, we focused on the differences of culture conditions on the micropatterned co-culture of rat hepatocytes with 3T3 cells, and the morphological and functional behaviors of the micropatterned co-culture compared to a random co-culture. Furthermore, the effects of co-culture timing and inoculated density of 3T3 cells on the expression of hepatocyte functions in the micropatterned co-culture were evaluated. This study aimed to demonstrate the advantages of micropatterned co-culture compared to random co-culture and to specify the optimum conditions for the micropatterned co-culture.

Materials and Methods

Preparation of a microstencil chip

Figure 1A shows a schematic diagram of the microstencil chip. A polydimethylsiloxane (PDMS; Sylgard 184, Dow Corning Co., Midland, MI, USA) microstencil, which consists of a thin membrane with 724 through-microholes, each 500 μm in diameter, in a triangular arrangement of 800-μm pitch, was fabricated by peeling the microstencil off from a microfabricated polymethylmethacrylate (PMMA) mould. The limitation of microstencil diameter that could make it by this

*Corresponding author: Kohji Nakazawa, Department of Life and Environment Engineering, The University of Kitakyushu,1-1 Hibikino, Wakamatsu-ku, Kitakyushu, Fukuoka 808-0135, Japan

method was 500 μm. Furthermore, previous studies including us have reported that the micropatterned culture with hepatocyte islands of 500 μm in diameter maintains high expression of liver-specific functions as compared to the conventional monolayer culture [20,25,26]. Therefore, we adopted this microstencil condition. To prepare the microstencil chip, the fabricated microstencil was sealed onto the surface of a glass plate (24 × 24 mm), and the surface of the microstencil chip was coated with type IV collagen to promote cell adhesion Figure 1B. The chip was sterilized by immersion in 70% ethanol, thoroughly rinsed with distilled water, and immersed in the culture medium until use.

Micropatterned culture of rat hepatocytes

This experiment was reviewed by the Committee of Ethics on Animal Experiments of our institute and conducted as per the Guidelines for Animal Experiments at our institute.

Hepatocytes were isolated from the whole liver of an adult Wistar rat (male, 7–8 weeks old, and weighing approximately 200 g) by perfusion with 0.05% collagenase (Wako Pure Chemical Industries Ltd., Osaka, Japan) [27]. Cell viability was determined by the trypan blue exclusion method, and cells that exhibited >85% viability were used for subsequent experiments.

The culture medium was Dulbecco's modified Eagle's medium (DMEM; Invitrogen Corp., Carlsbad, CA, USA) supplemented with 10% fetal bovine serum (FBS), 10 mg/L insulin (Sigma, St. Louis, MO, USA), 7.5 mg/L hydrocortisone (Wako), 50 μg/L epidermal growth factor (Biomedical Technologies Inc., Stoughton, MA, USA), 60 mg/L proline (Wako), 50 μg/L linoleic acid (Sigma), 0.1 μM $CuSO_4 \cdot 5H_2O$, 3 μg/L H_2SeO_3, 50 pM $ZnSO_4 \cdot 7H_2O$, 58.8 mg/L penicillin, and 100 mg/L streptomycin.

Figure 1B shows the process of creating micropatterned culture of hepatocytes using the microstencil chip. Hepatocytes ($1.0 × 10^6$) were inoculated onto the microstencil chip that was placed in a polystyrene dish (diameter, 35 mm) containing 2 mL of culture medium. After 24 h, the stencil was peeled off the chip, and the chip with micropatterned hepatocytes was transferred to another polystyrene dish containing 2 mL of fresh culture medium (micropatterned mono-culture). By this

procedure, approximately $7.0 × 10^4$ cells were immobilized on the chip after 24 h of culture. To obtain a random mono-culture, hepatocytes ($2.5 × 10^5$) were inoculated onto a 35-mm dish coated with type IV collagen. Approximately $9.0 × 10^4$ cells were immobilized on the dish after 24 h of culture.

Micropatterned co-culture of rat hepatocytes with 3T3 cells

NIH/3T3 cells (JCRB0615; Health Science Research Resources Bank, Osaka, Japan) were subcultured as a continuous monolayer in a 100-mm tissue culture dish (Corning, Corning, NY, USA) containing 10 mL DMEM supplemented with 10% FBS, 100 U/mL penicillin, and 100 μg/mL streptomycin. For co-culture experiments, the 3T3 cells were dispersed by treating the confluent monolayer formed on the tissue culture dish with 0.25% trypsin (Invitrogen), and the cells were suspended in the same medium with rat hepatocyte culture.

To compare the effect between micropatterned co-culture and random co-culture, $5.0 × 10^5$ 3T3 cells were inoculated onto the micropatterned- and random-hepatocyte cultures at day 1. To evaluate the effect of co-culture timing, $5.0 × 10^5$ 3T3 cells were inoculated onto the micropatterned hepatocyte culture at days 1, 3, and 5 of culture. To evaluate the effect of inoculated density of 3T3 cells, cell suspensions of densities $1.0 × 10^5$ and $5.0 × 10^5$ cells were inoculated onto the micropatterned hepatocyte cultures at 3 days of culture. The culture medium was changed at 2-day intervals. All cells were cultured at 37°C in a humidified atmosphere of 5% CO_2 and 95% air.

Albumin secretion activity

Albumin secretion activity was evaluated as a typical liver-specific function. The concentrations of albumin secreted into the culture medium during the 24 h of culture were determined by performing an enzyme-linked immunosorbent assay (ELISA). We used two-antibody sandwich method. Briefly, goat anti-rat albumin antibody (MP Biomedicals, Capple Products, USA) was bound to each well of a 96-well microtiter plate, and then the samples were applied to the wells. Subsequently, peroxidase-conjugated sheep anti-rat albumin antibody (MP Biomedicals) was added to each well. ABTS (KPL; Kirkegaard & Perry Laboratories, Inc., USA) was used as a chromogenic substrate, and a microplate reader (Model 550, Bio-Rad Laboratories, USA) was used for the measurement. The albumin concentration of each sample was calculated from the standard curve of rat albumin (MP Biomedicals). The activity was evaluated on days 3, 5, 10, 15, and 20 of the culture, and the values were normalized with the immobilized cells at day 1 of culture.

Statistical analysis

Data obtained from the albumin secretion of rat hepatocytes are represented as mean ± SD of 3 points. Statistical analysis of the numerical variables was performed using a repeated-measures ANOVA test. A p value of <0.05 was considered significant.

Results and Discussion

Cell morphology

In the random mono-culture, rat hepatocytes adhered to the collagen-coated plate and formed a monolayer; however, cell distribution was heterogeneous throughout during the culture Figure 2A. In the random co-culture, the inoculated 3T3 cells spread onto the interstitial space between the adhered hepatocytes, and the hepatocytes exhibited heterogeneous colony-like morphology by the proliferation

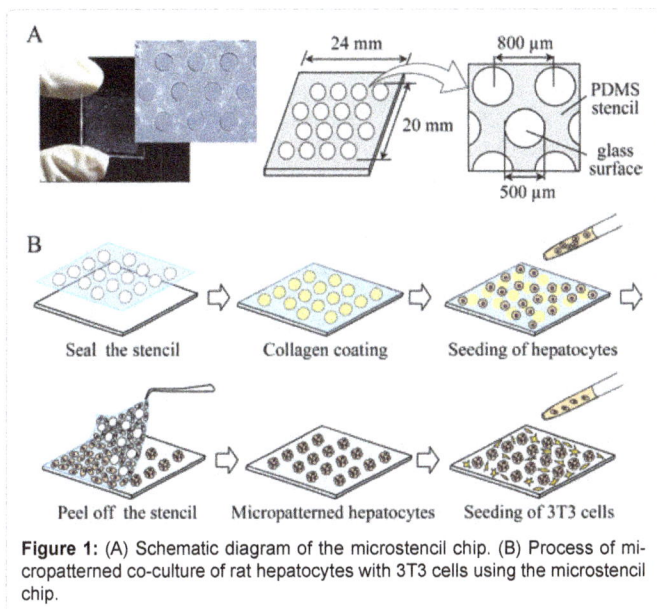

Figure 1: (A) Schematic diagram of the microstencil chip. (B) Process of micropatterned co-culture of rat hepatocytes with 3T3 cells using the microstencil chip.

of 3T3 cells Figure 2B. Although hepatocytes in the random mono-culture rapidly lost cuboidal morphology by elongation of cells, they maintained bright intercellular borders with distinct nuclei after long-term culture in the random co-culture, indicating that the co-culture is effective in maintaining the hepatocellular morphology.

Although the hepatocytes inoculated onto the microstencil chip adhered to the surface of the microstencil and to the glass surface in through-microholes of the microstencil and formed a monolayer within 1 day of culture, the micropatterned hepatocyte islands were formed by peeling off the stencil Figure 2C. The configuration of micropatterned hepatocytes was maintained until 5 days of culture; however, the hepatocytes were elongated at the chip surface after it was removed by the stencil. Consequently, the circular pattern of hepatocytes gradually decayed after 5-7 days of culture Figure 2E. In contrast, in the micropatterned co-culture, the 3T3 cells proliferated in the gap between the hepatocyte islands, thus forming the clear heterotypic interface between the hepatocytes and 3T3 cells Figure 2D. The hepatocytes exhibited hepatocellular morphology, and the configuration of micropatterned hepatocyte islands was also maintained for several weeks of culture by supporting the heterotypic interface Figure 2F. The result that the co-culture maintained the hepatocellular morphology irrespective of the random or micropatterned culture corresponds well with those of previous studies [19,20,25], probably because hepatocyte elongation is inhibited when the substratum surface is covered by 3T3 cells.

Effect of the micropatterned co-culture

Among the various liver-specific functions, we chose albumin secretion activity as an index for the effect of co-culture because albumin secretion is the hepatocyte function that is known to be elevated by co-culture [10-25].

Figure 3 shows changes in the albumin secretion activities of hepatocytes under 4 culture conditions: random mono-culture, random co-culture, micropatterned mono-culture, and micropatterned co-culture. The albumin secretion activity was maintained at a higher level in co-cultures than in mono-cultures, irrespective of whether the culture was random or micropatterned. Furthermore, the secretion activity was maintained at a higher level in micropatterned co-culture than in random co-culture, indicating that micropatterned co-culture is most useful method of maintaining liver-specific functions. Our previous study of mono-culture (with only hepatocytes) revealed

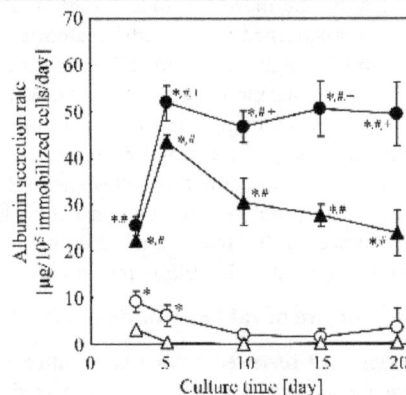

Figure 3: Changes in albumin secretion activity of hepatocytes on random mono-culture (open triangles), random co-culture (closed triangles), micropatterned mono-culture (open circles), and micropatterned co-culture (closed circles). Error bars represent SD; *, $p < 0.05$ compared with the value of random mono-culture; $^\#$, $p < 0.05$ compared with the value of micropatterned mono-culture; and $^+$, $p < 0.05$ compared with the value of random co-culture.

that hepatocytes in a micropatterned culture maintained normal hepatocellular morphology and stable homotypic cell–cell contacts (cell adhesion and gap junction in the hepatocytes); consequently, the expression of liver-specific functions was well retained as compared with the random monolayer [26]. This difference may be reflected to the difference of function between the random and micropatterned co-cultures. Although the maintenance of high functional expression in the mono-culture was difficult even if it was the micropatterned culture, the functional expression was drastically improved by the co-culture. Thus, hepatocytes in the micropatterned co-culture may be able to develop higher functions due to the synergistic effects of improving the culture environment by co-culture and stabilizing the hepatocyte functions by micropatterned culture.

Effect of the co-culture conditions

To evaluate the effect of the micropatterned co-culture conditions, the albumin secretion activities of hepatocytes were compared before and after changing the co-culture timing and inoculation density of 3T3 cells.

The albumin secretion activity of hepatocytes in the micropatterned co-culture decreased on day 3 as compared to day 1 and on day 5 as compared to day 3 Figure 4. Thus, earlier formation of co-culture induces higher functional activities than does later formation. Although recovery of hepatocyte functions by co-culture has been reported [10], to the best of our knowledge, this is the first study that reports that the expression of hepatocyte functions differs depending on the co-culture timing. This finding shows that early formation of co-culture is necessary to maintain a high functional expression of hepatocytes. The activity of micropatterned mono-culture (with only hepatocytes) gradually decreases with the increase in the culture time. This phenomenon may occur by the lack of stimulations such as cell signaling factors, extracellular matrixes, and cell-cell contacts from other cells. Therefore, earlier formation of co-culture may shorten those lack periods, and consequently, the expression of hepatocyte functions may be well maintained.

Furthermore, the albumin secretion activity of hepatocytes in the micropatterned co-culture inoculated at a density of 5×10^5 3T3 cells was higher than that in a co-culture inoculated at a density of 1×10^5

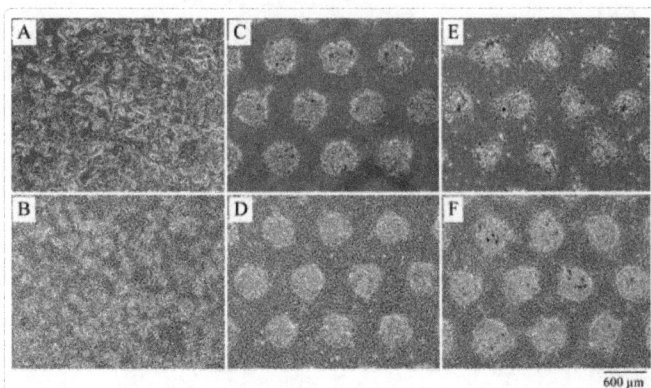

Figure 2: Phase-contrast micrographs of different culture conditions. (A) Random mono-cultures; (B) random co-culture; (C and E) micropatterned mono-culture; and (D and F) micropatterned co-culture. A–D, day 3 of culture; E, day 7 of culture; and F, day 15 of culture.

3T3 cells, indicating that forming a co-culture with a high density of 3T3 cells at the initial stage of the culture is superior for maintaining the high functional expression of hepatocytes Figure 5. Bhatia et al. reported that hepatocyte functions in random co-culture were induced depending on an increase of inoculated 3T3 cell density [19]; similarly, we have showed that the same phenomena exist in the micropatterned co-culture.

It is known that co-culture improves culture environment, since it induces production of cell signaling factors and extracellular matrixes, and/or formation of heterotypic cell–cell interaction. In particular, recent studies of co-culturing hepatocytes with 3T3 cells revealed that the expression of hepatocyte functions is stabilized by an increase of direct heterotypic cell–cell contacts (cell adhesion and gap junction) [19–22,25,28]. This information corresponds with our results that the earliest formation of co-culture under high cell density of 3T3 cells induced the highest functional activities of hepatocytes. The direct heterotypic cell–cell contacts are important factors for hepatocellular

polarity and organization [21,25]. Although the detailed mechanism is not clear, their development may operate to normal hepatocellular structure that includes cuboidal cell shape and abundant cytoplasmic organelles; consequently, the expression of hepatocyte functions may be well maintained. Although the sufficient heterotypic cell-cell contacts are formed even if it was the random co-culture, the function of micropatterned co-culture was higher than that of random monolayer. This fact suggests that the combination of the heterotypic contacts (hepatocyte-3T3 cell) and the homotypic interactions (hepatocyte-hepatocyte) is important for maintaining the liver-specific functions. Further studies of heterotypic and homotypic interactions may help to better understand the mechanism.

Conclusions

In this study, we demonstrated the effect of culture conditions on the micropatterned co-culture of rat hepatocytes with 3T3 cells by using the microstencil method. The micropatterned configuration of hepatocyte islands was maintained for several weeks in culture by supporting the heterotypic interface. The albumin secretion activities of hepatocytes in the micropatterned co-culture were maintained at higher levels than those observed in random co-culture. Furthermore, the earlier formation of co-culture expressed higher functional activities of hepatocytes than the later formation, and better induction of hepatocyte functions depended on increasing 3T3 cell density of inoculation. These results suggest that the formation of a heterotypic interaction between the hepatocytes and 3T3 cells at an early stage is important to maintain high levels of hepatocyte functions. The findings of this study may provide useful information for designing liver tissue engineering as well as pharmacological and toxicological studies.

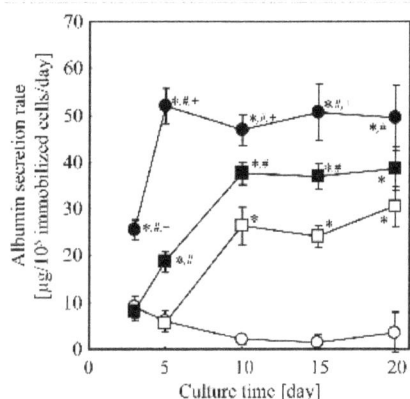

Figure 4: Effect of the co-culture timing on the albumin secretion activity of hepatocytes on micropatterned mono-culture (open circles), micropatterned co-culture inoculated on day 1 (closed circles), micropatterned co-culture inoculated on day 3 (closed squares), and micropatterned co-culture inoculated on day 5 (open squares) of culture. Error bars represent SD. *, $p < 0.05$, compared with the value of mono-culture. $^#$, $p < 0.05$, compared with the value of co-culture at day 5. $^+$, $p < 0.05$, compared with the value of co-culture at day 3.

Acknowledgments

This work was partly supported by the Regional Innovation Cluster Program (Global Type–2nd Stage) and by the Nanotechnology Network Project implemented by the Ministry of Education, Culture, Sports, Science, and Technology (MEXT).

References

1. Enat R, Jefferson DM, Opazo NR, Gatmaitan Z, Leinward LA, et al. (1984) Hepatocyte proliferation in vitro: Its dependence on the use of serum-free hormonally defined medium and substrata of extracellular matrix. Proc Natl Acad Sci 81: 1411–1415.

2. Hamilton GA, Westmorel C, George AE (2001) Effects of medium composition on the morphology and function of rat hepatocytes cultured as spheroids and monolayers. In Vitro Cell Dev Biol Anim 37: 656–667.

3. Mizumoto H, Ishihara K, Nakazawa K, Ijima H, Funatsu K, et al. (2008) A new culture technique for hepatocyte organoid formation and long-term maintenance of liver-specific functions. Tissue Eng Part C 14: 167–175.

4. Bissell DM, Stamatoglou SC, Nermut MV, Hughes RC (1986) Interactions of rat hepatocytes with type IV collagen, fibronectin and laminin. Distinct matrix-controlled modes of attachment and spreading. Eur J Cell Bio 40: 72–78.

5. Bissell DM, Arenson DM, Maher JJ, Roll FJ (1987) Support of cultured hepatocytes by a laminin-rich gel. J Clin Invest 79: 801–812.

6. Sawada N, Tomomura A, Sattler CA, Sattler GL, Kleinman HK, Pitot HC (1987) Effects of extracellular matrix components on the growth and differentiation of cultured rat hepatocytes. In Vitro Cell Dev Biol 23: 267–273.

7. Koide N, Sakaguchi K, Koide Y, Asano K, Kawaguchi M, et al. (1990) Formation of multicellular spheroids composed of adult rat hepatocytes in dishes with positively charged surfaces and under other nonadherent environments. Exp Cell Res 186: 227–235.

8. Tong JZ, Lagausie PD, Furlan V, Cresteil T, Bernard O, et al. (1992) Long-term culture of adult rat hepatocyte spheroids. Exp Cell Res 200: 326–332.

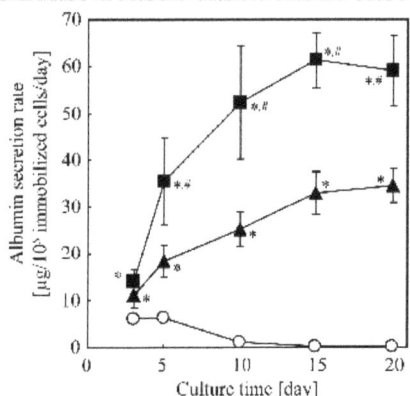

Figure 5: Effect of the inoculated density of 3T3 cells on albumin secretion activity of hepatocytes. Micropatterned mono-culture (open circles); micropatterned co-culture inoculated with a 3T3 cell density of 1×10^5 cells (closed triangles); and micropatterned co-culture inoculated with a 3T3 cell density of 5×10^5 cells (closed squares). Error bars represent SD; *, $p < 0.05$ compared with the value of the mono-culture; $^#$, $p < 0.05$ compared with the value of the co-culture at 1×10^5 3T3 cells.

9. Sakai Y, Yamagami S, Nakazawa K (2010) Comparative analysis of gene expression in rat liver tissue and monolayer- and spheroid-cultured hepatocytes. Cells Tissues Organs 191: 281–288.

10. Clement B, Guguen-Guillouzo C, Campion JP, Glaise D, et al. (1984) Long-term co-cultures of adult human hepatocytes with rat liver epithelial cells: modulation of albumin secretion and accumulation of extracellular material. Hepatology 4: 373–380.

11. Shimaoka S, Nakamura T, Ichihara A (1987) Stimulation of growth of primary cultured adult rat hepatocytes without growth factors by coculture with nonparenchymal liver cells. Exp Cell Res 172: 228–242.

12. Koike M, Matsushita M, Taguchi K, Uchino J (1996) Function of culturing monolayer hepatocytes by collagen gel coating and coculture with nonparenchymal cells. Artif Organs 20: 186–192.

13. Bhandari RN, Riccalton LA, Lewis AL, Fry JR, Hammond AH, et al. (2001) Liver tissue engineering: a role for co-culture systems in modifying hepatocyte function and viability. Tissue Eng 7: 345–357.

14. Lu HF, Chua KN, Zhang PC, Lim WS, Ramakrishna S, et al. (2005) Three-dimensional co-culture of rat hepatocyte spheroids and NIH/3T3 fibroblasts enhances hepatocyte functional maintenance. Acta Biomater 1: 399–410.

15. Thomas RJ, Bhandari R, Barrett DA, Bennett AJ, Fry JR, et al. (2005) The effect of three-dimensional co-culture of hepatocytes and hepatic stellate cells on key hepatocyte functions in vitro. Cells Tissues Organs 181: 67–79.

16. Revzin A, Rajagopalan P, Tilles AW, Berthiaume F, Yarmush ML, Toner M (2004) Designing a hepatocellular microenvironment with protein microarraying and poly(ethylene glycol) photolithography. Langmuir 20: 2999–3005.

17. Otsuka H, Hirano A, Nagasaki Y, Okano T, Horiike Y, et al. (2004) Two-dimensional multiarray formation of hepatocyte spheroids on a microfabricated PEG-brush surface. Chembiochem 5: 850–855.

18. Kikuchi K, Sumaru K, Edahiro J, Ooshima Y, Sugiura S, et al. (2009) Stepwise assembly of micropatterned co-cultures using photoresponsive culture surfaces and its application to hepatic tissue arrays. Biotechnol Bioeng 103: 552–561.

19. Bhatia SN, Balis UJ, Yarmush ML, Toner M (1998) Microfabrication of hepatocyte/fibroblast co-culture: role of homotypic cell interactions. Biotechnol Prog 14: 378–387.

20. Bhatia SN, Balis UJ, Yarmush ML, Toner M (1998) Probing heterotypic cell interactions: hepatocyte function in microfabricated co-cultures. J Biomater Sci Polym Ed 9: 1137–1160.

21. Khetani SR, Szulgit G, Del Rio JA, Barlow C, Bhatia SN (2004) Exploring interactions between rat hepatocytes and nonparenchymal cells using gene expression profiling. Hepatology 40: 545–554.

22. Takahashi S, Yamazoe H, Sassa F, Suzuki H, Fukuda J (2009) Preparation of coculture system with three extracellular matrices using capillary force lithography and layer-by-layer deposition. J Biosci Bioeng 108: 544–550.

23. Zinchenko YS, Coger RN (2005) Engineering micropatterned surfaces for the coculture of hepatocytes and Kupffer cells. J Biomed Mater Res A 75: 242–248.

24. Cho CH, Berthiaume F, Tilles AW, Yarmush ML (2008) A new technique for primary hepatocyte expansion in vitro. Biotechnol Bioeng 101: 345–356.

25. Cho CH, Park J, Tilles AW, Berthiaume F, Toner M, Yarmush ML (2010) Layered patterning of hepatocytes in co-culture systems using microfabricated stencils. Biotechniques 48: 47–52.

26. Nakazawa K, Shinmura Y, Yoshiura Y, Sakai Y (2010) Effect of cell spot sizes on micropatterned cultures of rat hepatocytes. Biochem Eng J 53: 85–91.

27. Seglen PO (1976) Preparation of isolated rat liver cells. Methods Cell Biol 13: 29–83.

28. Hui EE, Bhatia SN (2007) Micromechanical control of cell-cell interactions. Proc Natl Acad Sci 104: 5722–5726.

23

The Comparison of Chemiluminescent- and Colorimetric-detection Based ELISA for Chinese Hamster Ovary Host Cell Proteins Quantification in Biotherapeutics

Fengqiang Wang*, Dennis Driscoll, Daisy Richardson and Alexandre Ambrogelly

Bioprocess Development, Merck Research Laboratories, Union, NJ, USA

Abstract

Biologics manufacturing requires the clearance of Host Cell Proteins (HCPs) from recombinant therapeutic protein to acceptable low levels to ensure product purity and patient safety. To ensure adequate removal, a highly sensitive method, commonly in the form of Enzyme-Linked Immunosorbent Assay (ELISA), is necessary to quantify the HCPs amount in process intermediates and drug substance. We report the development of a chemiluminescent detection based ELISA (luminescent ELISA) in lieu of previously used colorimetric method (colorimetric ELISA) to improve assay sensitivity for the quantification of Chinese Hamster Ovary (CHO) HCPs in a monoclonal antibody product (mAb-A). For luminescent ELISA, Pierce Supersignal ELISA Femto was chosen as the substrate to replace colorimetric substrate TMB. The assay performance of luminescent and colorimetric ELISA was directly compared side-by-side. Our data show that luminescent ELISA has better signal/background ratio, broader linear range over logarithmic scales, and better linearity within the same linear range than colorimetric ELISA. Luminescent ELISA also demonstrates better low-end linearity, greater accuracy and precision. In addition, the Limit of Detection (LOD) and Limit of Quantification (LOQ) are significantly improved with luminescent ELISA as compared to colorimetric ELISA. In summary, luminescent ELISA is a more sensitive method and demonstrates superiority over colorimetric method for CHO HCP quantification.

Keywords: Host cell proteins; ELISA; Chemiluminescence; Chinese hamster ovary; Biotherapeutics; Monoclonal antibody

Introduction

Monoclonal antibodies (mAbs) have become a significant focus of the pharmaceutical industry due to their high specificity and their ability to engage a wide variety of targets [1,2]. While mAb therapeutics have been produced in a variety of genetically engineered host cell of non-human origin such as bacteria, yeast, plant, insect and mammalian cells, they are most commonly expressed in immortalized Chinese hamster ovary (CHO) cell lines [3-5]. CHO is a robust host that offers high productivity and glycosylation patterns similar to those found in endogenous human antibodies. Harvest of therapeutic antibodies of interest is relatively straight forward since the recombinant product is often secreted in the media. However, the harvest also contains significant amounts of proteins originated from the host, namely host cell proteins (HCPs), which are either secreted during fermentation or released into culture fluid as a result of cell lysis. Due to their non-human origin and thus potential immunogenic nature, HCPs can pose significant safety risk for patients and are part of process-related impurities that need to be controlled during bioprocess development [6-8]. Since after the purification steps, the residual HCPs amount in final drug substance is often very low in the parts per million (ppm) level, a highly specific, highly sensitive, and quantitative assay is desired to ensure their adequate removal and patient safety [9,10]. Due to its high specificity and sensivity, enzyme-linked immunosorbent assay (ELISA) is the most commonly accepted method by regulators for HCPs quantification [10]. Alternative immunospecific methods such as a quantitative slot blot assay [11] and solid-phase proximity ligation assay [12] as well as non-specific methods including mass spectrometry (MS) and 2D liquid chromatography (LC)-MS are also being developed or explored [8,10]. However, none of these methods are robust enough or can achieve the same level of sensitivity as ELISA, which remains the gold standard for HCP quantification. Commercially available HCP ELISA kits, commonly used as generic HCP assays in the early phase of development (Phase I/II) as well as previously reported late stage

process-specific HCP ELISA often use colorimetric detection for signal generation [13,14]. Colorimetric detection limits the assay sensitivity for low levels of HCP especially in final drug product [13,15]. At Merck Research Laboratories, we have initially developed a process-specific colorimetric ELISA assay for one of our late stage CHO-produced monoclonal antibodies, mAb-A. Genetically engineered CHO cell line is used to manufacture mAb-A and thus a process-specific HCP ELISA using proprietary antibodies raised against the null CHO cells has been developed in-house for Phase III mAb-A to measure HCP components in the drug substance (DS).

The process-specific ELISA in its current format has a limit of quantification (LOQ) of 7.6 ng/ml in 5 mg/ml of drug substance (equivalent to 1.5 ppm). While this LOQ value is sufficient to demonstrate process clearance of HCPs, improvements can be made to increase the assay sensitivity to measure HCP concentration < 7.6 ng/ml. Since colorimetric detection limits the assay sensitivity for low levels of HCPs especially in the final drug susbtance, alternative method using chemiluminescent detection has been explored. Since its introduction in the late 1970s, chemiluminescence has been used in a variety of analytical and immunological tests such as high performance liquid chromatography [16], capillary electrophoresis

*Corresponding author: Fengqiang Wang, Ph.D., Associate Principal Scientist, Bioprocess Development, Merck Research Laboratories, Kenilworth, NJ, USA

[17], immunoassays and DNA analyses [18,19]. Analytical methods using chemiluminescent detection are often characterized by their high sensitivity, broader dynamic range, and high signal-to-noise ratio [19]. In immunoassays, chemiluminescent horse radish peroxidases (HRP) substrates have shown improved sensitivity over colorimetric substrates [20]. Commercially available Thermo Scientific Super Signal ELISA Femto Maximum Sensitivity substrate uses an improved enhancer system with much greater sensitivity and has been successfully used in high throughput enzyme immunoassay [21], antibody microarrays [22], and blood-based diagnostic assays [23]. With its known advantage of fast light generation, high sensitivity (1.7 pg/ml), and improved low-end linearity, the ELISA Femto substrate was adopted for assay development and its assay performance was compared side-by-side with 1-step turbo TMB (sensitivity 70 pg/ml) based colorimetric detection method. The signal/noise ratio, linear range and linearity over logarithmic scales, precision and accuracy as well as the limit of detection (LOD) and limit of quantification (LOQ) of both methods were assessed and compared following ICH guidelines-Q2 (R1).

Materials and Methods

Commercial reagents and consumables

Hyclone phosphate buffer saline (PBS, 10×), carbonate-bicarbonate buffer packet PK40, blocker BSA in PBS (10×), neutrAvidin-horseradish peroxidase (HRP) conjugate, 1-step Turbo TMB ELISA substrate, and Supersignal ELISA Femto Maximum Sensitivity substrate were purchased from Thermo Fisher Scientific Inc.(Waltham, MA, USA); Tween-20 was purchased from Sigma Aldrich (St. Louis, MO, USA); Costar EIA/RIA ELISA clear bottom 96-well plate was purchased from Corning Inc (Corning, NY, USA); NUNC white opaque 96-well plate was purchased from Thermo Fisher Scientific Inc. (Waltham, MA, USA). ImmunoWare tubes and ImmunoWare reagent reservoirs were product of Thermo Scientific Pierce Inc (Waltham, MA, USA).

Merck proprietary reagents

Anti-HCP polyclonal antibodies were raised in goat by Pocono Farms& Laboratory, Inc. (Tobyhanna, PA, USA) against CHO null cell culture (mock) that doesn't have the gene encoding mAb-A and the anti-sera was affinity purified by a self-prepared mock HCP affinity column. Affinity-purified goat anti-HCP IgG was then aliquoted and partially labeled with biotin using EZ-link Sulfo-NHS-LC-Biotin kit from Thermo Scientific Inc (Waltham, MA, USA). Unlabeled anti-CHO HCP antibody (lot# 68780/140, 1.69 mg/mL) was used as coating antibody and biotinylated goat anti-HCP IgG (lot# 68780/147, 1.69 mg/mL) was used as detecting antibody in a sandwich ELISA format (Figure 1). CHO HCP stock generated from mock cell fermentation was used as standard (5.70 mg/mL, lot# 68383/106). The reagents are stored at -20°C, with one working aliquot stored at 2-8°C.

Instruments

ELISA plate wash was done using a BioTek Elx 405 Select semi-automatic plate washer (BioTek USA,Winooski, VT), and the absorbance/luminescence signal was read by a Molecular Devices SpectraMax M5 plate reader (Molecular Devices, Sunnyvale, CA).

ELISA working solutions

Coating buffer was prepared by dissolving one packet of the carbonate-bicarbonate buffer concentrate in 500 ml of deionized water; washing solution was prepared by adding Tween-20 to 1× PBS to a final concentration of 0.1%; blocking solution was prepared by adding

Tween-20 and 10× Pierce Blocker to PBS to a final solution containing 1% BSA and 0.05% Tween-20; assay diluents was prepared by adding Tween-20 and 10× Pierce Blocker to PBS to a final solution containing 0.1% BSA and 0.05% Tween-20.

Performing CHO colorimetric ELISA and luminescent ELISA in Corning Costar clear EIA/RIA 96-well plate

The CHO HCP ELISA was performed using established protocols for colorimetric detection with the adaption on the substrate addition step for chemiluminescent detection (Figure 1). For chemiluminescent ELISA, SuperSignal ELISA Femto substrate, instead of 1-step Turbo TMB substrate, was added to the appropriate wells and light emission was measured at 425 nm. Briefly, the ELISA plate was coated by 100 µL/well of coating antibody solution (1 µg/ml) prepared in carbonate-bicarbonate buffer and incubated at room temperature with gentle shaking for 2 hrs. After 4 washes with 300 µL/well washing buffer (10 second incubation/wash) using BioTek Elx 405 Select plate washer, the plate was blot dried using tissue paper and incubated with 300 µL/well of blocking solution at room temperature for 1 hr. After wash, the ELISA reactions were performed at room temperature with the addition of CHO HCP standards or unspiked/spiked mAb-A samples (100 µL/well), followed by the subsequent incubation with biotinylated anti-CHO HCP antibody (1 µg/mL, 100 µL/well), NeutrAvidin-HRP conjugate (1:15,000 dilution in assay diluents, 100 µL/well), and 1-step TMB turbo ELISA substrate (100 µL/well) or Supersignal ELISA Femto Maximum Sensitivity substrate (100 µL/well). Plate was washed 4 times with 300 µL/well washing buffer between each incubation steps. For those wells with TMB as a substrate, plate was incubated in dark for 10 min and the reactions were terminated by the addition of 100 µL/well of 1 M sulfuric acid, and then the absorbance of those wells were read at 450 nm using a Molecular Devices plate reader; for wells with ELISA Femto as a substrate, the light emission was measured at 425 nm 10 min after the addition of substrate using the same plate reader.

Performing CHO HCP colorimetric and luminescent ELISA in Nunc opaque 96-well plate

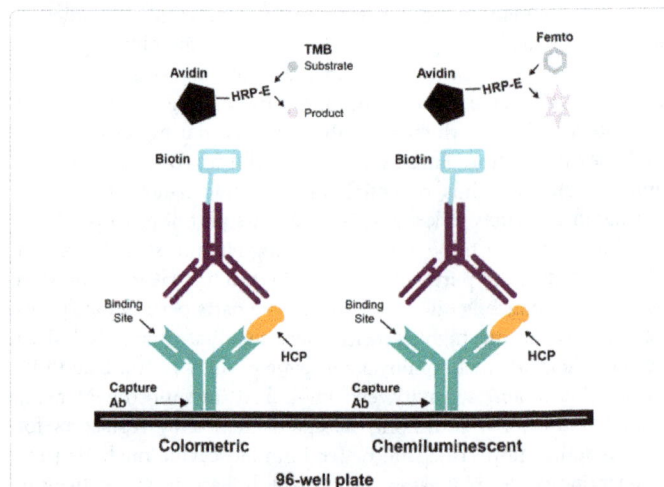

Figure 1: Schematic illustration of colorimetric and luminescent ELISA performed. Both ELISA methods are performed on the same 96-well plate and following the same procedure of HCPs capturing by coated anti-HCP antibodies, primary detection by biotinylated anti-HCP antibodies, and secondary detection by NeutrAvidin-HRP conjugates. The signal detection is either obtained by measuring absorbance at 450 nm generated from catalyzing TMB substrate (left) or by measuring the luminescence at 425 nm generated from oxidizing of ELISA Femto substrate (right).

To avoid luminescence signal cross-interference from adjacent wells, NUNC white opaque 96-well plate that provides maximum reflection and low cross-talk was chosen for luminescent ELISA development and optimization. The ELISA procedure follows the same steps of coating, washing, blocking, standards and samples incubation, primary detection (biotin-antibody conjugate), secondary detection (NeutrAvidin-HRP conjugate), and substrate incubation steps as described previously. The ELISA Femto substrate solution was prepared fresh on day of use by mixing equal volume of the signal enhancer with the Femto substrate. The steps and reagent volumes used in each step are summarized in Supplemental Table 1. For colorimetric assay, the reaction was stopped with the addition of 1 M sulfuric acid after 10 min incubation in TMB substrate, and then the reaction mixtures were transferred to a Costar clear 96-well plate using multi-channel pipette or and the absorbance of each well was measured at 450 nm using Molecular Devices SpectraMax microplate reader.

Standards and samples preparation for evaluating colorimetric and luminescent ELISA assay performance in Corning Costar 96-well clear plate

The ELISA assay performances including signal/noise ratio, dynamic range, low-end linearity, accuracy and precision were compared side-by-side using CHO HCP standard in triplicate prepared in a series of 3-fold dilution over the range from 2000 ng/ml to 0.034 ng/ml. Additional experiments were performed using standards in a series of 2-fold dilution ranging from 200 to 0.195 ng/ml or 50 to 0.049 ng/ml. Different amount of CHO HCPs (25 ng/ml, 5 ng/ml, 2 ng/ml and 1 ng/ml) were also spiked into mAb-A drug substance (final concentration 5 mg/ml) to assess the accuracy and precision of each assay.

Standards and samples preparation for evaluating colorimetric ELISA and luminescent ELISA performance in NUNC white opaque 96-well plate

The ELISA assay performances including signal/noise ratio, the assay linearity, accuracy, precision, LLOD and LLOQ were compared over the standards range from 1.56 ng/ml to 100 ng/ml using the same amount of reagents (Supplemental Table 1) optimized for luminescent ELISA in NUNC white opaque plate.

Data analysis

All experiments were done in triplicate. The comparison of colorimetric and luminescent ELISA was performed side-by-side in three repeated experiments with slightly variation on standard range. The data analysis was performed using Molecular Devices Soft max Pro v 5.3 software and Microsoft Excel. The chemiluminescence light emission at 425 nm (E425) or absorbance values at 450 nm (A450) are plotted against standard concentrations using the software's built in 4-parameter fit, linear fit, or log-log curve fitting. The signal/noise ratio of each standard data point is determined using the signal generated by a standard at a given concentration to that of concentration 0 (blank). The linear range over logarithmic scales is determined as where the correlation coefficient (R^2) has a value >0.99. The concentrations of standards are then back-calculated from the standard curve fit equation to assess the precision (CV%) and accuracy (recovery%) of the assay. In all conditions, the same curve-fitting method is applied for both colorimetric and luminescent ELISA for direct comparison. The recovery% of standards is calculated as the ratio of back-calculated concentration to the expected concentration × 100%.

Results and Discussion

The assay performance of luminescent ELISA and colorimetric ELISA was directly compared by a variety of assay parameters such as signal/noise ratio, linear range and linearity over logarithmic scales, accuracy and precision, lower limit of detection (LLOD) and quantification (LLOQ). To reduce the variations caused by assay plate type, the comparison of assay performance was carried out in both 96-well clear plate and opaque plate. Several standard curve-fitting methods built in the Softmax Pro v5.4.1 were applied to determine the optimum method for HCP quantification.

Assay performance comparison for colorimetric ELISA and luminescent ELISA in Corning Costar clear 96-well plate using conditions optimized for colorimetric ELISA

ELISA assays using 1-step Turbo TMB (colorimetric) or ELISA Femto (luminescent) as a substrate for CHO HCP testing were performed side-by-side on the same 96-well Costar clear EIA/RIA plate following the procedure described in materials and methods. The standard curves were first fit using a 4-parameter non-linear regression. As seen in Figure 2A, colorimetric ELISA standard curve (bottom) displayed a sigmoid shape with the absorbance value at 2000 ng/ml reaching a plateau, in contrast, luminescent ELISA standard curve at 2000 ng/ml remains in the rising phase (top). The C values

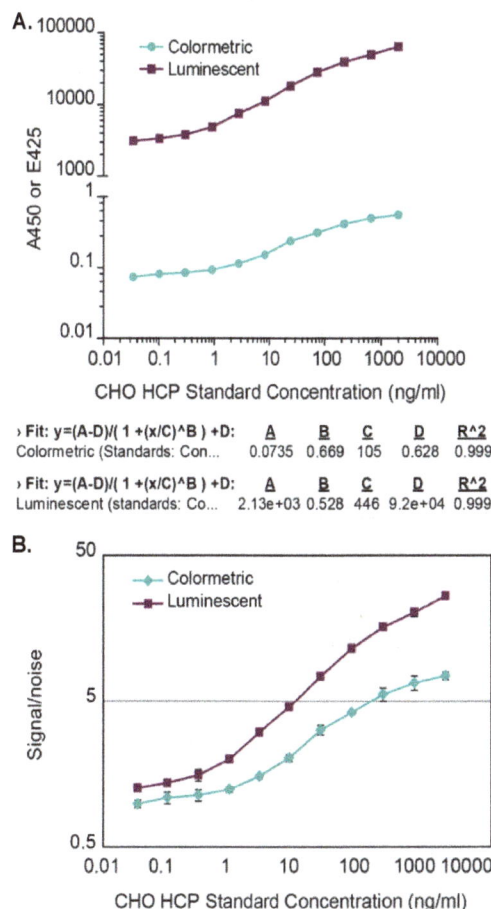

› Fit: y=(A-D)/(1 +(x/C)^B) +D:

	A	B	C	D	R^2
Colormetric (Standards: Con...	0.0735	0.669	105	0.628	0.999

› Fit: y=(A-D)/(1 +(x/C)^B) +D:

	A	B	C	D	R^2
Luminescent (standards: Co...	2.13e+03	0.528	446	9.2e+04	0.999

Figure 2: Standard curves from colorimetric and luminescent ELISA performed in clear 96-well plate. A. Absorbance values (A450) or luminescence emissions at 425 nm (E425) were plotted against standard concentrations on log-log scales and the curve equations were fit using 4-parameter non-linear regression. B. Signal/noise ratio values plotted against CHO HCP standard concentration over logarithmic scales.

from the 4-parameter fit standard curve equation of colorimetric and luminescent ELISA, which represent the standard concentrations where the signal response is ~½ of the maximum signal response, are 105 and 446 respectively, indicating that luminescent detection has a much broader range. To assist the direct comparison between the two detection methods, relative signal response or signal/noise ratio of each standard was also plotted against its concentration on logarithmic scales (Figure 2B). Luminescent detection showed significant higher signal/noise ratio at all standard concentrations and its standard curve has steeper slope than colorimetric method. As shown in Table 1, the absorbance values at 450 nm (A450) generated by colorimetric method ranged from 0.075 to 0.561 at standard concentrations from 0 to 2000 ng/ml; in contrast, light emissions (425 nm) from luminescent method have a reading from 2461 to 64780 at the same standard range. At 2000 ng/ml standard concentration, luminescent ELISA has a signal/noise ratio of 26.32, which is equivalent to 3.5 fold of the S/N for colorimetric ELISA (7.467, Table 1). The mean S/N for luminescent detection method (8.677 ± 8.736) is also significantly higher than the mean S/N for colorimetric method (3.191 ± 2.348) as analyzed by paired t-test (p=0.017, n=11, Table 1).

To accurately quantify the HCPs in drug substance, the ELISA standard curves need to be fit using the appropriate mathematical models. Several models are available in SoftMax software that include linear, semi-log, log-log, 4-parameter and 5-parameter fit. The method used to form the calibration curve dictates the working range and overall accuracy of the assay [24]. For broad range CHO HCP standard curves as shown in Figure 2, 4-parameter non-linear regression fit gave the best correlation coefficient of R^2=0.999. Using the 4-parameter fit equation, the concentration of each CHO HCP standard was back-calculated according to its corresponding A450 or E425 value. The precision and accuracy of the back-calculated concentration was demonstrated by its relative standard deviation calculated from triplicate (CV%) and the recovery% (the ratio of back-calculated concentration to the expected concentration) (Table 2). Using CV% ≤ 30 and 70 ≤ Recovery% ≤ 130 as acceptance criteria for accuracy and precision, luminescent ELISA has a working/dynamic range from 0.914 ng/ml to 666.7 ng/ml, which is 27 fold of that for colorimetric ELISA (2.743 ng/ml to 74.07 ng/ml, Table 2). In addition, we assessed the linearity of both colorimetric and luminescent ELISA at various standard concentration ranges using several mathematic models and determined their linear range with a log-log fit curve. For luminescent detection, the linear range over logarithmic scales is 0.31-2000 ng/ml (R^2=0.991), 27 fold of that for colorimetric method (0.91-222.22 ng/ml, R^2=0.992). Moreover, at the 0.91-222.22 ng/ml standard range, the R^2 of luminescent detection for log-log linear fit is 0.998, showing better linearity than colorimetric method at the same range (R^2=0.992, Figure 3). The better linearity within the same concentration range was also observed in experiment covering the standard range of 0.781-200 ng/ml, where luminescent

CHO HCP Standard (ng/ml)	Colorimetric ELISA A450	Luminescent ELISA E425	Colorimetric ELISA S/N	Luminescent ELISA S/N
0	0.075 ± 0.003	2461 ± 265.3	1	1
0.034	0.075 ± 0.004	3124 ± 140.6	0.993	1.269
0.102	0.082 ± 0.008	3389 ± 151.6	1.091	1.377
0.305	0.086 ± 0.008	3842 ± 351.9	1.139	1.561
0.914	0.093 ± 0.004	4924 ± 216.0	1.244	2.001
2.743	0.115 ± 0.004	7537 ± 432.9	1.533	3.063
8.23	0.153 ± 0.008	11179 ± 536.5	2.039	4.543
24.69	0.238 ± 0.018	18228 ± 335.9	3.17	7.407
74.07	0.313 ± 0.011	28313 ± 1656	4.17	11.51
222.2	0.419 ± 0.044	39576 ± 1913	5.577	16.08
666.7	0.502 ± 0.056	49976 ± 3293	6.683	20.31
2000	0.561 ± 0.034	64780 ± 3319	7.467	26.32
Mean			3.191	8.677
Stdev			2.418	8.736
P value (paired t-test on mean S/N, n=11)				0.017

Table 1: Comparison of the signal/noise (S/N) ratios of CHO HCP standards from colorimetric ELISA and luminescent ELISA performed in Corning Costar clear 96-well plate (n=3 for each standard concentration).

CHO HCP standard (ng/ml)	Colorimetric ELISA		Luminescent ELISA	
	%CV	%Rec	%CV	%Rec
2000	59.97	137.66	33.69	113.15[b]
666.7[l]	105.06	160.25	26.93[a]	87.78[b]
222.2[l]	51.35	111.11[b]	16.72[a]	106.95[b]
74.07[l, c]	11.90[a]	94.67[b]	16.75[a]	112.54[b]
24.69[l, c]	21.87[a]	118.46[b]	4.86[a]	100.98[b]
8.230[l, c]	17.90[a]	88.77[b]	12.54[a]	85.78[b]
2.743[l, c]	18.16[a]	89.84[b]	15.76[a]	89.39[b]
0.914[l]	31.8	84.94[b]	15.29[a]	72.43[b]
0.305	93.76	129.78[b]	37.1	85.82[b]
0.102	125.17	247.84	22.16[a]	140.18
0.034	57.38	171.49	27.08[a]	267.85
0	75.77		107.53	

Table 2: Comparison of the precision (%CV) and accuracy (%Rec) of back-calculated concentration using 4-parameter fit for colorimetric and luminescent ELISA performed in Corning Costar clear 96-well plate.

A.

Standard Range 0.31-2000 ng/ml

Log-Log Fit: Log(y) = A + B * Log(x):	A	B	R^2
△ Colormetric ELISA (Standards@Ex...	0.134	0.241	0.98
○ Luminescent ELISA (Standards@E...	0.278	0.341	0.991

B.

Standard Range 0.91-222.22 ng/ml

Log-Log Fit: Log(y) = A + B * Log(x):	A	B	R^2
△ Colormetric ELISA (Standards@Ex...	0.0818	0.285	0.992
○ Luminescent ELISA (Standards@E...	0.237	0.387	0.998

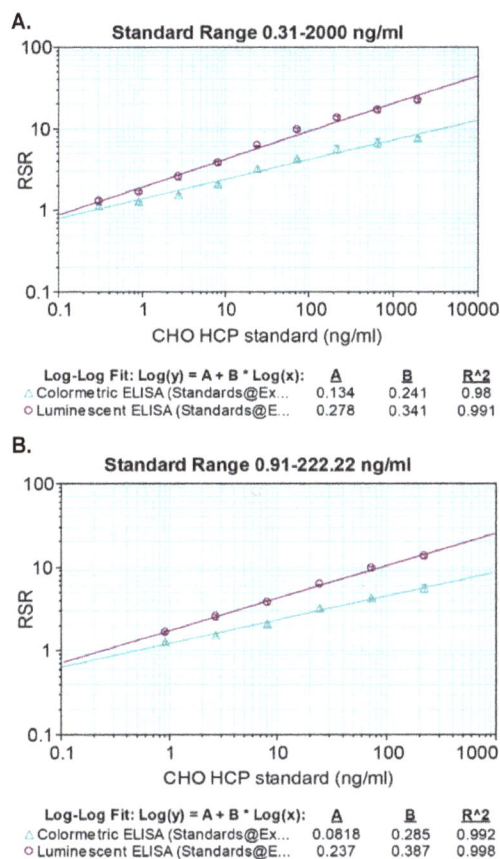

Figure 3: The linear range and linearity comparison over logarithmic scales between colorimetric ELISA and luminescent ELISA. A. Log-log linearity comparison over the range of 0.31-2000 ng/ml. B. Log-log linearity comparison over the range of 0.91-222.22 ng/ml.

Standard Range 0.034-2.743 ng/ml

Linear Fit: y = A + Bx:	A	B	R^2
△ Colorimetric ELISA (Standards@Ex...	1.05	0.18	0.961
○ Luminescent ELISA (Standards@E...	1.1	0.536	0.992

Figure 4: Low-end linear curves of colorimetric and luminescent ELISA at concentrations ranging from 0.034 to 2.743 ng/ml. The mean signal value ± standard deviation from triplicate wells was plotted against standard concentration by a linear fit method. Larger variations are observed in colorimetric method at this concentration range. The standard curve is also less linear than luminescent ELISA ($R2=0.961$ vs. 0.992).

method has a correlation coefficient of R=0.999 for log-log linear fit curve, as compared to $R^2=0.994$ for colorimetric method. In addition to the broader log-log fit linear range and better linearity within the same log-log fit range, luminescent detection also showed better low-end linearity at standard concentrations from 0.034 to 2.743 ng/ml, where the linear correlation has a $R^2=0.992$ in comparison to $R^2=0.961$ for the colorimetric ELISA (Figure 4). Using the low-end linear curve, the concentration of CHO HCPs at 0.914 ng/ml can be more precisely determined by luminescent ELISA, but not by colorimetric method (Supplemental Table 2).

Within the log-log linear range of 0.91-222.22 ng/ml, the accuracy and precision of both ELISAs were compared. Our results show that within this range luminescent method is significantly more precise (CV%, $p=0.03$, single tail t-test) than colorimetric method (Table 3). Using the log-log fit curves within this range to back-calculate standard concentrations from 0.03 to 2000 ng/ml, all standard concentrations exhibited comparable or better CV and recovery values than colorimetric method, which also holds true when back-calculating HCP concentration using 4-parameter fit (data not shown). Despite the standard curve fit methods, luminescent ELISA always has smaller CV values (~1/2) than colorimetric method. However, when using log-log fit curve to determine the concentration within this range, luminescent method also exhibit better accuracy (100.65 ± 9.12% recovery) than colorimetric method (102.66 ± 18.77% recovery), although not statistically different ($p=0.41$, single tail t-test).

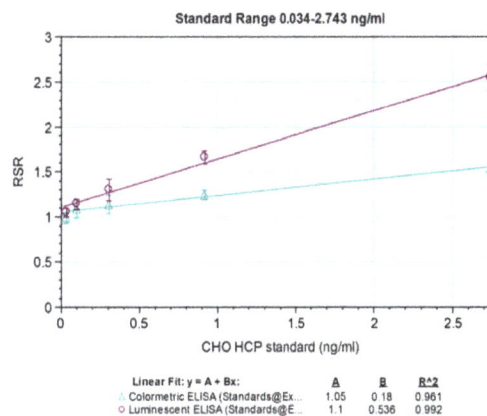

The limit of detection, normally refers to the lower limit of detection (LLOD) is the lowest quantity of a substance that can be distinguished from the absence of that substance (i.e., a blank value) within a stated confidence limit. The lower limit of quantification (LLOQ) is the lowest quantity of a substance that can be accurately quantified. There are a variety of ways to determine the LLOD and LLOQ of an analytical method, here we calculated the theoretical LLOD and LLOQ of ELISA according to the guidelines of ICH-Q2 (R1) using the formula: LLOD=$3.3 \times \theta/S$, and LLOQ=$10 \times \theta/S$, where θ is the standard deviation of blank concentration back-calculated from the calibrate curve, S is the slope of the standard curve. We also determined the experimental LOQ as the lowest and highest standard concentration which can be accurately determined (CV<20% and recovery within the range of 100 ± 20%). Using the log-log fit curve within linear range of 0.31-2000 ng/ml ($R^2=0.991$), the calculated LLOD and LLOQ for luminescent ELISA is 0.245 and 0.816 ng/ml, respectively; in contrast, within its log-log linear range ($R^2=0.992$), colorimetric method has a LLOD and LLOQ of 0.819 and 2.483 ng/ml, respectively, which is ~2 times higher than that for luminescent method. When calculated using log-log linear curve fitting in the same range of 0.91-222.22 ng/ml ($R^2=0.998$), luminescent ELISA has a LLOD=0.365 ng/ml and LLOQ=1.216 ng/ml, remaining much lower than that of colorimetric method (Table 4). In addition, lower LLOD and LLOQ is consistently observed in luminescent ELISA from replicated experiments (data not shown) in spite of the methodology used for calculating LLOD and LLOQ, indicating that luminescent detection is indeed more sensitive than colorimetric ELISA for CHO HCP detection and quantification.

One common method to validate the precision and accuracy of ELISA is using the standard curve to measure mAb-A DS with known amount of spike. In our experiments, different amount of CHO HCP was spiked into 5 mg/ml drug substance and spike recovery was calculated to indicate the accuracy of quantification. Using clear 96-well plate, we observed low recovery on the back-calculated concentration in the spiked samples when the unspiked sample wells are adjacent to the high concentration standards, which lead us to change the assay plate for luminescent ELISA from Costar clear plate to NUNC white opaque plate for assay optimization to avoid cross-interference of luminescence signals.

		CV%		Recovery%	
		Colorimetric	Luminescent	Colorimetric	Luminescent
Log-log fit	Mean ± SD	20.29 ± 9.12	11.79 ± 3.67	102.66 ± 18.77	100.65 ± 9.12
	P value (one tail)	0.03		0.41	
4-parameter fit	Mean ± SD	25.50 ± 14.27	13.65 ± 4.58	97.97 ± 13.59	94.68 ± 14.91
	P value (one tail)	0.04		0.35	

Table 3: Comparison of mean CV% and recovery% of colorimetric method to that of luminescent method using different fitting method (n=6, for standards ranging from 0.91 to 222.22 ng/ml).

Detection method	Log-log fit curve range (R^2)	LLOD (ng/ml)	LLOQ (ng/ml)
Colorimetric	0.91-222.22 ng/ml (0.992)	0.819	2.483
Luminescent	0.91-222.22 ng/ml (0.998)	0.365	1.216
Luminescent	0.31-2000 ng/ml (0.991)	0.245	0.816

Table 4: The LLOD and LLOQ for luminescent and colorimetric ELISA.

The comparison of colorimetric method to luminescent method using conditions optimized for luminescent ELISA in NUNC white opaque 96-well plate

To avoid cross-talk among adjacent wells, chemiluminescence detection based assays are often carried out in white or black opaque plate. In this study, we also compared the assay performance between luminescent ELISA and colorimetric ELISA using conditions optimized for luminescent ELISA in Nunc white opaque 96-well plate, which provides maximum reflection and low cross-talk. Performing luminescent ELISA in white opaque plate using conditions previously described in clear plate dramatically raised the signal response or relative luminescence unit by 6 to 15 fold. Thus, the reagents concentration used in ELISA assays performed in opaque 96-well plate were re-optimized and finalized as shown in Supplemental Table 1. The assay performance of luminescent and colorimetric ELISA was also compared side-by-side using the same amount of assay reagents. The signal/noise ratio, linearity over logarithmic scales, accuracy and precision, as well as the LLOD and LLOQ of both assays were re-assessed.

Under the assay conditions shown in Supplemental Table 1, colorimetric ELISA has a signal response range from 0.056 to 0.121, the log-log fit linear standard curve has a correlation coefficient R^2=0.95, considerably lower than that of luminescent ELISA, which has a signal response range from 4036 to 47731 and a linearity of R^2=0.998 (Figure 5). The accuracy and precision of both colorimetric and luminescent ELISA assays were evaluated using the back-calculated standard concentration and the measured spiked DS concentration. For standards, luminescent has significantly lower CV values (9.76 ± 8.87 vs. 28.17 ± 15.96, p=0.006, n=8) than colorimetric method, indicating improved precision. For accuracy, the recovery of standards at all concentrations in luminescent ELISA ranges from 94.43% to 109.73% (100 ± 6.15%); in contrast standard recovery in colorimetric ELISA ranges from 68.30 to 162.62%, showing much higher variation than luminescent method (Supplemental Table 3). For spiked DS recovery, luminescent ELISA was capable of accurately measure 25 ng/ml and 5 ng/ml HCP spike in 5 mg/ml mAb-A drug substance, with much lower CV values and closer to 100% recovery than colorimetric method; for 1 ng/ml HCP spiked in mAb-A DS (0.2 ppm) recovery, luminescent method was relatively accurate with a CV=30.5% and a recovery=77.35%, while colorimetric method only showed 39.39% recovery (Table 5). The LLOD and LLOQ of luminescent ELISA and colorimetric ELISA were calculated according to the standard deviation of blank (θ) and the slope of calibration curve (S) following the formula LLOD=3.3 × θ/S and LLOQ=10 × θ/S (Table 6). As shown in Table 6, luminescent ELISA is ~9 times more sensitivity than colorimetric ELISA when using 0.25 µg/ml biotin-anti-

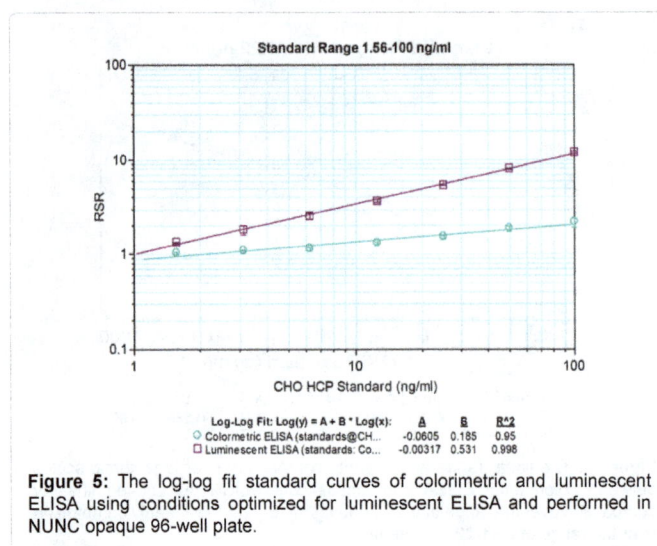

Figure 5: The log-log fit standard curves of colorimetric and luminescent ELISA using conditions optimized for luminescent ELISA and performed in NUNC opaque 96-well plate.

HCP, 1:120,000 dilution of NeutrAvidin-HRP and 50 µL of substrate. Additionally, colorimetric ELISA was also repeated in Costar clear 96-well plate using concentrations of reagents described above, however, the signal response only ranged from 0.056 to 0.074 at the standards concentrations from 1.563 to 100 ng/ml. The signal response was not sensitive enough to differentiate concentration changes, confirming the much lower sensitivity of colorimetric substrate than luminescent substrate.

Conclusion

The sensitivity of luminescent detection was compared side-by-side with colorimetric detection in a sandwich ELISA format under conditions optimized for either colorimetric ELISA or luminescent ELISA. The performance of both assays including parameters such as signal/noise ratio, linear range and linearity over logarithmic scales, precision and accuracy, as well as LLOD and LLOQ were fully evaluated and compared. Our results show that luminescent detection has enhanced signal/noise ratio, broader linear range on logarithmic scales and better linearity within the same range. In addition, luminescent detection also shows better lower end linearity of its standard curve, which allows the accurate quantification of HCP concentration at as low as 1 ng/ml. Under the same experimental conditions, luminescent ELISA is able to detect and accurately quantify lower amount of HCP than the colorimetric method. Moreover, when calculating HCP concentration from its standard curve, better recovery (closer to

	Colorimetric method		Luminescent method	
	CV%	Recovery%	CV%	Recovery%
Unspiked DS	-		20.9	
25 ng/ml spiked DS	9.8	109	5.5	98.75
5 ng/ml spiked DS	14.4	121.04	6.3	94.52
1 ng/ml spiked DS	-	39.39	30.5	77.35

Table 5: The CV and recovery of spiked DS as back-calculated from the log-log fit calibration curve of luminescent or colorimetric ELISA.

	Colormetric method	Luminescent method
LLOD (ng/ml)	12.47	1.34
LLOQ (ng/ml)	37.78	4.06

Table 6: The LLOD and LLOQ of colormetric and luminescent ELISA performed in Nunc white opaque plate using conditions optimized for luminescent ELISA.

100%) and lower CV values were observed with luminescent ELISA. Lastly, chemiluminescent ELISA using the Femto substrate doesn't require a stopping step. Chemiluminescent signals can be measured immediately after the addition of substrate and for 10 min after, since the signal will remain at a plateau for that amount of time. In summary, chemiluminescent ELISA proves to be more sensitive than its colorimetric counterpart for mAb-A HCP quantification. Although the experiments were performed using polyclonal antibodies raised against mAb-A null culture HCP, the same principal should apply to the quantitation of HCP in other drug substance samples with minimal assay development work needed. In conclusion, improvement of the HCP ELISA detection limit without compromising assay robustness offers the opportunity to better control HCP clearance during each of the purification steps and reduces the risk of HCP associated immunogenic response in patients.

Acknowledgements

We thank Xiaoyu Yang and Kimberly May for their support and critical review of this manuscript. The study is funded by the Bioprocess Development Department in Merck Research Laboratories.

References

1. Brekke OH, Sandlie I (2003) Therapeutic antibodies for human diseases at the dawn of the twenty-first century. Nat Rev Drug Discov 2: 52-62.

2. Leader B, Baca QJ, Golan DE (2008) Protein therapeutics: a summary and pharmacological classification. Nat Rev Drug Discov 7: 21-39.

3. Hudson PJ, Souriau C (2003) Engineered antibodies. Nat Med 9: 129-134.

4. Wurm FM (2004) Production of recombinant protein therapeutics in cultivated mammalian cells. Nat Biotechnol 22: 1393-1398.

5. Nelson AL, Dhimolea E, Reichert JM (2010) Development trends for human monoclonal antibody therapeutics. Nat Rev Drug Discov 9: 767-774.

6. Wang X, Hunter AK, Mozier NM (2009) Host cell proteins in biologics development: Identification, quantitation and risk assessment. Biotechnol Bioeng 103: 446-458.

7. Briggs J, Panfili PR (1991) Quantitation of DNA and protein impurities in biopharmaceuticals. Anal Chem 63: 850-859.

8. Schenauer MR, Flynn GC, Goetze AM (2012) Identification and quantification of host cell protein impurities in biotherapeutics using mass spectrometry. Anal Biochem 428: 150-157.

9. Eaton LC (1995) Host cell contaminant protein assay development for recombinant biopharmaceuticals. J Chromatogr A 705: 105-114.

10. Tscheliessnig AL, Konrath J, Bates R, Jungbauer A (2013) Host cell protein analysis in therapeutic protein bioprocessing - methods and applications. Biotechnol J 8: 655-670.

11. Zhu D, Saul AJ, Miles AP (2005) A quantitative slot blot assay for host cell protein impurities in recombinant proteins expressed in E. coli. J Immunol Methods 306: 40-50.

12. Liu N, Brevnov M, Furtado M, Liu J (2012) Host Cellular Protein Quantification. BioProcess International 10: 44-50.

13. Savino E, Hu B, Sellers J, Sobjak A, Majewski N, et al. (2011) Development of an In-House, Process-Specific ELISA for Detecting HCP in a Therapeutic Antibody, Part 2. BioProcess International 9: 68-75.

14. Nicholson P, Storm E (2011) Single-Use Tangential Flow Filtration in Bioprocessing. BioProcess International 9: 38-47.

15. Wang X, Schomogy T, Wells K, Mozier NM (2010) Improved HCP Quantitation By Minimizing Antibody Cross-Reactivity to Target Proteins. BioProcess International 8: 18-24.

16. Gamiz-Gracia L, Garcia-Campana AM, Huertas-Perez JF, Lara FJ (2009) Chemiluminescence detection in liquid chromatography: Applications to clinical, pharmaceutical, environmental and food analysis-A review. Anal Chim Acta 640: 7-28.

17. Lara FJ, Garcia-Campana AM, Velasco AI (2010) Advances and analytical applications in chemiluminescence coupled to capillary electrophoresis. Electrophoresis 31: 1998-2027.

18. Fan A, Cao Z, Li H, Kai M, Lu J (2009) Chemiluminescence Platforms in Immunoassay and DNA Analyses. Anal Sci 25: 587-597.

19. Roda A, Guardigli M (2012) Analytical chemiluminescence and bioluminescence: latest achievements and new horizons. Anal Bioanal Chem 402: 69-76.

20. Dotsikas Y, Loukas YL (2012) Improved Performance of Antigen-HRP Conjugate-based Immunoassays after the Addition of Anti-HRP Antibody and Application of a Liposomal Chemiluminescence Marker. Anal Sci 28: 753-757.

21. Roda A, Manetta AC, Portanti O, Mirasoli M, Guardigli M, et al. (2003) A rapid and sensitive 384-well microtitre format chemiluminescent enzyme immunoassay for 19-nortestosterone. Luminescence 18: 72-78.

22. Wolter A, Niessner R, Seidel M (2007) Preparation and Characterization of Functional Poly(ethylene glycol) Surfaces for the Use of Antibody Microarrays. Anal Chem 79: 4529-4537.

23. Edgeworth JA, Farmer M, Sicilia A, Tavares P, Beck J, et al. (2011) Detection of prion infection in variant Creutzfeldt-Jakob disease: a blood-based assay. Lancet 377: 487-493.

24. Plikaytis BD, Turner SH, Gheesling LL, Carlone GM (1991) Comparisons of standard curve-fitting methods to quantitate Neisseria meningitidis group A polysaccharide antibody levels by enzyme-linked immunosorbent assay. J Clin Microbiol 29: 1439-1446.

24

Optimization of Submerged Culture Conditions for Mycelial Growth and Extracellular Polysaccharide Production by *Coriolus versiolor*

Feifei Wang[1,3], Jianchun Zhang[1,2*], Limin Hao[1,2*], Shiru Jia[3], Jianming Ba[4] and Shuang Niu[1,3]

[1]*The Research Center of China Hemp Materials, Beijing 100027, P. R. China*
[2]*The Quartermaster Equipment Institute of GLD of PLA, Beijing 100010, P. R. China*
[3]*Key Laboratory Industrial Fermentation Microbiology, University of Science & Technology, Tianjin 300457, P. R. China*
[4]*Department of Endocrinology, Chinese PLA General Hospital, Beijing 100853, P. R. China*

Abstract

This paper is concerned with optimization of submerged culture conditions for mycelial growth and Exopolysaccharides (EPS) yield with *Coriolus versiolor* by both one-factor-at-a-time and orthogonal matrix methods. Glucose and yeast-Extracts were identified to be the most suitable carbon and nitrogen sources, respectively. The optimal initial pH, inoculum size and liquid volume for mycelial growth and EPS yield were 5.0, 8% and 150mL/500mL, respectively. Subsequently, the concentration of glucose, yeast-extract, KH_2PO_4, and $MgSO_4 \cdot 7H_2O$ were optimized using the orthogonal matrix method. The effects of media composition on the mycelial growth of *Coriolus versiolor* were in the order of glucose >KH2PO4 > yeast-extract > MgSO4•7H2O, and those on EPS yield were in the order of glucose >$MgSO_4 \cdot 7H_2O$> yeast-extract >KH_2PO_4. The optimal concentrations for enhanced yield were determined as 30 g/L glucose, 7.0 g/L yeast-extract, 1.0 g/L KH_2PO_4, 1.0 g/L $MgSO_4 \cdot 7H_2O$, and 40 g/L glucose, 6.0 g/L yeast-extract, 1.0 g/L KH_2PO_4, 1.5 g/L MgSO4•7H2O for mycelial and EPS yield, respectively. The verification experiments confirmed the final medium. This optimization strategy (30 g/L glucose, 7.0 g/L yeast-extract, 2.0 g/L KH_2PO_4 and 0.5 g/L $MgSO_4 \cdot 7H_2O$) in shake flask culture lead to a mycelial yield of 5.18 g/L, and EPS yield of 0.64 g/L, respectively, which were considerably higher than those obtained in preliminary studies. Under optimal culture conditions, the maximum EPS concentration in a 5-L stirred-tank bioreactor was 0.75 g/L, while the maximum mycelial yield was 8.55 g/L. This also corresponded to 14.67% and 39.42% enhancement in EPS yield and mycelial dry weight, respectively, compared with the verification test results.

Keywords: *Coriolus versiolor*; Exopolysaccharide; Mycelial; Submerged culture; Optimization

Introduction

Coriolus versiolor (CV), known as Yunzhi in China, is a mushroom belonging to species of the Basidiomycetes class of fungi, and its polysaccharide has been widely used as a magic drug to treat cancer and immune deficiency related illnesses [1]. *Coriolus versicolor* was recorded in the Compendium of Materia Medica by Li Shizhen during the Ming Dynasty in China, as being beneficial to health and able to bring longevity if consumed regularly [2].

However, it is time-consuming to harvest the fruiting body from cultivated mushrooms for commercial purposes [3]. Submerged culture has a number of advantages including higher mycelial yield in a more compact space and shorter time, with fewer chances of contamination [4,5-7], therefore it is now attracting attention as an alternative for efficient yield of mycelia and polysaccharide. In addition to polysaccharide obtain from the biomass, mycelial cultures also excrete EPS into the fermentation broth.

EPS is easier to obtain from submerged culture than the internal polysaccharide localized within the mycelia, but exhibits similar biological activities [8]. In other species, mycelial growth rate, EPS yield rate, and EPS productivity have been shown to vary with environmental conditions and medium composition, including carbon source, nitrogen source, pH, etc. [9].

Medium optimization by the one-factor-at-a-time method involves changing one independent variable (i.e. nutrient, pH, inoculum size, etc.) while fixing the others at certain levels. This single-dimensional search is laborious and time-consuming, especially for a large number of variables, and frequently does not guarantee the determination of optimal conditions. Hence, as a more practical method, the orthogonal matrix method was employed to study the relationships between the medium components and their effects on mycelial growth and EPS yield.

The purpose of this study was to optimize the submerged culture conditions to simultaneously produce mycelial biomass and EPS by *Coriolus versiolor* using a statistically based experimental design. In the first step, the one-factor-at-a-time method was used to investigate the effects of variables of medium composition (i.e. carbon, and nitrogen) and environmental factors (i.e. pH and inoculum size) on mycelial growth and EPS yield. Subsequently, the concentration of the medium components was optimized using an orthogonal matrix method.

Materials and Methods

Microorganism and culture conditions

Coriolus versiolor was kindly provided by the Quartermaster Equipment Institute of the General Logistics Department, China. The strain was maintained on Potato Dextrose Agar (PDA) at 4ºC and subcultured every 3 months.

*Corresponding author: Jianchun Zhang, The Research Center of China Hemp Materials, Beijing 100027, China

Cultivation was performed in two stages. The seed culture medium was consisted of the following components: 30.0 g/L glucose, 4.0 g/L peptone, 1.0 g/L KH_2PO_4 and 0.5 g/L $MgSO_4 \cdot 7H_2O$. The initial pH was not adjusted (pH 5.0–5.5). The solution was sterilized at 121°C for 15 min. The preculture which was inoculated with 3~4 cm^2 mycelium was incubated on a rotary shaker at 160 r/min and 27°C for 6 days. The flask culture experiments were performed in 500 mL flasks containing 150 mL of fermentation medium, which was inoculated with 10 % (v/v) of the seed culture. The flasks were cultured under the same conditions as above.

Analysis of mycelial yield and EPS yield

The biomass was obtained by vacuum filtration, washed twice with distilled water and then dried overnight to a constant weight at 60°C. EPS was precipitated from the remaining filtrate by mixing 5mL with three volumes of 95% (v/v) ethanol. It was standing at 4°C overnight to precipitate crude EPS. The precipitated EPS was collected by centrifugation at 15,000×g for 40 min at 4°C and the supernatant was discarded. The precipitate was then resuspended in an equal volume of 75 % ethanol [10,11] to remove oligosaccharides and centrifuged again as above. The precipitate of EPS was dried at 40°C to remove residual ethanol. The EPS content was determined by a phenol–sulfuric acid method using glucose as standard [12].

One-factor-at-a-time

In each experiment, one factor was varied, while all other factors were holding constant. Different carbon sources (glucose, sucrose, maltose, lactose, malt extract (ME), corn starch (CS)), nitrogen sources (ammonium chloride (AC), ammonium sulfate (AS), potassium nitrate (PN), peptone (PT), yeast-extract (YE)), at different concentrations, major components, and different conditions (pH, inoculum size, liquid volume) were initially studied by single factor experiments.

Orthogonal matrix method

To investigate the relationships between variables of medium components and optimize their concentrations for mycelial growth and EPS yield, the orthogonal matrix experimental design $L_9(3^4)$ method can be used. According to preliminary experiments, with only 9×2 replicates (=18) experiments of $L_9(3^4)$ orthogonal projects, three varied levels of each factor were selected, as shown in Table 1.

Results and Discussion

One-factor-at-a-time

Screening of carbon sources and its concentrations: Carbohydrates are a major component of the cytoskeleton and they are an important nutritional requirement for growth and development of higher fungi [13]. All of the selected carbon sources resulted in high mycelial growth and product yield. The mycelial dry weight and EPS yield in ME and CS medium which contained complex polysaccharides were higher than those in other carbon sources (shown in Figure 1a). However,

glucose was low priced and biologically was the most effective energy source, so glucose was chosen as the carbon source for analyzing easily of the EPS from *Coriolus versiolor*.

It can be seen (Figure 1b) that the maximum concentration of 4.10 g/L for mycelia yield was achieved with 3% glucose concentration. However, EPS increased with the increase of glucose concentration which maybe resulted from residual sugar not cleaned completely by alcohol precipitation.

Screening of nitrogen sources and its concentrations

The effect of nitrogen sources on secondary metabolism is conditioned by many factors [14], including the producing organism, the type and concentration of the nitrogen sources and culture method (stationary or submerged).

Coriolus versiolor could grow on a number of different nitrogen sources, but the nitrogen sources effects on EPS and biomass yield were quite distinct (Figure 2a). Among the five different nitrogen sources examined, yeast-extract was the most effective for enhancing the EPS yield (0.40 g/L) and biomass (4.43 g/L) by *Coriolus versiolor*. Inorganic nitrogen was not effective for both EPS and biomass production. Yeast-extract has often been used to provide necessary growth factors; however, too high a concentration of yeast extract would lower the use of other carbon sources and cause the reduction of metabolites (Figure 2b). Besides carbon and nitrogen sources, many growth factors also have positive impact on the yields of EPS such as sodium carboxymethylcellulose, L-glutamic acid, VB1, naphthalene acetic acid, oleic acid, and Tween 80, which has been proved by one-factor-at-a-time and the orthogonal matrix method [15].

It can be determined that the optimal conditions is pH 5.0, 8% of the inoculum size and 150mL/500mL of liquid volume(Figure 3).

Orthogonal matrix design to determine the optimum medium

The experimental conditions for each project are listed in Table 2, and experimental results are included in the last two columns.

Order of effects of factors

According to the magnitude order of R (Max Dif), the order of effects of all factors on mycelial growth and EPS yield could be determined. The order of effects of factors on mycelial growth was glucose > KH_2PO_4 > yeast-extract > $MgSO_4 \cdot 7H_2O$. Applying the same method, the order of effects of factors on EPS yield was glucose > $MgSO_4 \cdot 7H_2O$ > yeast-extract > KH_2PO_4. This result pointed out that the effect of glucose was more important than that of other nutrients.

Optimum levels of each factor

To obtain the optimization levels or composition of each factor, the intuitive analysis based on statistical calculation using the data in Table 2, is shown in Table 3. The results were as follows: (1) to obtain a high mycelial growth, the optimum composition was 30 g/L glucose, 7.0 g/L yeast-extract, 2 g/L KH_2PO_4 and .0.5 g/L $MgSO_4 \cdot 7H_2O$; (2) to obtain a high EPS yield, the optimum composition was 40 g/L glucose, 7.0 g/L yeast-extract, 1.0 g/L KH_2PO_4 and 1.5 g/L $MgSO_4 \cdot 7H_2O$.

Verification Test

Further experiments were carried out using these nutrient concentrations due to the different results between the one-factor-at-a-time and the orthogonal matrix methods. Finally, the medium composition

Level	Glucose A(g/L)	Yeast-extract B(g/L)	KH_2PO_4 C(g/L)	$MgSO_4 \cdot 7H_2O$ D(g/L)
1	20	5	1	0.5
2	30	6	2	1
3	40	7	3	1.5

Symbols A, B, C, and D represent factors of glucose, yeast-powder, KH2PO4 and MgSO4•7H2O, respectively. Symbols 1, 2, and 3 represent concentration levels of each factor.

Table 1: Experimental factors and their levels for orthogonal projects

(a)   (b)

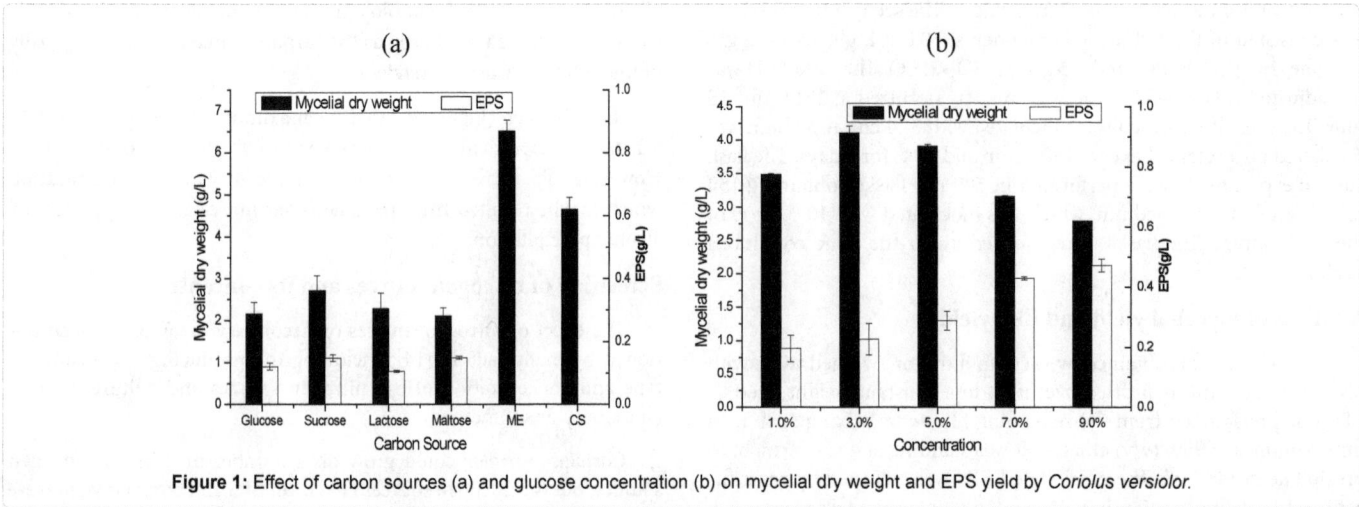

Figure 1: Effect of carbon sources (a) and glucose concentration (b) on mycelial dry weight and EPS yield by *Coriolus versiolor*.

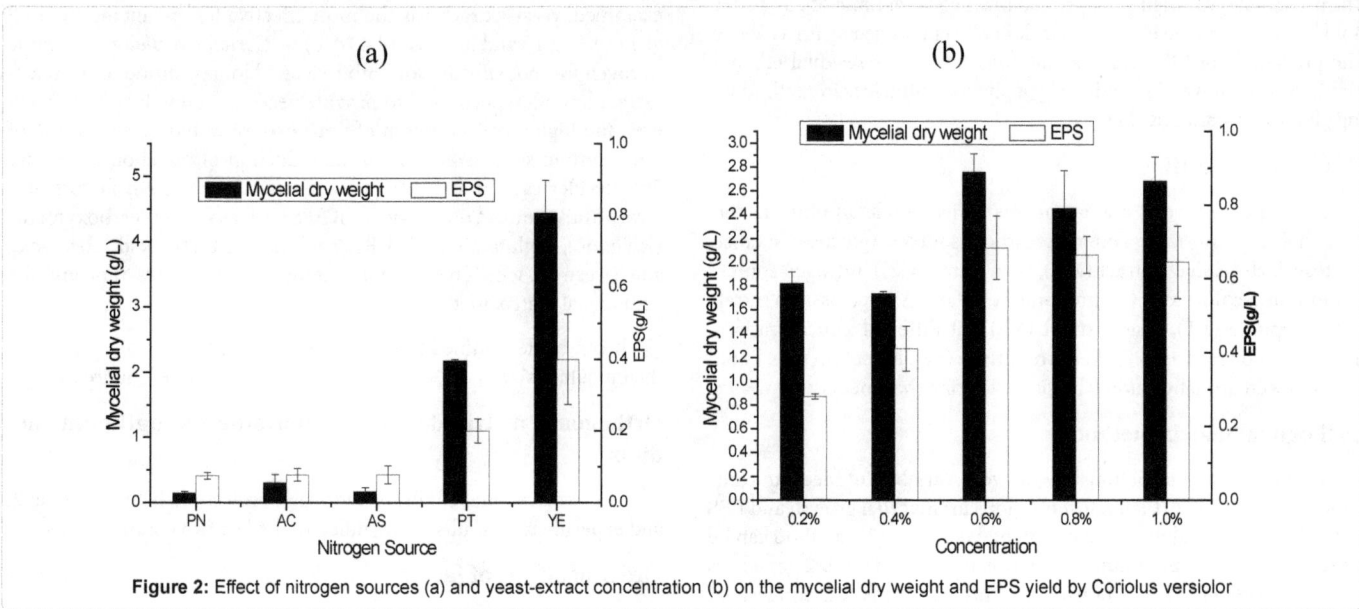

(a)   (b)

Figure 2: Effect of nitrogen sources (a) and yeast-extract concentration (b) on the mycelial dry weight and EPS yield by Coriolus versiolor

Effects of different fermentation conditions

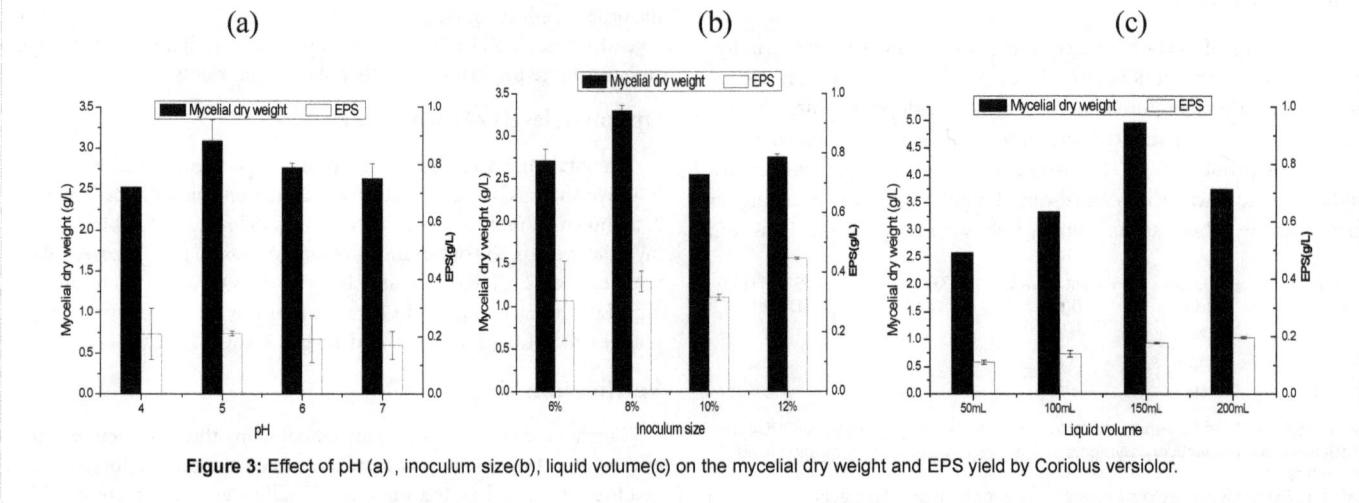

(a)   (b)   (c)

Figure 3: Effect of pH (a) , inoculum size(b), liquid volume(c) on the mycelial dry weight and EPS yield by Coriolus versiolor.

Run	Experimental factor					Result
	A Glucose (g/L)	B Yeast-extract (g/L)	C KH$_2$PO$_4$ (g/L)	D MgSO$_4$•7H$_2$O (g/L)	Mycelia (g/L)	EPS (g/L)
1	1 (20)	1 (5)	1 (1)	1 (0.5)	5.25	0.71
2	1	2 (6)	2 (2)	2 (1)	5.30	0.65
3	1	3 (7)	3 (3)	3 (1.5)	5.24	0.75
4	2 (30)	1	2	3	6.06	0.67
5	2	2	3	1	5.77	0.74
6	2	3	1	2	6.15	0.70
7	3 (40)	1	3	2	5.39	0.74
8	3	2	1	3	5.41	0.87
9	3	3	2	1	5.88	0.84

The arrangements of column A, B, C, and D were decided by orthogonal design for 4 (factor)×9 (run number).
Table 2: Application of L$_9$(3^4) orthogonal projects to the mycelial growth and EPS yield by *Coriolus versiolor*

	EPS(g/L)				Mycelia(g/L)			
	A	B	C	D	A	B	C	D
K1	0.703	0.703	0.760	0.761	5.265	5.567	5.602	5.635
K2	0.704	0.756	0.720	0.697	5.992	5.494	5.746	5.613
K3	0.814	0.763	0.742	0.763	5.559	5.756	5.468	5.568
k$_1$	0.234	0.234	0.253	0.254	1.755	1.856	1.867	1.878
k$_2$	0.235	0.252	0.24	0.232	1.997	1.831	1.915	1.871
k$_3$	0.271	0.254	0.247	0.254	1.853	1.919	1.823	1.856
R	0.111	0.060	0.040	0.066	0.727	0.262	0.278	0.067
Optimal level	3	3	1	3	2	3	2	1

$K_i^A = \sum$ mycelial yield at Ai; $k_i^A = K_i^A / 3$; $R_i^A = \max\{k_i^A\} - \min\{k_i^A\}$
Table 3: The range analysis of L9(34) orthogonal experiment of EPS yield

of 30 g/L glucose, 7.0 g/L yeast-extract, 2.0 g/L KH$_2$PO$_4$ and 0.5 g/L MgSO$_4$•7H$_2$O was chosen as the optimum formula to obtain the maximum EPS (0.64 g/L) while the mycelia dry weight is 5.18 g/L, which increased by 2.6 times and 1.6 times than those of previous report (0.18g/L EPS and 2.01g/L biomass) [16], respectively.

The verification experiment result applying in 5-L bioreactor

Figure 4 showed the typical time courses of mycelial growth and EPS yield in a 5-L stirred-tank bioreactor under optimal culture condition (30 g/L glucose, 7.0 g/L yeast-extract, 2.0 g/L KH$_2$PO$_4$ and 0.5 g/L MgSO$_4$•7H$_2$O) for EPS yield. The maximum EPS yield was achieved at 0.75 g/L after 6 d of fermentation, while the maximum mycelial yield was 8.55 g/L after 4 d. This also corresponded to 14.67% and 39.42% enhancement in EPS yield and mycelial dry weight, respectively, compared with the verification test results. The result implied that the selected conditions were the most suitable in practice. The variance in pH value during fermentation was not so significant. Optimization of operating parameter (e.g. agitation, aeration, and dissolved oxygen tension) in bioreactor fermentation deserves further investigation.

Although some research has found no relationship between the mycelial growth and EPS yield, other results have shown a positive correlation between them [17,18]. In this experiment, the data indicated the EPS began to accumulate rapidly after the biomass reached the maxima. The relationship between the mycelial and EPS yield is still unclear and is worthy of further study.

Both natural medium and semisynthetic medium contain complex sugar components which have an influence on the estimation for exopolysaccharide secreted by *Coriolus versiolor*. Thus we chose glucose as the sole carbon source out of consideration for analysis of the fermentation broth even though the yields were less than those of using natural medium [19].

It is not enough convincing only depending on one-factor-at-a-time and orthogonal matrix methods to determine the optimal medium for yield of EPS from *Coriolus versiolor*. Subsequent experiments can be combined a Box–Behnken design and response surface methodology which has been proved that the yields can be enhanced efficiently [20].

Conclusions

Coriolus versicolor polysaccharide has been widely studied as a medicinal fungus because of its anti-tumor, antioxidant and immunity improving activities. Most of studies focused on its Intracellular Polysaccharide (IPS), however we hope to improve the yield of EPS which also has various of functions such as antioxidant activity [16].

At first, one-factor-at-a-time was taken. Although it was tedious, defective and ignored in much research, this method was helpful in the selection of factor level in orthogonal matrix design. We then selected the main factors and found the preliminary vicinity of the optimums. Finally the optimal medium was obtained by the verification test. This optimized conditions (30 g/L glucose, 7.0 g/L yeast-extract, 2.0 g/L KH$_2$PO$_4$ and 0.5 g/L MgSO$_4$•7H$_2$O) in shake flask culture led to a mycelial yield of 5.18 g/L and EPS yield of 0.64 g/L, respectively, which increased by 1.6 times and 2.6 times than those of previous report [16], respectively.. The subsequent experiments in 5-L fermentor confirmed the results which obtained the maximum EPS of 0.75g/L and mycelia yield of 8.55 g/L. This also corresponded to 14.67% and 39.42% enhancement in EPS and mycelial yield, respectively, compared with the verification test results.

Two optimization techniques used in this work can be widely applied to other processes for optimization of submerged culture conditions for the mushrooms. The results obtained in this study may be useful for a highly effective yield of biomass and valuable bioactive metabolites.

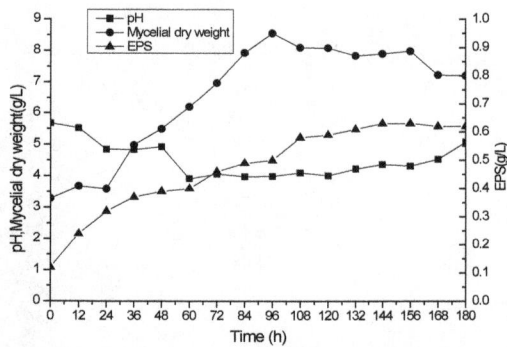

Figure 4: Typical time courses of the mycelial growth and EPS yield by *Coriolus versiolor* under optimal condition (30 g/L glucose, 7.0 g/L yeast-extract, 2.0 g/L KH_2PO_4 and 0.5g/L $MgSO_4 \cdot 7H_2O$) in a 5-L stirred-tank bioreactor at 27°C and 100 r/min for 8 days.

Acknowledgements

This work was financially supported by the National Natural Science Foundation of China (Grant No.31171662, No. 21006072). We thank Mr Zhi-cun Sheng for his kindly help in fermentation experiments.

References

1.   Yang Q (1999) History, present status and perspectives of the study of Yun Zhi polysaccharopeptide. Advance research in PSP.

2.   Ng TB (1998) A review of research on the protein-bound polysaccharide (polysaccharopeptide, PSP) from the mushroom *Coriolus versicolor* (basidiomycetes: Polyporaceae). General Pharmacology: The Vascular System 30: 1-4.

3.   Huang HC, Liu YC (2008) Enhancement of polysaccharide production by optimization of culture conditions in shake flask submerged cultivation of *Grifola umbellata*. Journal of the Chinese Institute of Chemical Engineers 39: 307-311.

4.   Pokhrel CP, Ohga S (2007) Submerged culture conditions for mycelial yield and polysaccharides production by *Lyophyllum decastes*. Food Chemistry 105: 641-646.

5.   Yang FC, Huang HC, Yang MJ (2003) The influence of environmental conditions on the mycelial growth of *Antrodia cinnamomea* in submerged cultures. Enzyme and Microbial Technology 33: 395-402.

6.   Shih IL, Chou BW, Chen CC, Wu JY, Hsieh C (2008) Study of mycelial growth and bioactive polysaccharide production in batch and fed-batch culture of *Grifola frondosa*. Bioresource Technology 99: 785-793.

7.   Liu RS, Li DS, Li HM, Tang YJ (2008) Response surface modeling the

8.   Kim DH, Yang BK, Jeong SC, Park JB, Cho SP, et al. (2001) Production of a hypoglycemic, extracellular polysaccharide from the submerged culture of the mushroom, *Phellinus linteus*. Biotechnology Letters 23: 513-517.

9.   Papagianni M (2004) Fungal morphology and metabolite production in submerged mycelial processes. Biotechnology Advances 22: 189-259.

10.   Lin ES, Sung SC (2006) Cultivating conditions influence exopolysaccharide production by the edible Basidiomycete *Antrodia cinnamomea* in submerged culture. International Journal of Food Microbiology 108: 182-187.

11.   Lin ES, Chen YH (2007) Factors affecting mycelial biomass and exopolysaccharide production in submerged cultivation of *Antrodia cinnamomea* using complex media. Bioresource Technology 98: 2511-2517.

12.   Dubois M, Gilles KA, Hamilton JK, Rebers PA, Smith F (1956) Colorimetric method for determination of sugars and related substances. Anal Chem 28: 350-356.

13.   Xiao JH, Chen DX, Wan WH, Hu XJ, Qi Y, et al. (2006) Enhanced simultaneous production of mycelia and intracellular polysaccharide in submerged cultivation of *Cordyceps jiangxiensis* using desirability functions. Process Biochemistry 41: 1887-1893.

14.   Xu F, Tao W, Cheng L, Guo L (2006) Strain improvement and optimization of the media of taxol-producing fungus *Fusarium maire*. Biochemical Engineering Journal 31: 67-73.

15.   Hao LM, Xing XH , Li Z, Zhang J, Sun J, et al. (2010) Optimization of Effect Factors for Mycelial Growth and Exopolysaccharide Production *Schizophyllum commune*. Applied Biochemistry and Biotechnology 160: 621-631.

16.   Arteiro JMS, Martins MR, Salvador C, Candeias MF, Karmali A, et al. (2011) Protein–polysaccharides of *Trametes versicolor* production and biological activities. Medicinal Chemistry Research 21: 937-943.

17.   Benkortbi O, Hanini S, Bentahar F (2007) Batch kinetics and modelling of Pleuromutilin production by Pleurotus mutilis. Biochemical Engineering Journal 36: 14-18.

18.   Jiao YC, Chen QH, Zhou JS, Zhang HF, Chen HY (2008) Improvement of exo-polysaccharides production and modeling kinetics by *Armillaria luteovirens Sacc.* in submerged cultivation. LWT-Food Science and Technology 41: 1694-1700.

19.   Cui J, Goh KKT, Archer R, Singh H (2007) Characterisation and bioactivity of protein-bound polysaccharides from submerged-culture fermentation of *Coriolus versicolor* Wr-74 and ATCC-20545 strains. Journal of Industrial Microbiology & Biotechnology 34: 393-402.

20.   Thirunavukkarasu A, Nithya R (2011) Response Surface Optimization of Critical Extraction Parameters for Anthocyanin from Solanum melongena. J Bioprocess Biotechniq 1:1-4.

25

Label-Free Shape-Based Selection of Cardiomyocytes with on-Chip Imaging Cell Sorting System

Fumimasa Nomura, Tomoyuki Kaneko, Akihiro Hattori and Kenji Yasuda*

Department of Biomedical Information, Institute of Biomaterials and Bioengineering, Tokyo Medical and Dental University, 2-3-10 Kanda-Surugadai, Chiyoda, Tokyo 101-0062, Japan

Abstract

We have examined the method for the label-free optical microscopic image-based selection of cardiomyocytes from the mixture of collagenase-digested embryonic mouse heart cells using the on-chip imaging cell sorter system, in which 1/2000 s real-time high-speed camera-based phase-contrast cell image recognition was performed. To separate the cardiomyocytes from other cells including fibroblasts, we distinguished the roughness of the cell surface quantitatively as the index of sorting. The selected cells with smooth surface (roughness index not more than 1.1) for cardiomyocytes, and confirmed that the more than 80% of the collected and recultivated cells were self-beating and the 99.2% of them were identified as cardiomyocytes by immunostaining. The results indicates the potential of the label-free cell purification using imaging cell sorting system, especially for cardiomyocytes by the index of their cell surface roughness, which might be applicable for purification of cardiomyocytes for regenerative medicine.

Keywords: On-chip imaging cell sorting system, Cardiomyocyte, Index of roughness, Agarose microchamber

Introduction

The non-invasive purification technique of the particular target cells from tissues, organs or cultivated cells is the essential and fundamental for cell-based research and even for regenerative medicine. For example, for the purification of cardiomyocytes, we need to remove more than half of the noncardiomyocytes from whole heart cells [1].

At present, the most commonly used method for cardiomyocyte purification is the pre-cultivation method exploiting the difference of the adhesion of cardiomyocytes and other cells including fibroblasts to the surface of cultivation dish [2]. However, the method was probabilistic way and is still regarded to be a rough purification. Another more precise purification method is the fluorescence-activated cell sorting (FACS) [3] with fluorescent cardiomyocyte marker labeling, whereas it involves the potential problems of cell damages caused by antibody labeling. To overcome this labeling problem, we have proposed and developed the on-chip imaging cell sorting system for the label-free phase-contrast imaging-based identification of cell types using the index of cell surface shapes. Different shapes with 1/2000s real time processing [4, 5]. In this paper, as a practical application of imaging cell sorting, we have examined the purification efficacy of cardiomyocytes cells with using of the index of cell surface roughness and reported.

Materials and Methods

Imaging cell sorter system

The on-chip phase-contrast imaging cell sorter system was consisted of the following five modules; [4, 5] a cell sorting chip, an optical microscope equipped with a phase-contrast image processing module, a 1/2000 s high-speed charge-coupled devise (CCD) camera (HAS-220; DITECT, Japan), image processing unit, and a computer-controlled switching DC power supply for cell separation (Figure 1a). The cell sorting chip was made of poly dimethylsiloxane (PDMS) attached on a 0.9 mm thin glass slide, in which two inlets and flow channels for sample solution and buffer solution were symmetrically arranged. The generated two laminar flow in the two channels were connected at the cell separation area in center of the chip, and the two laminar flows were divided to the symmetrical two flow channels again,

one is for waste drain and the other is for target cell collection. A pair of 1.5 % agarose gel electrodes containing 0.5 M NaCl (carrier of ionic current) was also arranged at the connecting point of two laminar flows for cell separation to shift target cells from the sample laminar flow to the buffer laminar flow through their boundary by electrophoretic force.

First for separation of target cells from the mixture of cells, the samples were applied to the sample inlet (Figure 1b) and put the air pressure both to the sample inlet and the buffer inlet to introduce the cells into the central separation area with keeping the boundary of two laminar flows by maintaining the same flow velocities of medium buffers (Figure 1c). The phase-contrast real-time images of the cells flew in the sample flow channel were directly monitored with 2000 pictures/s intervals, and were analyzed to acquire specific indexes (cell size, cross-sectional area, cell surface roughness) of cells using the field-programmable gate array (FPGA) real time image processing unit. The system can record both of the cell image pictures and FPGA analysis data simultaneously for the further checking and evaluation.

Then, when the target cells which satisfy the required values of the indexes were recognized, they were sorted by shifting them from left laminar flow to right laminar flow by the computer-controlled electrostatic force generation between two gel electrodes, which arranged perpendicular to the flow direction at the separation area, by the switching DC power supply.

Finally, only the recognized and shifted target cells were moved

*Corresponding author: Kenji Yasuda, Department of Biomedical Information, Institute of Biomaterials and Bioengineering, Tokyo Medical and Dental University, 2-3-10 Kanda-Surugadai, Chiyoda, Tokyo 101-0062, Japan

from the waste drain channel flow to the target sample collection channel flow, and the target cells were acquired at the target cell reservoir arranged at the end of the target sample collection channel (Figure 1b and 1c).

Sample preparation

The cardiomyocytes were isolated from 12- to 13-day-old mouse embryos (ICR) (Saitama Experimental Animals Supply, Japan) [6]. A mouse was anesthetized with diethyl ether and embryos were rapidly removed. The hearts of the embryos were removed and then washed with phosphatebuffered saline (PBS, 137 mM NaCl, 2.7 mM KCl, 8 mM Na_2HPO_4, 1.5 mM KH_2PO_4, pH 7.4) containing 0.9 mM $CaCl_2$ and 0.5 mM $MgCl_2$ to induce heart contraction and remove corpuscles. The hearts were then transferred to PBS without $CaCl_2$ and $MgCl_2$ and minced into 1-mm^3 pieces with fine scissors, and incubated in PBS containing 0.25% collagenase (Wako, Osaka, Japan) for 30 min at 37 °C to digest the ventricular tissue. After this procedure was repeated twice, the cell suspension was transferred to a cell culture medium (DMEM [Invitrogen Corp, Carlsbad, SC, USA] supplemented with 10% fetal bovine serum, 100 U/ml penicillin, and 0.1 mg/ml streptomycin) at 4 °C. The cells were filtered through a 40-µm-nylon mesh and were centrifuged at 180 g for 5 min at room temperature. Then, the cell pellet

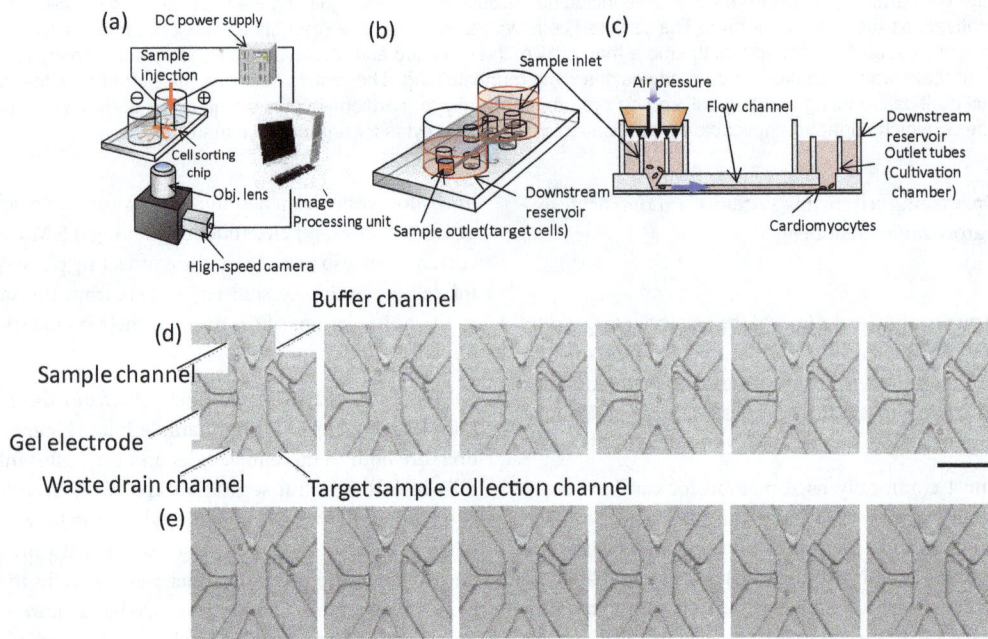

Figure 1: Selection of cardiomyocytes with on-chip cell sorting system.
(a) System design of on-chip cell sorting system. (b) Chip design of on-chip cell sorting system. (c) Cross sectional view of the chip. (d) The time course images of cell sorting area during the collection of selected cells. Each image was obtained 33ms. Bar, 100 µm. (e) The time course images of cell sorting area during the removing the non-selected cells. Each image was obtained 33ms. Bar, 100 µm.

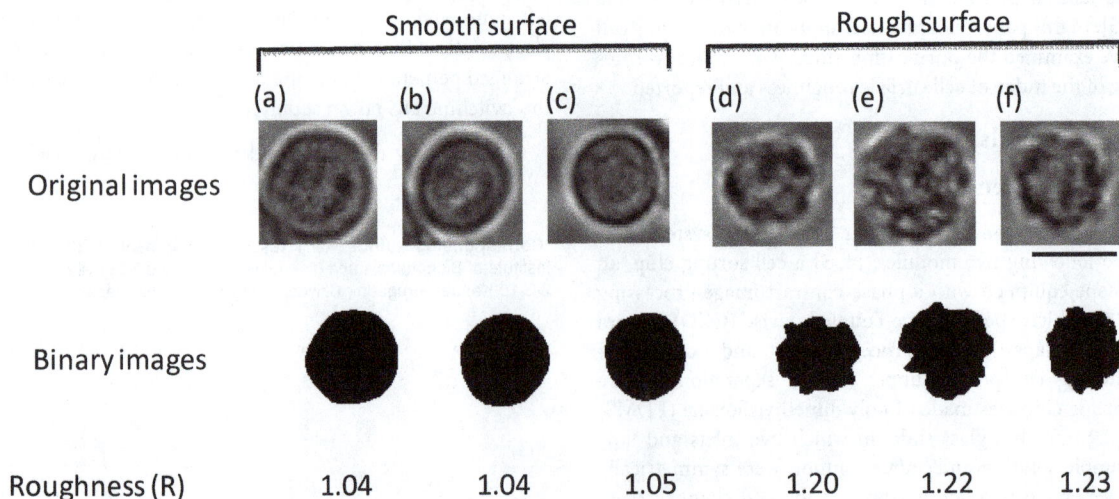

Figure 2: Selection of cardiomyocytes with on-chip cell sorting system.
(a)-(c) Acquired images and sorting indices of smooth surface cells, and (d)-(f) rough surface cells. Bar, 10 µm.

Sample (pixel)	Area (pixel)	Perimeter (pixel)	Radius (pixel)	Circumference (pixel)	
(a)	4232	240.6	36.7	230.6	1.04
(b)	3203	209.1	31.9	200.6	1.04
(c)	2767	195.2	29.6	186.4	1.04
(d)	2431	209.2	27.8	174.7	1.20
(e)	3787	265.0	34.7	218.1	1.21
(f)	2947	236.1	30.6	192.4	1.22

Table 1: Quantitative evaluation of cell surface roughness (Fig. 2).

was resuspended in a cell culture medium and incubated at least 20 min in room temperature.

Immunostaining of cardiomyocytes

The cells were fixed with 4% paraformaldehyde and permeabilized in 0.1% Triton X-100 and exposed for 2 hours to the mouse monoclonal antibody to heavy chain of cardiac myosin (abcam, Japan) dissolved in blocking buffer (PBS containing 1% BSA) and incubated for 1 hour at room temperature with Alexa Fluor 546, goat anti-mouse IgG (Molecular probes, Eugene, USA). To visualize the nuclei, cells were counterstained with Hoechst 33342 for 30 min at room temperature. The cell images were recorded using an inverted microscope equipped for epifluorescence (IX-70, Olympus, Japan) using cooled CCD camera (ORCA-ER, Hamamatsu photonics, Japan).

Results and Discussion

Index for distinguish the surface roughness

For the imaging cell sorting, the particular index of cell images is essential for their separation. Regarding separation of cardiomyocytes from the mixture, we have already examined and found that the most of the smooth round shaped cells should cardiomyocytes). Hence, we

have defined the index of roughness of cell surface R, which compare the real perimeter of cell surface and the estimated perimeter of cell surface by approximation of sphere model, to distinguish the surface roughness of cardiomyocytes and other cells quantitatively as,

$$R = \frac{l}{\sqrt{4\pi S}} \qquad (1),$$

Where l is the perimeter of cell surface, S is the cross-sectional area of cell (both of l and S were acquired from pixel-based digitized image analysis of cells).

For example, a sphere with radius r should have $R = [2\pi r / (4\pi \times \pi r2)1/2]$ equal to 1, and when the roughness increased, R should be increased in proportion to the increase of the real perimeter. Using the above index R, we can quantify the smoothness/roughness of cell surfaces. For example, as shown in Figure 2 and Table 1, the typical smooth cells have the index of roughness R less than 1.1 (Figures 2a, 2b and 2c), and the typical rough cells have R larger than 1.1 (Figure 2d, 2e, and 2f). Table 1 also shows the acquired values of each cell calculated by the FPGA unit calculation from the captured cell images.

Cardiomyocyte separation using the index of roughness R

There are two major points for accomplishing the non-labeled separation of cardiomyocytes from the mixture of enzyme-treated heart extracts. One is the development of the proper quantified index of surface roughness of cells for judging the target cells and others as described above, and the other is the proper enzymatical treatment for appearance of surface shape differences to be able to distinguish different cells. Regarding the cardiomyocyte purification, based on our experience, [11] the collagenase-digestion treatment was effective to appear the difference of smooth/rough surface shapes depending on their difference of cell types. Hence, in this experiment, the sample

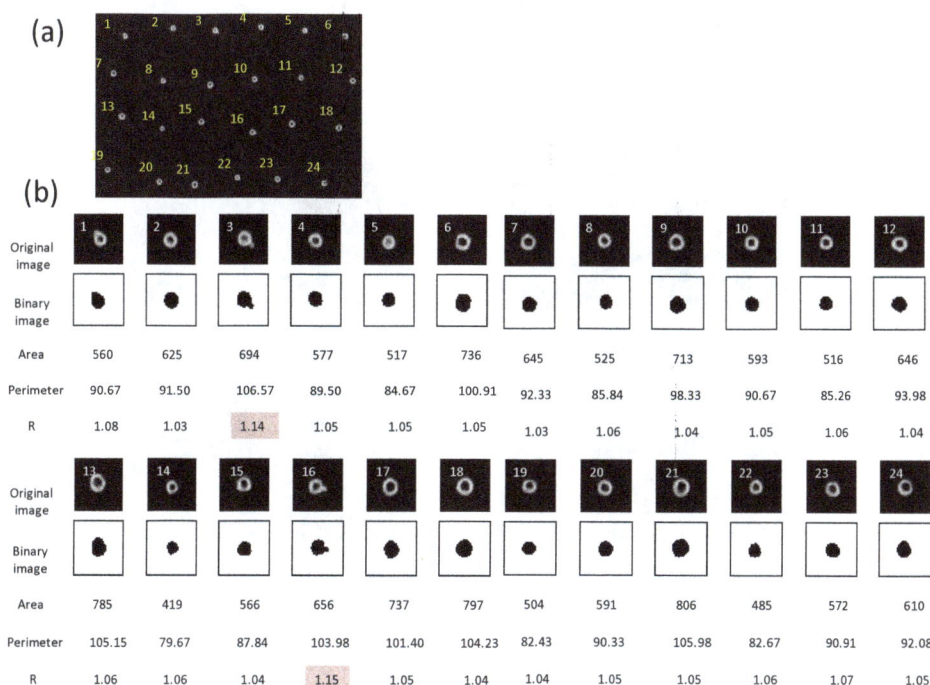

Figure 3: Representative images and indexes of selected smooth surface cells.
(a) Phase-contrast image of arrangement of the selected smooth surface cells settled on agarose microchamber at one by one manner using micropipette after cell sorting purification. (b) Phase-contrast image and FPGA calculated index of cells in Figure 3a.

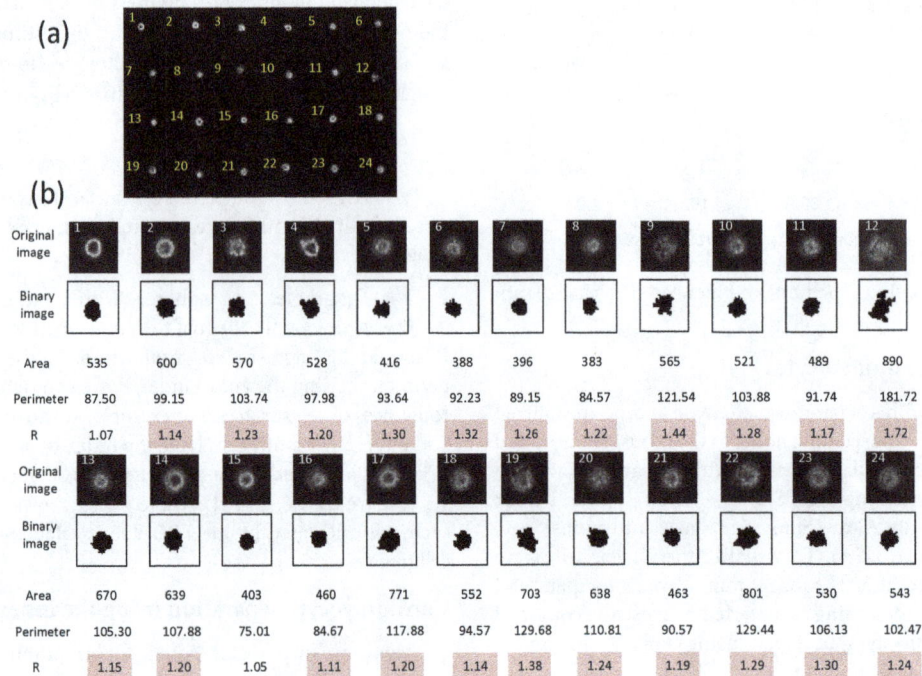

Figure 4: Representative images and indexes of removed rough surface cells.
(a) Phase-contrast image of arrangement of the selected rough surface cells settled on agarose microchamber at one by one manner using micropipette after cell sorting purification. (b) Phase-contrast image and FPGA calculated index of cells in Figure 4a.

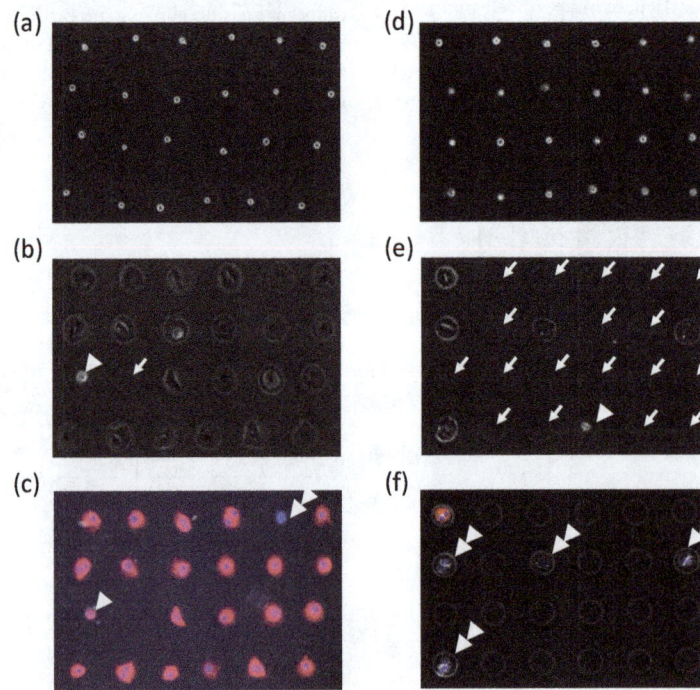

Figure 5. Representative images of selected cells in agarose microchambers.
(a) Phase-contrast image of arrangement of the selected smooth cells on agarose microchamber at one by one manner using micropipette (same image of Figure 3a).
(b) Phase-contrast image of selected cells after 2 days cultivation. White arrowhead shows a shrinked cell. White arrow shows the microchamber in which the cell was disappeared during cultivation caused by apoptosis or insufficient attachment. (c) Phase-contrast image superimposed on the fluorescence image of the selected cells (2 days cultivation) after fixation with 4% Formaldehyde solution. Nucleus (Hoechst33342; blue), heavy chain cardiac myosin (red). White double arrowheads show the no-stained cells by cardiac myosin (i.e., non cardiomyocytes). (d) Phase-contrast image of arrangement of the selected rough cells on agarose microchamber at one by one manner using micropipette (same image of Figure 4a). (e) Phase-contrast image of selected cells after 2 days cultivation. White arrowhead shows shrinkage cell. (f) Phase-contrast image superimposed on the fluorescence image of the selected cells (2 days cultivation) after fixation with 4% Formaldehyde solution. Nucleus (Hoechst33342; blue), heavy chain cardiac myosin (red). Bars, 100 µm.

	0 DIV		2 DIV			staining of cardiac myosin	
	selected cell	beating cell	no beating cell	shrinkage cell	removal cell	positive cell	negative cell
cell number	242	178	19	12	33	195	2
ratio (%)	100	73.6	7.9	5	13.6	80.6	0.8

Table 2: Quantitative evaluation of cardiomyocyte ratio

mixture of cells were acquired from collagenase-digested heart isolated from 13-day-old mouse embryos (ICR). It should be noted for this procedure that the collagenase-digestion procedure and following incubation before cell sorting is essential for appearing the difference of smooth/rough surface of cells. Otherwise, if we use the conventional trypsinization procedure for cell preparation, cells do not show apparent smooth/rough surface shape differences. And the incubation after the collagenase-treatment at least 20 min was also essential for appearing their shape differences. Without this incubation, the shape difference was not sufficiently appeared on the surface of cells.

The functional sorting procedure of cardiomyocytes using above on-chip imaging cell sorter system with this imaging index R was as follows; (1) when the target smooth cells ($R \leq 1.1$) flew into the sorting area, the target cell was recognized by high-speed camera and a pulse DC voltage was applied between a pair of agarose gel electrodes to shift the target cells from the sample flow channel to the buffer channel to introduce to the target sample collection channel (Figure 1d); (2) on the other hand, when other rough cells ($R > 1.1$) comes into this area, no electrostatic force was applied and the rough cells were introduced into the waste drain channel (Figure 1e).

Each of the collected smooth surface cell in the target cell reservoir was transferred into the 5-μm-thick 2 % (w/v) agarose (GenePure LowMelt ISC BioExpress) microchamber [7, 8] laid on the collagen coated polystyrene cultivation dish at one by one manner [9, 10] using micropipette (Figures 3, 4 and 5), and cultivated in a cell culture medium (Invitrogen DMEM supplemented with 10 % fetal bovine serum, 100 U/mL penicillin, and 100 mg/mL streptomycin) at 37 °C with a humidified atmosphere of 95 % air and 5 % CO_2.

Figure 3 shows the 24 samples of the typical target smooth cells acquired from cell sorting system. As shown in Figure 3a, each cell was put into the microchamber to isolate and identification of cell for identification of their characteristics after cultivation. Figure 3b shows the micrographs and the FPGA acquired indexes of the collected smooth cells. The index R of all the smooth cells except for the two coagulated smooth cells (see no. 3 and 16) showed less than 1.08. The samples of coagulated smooth cells (No. 3 and 16) were manually acquired from the rough sample cell reservoir to indicate the potential error of the index R for appropriate smooth cells. The results indicate that the visualized micrograph-based identification of smooth cells is suitable match to the index $R < 1.1$.

Figure 4 also shows the 24 samples of the typical rough surface cells acquired from cell sorting system. Just same as Figure 3, each cell was put into the microchamber (Figure 4a) and indicated the index R larger than 1.1 except for the samples no. 1 and no. 15, which were manually collected from the smooth sample reservoir to indicate the potential error of the index R for appropriate rough cells, too. In the results, some rough surface cells were judged $R < 1.1$ using FPGA procedure. One of the possible reasons and possible improvement for overcoming this problem is using of higher magnitude of objective lens for acquiring more precise perimeters.

Evaluation of the index of surface roughness for characterizing cardiomyocytes

After two days cultivation of the smooth surface cells (Figure 5b) and the rough surface cells (Figure 5e) laid in the agarose microchambers, we first categorized the cultivated cells by their ability of beating, and also by their adhesion abilities such as shrinking and stable expansion. For example, as shown in Figure 5b, a cell at the white arrowhead was shrinking, and a cell at the white arrow was disappeared. Moreover, as shown in Figure 5c, after the staining of myosin heavy chain of cells as the index of cardiomyocyte protein, a cell at the white double arrowheads was not stained, whereas all the remaining 21 cells in the 24 microchambers were well cultivated and showed their beating. In contrast, as shown in Figures 5d, 5e and 5f, most of rough surface cells were not attached on the surface of cultivation dish and disappeared (white arrows). Even the attached cells did not stained by myosin heavy chain probes (white double arrowheads) except for the rough surface sample no. 1, having the index R = 1.07.

In the experiment, from all of the 242 smooth cells, beating cells were 178 (73.6%), no-beating cells were 19 (7.9%), shrunk cells were 12 (5.0%), unattached cells were 33 (13.6%), respectively (Table 2).

As shown in Figure 5c, cardiac myosin positive-staining cells were 195 (99.99% of healthy cultivated cells; 80.6% of whole collected smooth cells), and the negative-staining cells were 2 cells (0.01% of healthy cultivated cells; 0.8% of whole cells) (Table 2). That means, even the cells which were not beating were cardiomyocytes except for two cells from 197 cells.

Hence, the results indicated that at least more than 80% of the collected cells were beating cardiomyocytes, and more than 99.99% of healthy cultivated cells were cardiomyocytes. The shrinkage and unattachment in our re-cultivation procedure might be caused by the insufficient removal of agarose coating of cultivation microchamber array from collagen layer. It should be noted that, although we did not count into the myosin-positive cells, even the shrinked cells in microchambers showed the positive-staining (see the white single arrowhead in Figure 5c).

In conclusion, we have succeeded in separating cardiomyocytes with label-free on-chip imaging cell sorting procedures with using of the index of their surface roughness, and the results indicated that the more than 80% of the smooth cells were beating cardiomyocyte cells, and also more than 99% of healthy re-cultivated cells were cardiac myosin positive cells. The results suggested that the non-labeling simple phase-contrast image-based cell sorting has a potential to separate cardiomyocytes precisely from the mixture of heart cells by the index of cell surface roughness with less than 1 % error rate.

Acknowledgements

This work was financially supported by New Energy Development Organization (NEDO).

References

1. Manabe I, Shindo T, Nagai R (2002) Gene expression in fibroblasts and fibrosis. Circ Res 91: 1103-1113.

2. Borg TK, Rubin K, Lundgren E, Borg K, Obrink B (1984) Recognition of extracellular matrix components by neonatal and adult cardiac myocytes. Dev Biol 104: 86-96.

3. Bonner WA, Hulett HR, Sweet RG, Herzenberg LA (1972) Fluorescence activated cell sorting. Rev Sci Instrum 43: 404-409.

4. Takahashi K, Hattori A, Suzuki I, Ichiki T, Yasuda K (2004) Non-destructive on-chip cell sorting system with real-time microscopic image processing. J Nanobiotechnol 2: 5.

5. Hattori A, Yasuda K (2010) Comprehensive study of microgel electrode for on-chip electrophoretic cell sorting. Jpn. JAppl Phys 49: 06GM04-1-4.

6. Kojima K, Kaneko T, Yasuda K (2006) Role of the community effect of cardiomyocyte in the entrainment and reestablishment of stable beating rhythms. Biochem Biophys Res Commun 351: 209-215.

7. Moriguchi H, Wakamoto Y, Sugio Y, Takahashi K, Inoue I, et al. (2002) An agar-microchamber cell-cultivation system: flexible change of microchamber shapes during cultivation by photo-thermal etching. Lab Chip 2: 125-132.

8. Kojima K, Moriguchi H, Hattori A, Kaneko T, Yasuda K (2003) Two-dimensional network formation of cardiac myocytes in agar microculture chip with 1480 nm infrared laser photo-thermal etching. Lab Chip 3: 292-296.

9. Kojima K, Kaneko T, Yasuda K (2005) Stability of beating frequency in cardiac myocytes by their community effect measured by agarose microchamber chip. J Nanobiotech 3: 4.

10. Kaneko T, Kojima K, Yasuda K (2007) An on-chip cardiomyocyte cell network assay for stable drug screening regarding community effect of cell network size. Analyst 132: 892-898.

11. KanekoT, Nomura F, Yasuda K (2011) On-chip constructive cell-network study (I): contribution of cardiac fibroblasts to cardiomyocyte beating synchronization and community effect. J Nanobiotech 9: 21.

26

Inhibitory Compounds in Lignocellulosic Biomass Hydrolysates during Hydrolysate Fermentation Processes

Ying Zha[1,2,*], Bas Muilwijk[3], Leon Coulier[2,4] and Peter J. Punt[1,2]

[1]Microbiology & Systems Biology, TNO, Utrechtsweg 48, 3704 HE, Zeist, The Netherlands
[2]Netherlands Metabolomics Centre (NMC), Einsteinweg 55, 2333 CC Leiden, The Netherlands
[3]TNO Triskelion BV, Utrechtseweg 48, 3700 AV, Zeist, The Netherlands
[4]Research group Quality & Safety, TNO, Utrechtsweg 48, 3704 HE, Zeist, The Netherlands

Abstract

To compare the composition and performance of various lignocellulosic biomass hydrolysates as fermentation media, 8 hydrolysates were generated from a grass-like and a wood biomass. The hydrolysate preparation methods used were 1) dilute acid, 2) mild alkaline, 3) alkaline/peracetic acid, and 4) concentrated acid. These hydrolysates were fermented at 30°C, pH 5.0 using *Saccharomyces cerevisiae* CEN.PK113-7D as model strain. The growth in different hydrolysates varied in the aspects of lag-phase, growth rate, glucose consumption rate and ethanol production rate. Subsequently, 11 potential inhibitory compounds as described in literature were selected for further analysis. The concentrations of these compounds were determined in the time-samples of the 8 fermentations, using a novel analytical method, ethyl-chloroformate derivatization-GC-MS. Some of these compounds, e.g. furfural, decreased during the fermentation process, while others, such as formic and benzoic acid, remained almost constant. Inhibitory effect analysis of individual compound revealed that most of these compounds exhibit little effect at the concentrations detected in hydrolysates. Only furfural and benzoic acid clearly affected the growth of the model yeast: furfural elongated the lag-phase, while benzoic acid reduced the growth rate and biomass yield.

Keywords: Lignocellulosic biomass hydrolysates; Bagasse; Oak; Pre-treatment; Fermentation; Inhibitors; Yeast

Abbreviations: Bag: Sugar cane Bagasse; Oak: Oak saw dust; MA: Mild Alkaline; DA: Dilute Acid; PAA: oxidative/PerAcetic Acid; CA: Concentrated Acid; HMF: 5-HydroxyMethyl Furfural; GC-MS: Gas chromatography–Mass Spectrometry; EC-GC-MS: Ethyl-Chloroformate derivatization-GC-MS; OD: Optical Density

Introduction

Lignocellulosic biomass is the feedstock for the production of 2nd generation biofuel. The biomass, such as bagasse, corn stover, wheat straw and willow wood, is structurally composed of cellulose, hemicellulose and lignin [1,2]. To transform the biomass to liquid fermentation medium, a pretreatment and a hydrolysis step are required to break down the biomass structure and to form monomer sugars, such as glucose and xylose [3,4]. The composition of this liquid medium, named hydrolysate, is determined by the biomass type and the pretreatment-hydrolysis method used.

During the pretreatment process, various degradation products of both sugar and lignin are formed, among which are some inhibitory compounds. These compounds negatively influence the hydrolysis as well as fermentation process [5,6]. Acetic acid, furfural and 5-hydroxymethyl furfural are the most studied inhibitory compounds in hydrolysates. These compounds were also used for toxicity studies in different microorganisms [1,7,8]. The other compounds that were reported inhibitory are mainly weak acids, phenolic and aromatic species. These compounds, for example, vanillin and syringic acid, are less well studied concerning their concentrations in hydrolysate and their effects [1,5].

The effects of inhibitory compounds in a fermentation process were shown as longer lag-phase, slower growth, lower cell density and decreased ethanol productivity [9,10]. To be able to use hydrolysate for biofuel production on an industrial level, these effects need to be reduced. Several detoxification methods have been developed and applied to different hydrolysates. Activated carbon, organic solvent absorbing and extracting inhibitory compounds were proven to be effective physical detoxification methods [11,12]. The chemical detoxification methods include over-liming, reacting with reducing agent and peroxide treatment [13,14].

Since hydrolysates were made from natural materials and the preparation methods are various, the composition and performance of different hydrolysates differ. These differences are of importance for both inhibitory compounds studies and detoxification method development. Therefore, studying the similarity and difference of various hydrolysates on their composition and fermentation performance is of considerable interest.

The results of these studies will provide information to analyze the relationship between hydrolysate composition and its fermentation performance as medium. For a proper study design, the selected hydrolysates should be different in their fermentation performance. This can be achieved by using different biomass types and diverse pretreatment-hydrolysis methods to prepare the hydrolysates. In this study, we generated 8 different hydrolysates from bagasse and oak sawdust to compare their performance as fermentation media. Hydrolysates and their fermentation time-series samples were taken to study their composition and dynamics during the fermentation process. These samples were analyzed with EC-GC-MS method. This analytical method was developed to remove the sugar content in the hydrolysates and detect sugar and lignin degradation products (Zha et al. [20]). Among all the

*Corresponding author: Ying Zha, Microbiology & Systems Biology, TNO, Utrechtsweg 48, 3704 HE, Zeist, the Netherlands

compounds detected, 11 were selected to be quantified in hydrolysates and their fermentation time-series samples. These selected compounds are formic acid, acetic acid, levulinic acid, furfural, furfuryl alcohol, 2-furanmethanol acetate, HMF, vanillin, syringic acid, benzoic acid and 4-hydroxybenzaldehyde. They were chosen because they were either reported as inhibitory compounds [15,16] or belong to the categories of potential inhibitors [17,18]. The concentrations of these compounds detected in the hydrolysates were used to analyze their dynamics during the fermentation process, and test their inhibitory effects individually using a screening method.

Materials and Methods

Biomass

Sugar cane bagasse was a kind gift from ZILOR, Brazil, and oak sawdust was obtained from ESCO, the Netherlands, a wood-flooring supplier. Both types of biomass were pre-dried at 80°C for 5 hours when received, and stored at room temperature. Sugar cane bagasse was ground to pieces with average length of 3 mm. Prior to pretreatment, the biomass was dried again at 80°C for minimum 16 hours.

Hydrolysate preparation method

Four pretreatment methods were used to prepare bagasse and oak sawdust for hydrolysis, namely dilute acid (2% H_2SO_4), mild alkaline (3% $Ca(OH)_2$), alkaline/peracetic acid [19], and concentrated acid (72% H_2SO_4). The biomass pretreated with the first three methods was hydrolyzed enzymatically while the concentrated acid pretreated biomass was hydrolyzed in acid. The detailed steps of these methods are described in Zha et al. [20].

Strain preculture

Saccharomyces cerevisiae CEN.PK113-7D (CBS8340) was used as model strain in this hydrolysate study. The strain was obtained from CBS Utrecht, the Netherlands.

The preculture for both fermentation and Bioscreen test was prepared in a 500 ml Erlenmeyer shake flask with 100 ml mineral medium and 20 g/l glucose. The mineral media was prepared according to van Hoek et al. [21]. The preculture was inoculated with 1 ml *S. cerevisiae* CEN.PK113-7D glycerol stock, and incubated at 30°C, 200 rpm for 20 hours.

Fermentation setup

The batch fermentation was carried out in a 2 l New Brunswick fermentor with working volume of 1 l. The fermentor, filled with 1l demineralized water, was sterilized at 121°C. After sterilization, the fermentor was connected to the console, emptied and filled with 950 ml filter-sterilized hydrolysate. For each hydrolysate, 1 fermentation run was conducted. The fermentation temperature was set at 30°C, pH at 5.0 by adding 2 M KOH or 1 M H_2SO_4, dissolved oxygen at 0 by flushing 0.5 l/min N_2 continuously. The fermentation began at the point of inoculation. The inoculum was prepared by harvesting the cells from 50 ml preculture and re-suspending the cells in 50 ml hydrolysate. Together with inoculum, 2 ml Tween 80-Ergosterol stock were added into the fermentor. The Tween 80-Ergosterol stock contained 5.0 g/l Ergosterol and 210.0 g/l Tween 80, which were dissolved in 95% ethanol. The whole fermentation process was monitored by continuously measuring the CO_2 percentage in the off-gas. The fermentation was considered finished when the CO_2 percentage value is 0 for 10 hours. During the fermentation process, samples were taken every 60 min or 99 min. The auto-samples were directly cooled to 4°C and later stored at 0°C.

Fermentation sample analysis

The monomer sugar concentrations in the hydrolysates were determined with DIONEX ICS 3000, equipped with CarboPac PA20 carbohydrate column and plused amperometric detector. The column was operated at 30°C, with 7.5 mM NaOH as eluent, and the flow rate was 0.5 ml/min.

The optical density, glucose and ethanol concentrations of the fermentation auto-samples were determined using ROCHE Cobas Mira Plus. Vortex was performed to each individual sample to reach a homogeneous cell distribution before measuring optical density at wavelength 600 nm. After optical density measurement, the samples were centrifuged at 4000 rpm for 15 min and the suspension was used for glucose and ethanol measurements. Glucose concentration was determined enzymatically, by adding reagent Glucose HK CP, purchased from ABX Pentra, and measuring formed NADH amount at wavelength 340 nm. The ethanol assay was performed by using NAD and aldehyde dehydrogenase in 0.4 M KH_2OP_4 buffer as the first reagent and alcohol dehydrogenase as the second reagent, and measuring NADH concentration at wavelength 340 nm (adapted from BIOCHEMICA© protocols) .

For each fermentation, 5 auto-samples were selected, representing the following time points: directly after inoculation, end of lag-phase, growth phase, end of growth phase and stationary phase. The concentrations of formic acid and acetic acid of these samples were measured with DIONEX ICS 3000, equipped with IonPac ICE-AS6 ion-exclusion column and suppressed conductivity detector. The column was operated at 30°C, with 1.6 mM perfluorobutyric acid as eluent, and the flow rate was 1.0 ml/min.

The concentrations of furfural, furfurylalcohol, 2-furanmethanol acetate, levulinic acid, benzoic acid, syringic acid, HMF, 4-hydroxybenzaldehyde and vanillin were analyzed with EC-GC-MS method. The method was conducted as follows. NaOH solution was added to 0.5 ml hydrolysate to bring the mixture pH above 10. Into the mixture, the labeled internal standard containing leucine-D3, succinic acid-D4 and cinnamic acid-D5 in pyridine was added. 300μl ethanol and the injection standard containing difluorobiphenyl and dicyclohexylphtalate in pyridine were also added. The formation of the ethylesters was done by two rounds of adding 40μl ethyl chloroformate then shaking vigorously by hand for 15 seconds. The reaction was stopped by adding 750μl dichloromethane and 500 μl of 1 M bicarbonate buffer. The formed derivates were extracted to the dichloromethane phase by shaking the mixture for 20 seconds. The dichloromethane phase was then transferred to another vial and dried with sodium sulfate. The dried dichloromethane phase was transferred to an auto-sample vial. The measurement was carried out by 1 μl splitless injection in the PTV injector of the AGILENT 7890A GC with AGILENT 5975C mass spectrometer as detector. A DB-1 30 m x 0.32 mm x 1 μm analytical column was used for the separation of the analytes.

Inhibitory effects test

The inhibitory effects of the selected compounds were examined by using growth tests in BIOSCREEN C Analyzer, LABSYSTEMS OY, Helsinki, Finland, as described in Zha et al. [20]. The compounds were added into mineral medium with 20 g/l glucose and 2 different hydrolysates, Oak-PAA and Bag-CA. The concentrations added were based on the highest levels detected in all hydrolysates, which are marked in bold in Table 5. The media pH was adjusted to 5.0±0.5 with either 3 M H_2SO_4 or 6 M KOH before inoculation. The tests were carried out in triplicates.

Results and Discussion

Biomass hydrolysates composition

Sugar cane bagasse and oak saw dust were chosen as the biomass for this study because they represent two distinct categories of biomass type, namely grass like and wood. More importantly, in a previous study, where the growth of the model yeast was screened in 24 different hydrolysates, bagasse and oak hydrolysates showed the largest diversity [20].

Both bagasse and oak were treated with the 4 different hydrolysate preparation methods. The resulting 8 hydrolysates were analyzed on their monosaccharide compositions, as shown in table 1. Glucose and xylose were the major monomer sugars in all 8 hydrolysates, and glucose had an approximately two fold higher concentration compare to xylose. Small amounts of galactose and arabinose were detected in both bagasse and oak hydrolysates, while in oak hydrolysates, also low levels of mannose were found.

Hydrolysate fermentation

For each of the 8 hydrolysates in table 1, a batch fermentation was carried out with the model yeast *S. cerevisiae* CEN.PK 113-7D, as described in section 2.4. The fermentation performance was determined by measuring the optical density, glucose concentration and ethanol concentration of the samples taken during the whole fermentation process. The growth of the yeast varied in these 8 hydrolysates, as shown in figure 1, as well as the glucose consumption (Figure 2) and the ethanol production (Figure 3).

Growth characteristics: The growth of the yeast cells in a fermentation process was monitored by measuring the optical density of the time samples at wavelength 600 nm. The optical density of a time sample was calculated by deducting the measured optical density value by the time-0 optical density value: $OD_{t-s} = OD_{t-m} - OD_{t-0}$.

The growth curves of the model yeast in 8 different hydrolysates and in mineral medium are shown in figure 1. The growth of the model yeast in mineral medium in this study was highly comparable to the growth reported by Kuyper et al. [22]. By comparing the growth in hydrolysates and in mineral medium, it can be seen that the growth in all hydrolysates were negatively affected. This was mainly shown as slower growth, longer lag-phase and lower OD yield. It can be seen that the growth of the model yeast was similar in the hydrolysates prepared with the same method, indicating that the hydrolysate performance was mainly dependent on the pretreatment-hydrolysis method. The hydrolysates prepared by mild alkaline method resulted in the shortest lag-phase and relatively high growth rate, while the concentrated acid method prepared hydrolysates had the longest lag-phase and slower growth. The performance of the hydrolysates made by dilute acid and peracetic acid methods was in between the other two (Figure 1).

To quantitatively compare these 8 fermentations, lag-phase, growth rate and OD yield were defined as parameters to describe the characteristics of the fermentation performance (Table 2). By comparing the growth rates in bagasse and oak hydrolysate prepared with the same method, it is noticed that only when prepared with mild alkaline method, oak hydrolysate was with a higher growth rate than bagasse. This indicates that hydrolysates prepared from bagasse are, in general, less inhibitory than those prepared from oak. Probably, if the pretreatment method was mild and the biomass structure was relatively more difficult to break down, such as oak, there would be little inhibitory compounds released or formed. In this case, the generated hydrolysate would be less toxic. As shown, different from the growth in mineral medium, the growth in some hydrolysates slowed down several hours after the growth started (Figure 1). This phenomenon is most illustrative in PAA hydrolysates, the model yeast started with fast growth, but the growth rate dropped at a specific point, 14 h for Bag-PAA and 20 h for Oak-PAA. The 2 different growth rates shown in table 2 are before and after the rate drop, respectively. A possible explanation for the phenomenon is that the amount of essential nutrients in these hydrolysates was limited, which could only support the growth in the first several hours. To continue growth, the yeast had to use different nutrients that were less efficient, which caused the growth rate to slow down. This explanation was consistent with the fact that the growth slowed down particularly in PAA hydrolysates. As during PAA pretreatment, 2 washing steps were involved, which removed dissolved nutrients at that moment. This possibly caused nutrient limitation in PAA hydrolysates, which lead to the reduction of the growth rate. In agreement with this, the analysis results of the hydrolysate fermentation time samples revealed that most of the amino acids present in the hydrolysates were consumed during the fermentation process (Zha et al. [20]).

As far as the OD yield is considered, it seems that it was related to lag-phase and growth rate, namely, long lag-phase and/or slow growth corresponded to low OD yield. For instance, the lowest OD yield was of Bag-CA and Oak-CA hydrolysate fermentations, which had the longest lag-phase (17 h) and lowest growth rate (0.035), respectively (Table 2). The differences in OD yield indicate that the yeast cells spent a higher percentage of the total energy on maintenance in hydrolysates, which maybe the result of overcoming inhibitory effect and/or using less efficient nutrients.

unit: g/l	glucose	xylose	galactose	arabinose	mannose
Bag-MA	57.9	33.6	0.1	3.3	0.0
Bag-DA	66.5	29.7	0.7	2.0	0.0
Bag-PAA	67.8	31.2	0.1	0.7	0.0
Bag-CA	107.3	62.9	1.5	4.2	0.0
Oak-MA	47.4	24.6	0.2	0.1	0.0
Oak-DA	42.1	25.0	1.0	0.6	1.3
Oak-PAA	61.0	28.1	0.2	0.1	0.0
Oak-CA	90.9	42.3	2.5	1.8	4.2

Table 1: Concentrations of monomer sugars in the 8 hydrolysates.

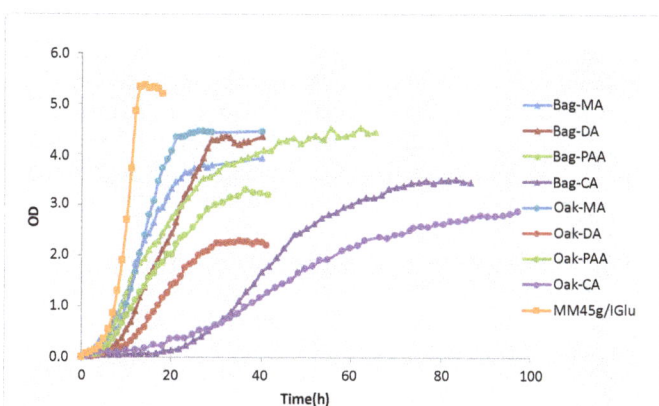

Figure 1: Growth curves of the model yeast, Saccharomyces cerevisiae CEN. PK113-7D, in the 8 different hydrolysates.

Glucose consumption and ethanol production profile: The 8 different hydrolysates differ in their initial glucose concentrations due to the diverse biomass types and hydrolysate preparation methods used,

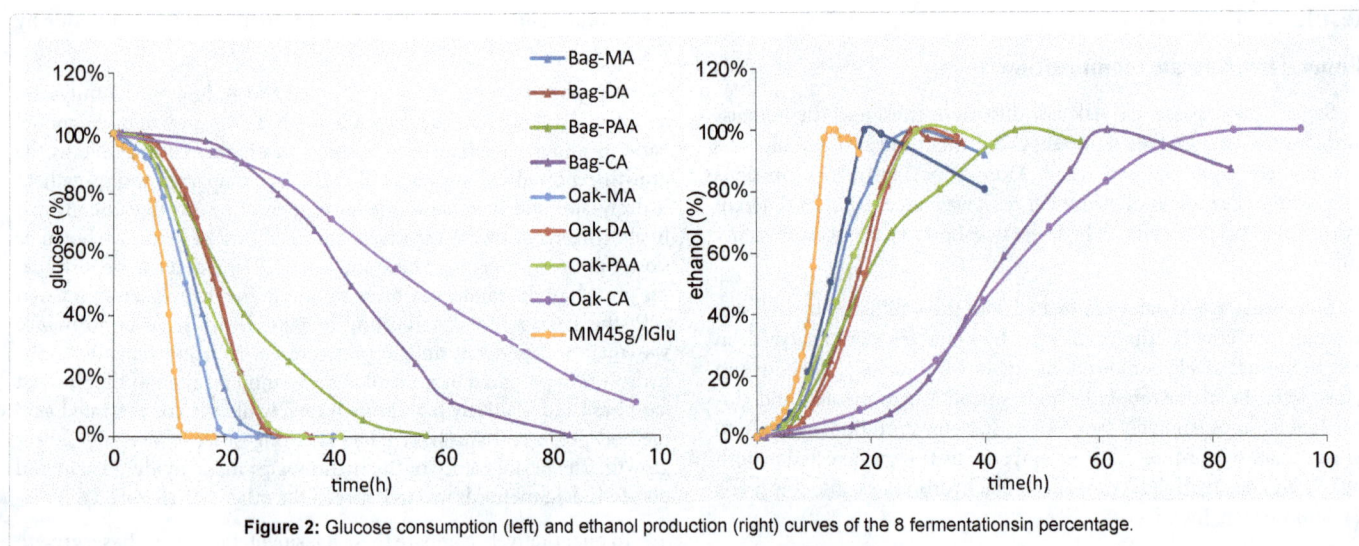

Figure 2: Glucose consumption (left) and ethanol production (right) curves of the 8 fermentationsin percentage.

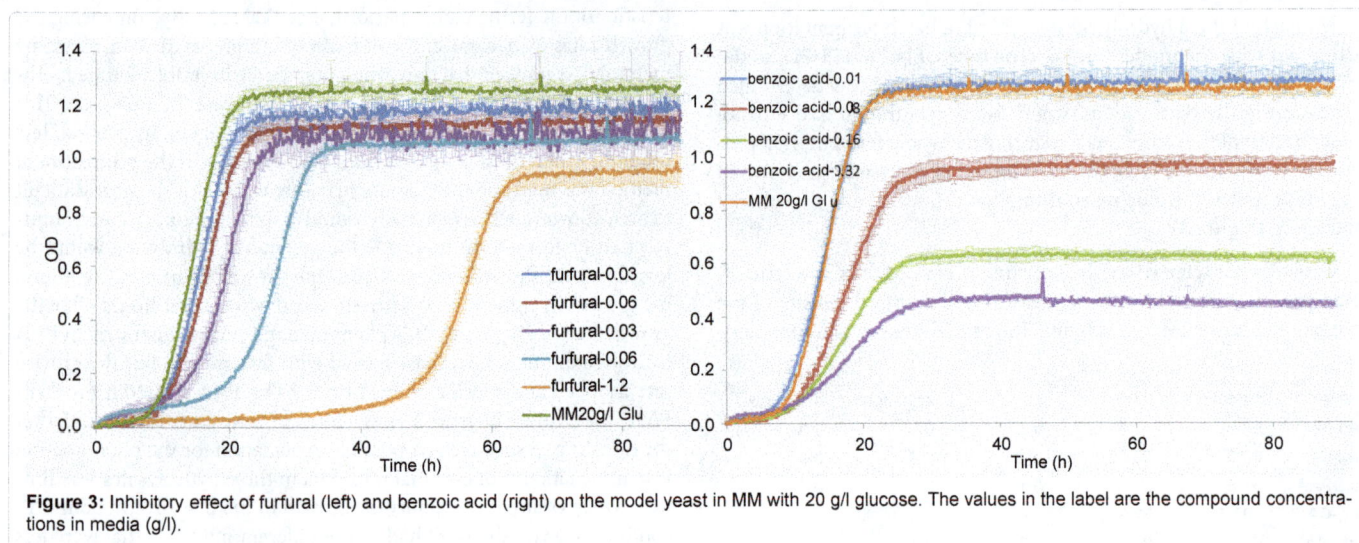

Figure 3: Inhibitory effect of furfural (left) and benzoic acid (right) on the model yeast in MM with 20 g/l glucose. The values in the label are the compound concentrations in media (g/l).

see table 3. To analyze the effect of initial glucose concentration on growth, mineral medium with glucose concentration 20 g/l, 40 g/l, 60 g/l and 80 g/l were used to test the model strain in Bioscreen. The results showed that with glucose concentration at this range, the model yeast did not show any difference in their growth, in terms of the 3 parameters listed in table 2 (data not shown). So it was assumed that the performance differences of the model strain in these hydrolysates were not caused by the variation of initial glucose concentration.

To present the glucose consumption and ethanol production of the 8 hydrolysate fermentations in a comparable manner, both glucose and ethanol concentrations were expressed as a percentage, with maximum value set as 100% and 0 g/l set as 0%, as shown in figure 2. It can be seen that the hydrolysates prepared with the same method had similar pattern in both glucose consumption and ethanol production curves. This is consistent with the observation of growth curves, confirming that the hydrolysate performance was mainly determined by pretreatment-hydrolysis method rather than biomass type.

The maximum ethanol concentration and ethanol yield of the 8 different hydrolysates are listed in table 3. The highest ethanol concentration in all fermentations was 34.9 g/l of Bag-CA hydrolysate, while the highest ethanol yield was of Oak-MA hydrolysate, 0.44 g ethanol per g glucose. This yield was 86% of the theoretical ethanol yield on glucose [23]. Furthermore, also the maximum glucose consumption rate and the maximum ethanol production rate of the 8 fermentations are compared in table 3. It can be seen that these two rates were closely related, in general, the faster the glucose was consumed the quicker the ethanol was produced, in other words, the ethanol yields of these 8 fermentations were quite similar. Additionally, these ethanol yields were not only similar to each other, but also comparable to the one of mineral media fermentation. This suggests that ethanol yield was only slightly influenced by the inhibitory compounds in the hydrolysates, which agrees with the effect of furans and phenols on yeasts performance [17]. Since the effects of inhibitory compounds in hydrolysates were mainly on growth rate, OD yield and glucose consumption rate, it is practical to use these parameters as indicators for studying the hydrolysate inhibitory effect.

Selection and quantification of the compounds in the 8 hydrolysates and their fermentation samples

To identify the role of specific inhibitory compounds on fermenta-

	lag-phase[1] (h)	growth rate[2]	OD yield[3] (OD/ g glucose)
Bag-MA	1	0.169	0.067
Bag-DA	6	0.187	0.069
Bag-PAA	4	0.243 / 0.052	0.064
Bag-CA	17	0.085	0.033
Oak-MA	2	0.200	0.100
Oak-DA	5	0.125	0.060
Oak-PAA	4	0.190 / 0.058	0.057
Oak-CA	10	0.035	0.032
MM 45 g/l Glucose	1	0.306	0.132

[1]: lag-phase is defined as the time needed to reach 2% of the maximum OD
[2]: growth rate is calculated as the slope of the linear part of the *logOD vs. time* plot
[3]: OD yield is calculated by dividing maximum OD by the amount of glucose consumed in the whole fermentation process.

Table 2: Growth characteristics of the model yeast in the 8 different hydrolysates.

	initial glucose concentration[1] (g/l)	maximum ethanol concentration[2] (g/l)	ethanol yield[3] (g/g)	maximum glucose consumption rate[4] (g/l/h)	maximum ethanol production rate[5] (g/l/h)
Bag-MA	58.8	22.7	0.39	3.3	1.3
Bag-DA	63.3	24.2	0.38	3.1	1.2
Bag-PAA	69.8	24.0	0.34	1.6	0.6
Bag-CA	104.4	34.9	0.33	2.3	0.9
Oak-MA	44.4	19.5	0.44	3.1	1.4
Oak-DA	38.2	15.3	0.40	2.0	0.8
Oak-PAA	58.0	24.2	0.42	2.4	1.0
Oak-CA	88.9	22.0	0.35	1.1	0.4
MM 45 g/l Glucose	42.5	16.0	0.38	5.5	2.0

[1]: the glucose concentration of the time-0 fermentation sample;
[2]: the highest ethanol concentration among all fermentation samples;
[3]: maximum ethanol concentration divided by the total amount of glucose consumed
[4]: the slope of the linear part of the *glucose concentration vs. time* plot
[5]: the slope of the linear part of the *ethanol concentration vs. time* plot

Table 3: Glucose consumption and ethanol production in the 8 fermentations.

tion performance of the various hydrolysates, exo-metabolomics analysis was carried out. As a first step in interpreting this type of analysis, a group of 11 compounds were selected and quantified for the hydrolysates and their fermentation samples. Base on quantification results, the dynamics of these compounds during a fermentation process could be determined. This analysis will also allow tests of inhibitory effects of these compounds at concentrations present in the hydrolysates. The selected compounds were formic acid, acetic acid, levulinic acid, furfural, furfurylalcohol, 2-furanmethanol acetate, HMF, vanillin, syringic acid, 4-hydroxybenzaldehyde and benzoic acid. The selection was made based on data reported in literature and observations made in our preliminary studies (Table 4).

Furfural and HMF are both furan compounds, and were identified as potential inhibitors in biomass hydrolysates [17,24,25]. Furfural was pointed to be the key inhibitor in hydrolysates by Heer and Sauer [15] in 2008 [15]. It was known that furfural was converted to furfurylalcohol by yeast as a detoxification mechanism [26]. The inhibitory effect of furfural was reported as increasing lag-phase [15] and reducing specific growth rate [6].

Formic acid, acetic acid and levulinic acid are the weak acids formed in most of the biomass hydrolysis preparation process [1,5] and their inhibitory effects and mechanism on yeasts have been studied in the past several years [6,27]. It was suggested that these weak acids re-

duce yeast growth and ethanol yield by causing intracellular anion accumulation, which is pH dependent [1,16]. Recently, Sanda et al. [27] reported that both formic acid and acetic acid affect the utilization of xylose in recombinant xylose-fermenting strain [27].

Vanillin, syringic acid and 4-hydroxybenzoic acid were characterized phenolic compounds in hydrolysates [5,17]. The inhibitory effects of 4-hydroxybenzoic acid have been studied with several different yeast strains [18], the study concluded that the compound showed little effects on the yeasts used. In this study, the two closely related compounds, benzoic acid and 4-hydroxybenzaldehyde were chosen to be quantified in hydrolysates and tested on their effects on the growth of the model yeast.

As summarized in table 4, formic acid, acetic acid, levulinic acid, furfural, HMF, vanillin, syringic acid, 4-hydroxybenzaldehyde and benzoic acid were the characterized degradation products in biomass hydrolysates, with inhibitory effects on *S. cerevisiae*. Furfurylalcohol and 2-furanmethanol acetate are the possible conversion products of furfural and/or HMF.

These selected compounds were analyzed and quantified in both hydrolysates and their fermentation samples. For each fermentation, 5 samples were chosen according to the following criteria: 1) directly after inoculation, 2) end of lag-phase, 3) growth phase, 4) end of growth phase, 5) stationary phase. The concentrations of these selected compounds in the fermentation samples are listed in table 5.

Formic acid and acetic acid were detected in all hydrolysates and their fermentation samples. In general, acetic acid concentrations were 10-15 times higher than that of formic acid. The highest concentrations of these two acids were found in CA hydrolysates, 0.57 g/l of formic acid and 8.0 g/l of acetic acid. These concentrations are comparable with the ones detected in acid pretreated spruce and bagasse hydrolysates [14]. During fermentation processes, no obvious consumption of either acid was observed, though both fluctuated slightly. Unlike formic acid and acetic acid, levulinic acid was only present in CA hydrolysates with a concentration of 1.2 g/l, without a decrease during fermentation.

Vanillin, syringic acid and 4-hydroxybenzaldehyde were present mainly in MA and DA hydrolysates although in rather low amounts, 30-50 mg/l. These concentrations are similar with those detected previously [1,5]. In contrast to syringic acid, both vanillin and 4-hydroxybenzaldehyde decreased during fermentation, suggesting their conversion or consumption. Due to its presence as a preservative in the enzyme cocktails of about 2.0 g/l [28], benzoic acid was detected in all enzymatic hydrolyzed hydrolysates, namely MA, DA and PAA hydrolysates, with a similar concentration of 150 mg/l. During the whole fermentation process, the level of benzoic acid did not change (Table 5). Surprisingly, in PAA treated bagasse hydrolysate, benzoic acid was apparently converted into its corresponding ethanol before the starting of the fermentation. It is unclear why this conversion took place specifically in Bag-PAA hydrolysate.

It can be seen in table 5 that furfural was found at considerable levels in CA hydrolysates, and at low amounts in DA hydrolysates. In both DA and CA hydrolysate fermentations, the furfural concentration rapidly decreased at the onset of the fermentation until levels of about 30 mg/l, with exception of Bag-DA hydrolysate. Correspondingly, the concentration of furfurylalcohol increased in the same time frame. This suggests that furfural was converted to furfurylalcohol in the lag-phase of the fermentation, which agrees with the report of Palmqvist et al. [26]. Different from the observation in *A. niger* [29], furfurylalcohol was not further converted into furoic acid. Furfural was also found in

Compound	Structure	Hydrolysates		Concentrations in hydrolysates		Concentrations showed effects on *S.cerevisiae*	
		biomass	preparation method	mg/l	ref	mg/l	ref
formic acid		Corn stover	Acid/temperature	130-310	[12]	4000	[16]
		Spruce/ bagasse	Acid/temperature	600-800	[14]	2700	[32]
		Hardwood chips	Autohydrolysis/ temperature	4000-4600	[33]	5000	[34]
		Corn stover	Steam explosion	6800	[16]	6000	[16]
acetic acid		Bagasse	Acid hydrolysis	2400	[34]	7500	[32]
		Bagasse	Enzymehydrolysis	2100	[34]	>10000	[35]
		Wheat straw	Acid hydrolysis	1300	[34]		
		Wheat straw	Enzymehydrolysis	900	[34]		
		Corn stover	Acid hydrolysis	2300	[34]		
		Willow wood	Acid hydrolysisa	2200	[34]		
		Yellow polar wood	Organosolv	900-4900	[36]		
		Spruce/ bagasse	Acid/temperature	3100-5200	[14]		
		Corn stover	Acid/temperature	2270-3740	[12]		
		Hardwood chips	Autohydrolysis/ temperature	4500-5800	[33]		
		Corn stover	Steam explosion	7800	[16]		
levulinic acid		Spruce/ bagasse	Acid/temperature	200-300	[14]		
		Corn stover	Acid/temperature	130-410	[12]		
furfural		Yellow polar wood	Organosolv	0.2-35.2	[36]	1000	[34]
		Wheat straw	Alkaline/oxidation	0-146*	[5]	>800	[35]
		Bagasse	Acid hydrolysis	410	[34]	>4000	[16]
		Wheat straw	Acid hydrolysis	270	[34]		
		Corn stover	Acid hydrolysis	510	[34]		
		Willow wood	Acid hydrolysis	500	[34]		
		Hardwood chips	Autohydrolysis/ temperature	510-780	[33]		
		Corn stover	Acid/temperature	570	[12]		
		Corn stover	Steam explosion	710	[16]		
		Wheat straw	Acid steam explosion	480-680	[15]		
		Spruce	Acid steam	1100	[15]		
		Barley straw	Acid steam	2880	[15]		
		Spruce/ bagasse	Acid/temperature	600-1200	[14]		
furfuryl alcohol		N.A.					
2-furan methanol acetate		N.A.					
HMF		Wheat straw	Alkaline/oxidation	0-16*	[5]	1000	[34]
		Yellow polar wood	Organosolv	0-56.5	[36]	>3000	[35]
		Hardwood chips	Autohydrolysis/ temperature	80-130	[33]	>4000	[16]
		Bagasse	Acid hydrolysis	70	[34]		
		Wheat straw	Acid hydrolysis	60	[34]		
		Corn stover	Acid hydrolysis	100	[34]		
		Willow wood	Acid hydrolysis	140	[34]		
		Corn stover	Acid/temperature	50-140	[12]		
		Wheat straw	Acid steam explosion	177-277	[15]		
		Spruce	Acid steam	2140	[15]		
		Barley straw	Acid steam	996	[15]		
		Corn stover	Steam explosion	560	[16]		
		Spruce/ bagasse	Acid/temperature	1600-3400	[14]		
vanillin		Wheat straw	Alkaline/oxidation	8-96*	[5]	4000	[16]
		Wheat straw	Acid steam explosion	91-122	[15]		
		Spruce	Acid steam	152	[15]		
		Barley straw	Acid steam	106	[15]		
		Willow	Dilute acid	430	[1]		
		Corn stover	Steam explosion	4000	[16]		

syringic acid		Wheat straw	Wet oxidation	22	[1]		
		Wheat straw	Alkaline/oxidation	6-52*	[5]		
4-hydroxy benzaldehyde		Wheat straw	Wet oxidation	21	[1]		
		Willow	Dilute acid	10	[1]		
		Wheat straw	Alkaline/oxidation	12-59*	[5]		
benzoic acid		Corn stover	Steam explosion	900	[16]	2000	[16]

*: these values are expressed as g/100g straw
N.A.: Not Available

Table 4: A summary of the selected compounds: their concentrations detected in various hydrolysates, and the concentrations at which inhibitory effects were shown on *S.cerevisiae*.

mg/l	formic acid	acetic acid	levulinic acid	furfural	furfuryl alcohol	2-furanmethanol acetate	HMF	vanillin	syringic acid	4-hydroxybenzaldehyde	benzoic acid
Bag-MA 1	93	1342	<25	<10	<10	<1	<8	**34**	7	**42**	139
Bag-MA 2	83	1190	<25	<10	<10	<1	<8	26	6	40	143
Bag-MA 3	77	1164	<25	<10	<10	<1	<8	3	6	5	143
Bag-MA 4	63	1026	<25	<10	<10	<1	<8	3	6	5	140
Bag-MA 5	56	978	<25	<10	<10	<1	<8	3	6	5	143
Bag-DA 1	184	1816	<25	30	21	<1	11	19	8	23	141
Bag-DA 2	173	1750	<25	<10	42	1	10	5	9	13	141
Bag-DA 3	165	1661	<25	<10	44	1	<8	4	9	6	145
Bag-DA 4	147	1535	<25	<10	46	1	<8	5	9	10	149
Bag-DA 5	153	1520	<25	<10	48	2	<8	5	9	10	153
Bag-PAA 1	16	241	<25	27	<10	<1	<8	3	<5	4	<10*
Bag-PAA 2	16	273	<25	27	<10	<1	<8	3	<5	4	<10*
Bag-PAA 3	0	128	<25	28	<10	<1	<8	3	<5	6	<10*
Bag-PAA 4	0	28	<25	29	<10	<1	<8	3	<5	9	<10*
Bag-PAA 5	0	48	<25	30	<10	<1	<8	3	<5	10	<10*
Bag-CA 1	568	7234	1148	**579**	97	12	57	<1	<5	<1	<10
Bag-CA 2	528	7049	1159	32	**750**	98	29	<1	<5	<1	<10
Bag-CA 3	552	6922	1206	28	730	99	<8	<1	<5	<1	<10
Bag-CA 4	533	6460	1297	30	739	97	<8	<1	<5	<1	<10
Bag-CA 5	534	6469	1314	29	747	99	<8	<1	<5	<1	<10
Oak-MA 1	133	1198	<25	<10	<10	<1	<8	13	7	<1	128
Oak-MA 2	135	1310	<25	<10	<10	<1	<8	3	7	<1	130
Oak-MA 3	173	1679	<25	<10	<10	<1	<8	2	6	<1	129
Oak-MA 4	151	1547	<25	<10	<10	<1	<8	2	7	3	132
Oak-MA 5	154	1562	<25	<10	<10	<1	<8	2	7	3	130
Oak-DA 1	330	3490	<25	60	30	1	17	14	46	<1	157
Oak-DA 2	318	3420	<25	28	94	6	14	4	44	<1	159
Oak-DA 3	302	3228	<25	26	106	7	8	3	46	<1	163
Oak-DA 4	282	3003	<25	26	107	7	<8	3	**47**	<1	162
Oak-DA 5	278	3051	<25	27	112	7	<8	3	45	<1	**164**
Oak-PAA 1	36	560	<25	<10	<10	<1	<8	4	<5	<1	144
Oak-PAA 2	39	592	<25	<10	<10	<1	<8	3	<5	<1	139
Oak-PAA 3	33	603	<25	<10	<10	<1	<8	3	<5	<1	141
Oak-PAA 4	34	474	<25	<10	<10	<1	<8	3	<5	<1	144
Oak-PAA 5	37	500	<25	<10	<10	<1	<8	3	<5	<1	141
Oak-CA 1	492	**7994**	1082	431	95	12	55	<1	7	<1	<10
Oak-CA 2	454	7877	1119	50	603	84	37	<1	6	<1	<10
Oak-CA 3	499	7869	1198	34	640	93	10	<1	7	<1	<10
Oak-CA 4	479	7591	1324	33	684	97	<8	<1	7	<1	<10
Oak-CA 5	509	7901	**1360**	34	698	**100**	<8	<1	7	1	<10

*: Instead of benzoic acid, benzylalcohol peak was found in Bag-PAA hydrolysate samples. Since benzylalcohol and several unknown peaks that may relate to benzoic acid were unique to Bag-PAA samples, it is possible that the benzoic acid presented in Bag-PAA hydrolysate was converted to several related compounds.

Table 5: The concentrations of selected 11 compounds in the 5 time-point samples of the 8 fermentations (mg/l).

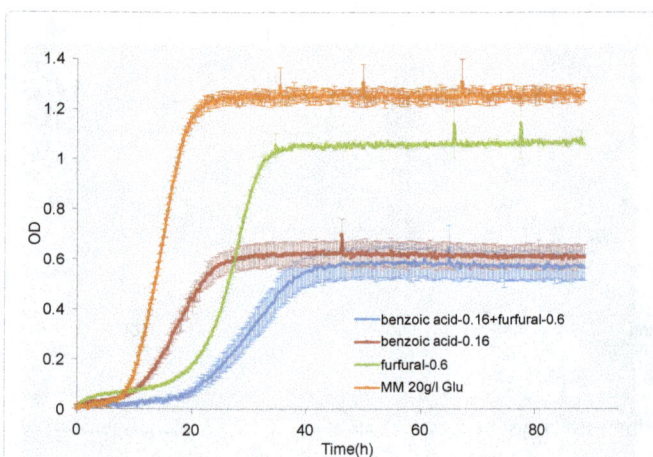

Figure 4: Combined inhibitory effects of furfural and benzoic acid on the model yeast in MM with 20 g/l glucose. The values in the label are the compound concentrations in media (g/l).

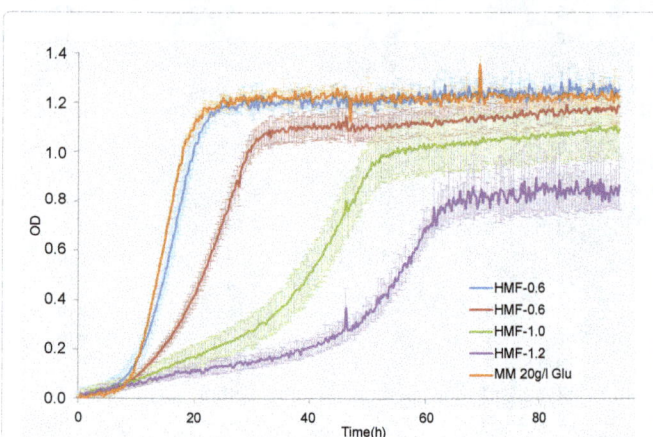

Figure 5: Inhibitory effects of HMF on the model yeast in MM with 20 g/l glucose. The values in the label are the compound concentrations in media (g/l).

Bag-PAA hydrolysate at 30 mg/l, but it was not converted during the whole fermentation process. Similar to furfural, HMF was also found in DA and CA hydrolysates, but with a much lower amount. The HMF concentration reduced gradually in both lag-phase and growth-phase of these hydrolysate fermentations.

Interestingly, 2-furanmethanol acetate showed similar pattern as furfurylalcohol, the compound increased with the decrease of furfural and HMF in the fermentation lag-phase (Table 5). Based on the structure of 2-furanmethanol acetate, it is suspected that the compound was the reaction product of furfurylalcohol and acetic acid. From this result, we suggest that furfurylalcohol was possibility partially converted to 2-furanmethanol acetate by reacting with acetic acid.

Inhibitory effect of the selected compounds tested in mineral medium

The quantification results of the selected compounds in the hydrolysates provided reference concentrations to test their inhibitory effects. For each compound, the highest concentration detected among all samples, marked as bold in table 5, was used as the initial testing value. Based on initial test results, the concentrations were increased or decreased up to 5-10 folds for the actual test. The medium used here was mineral medium with 20 g/l glucose.

Furfural and benzoic acid: Furfural and benzoic acid clearly affected the growth of the model yeast at concentrations as were present in the hydrolysates, see figure 3. The inhibitory effect of furfural displayed mainly as longer lag-phase. The lag-phase started to elongate already at a very low furfural concentration, 0.06 g/l, and increased from about 5 hours to 15 hours at a concentration of 0.6 g/l, which was about the concentration in CA hydrolysates (Table 5). It was observed that furfural concentration reduced mainly in the lag-phase during the fermentation process. This suggests that the presence of furfural obstructed the growth of the model yeast, and only when its concentration in the medium dropped below a threshold, the growth could start. It is suspected that this threshold was 0.03 g/l, as the growth commenced in most hydrolysates at this furfural concentration.

Unlike furfural, the inhibitory effect of benzoic acid was lowering the growth rate and final optical density level of the model yeast, as

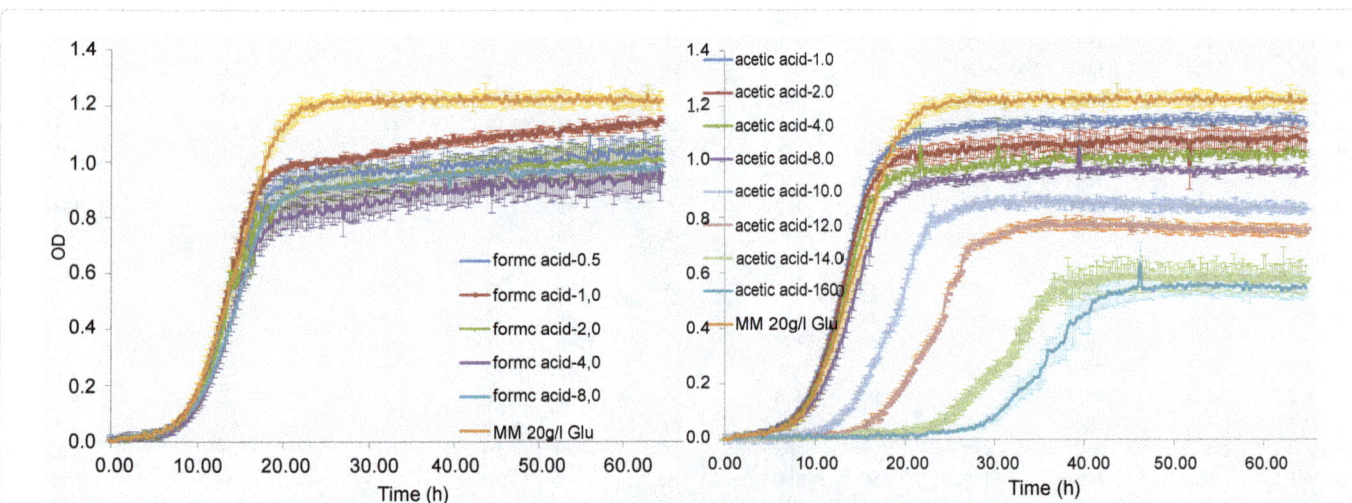

Figure 6: Inhibitory effects of formic acid (left) and acetic acid (right) on the model yeast in MM with 20 g/l glucose. The values in the label are the compound concentrations in media (g/l).

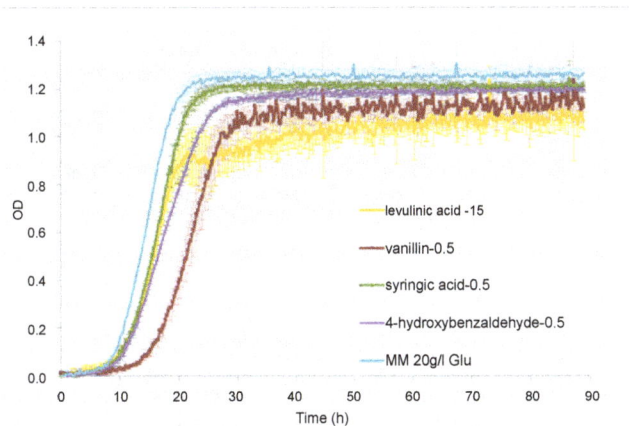

Figure 7: Inhibitory effects of levulinic acid, vanillin, syringic acid and 4-hydroxybenzaldehyde on the model yeast in MM with 20 g/l glucose. The values in the label are the compound concentrations in media (g/l).

shown in figure 3. At the concentration of 0.16 g/l, which was also the highest benzoic acid concentration detected in the hydrolysates, the growth rate decreased more than 60% compared to the reference medium, and the final optical density level dropped from 1.28 to 0.65. It seems that the inhibitory effect of benzoic acid was closely related to its concentration present in the medium.

The combination effect of furfural and benzoic acid on the model yeast seems to be addable, as shown in figure 4. That is to say, the lag-phase and the growth rate in the medium with both furfural and benzoic acid were very similar to which in the medium with furfural and with benzoic acid, respectively. Apparently, the inhibition by furfural and benzoic acid takes place at different stages of the growth process, namely, furfural before growth started and benzoic acid after. This indicates the inhibitory mechanisms of furfural and benzoic acid was different.

5-hydroxymethyl furfural (HMF): HMF was frequently mentioned as a inhibitor next to furfural in hydrolysates [24,30] , but seemed to have a milder inhibitory effect [7,23]. The highest HMF con-

centration present in the 8 hydrolysates in this study was 0.06 g/l, which did not give any effect on growth when added into mineral medium (data not shown). The inhibitory effect of HMF only became visible when its concentration reached 0.6 g/l and enhanced strongly when it was increased to 1.2 g/l, see figure 5. In contrast to furfural, the inhibitory effect of HMF was mainly shown as slower growth and lower final optical density, next to elongated lag-phase.

Formic acid and acetic acid: The presence of formic acid at 0.5 g/l had little effect on the growth rate of the model yeast, but reduced final optical density slightly, see figure 6. Increasing the formic acid concentration in mineral medium from 0.5 g/l to 8.0 g/l hardly enhanced this effect. The influence of acetic acid on the growth of the model yeast was similar to formic acid up to 8.0 g/l. Only when acetic acid concentration exceeded 8.0 g/l, both growth rate and final optical density were reduced significantly, and the lag-phase was clearly elongated, similar as described previously [16,31]. The highest concentrations of formic acid and acetic acid found in hydrolysates were 0.6 g/l and 8.0 g/l, respectively (Table 5). At these concentrations, the inhibitory effects of both acids were only marginal. To reach severe inhibitory effect, the level of formic acid needs to be enhanced by more than 13 folds, while the acetic acid levels are close to the inhibiting concentration. From this point of view, acetic acid is more likely to be an inhibitor in hydrolysates than formic acid.

Levulinic acid, syringic acid, vanillin and 4-hydroxybenzaldehyde: Though reported as inhibitory compounds in the hydrolysates [17], the inhibitory effects of levulinic acid, syringic acid, vanillin and 4-hydroxybenzaldehyde were only marginal. The four compounds were tested by adding them individually into mineral medium according to their highest concentrations detected in the hydrolysates. Only at 10 fold increased levels, effects became visible for these compounds, although still mild, see figure 7.

Among the 4 compounds, vanillin with concentration 0.5 g/l gave the most inhibitory effect, which was mainly on lag-phase. Levulinic acid showed similar effect on growth as formic and acetic acid, but at a much higher concentration, 15.0 g/l. Since these 4 compounds only started to affect the growth of the model strain at a 10-fold concentra-

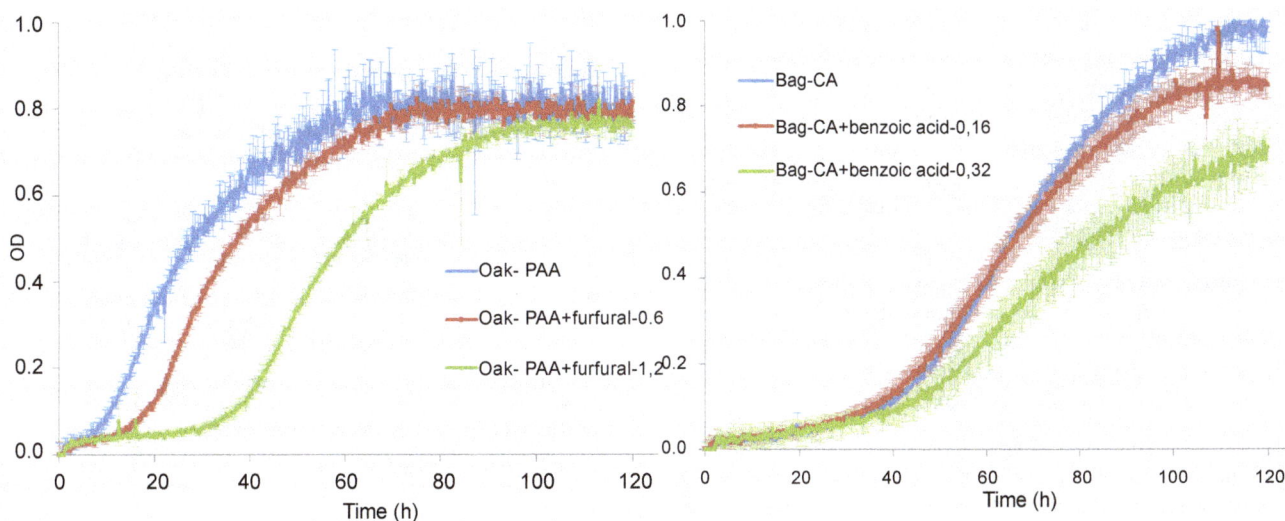

Figure 8: Inhibitory effects of furfural in Oak-PAA hydrolysate and benzoic acid in Bagasse-CA hydrolysate, on the model yeast. The values in the label are the compound concentrations in media (g/l).

tion compare to their highest concentrations in hydrolysates, they are thought to be none-inhibitory in the hydrolysates.

Inhibitory effect test in hydrolysates

The inhibitory effect tests of the selected compounds in mineral medium suggested that furfural and benzoic acid were the most important inhibitory compounds. They affected the growth of the model strain considerably at their concentrations presented in the hydrolysates. As hydrolysates have a total different matrix compare to mineral medium, it is interesting to test if these two compounds display similar inhibitory effect in hydrolysates.

Furfural and benzoic acid were tested in Oak-PAA and Bagasse-CA hydrolysate, respectively. The reason for using these two hydrolysates was that Oak-PAA was a furfural free hydrolysate and Bagasse-CA was benzoic acid free. The testing concentrations of both compounds were one and twofold of their highest levels in the hydrolysates, 0.6 g/l and 1.2 g/l of furfural, and 0.16 g/l and 0.32 g/l of benzoic acid (Figure 8).

As shown in figure 8, the inhibition effects of both compounds were also observed in hydrolysates and similar to those seen in mineral medium. The presence of furfural lengthened the lag-phase of the growth in Oak-PAA hydrolysate, and benzoic acid affected the growth rate and the final optical density level in Bagasse-CA hydrolysate. However, these effects were milder than in mineral medium, as by adding 1.2 g/l furfural in mineral medium, the lag-phase increased to 40 h, while the lag-phase enhanced to only 30 h when added into hydrolysate. For benzoic acid, in mineral medium the optical density dropped to half of that in reference medium when 0.16 g/l was present, while in hydrolysate the optical density decreased less than 10% (Figure 2,8). These results indicate that the hydrolysate matrix buffers inhibitory effects. It can also be seen in figure 8 that the pattern of the growth curve of both hydrolysates changed little by adding either furfural or benzoic acid. This suggests that the growth curve pattern of a hydrolysate is determined by the combined structure of most or all the compounds present in it.

Conclusion

As shown in this study, the fermentation performance of different hydrolysates varied in lag-phase, growth rate and biomass yield (Table 2), as well as their composition as far as the selected 11 compounds is considered (Table 5). These differences among hydrolysates seem to be caused mainly by hydrolysate preparation method, and secondly by biomass type. The detection of the 11 selected compounds in fermentation samples revealed that the levels of most compounds changed during fermentation process. Remarkably, furfural was converted to furfurylalcohol and possibly also 2-furan methanol acetate in the fermentation lag-phase.

The toxicity test of the 11 selected compounds showed that furfural and benzoic acid exhibited clear inhibitory effects on model yeast *S. cerevisiae* CEN.PK 113-7D at their concentrations detected in hydrolysates, while the effects of acetic acid and HMF were minor, but enhanced dramatically at the increase of concentration.

The fact that concentrated acid (CA) prepared hydrolysates performed the worst as fermentation media but were absent of benzoic acid, indicates that more compounds are involved in causing the inhibitory effects of the hydrolysates. The identification of these compounds could be conducted by using a systematic approach and metabolomics tools.

Acknowledgement

This project was (co) financed by the Netherlands Metabolomics (NMC) which is part of the Netherlands Genomics Initiative / Netherlands Organisation for Scientific Research.

References

1. Almeida JRM, Modig T, Petersson A, Hähn-Hägerdal B, Lidén G, et al. (2007) Increased tolerance and conversion of inhibitors in lignocellulosic hydrolysates by *Saccharomyces cerevisiae*. J Chem Technol Biotech 82: 340-349.

2. da Silva AS, Inoue H, Endo T, Yano S, Bon EPS (2011) Milling pretreatment of sugarcane bagasse and straw for enzymatic hydrolysis and ethanol fermentation. Bioresour Technol 101: 7402-7409.

3. Sun Y, Cheng J (2002) Hydrolysis of lignocellulosic materials for ethanol production: A review. Bioresour Technol 83: 1-11.

4. Alvira P, Tomás-Pejó E, Ballesteros M, Negro MJ (2010) Pretreatment technologies for an efficient bioethanol production process based on enzymatic hydrolysis: A review. Bioresour Technol 101: 4851-4861.

5. Klinke HB, Ahring BK, Schmidt AS, Thomsen AB (2002) Characterization of degradation products from alkaline wet oxidation of wheat straw. Bioresour Technol 82: 15-26.

6. Palmqvist E, Hahn-Hägerdal B (2000) Fermentation of lignocellulosic hydrolysates. II: Inhibitors and mechanisms of inhibition. Bioresour Technol 74: 25-33.

7. Martin C, Jönsson LJ (2003) Comparison of the resistance of industrial and laboratory strains of *Saccharomyces* and *Zygosaccharomyces* to lignocellulose-derived fermentation inhibitors. Enzy Microb Technol 32: 386-395.

8. Franden MA, Pienkos PT, Zhang M (2009) Development of a high-throughput method to evaluate the impact of inhibitory compounds from lignocellulosic hydrolysates on the growth of *Zymomonasmobilis*. J Biotechnol 144: 259-267.

9. Albers E, Larsson C (2009) A comparison of stress tolerance in YPD and industrial lignocellulose-based medium among industrial and laboratory yeast strains. J Ind Microbiol Biotechnol 36: 1085-1091.

10. Modig T, Almeida JR, Gorwa-Grauslund MF, Lidén G (2008) Variability of the response of *Saccharomyces cerevisiae* strains to lignocellulose hydrolysate. Biotechnol Bioeng 100: 423-429.

11. Mussatto SI, Roberto IC (2004) Alternatives for detoxification of diluted-acid lignocellulosic hydrolyzates for use in fermentative processes: A review. Bioresour Technol 93: 1-10.

12. Zhu J, Yong Q, Xu Y, Yu S (2011) Detoxification of corn stover prehydrolyzate by trialkylamine extraction to improve the ethanol production with *Pichia stipitis* CBS 5776. Bioresour Technol 102: 1663-1668.

13. Jönsson LJ, Palmqvist E, Nilvebrant NO, Hahn-Hägerdal B (1998) Detoxification of wood hydrolysates with laccase and peroxidase from the white-rot fungus *Trametesversicolor*. Appl Microbio Biotechnol 49: 691-697.

14. Alriksson B, Cavka A, Jönsson LJ (2011) Improving the fermentability of enzymatic hydrolysates of lignocellulose through chemical in-situ detoxification with reducing agents. Bioresour Technol 102: 1254-1263.

15. Heer D, Sauer U (2008) Identification of furfural as a key toxin in lignocellulosic hydrolysates and evolution of a tolerant yeast strain. Microb Biotechnol 1: 497-506.

16. Huang H, Guo X, Li D, Liu M, Wu J, et al. (2011) Identification of crucial yeast inhibitors in bio-ethanol and improvement of fermentation at high pH and high total solids. Bioresour Technol 102: 7486-7493.

17. Klinke HB, Thomsen AB, Ahring BK (2004) Inhibition of ethanol-producing yeast and bacteria by degradation products produced during pre-treatment of biomass. Appl Microbiol Biotechnol 66: 10-26.

18. Palmqvist E, Grage H, Meinander NQ, Hahn-Hägerdal B (1999) Main and interaction effects of acetic acid, furfural, and p-hydroxybenzoic acid on growth and ethanol productivity of yeasts. Biotechnol Bioeng 63: 46-55.

19. Zhao X, Peng F, Cheng K, Liu D (2009) Enhancement of the enzymatic digestibility of sugarcane bagasse by alkali-peracetic acid pretreatment. Enzy Microb Technol 44: 17-23.

20. Zha Y, Slomp R, Groenestijn JV, Punt JP (2012) Preparation and evaluation of lignocellulosic biomass hydrolysates for growth by ethanologenic yeasts. Microb Metabol Eng 834: 245-259.

21. van Hoek P, van Dijken JP, Pronk JT (1998) Effect of specific growth rate on fermentative capacity of baker's yeast. Appl Environ Microbiol 64: 4226-4233.

22. Kuyper M, Winkler AA, van Dijken JP, Pronk JT (2004) Minimal metabolic engineering of Saccharomyces cerevisiae for efficient anaerobic xylose fermentation: a proof of principle. FEMS Yeast Research 4: 655-664.

23. Kwon YJ, Ma AZ, Li Q, Wang F, Zhuang GQ, et al. (2011) Effect of lignocellulosic inhibitory compounds on growth and ethanol fermentation of newly-isolated thermo tolerant Issatchenkiaorientalis. Bioresource Technology 102: 8099-8104.

24. Wierckx N, Koopman F, Bandounas L, de Winde JH, Ruijssenaars HJ (2009) Isolation and characterization of Cupriavidusbasilensis HMF14 for biological removal of inhibitors from lignocellulosic hydrolysate. Microbial Biotechnol 3: 336-343.

25. Palmqvist E, Hahn-Hägerdal B (2000) Fermentation of lignocellulosic hydrolysates. I: inhibition and detoxification. Bioresour Technol 74: 17-24.

26. Palmqvist E, Almeida JS, Hahn-Hägerdal B (1999) Influence of furfural on anaerobic glycolytic kinetics of Saccharomyces cerevisiae in batch culture. Biotechnol Bioeng 62: 447-454.

27. Sanda T, Hasunuma T, Matsuda F, Kondo A (2011) Repeated-batch fermentation of lignocellulosic hydrolysate to ethanol using a hybrid Saccharomyces cerevisiae strain metabolically engineered for tolerance to acetic and formic acids. Bioresour Technol 102: 7917-7924.

28. Hazan R, Levine A, Abeliovich H (2004) Benzoic acid, a weak organic acid food preservative, exerts specific effects on intracellular membrane trafficking pathways in Saccharomyces cerevisiae. Appl Environ Microbiol 70: 4449-4457.

29. Rumbold K, van Buijsen HJJ, Gray VM, van Groenestijn JW, Overkamp KM , et al. (2010) Microbial renewable feedstock utilization: A substrate-oriented approach. Bioeng Bugs 1: 359-366.

30. da Cunha-Pereira F, Hickert LR, Sehnem NT, de Souza-Cruz PB, Rosa CA, et al. (2011) Conversion of sugars present in rice hull hydrolysates into ethanol by Spathasporaarborariae, Saccharomyces cerevisiae, and their co-fermentations. Bioresour Technol 102: 4218- 4225.

31. Casey E, Sedlak M, Ho NW, Mosier NS (2010) Effect of acetic acid and pH on the cofermentation of glucose and xylose to ethanol by a genetically engineered strain of Saccharomyces cerevisiae. FEMS Yeast Res 10: 385-393.

32. Maiorella B, Blanch HW, Wilke CR (1983) By-product inhibition effects on ethanolic fermentation by Saccharomyces cerevisiae. Biotechnol Bioeng 25: 103-121.

33. Sannigrahi P, Pu Y, Ragauskas A (2010) Cellulosic biorefineries--unleashing lignin opportunities. Curr Opin Envir Sustain 2: 383-393.

34. Rumbold K, van Buijsen HJJ, Overkamp KM, van Groenestijn JW, Punt PJ, et al. (2009) Microbial production host selection for converting second-generation feedstocks into bioproducts. Microb Cell Fact 8: 64.

35. Keating JD, Panganiban C, Mansfield SD (2006) Tolerance and adaptation of ethanologenic yeasts to lignocellulosic inhibitory compounds. Biotechnol Bioeng 93: 1196-1206.

36. Koo BW, Park N, Jeong HS, Choi JW, Yeo H, et al. (2011) Characterization of by-products from organosolvpretreatments of yellow poplar wood (Liriodendron tulipifera) in the presence of acid and alkali catalysts. J Indust Eng Chem 17: 18-24.

27

Fermentation Monitoring of a Co-Culture Process with *Saccharomyces cerevisiae* and *Scheffersomyces stipitis* Using Shotgun Proteomics

Eric L Huang and Mark G Lefsrud*

Bioresource Engineering, McGill University, 21111 Lakeshore Road, Ste-Anne-de-Bellevue, Quebec, H9X 3V9, Canada

Abstract

Co-culture processes present the opportunity to produce value-added products from economical raw materials, but there lacks a high-throughput fermentation monitoring technique to study the temporal physiology of fermentation organisms in co-culture processes. In this study, we applied shotgun proteomics to investigate a co-culture process of *Saccharomyces cerevisiae* and *Scheffersomyces stipitis*, and we monitored the fermentation until glucose depletion. Three time points were taken for proteomics analysis at 11.5 hour, 18.5 hour and 32 hour, representing transition into diauxic shift. Using label-free quantitation, we observed cellular dynamic within 20-hour time frame. We distinguished the proteome between two yeasts, and the most abundant proteins of *S. stipitis* and *S. cerevisiae* contained expected processes of glycolytic enzymes, histones, heat shock proteins, ribosomal proteins and F_1F_0-ATPase. After glucose depletion, we identified up-regulations of *S. stipitis* malate synthase and isocitrate lyase as key enzymes in glyoxylate cycle and gluconeogenesis. Increased expression of *S. stipitis* histone 2B was observed in diauxic shift, and histone acetylation was suggested by up-regulation of acetyl-CoA synthetase. Without the presence of xylose, we observed induction of NAD(P)H-dependent D-xylose reductase (Xyl1p) as early as 11.5-hour before glucose depletion. We also observed the expression of D-xylulose reductase after glucose depletion without xylose induction. Further study is needed to investigate the cause of derepression signals for xylose oxo-reductive pathway. Without cellulose induction, the up-regulation of *S. stipitis* endo-1,4-beta-glucanase suggested *S. stipites'* strategy in diversifying carbon choices after glucose repression. This research demonstrated the application of shotgun proteomics in high-throughput monitoring of complex co-culture system and able to elucidate the temporal physiology of *S. stipitis*.

Keywords: *Saccharomyces cerevisiae*; *Scheffersomyces stipitis*; *Pichia stipitis*; Shotgun proteomics; Co-Culture; Mudpit; Fermentation monitoring

Introduction

Co-culture processes present the opportunity to establish stable and profitable biotechnological bioprocesses and produce value-added products from economical raw materials such as agricultural residues [1]. Mixed culture systems have demonstrated promise in hydrogen, methane, ethanol and polyhydroxyalkanoates productions from renewable resource, increasing potential revenue and reducing environmental impacts [2-4]. The use of controlled mixed fermentation using *Saccharomyces* and non-*Saccharomyces* yeasts has been implemented in winemaking to alter both chemical and the aromatic composition of wines [5]. The synergistic interaction between different yeasts provides a tool for the implementation of new fermentation technologies [5]. In biofuel research, the co-culture process of *Saccharomyce cerevisiae* and *Scheffersomyces stipitis (Pichia stipitis)* has been extensively studied with attempts to achieve a simple, one-batch process to ferment glucose and xylose using lignocellulosic feedstock [6,7]. The co-culture process poses technical obstacles such as low ethanol yields associated with different optimal oxygen transfer rate required by each organism. Novel fermentation apparatus using two fermentors and two microfiltration modules to achieve optimal fermentation condition for each yeast [7]. However, before optimizing a co-culture process such as ones described by Taniguchi et al. [7], there lacks a robust, high-throughput fermentation monitoring technique to monitor the temporal physiology of each organism in more complex process such as co-culture fermentation. In recent years, fermentation technology has turned '-omics' for solutions in process development [8]. Proteomics quantification of yeast high gravity ethanol fermentations was investigated to determine the dynamic of metabolic pathways (e.g. the deactivation of secondary metabolites pathway) [9]. Shotgun proteomics can provide the physiological backgrounds needed for non-model organisms such as *S. stipitis* and allow observation of the global profile of the organism's proteome. The analysis of the proteome allows the evaluation of cellular processes at greater depth with a much simpler technique than 2D-PAGE-MS [10]. In this study we demonstrated the application of shotgun proteomics in monitoring a co-culture process using *S. cerevisiae* S288C and *S. stipitis* CBS 6054. The yeast proteome of three time points were analyzed based on metabolic and regulatory pathways. Without the presence of xylose, we observed the induction of xylose reductase (Xyl1p, *S. stipitis*) as early as 11.5-hour before glucose depletion, and D-xylulose reductase (Xyl2p, *S. stipitis*) as early as 18.5-hour after glucose depletion. . We distinguished the proteome between two yeasts, identified the most abundant proteins of *S. stipitis* and *S. cerevisiae*, and determine the dynamic of *S. stipitis* proteome during the co-culture process.

Materials and Methods

Microorganism and culture condition

Saccharomyces cerevisiae S288C and *Scheffersomyces stipitis (syn. Pichia stipitis* CBS 6054) were obtained from American Type Culture

*Corresponding author: Mark G Lefsrud, Bioresource Engineering, McGill University, 21111 Lakeshore Road, Ste-Anne-de-Bellevue, Quebec, H9X 3V9, Canada

Collection (ATCC, Manassas, VA, USA) and maintained on YPD or YPX agar slants accordingly. YPX substituted xylose with glucose in YPD. Pre-inoculums of S. cerevisiae and S. stipitis were grown separately at 26°C, 100 rpm and harvested at mid-exponential phase. Two biological replicates of the co-culture process were carried out in a 250mL Erlenmeyer flask with 100mL working volume at 26°C and 100rpm. The medium composition for the co-culture process was 20g/L glucose, 20g/L peptone and 10g/L yeast extract.

Fermentation parameters analysis

Glucose, ethanol, acetic acid and glycerol were measured using enzymatic kits (Megazyme, Wicklow, Ireland). OD610 was monitored throughout growth with a UV-VIS spectrophotometer. Colony-forming unit (CFU) was used to obtain viable yeast counts of S. stipitis and S. cerevisiae. S. stipitis and S. cerevisiae were enumerated by adapting methods as described in Laplace and colleagues [11]. In brief, S. stipitis and S. cerevisiae were distinguished from each other by plating on medium containing erythromycin (1g/L), 1% yeast extract, 2% peptone, 2% glycerol or 2% glucose. S. stipitis was able to grow on media containing erythromycin (1 g/L) with 2% glycerol, whereas S. cerevisiae was inhibited by the mitochondrial inhibitor.

Cells lysis and protein extraction

Three protein samples from each biological replicate were taken at 11.5 hour, 18.5 hour and 32 hour. ~10 mg of wet-mass cell pellet for each sample was obtained and processed through single tube cell lysis and protein digestion [12,13]. The protein concentrations after the extraction were ~6 mg/mL for each sample after extraction. Tris/10mM $CaCl_2$ at pH 7.6 and 6M Guanidine/10mM Dithiothreitol (DTT) (Sigma-Aldrich Canada, Oakville, Ontario) were used to lyse cells and extrude proteins. In brief, the mixture was placed on the rocker for the first hour and vortexed every 10 minutes, and then incubated for 12 hours at 37°C. The mixture was spun down at 10,000 g. The supernatant was discarded and the pellet was diluted with 6-fold 50mM Tris/10mM $CaCl_2$. ~5 µg of sequencing grade trypsin (Promega, Madison, WI, USA) was added to each sample and digested for 12 hours at 37°C by gentle rocking. The same amount of trypsin was added a second time and incubated at 37°C for another 12 hours. 1M DTT was added to a final concentration of 20mM and incubated for another hour with gentle rocking at 37°C followed by centrifugation at 10,000 g for 10 minutes. The supernatant was kept and samples were cleaned and de-salted via solid phase extraction with Sep-Pak Plus cartridges (Waters Limited, Mississauga, Ontario), concentrated, solvent exchanged, filtered, aliquoted and frozen at -80°C [14].

2D-LC MS/MS

Each extracted and digested sample was analyzed twice for technical replicate in 2D-LC MS/MS. Samples were loaded onto the back column of split phase 2D column [15] (~3-5cm Luna SCX 5 µm and ~3- 5cm C_{18} Aqua 5µm 100A) (Phenomenex, Torrance, CA) using a pressure cell and then connected to a front column with integrated nanospray tip (~15 cm of Aqua C_{18} Aqua 5µm 100A, Phenomenex, Torrance, CA; 150 µm with 15 µm tip, New Objective, Woburn, MA). Samples were analyzed via 2-D LC-nanoESI-MS/MS on a LTQ XL (ThermoFisher Scientific, San Jose, CA) as previously described [13]. Samples were analyzed via 13-hr runs (6 steps with 1 wash step) [14,16].

Proteome informatics

Proteome Discoverer 1.1 (ThermoFisher Scientific, San Jose, CA) and SEQUEST algorithm [17] was used to correlate peptides spectra with S. stipitis protein sequence (filtered models containing 5839 entries downloaded from the Joint Genome Institute; accessed on 05/15/2011) and S. cerevisiae protein sequence (5884 entries downloaded from the Saccharomyces Genome Database; accessed on 05/15/2011) [6,16]. Proteome Discoverer 1.1 was used to filter SEQUEST results and sort peptides into protein identification based on modified method of Verberkmoes et al. [17] (minimum Xcor of at least 1.5 [+1], 2.3 [+2], 2.8 [+3]; minimum probability score of at least 2 [+1], 7 [+2] and 7.5 [+3]; minimum two fully tryptic peptides) [13]. We enabled protein grouping, counted only rank 1 peptides and only in the top scored proteins. Each protein identified had at least one unique peptide. Reverse protein sequences were included in the search database to estimate the overall false-positive rates of protein identification (false-positive rate = $2[n_{rev}/(n_{rev} + n_{real})]*100$; n_{rev} = the number of peptides identified from the reverse database; n_{real} = the number of peptides identified from the real database) [18].

Data analysis

KEGG and GO annotations of S. cerevisiae and S. stipitis (also KOGs) were downloaded from Saccharomyces Genome Database and Joint Genome Institute [19,20]. The identified proteomes were examined manually with each biological function. In-house Perl scripts were written to perform the functional counts of GOs and KOGs. The metabolic pathways were visualized using KEGG Mapper [21]. For comparison of protein abundance levels, normalized spectral abundance factors (as label free estimation of relative protein abundance) (NSAF) were determined as described in Florens et al. [22]. This method estimates protein abundance by first dividing the spectral count for each protein by the protein length and then dividing this number by the sum of all length-normalized spectral counts for each organism and multiplying by 100 [22,17]. Differentially expressed proteins between 11.5-hour and 32-hour were determined using spectral count and the software QSpec with log. fold change of 3 and up-regulated FDR = 0.01 [23].

Results and Discussions

Overview of proteome

This study represented the first attempt to monitor the temporal physiological profile of a co-culture process using S. cerevisiae and S. stipitis with shotgun proteomics. Key metabolites, including ethanol, glucose, acetic acid and glycerol were measured and monitored throughout the co-culture fermentation (Figure 1 and 2). The fermentation lasted for 36 hour in both biological replicates. The highest ethanol yielded ~9 g/L. We observed low acetic acid (< 0.5 g/L) throughout the process, but glycerol was observed at moderate level (<1.15 g/L). Glucose was consumed to near depletion (< 1 g/L) at ~16 hour, indicating diauxic shift for both organisms. We inoculated batch co-culture fermentation with the pre-inoculums and obtained initial cell density of 8.07x10^5 CFU for S. cerevisiae and 7.23x10^5 CFU for S. stipitis in biological replicate 1 at time zero. In biological replicate 2 we obtained initial cell density of 7.87 x10^5 CFU for S. cerevisiae and 1.10 x10^6 CFU for S. stipitis. Before 11.5 hour, both S. cerevisiae and S. stipitis multiplied at near equal rate. In biological replicate 1, S. stipitis outgrew S. cerevisiae, reaching 1.06x10^8 CFU for S. stipitis and 6.20x10^7 CFU for S. cerevisiae at the end of the fermentation. However, in biological replicate 2, the population of S. stipitis and S. cerevisiae fluctuated back and forth, failing to determine the dominant specie. This study further demonstrated the challenge in producing consistent batches in co-culture process. Based on the growth curve of each organism, the maximal growth rates of both S. cerevisiae and S. stipitis occurred at 11.5 hour and the maximal glucose consumption rate was also observed

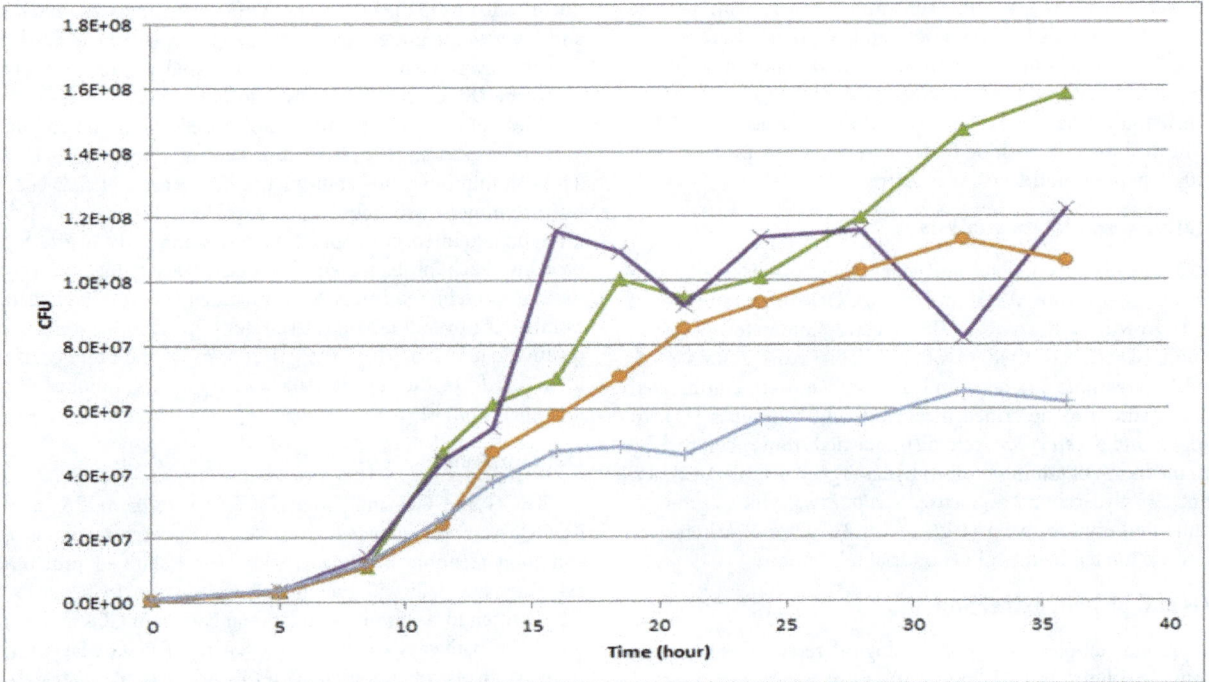

Figure 1: Viable cell counts of the co-culture process of *Saccharomyces cerevisiae* and *Scheffersomyces stipitis* in both biological replicate 1 and 2. Proteomics samples were taken at 11.5 hour, 18.5 hour, 32 hour, representing transition into diauxic shift. Fermentation was terminated at 36-hour (● *S. stipitis* CFUs Biological 1; ▲ *S. stipitis* CFUs Biological 2; + *S. cerevisiae* CFUs Biological 1; X *S. cerevisiae* CFUs Biological replicate 2).

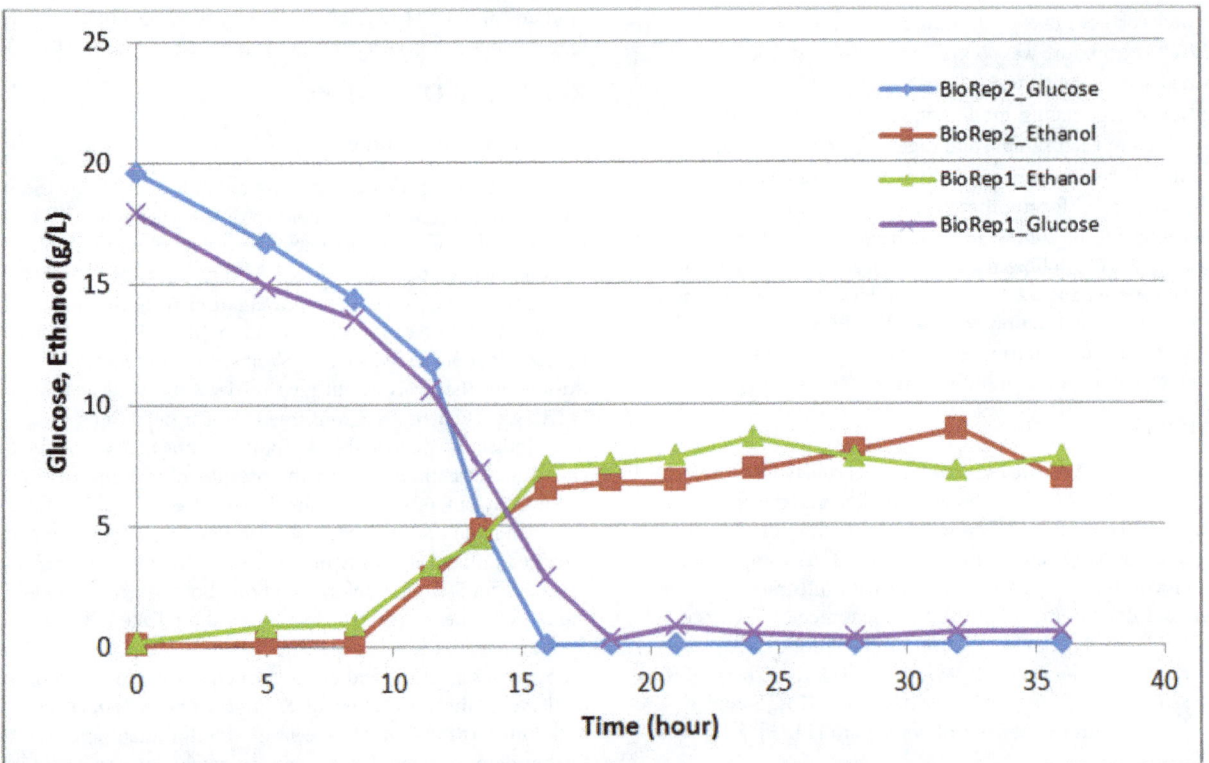

Figure 2: Monitoring glucose and ethanol concentrations during the co-culture process of *Saccharomyces cerevisiae* and *Scheffersomyces stipitis*. Data shown included both biological replicate 1 and 2.

at 11.5 hour, corresponding to the exponential growth phase of both yeasts. We collected proteomics samples from two flask fermentations at three time points (11.5 hour, 18.5 hour and 32 hour) to monitor the co-culture process. The shotgun approach was used to identified and matched peptide mass data to the sequenced S. stipitis and S. cerevisiae sequence databases. In biological replicate 1, combining all time points and technical replicates for each time point, we identified a total number of 1390 proteins throughout the co-culture process (650 proteins from S. stipitis and 740 proteins from S. cerevisiae (Table 1). In biological replicate 2, combining all time points and technical replicates for each time point, we identified a total number of 1499 proteins throughout the co-culture process (666 proteins from S. stipitis and 833 proteins from S. cerevisiae; Table 1). The false discovery rates of 2D-LC-MS/MS runs were calculated between 0.16% to 0.57%. Table 1 summarized spectral counts and the CFU for each organism in each time point in both biological replicates. We observed that although both S. cerevisiae and S. stipitis CFU were at approximately 1:1 ratio (~2.5x10^7 CFU each) at 11.5 hour, the total peptides counts were not equal between two yeasts. The difference between total peptides counts demonstrated the difference in cell size and mass between S. stipitis and S. cerevisiae, where S. stipitis is smaller than S. cerevisiae in cell mass per cell. S. stipitis exhibited higher CFU than S. cerevisiae during the fermentation, but higher spectral counts in S. cerevisiae were observed. Looking at the spectral counts, CFU and protein identified, we found an increase in S. cerevisiae CFU did not necessary result in identifications of higher proteins and spectral counts, but an increase in S. stipitis CFU resulted in higher spectral counts and minor increase in numbers of protein identification. This implied other variables, which could be time dependent, interacted with spectrum identification capacity of the mass spectrometry. One possible explanation could be the competition for electro-spray ionization between S. cerevisiae and S. stipitis peptides. Another explanation was the difference in extraction and digestion efficiency between two organisms. We calculated normalized spectral abundance factors (NSAFs) to estimate the relative protein abundance. By comparing the technical replicates of each dataset, we obtained highly reproducible results in each biological replicate, with the linear regression line fell between $R^2 = 0.81$ and 0.89 for S. cerevisiae in biological replicate 1(as illustrated in Figure 3 for 18.5 hour time point) and between $R^2 = 0.85$ and 0.90 for S. stipitis (data not shown).

Most abundant proteins in s. cerevisiae and s. stipitis

Proteomics data was further analyzed below by using both biological replicates if not stated. Similar to the most abundant proteins found using GFP fusion library and high-throughput flow cytometry, S. cerevisiae glycolytic enzymes were among the most abundant proteins in the cell from exponential to diauxic shift [24]. These included pyruvate kinase (Cdc19p), 3-phosphoglycerate kinase (Pgk1p), glyceraldehyde-3-phosphate dehydrogenase (Tdh1p,

Tdh2p, Tdh3p), enolase II (Eno2p) and fructose 1,6-bisphosphate aldolase (Fba1p). Alcohol dehydrogenase (Adh1p) and F1 sector of mitochondrial F1F0 ATP synthase (Atp2p) were also highly abundant in S. cerevisiae proteome, indicating the importance of oxidative phosphorylation. The fact that the other enzymes of the tricarboxylic acid cycle were not switched off during fermentative growth can be explained by their involvement in amino acid biosynthesis, gluconeogenesis, heme biosynthesis or de novo fatty acid synthesis.

The most abundant proteins identified in S. stipitis also came from glycolysis and gluconeogenesis. Similar to S. cerevisiae, the glycolytic enzymes observed in most abundance included glyceraldehyde-3-phosphate dehydrogenase (Tdh2p and Tdh3p), enolase I (Eno1p), phosphoglycerate mutase (Gpm1.1p), fructose-bisphosphate aldolase (Fba1p), phosphoglycerate kinase (Pgk13p) and pyruvate kinase (Pyk1p). S. stipitis cells also produced high amount of alcohol dehydrogenase (Adh1p), which reduces acetaldehyde to ethanol. We observed high abundance of F_1F_0-ATPase complex alpha and beta subunits (Atp1p, Atp2p) located in mitochondrial cellular components. We also identified high abundance of mitochondrial malate dehydrogenase (Mdh1p) involved in tricarboxylic acid (TCA) cycle, indicating respiro-fermentative. Proteins involved in chromatin assembly, protein translation and folding were also high in abundance. These included heat shock protein (HSP70 family Ssa2p), histone H4 and 2B (Hhf1p and Htb2.1p) and ribosomal proteins (L14B, 40S S14-B, S7-A and 60S L4360S). The resemblance in both S. cerevisiae and S. stipitis proteomes showed that highly abundant proteins were housekeeping proteins and essential for viability. In yeast cells, glycolytic enzymes constitute a major fraction (30 to 60%) of the soluble proteins [25], and in rapidly growing cells, 60% of total transcription is devoted to ribosomal RNA [26].

Functional categorization and analysis of s. cerevisiae and s. stipitis proteome

A comparison between S. stipitis and other more related yeast such as C. albicans would almost certainly show a higher degree of conservation, but because regulation has been best studied in S. cerevisiae, that is a standard frame of reference. Therefore in this study regulation of S. stipitis was inferred and compared with S. cerevisiae if not otherwise stated. The proteome of S. cerevisiae and S. stipitis observed in the co-cultural process contained housekeeping processes such as oxidative phosphorylation, glycolysis, non-oxidative branch of the pentose phosphate pathway, gluconeogenesis, biosynthesis of amino acids and aminoacyl-tRNA, protein synthesis and proteolysis, fatty acid metabolism and cell division. An attempt was made to categorize S. stipitis proteome based on GO terms for direct comparison between S. stipitis and S. cerevisiae, but the incomplete nature of the GO entries for S. stipitis proved impractical (data not shown). The incomplete

	Number of Proteins Identified Biological Replicate 1			Total proteins measured from each organism
	11.5 hour	18.5 hour	32 hour	
Scheffersomyces stipitis	322	456	574	650
Saccharomyces cerevisiae	535	594	539	740
Total proteins in each time-point	857	1050	1113	1390
	Number of Proteins Identified Biological Replicate 2			Total proteins measured from each organism
	11.5 hour	18.5 hour	32 hour	
Scheffersomyces stipitis	444	412	577	666
Saccharomyces cerevisiae	626	637	660	833
Total proteins in each time-point	857	1050	1113	1499

Table 1: Numbers of Scheffersomyces stipitis and Saccharomyces cerevisiae proteins identified from shotgun proteomics of the co-culture process in both biological replicates.

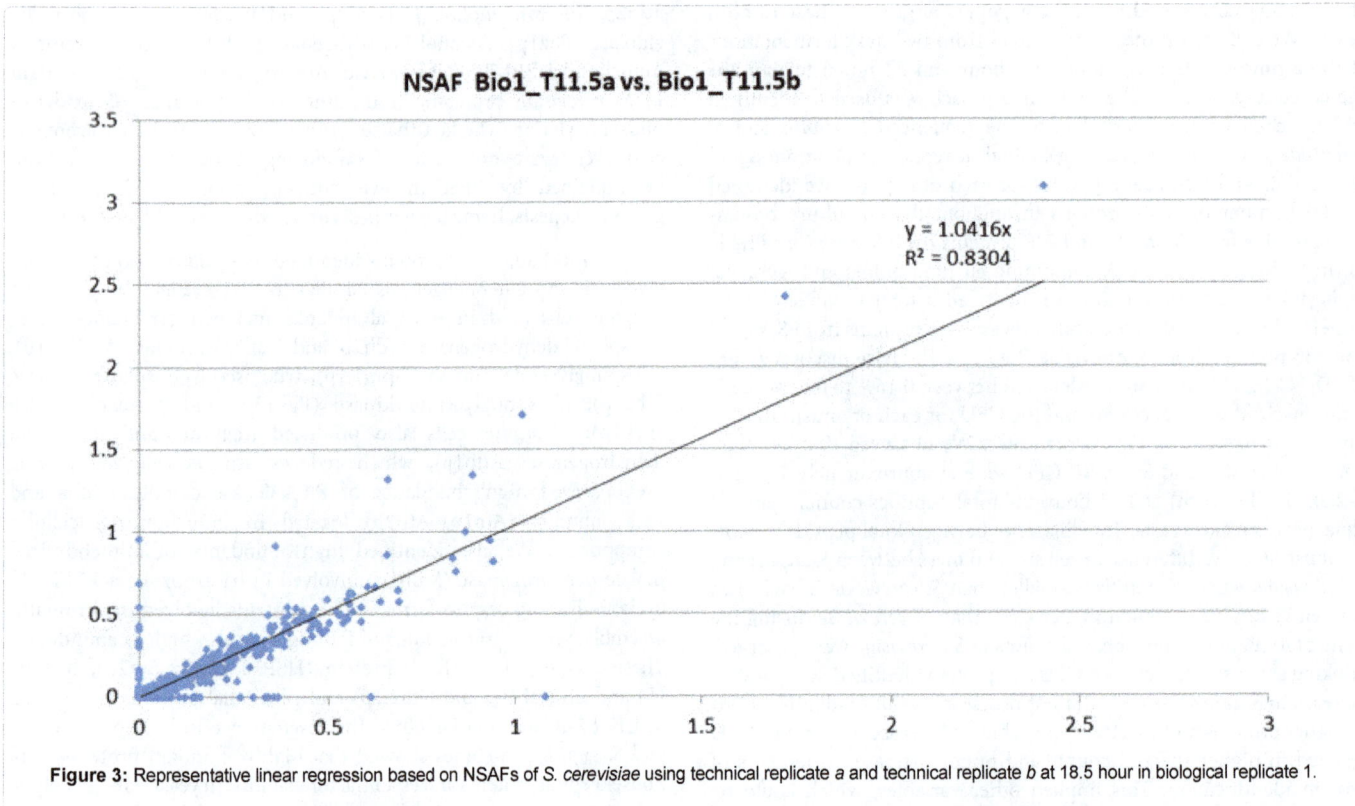

Figure 3: Representative linear regression based on NSAFs of *S. cerevisiae* using technical replicate *a* and technical replicate *b* at 18.5 hour in biological replicate 1.

nature of the GO annotations for *S. stipitis* directed us to categorize *S. stipitis* proteome based on KOGs. In analyzing the *S. stipitis* proteome, the majority of KOGs functions fell into the categories of carbohydrate metabolism, protein synthesis and modification. Technical replicates of each time point showed reproducible counts of each KOGs term. Even with the higher counts observed in later time points, each KOGs category still constituted the same relative ratio to the whole data set (data now shown). The majority of GO terms also fell into protein synthesis and carbohydrate metabolism in *S. cerevisiae*. The ratio of each category relative to the whole data set was consistent between time points. Technical replicate of each time point also showed reproducibility. All time points shared similar distribution of biological function in terms of KOGs for *S. stipitis* and GO Biological terms for *S. cerevisiae*.

Distinctive physiology in temporal proteome

In analyzing the proteome of *S. stipitis*, certain stress proteins were observed before diauxic shift, and these included thioredoxin peroxidase (acts as antioxidant) and glycogen synthase (Gsy1p). The expression of Gsy1p was observed as early as 11.5 hour in both *S. cerevisiae* and *S. stipitis*. It was possible that heat shock, nitrogen starvation or the 2% glucose medium (insufficient glucose) caused the de-repression of Gsy1p [27,28]. Earlier studies suggested *S. cerevisiae* and *S. stipitis* shared similar amino acid biosynthesis pathway. In *S. stipitis* proteome, we observed proteins involved in a list of amino acid biosynthesis pathways that are under general amino acid controls before diauxic shift. These included acetolactate synthase (Ilv2p), dihydroxyacid dehydratase (Ilv3p, 32 hour only), mitochondrial ketol-acid reductoisomerase (Ilv5p) and 3-deoxy-D-arabino-heptulosonate 7-phosphate synthase isoenzyme (Aro4p). Hans and colleagues observed the most drastic alternations of intracellular amino acid pools occurred during the diauxic shift and at the entry into the stationary

phase [29]. We also observed principal enzyme of glyoxylate cycle before diauxic shift, including malate synthase (Mls1.1p) and isocitrate lyase (Icl1p), a key enzyme for growth on ethanol and acetate in yeast. Earlier studies in *S. cerevisiae* showed that *ICL1* expression is tightly regulated and growth on ethanol requires the glyoxylate pathway for replenishing C4 compounds to the tricarboxylic acid cycle. *ICL1* was also induced by growth on ethanol and repressed by growth on glucose [30]. The most peculiar finding in our co-culture process was the observation of the enzymes involved in oxo-reductive pathway for xylose utilization. We observed NAD(P)H-dependent D-xylose reductase (Xyl1p) as early as 11.5- hour before glucose depletion. We also observed D-xylulose reductase as early as 18.5-hour after glucose depletion. Xylose was not presented in the media (confirmed by enzymatic assay). We observed aldehyde dehydrogenase (Ald5p) at 32 hour, but *S. stipitis* was not known to produce acetate. The function of Ald5p remained to be examined. Our metabolic data supported the expression of DL-glycerol-3-phosphatase (Gpp1p) and suggested anaerobic and osmotic stress for *S. stipitis*. Other proteins which suggested stress included Gph1p, glycogen (starch) synthase (Gsy1p), thiol-specific antioxidant protein (Tsa1p), NADPH-dependent methylglyoxal reductase (Gre2p) and thioredoxin reductase (Prx1p) [31,32].

Dynamic of *S. stipitis* during fermentation

Using the spectral counts and the software QSpec (log. fold change cutoff = 3; FDR up-regulated = 0.01) [23], we identified up-regulations of six *S. stipitis* proteins and one *S. cerevisiae* protein from 11.5 hour to 32 hour (Table 2). During the diauxic shift at 32-hour, we observed up-regulation of *S. stipitis* malate synthase and isocitrate lyase, the key enzymes involved in glyoxylate cycle. The depletion of glucose and the transition to growth on ethanol and acetate also triggered the up-regulation of *S. stipitis* phosphoenolpyruvate carboxykinase in gluconeogenesis. Interestingly, *S. stipitis* histone 2B did not exhibit

Organism Accession	Bio1 11.5hr	Bio2 11.5hr	Bio1 32hr	Bio2 32hr	Fold Change (log)	Z statistics	FDR	FDR Up-regulated	Description
S. stipitis 89873	0	0	41	26	3.185	4.94	0.021	0.006	Acetyl-coenzyme A synthetase
66312	0	0	10	40	3.228	5.28	0.01	0.002	Endo-1,4-beta-glucanase
73488	0	0	49	32	3.465	5	0.018	0.005	Histone 2B
53620	0	0	45	64	3.78	6.27	0.001	0	Malate synthase, glyoxysomal
62080	0	0	96	88	4.411	6.47	0	0	Isocitrate lyase
71471	0	0	58	123	4.941	5.32	0.009	0.002	Phosphoenolpyruvate carboxykinase
S. cerevisiae YBL075c	0	0	0	122	3.989	5.95	0.002	0	Stress-Seventy subfamily A

Table 2: Up-regulated proteins of *S. stipitis* and *S. cerevisiae* from 11.5 hour to 32 hour. Differentially expressed proteins were determined using spectral counts and the software QSpec with the cutoff of 3 log fold change and FDR Up-regulated of 0.01.

down-regulation and increased histone acetylation was suggested by up-regulation of acetyl-coA synthetase. With glucose repression no longer existed at 32 hour, we also observed up-regulation of *S. stipitis* endo-1,4-beta-glucanase, a family 5 glycoside hydrolase, without cellulose in the growth media. Another unexpected finding in our proteomics data was the observation of the expressed enzymes involved in oxo-reductive pathway of xylose utilization in *S. stipitis*. NAD(P)H-dependent D-xylose reductase (Xyl1p) and D-xylose reductase (Xyl2p) are repressed under glucose culture [33]. However, we observed *S. stipitis* Xyl1p expression as early as 11.5- hour before glucose depletion. We also observed *S. stipitis* Xyl2p expression as early as 18.5 hour after glucose depletion. Xylose was not present in the media (confirmed by enzymatic assay) and the mechanism for signal derepression of these two enzymes is unknown. Earlier studies have suggested that reducing ATP production can also bring about a partial derepression of xylose assimilation [34]. Further study is needed to investigate and verify the derepression signal for Xyl1p and Xyl2p. We hypothesize that the derepression signal came from the effect of co-culture, which could be a competitive advantage for *S. stipitis* [35] as a hedging bet to utilize other potential carbon sources. Along with the up-regulation of *S. stipitis* endo-1,4-beta-glucanase without cellulose induction, our data demonstrated S. stipites strategy in diversifying potential carbon sources once glucose no longer existed.

Conclusions

In this study, we demonstrated a shotgun proteomics to monitor a co-culture processing with *S. cerevisiae* and *S. stipitis*. Shotgun proteomics successfully distinguished, identified distinct physiology of both yeast during the co-culture process, and able to elucidate the temporal profile of *S. stipitis* proteome. The most abundant proteins of *S. stipitis* and *S. cerevisiae* came from the expected process such as glycolytic enzymes, histones, heat shock proteins, ribosomal proteins and F_1F_0-ATPase. Without the presence of xylose, the induction of NAD(P)H-dependent D-xylose reductase and D-xylulose reductase were unexpected. Up-regulation of *S. stipitis* endo-1,4-beta-glucanase, histone 2B and acetyl-coA synthetase provided insight into the dynamics of non-model yeast *S. stipitis* during the co-culture process.

Acknowledgments

We thank Brian Dill and Collin Olivier for giving us detailed parameters of data-depended and LTQ XL setup. Funding of the project came from McGill University start up fund and Canadian Foundation for Innovation (CFI-IOF grant #23635).

References

1. Kleerebezem R, Van Loosdrecht MCM (2007) Mixed Culture Biotechnology for Bioenergy Production. Curr Opin Biotechnol 18: 207-212.

2. Florens L, Carozza MJ, Swanson SK, Fournier M, Coleman MK, et.al., (2006) Analyzing Chromatin Remodeling Complexes using Shotgun Proteomics and Normalized Spectral Abundance Factors. Methods 40: 303-311.

3. Reis MAM, Serafim LS, Lemos PC, Ramos AM, Aguiar FR, et.al., (2003) Production of Polyhydroxyalkanoates by Mixed Microbial Cultures. Bioprocess Biosyst Eng 25: 377-385.

4. Rodrguez J, Kleerebezem R, Lema JM, Van Loosdrecht M (2006) Modeling Product Formation in Anaerobic Mixed Culture Fermentations. Biotechnol Bioeng 93: 592-606.

5. Ciani M, Comitini F, Mannazzu I, Domizio P (2010) Controlled Mixed Culture Fermentation: A New Perspective on the use of Non-*Saccharomyces* Yeasts in Winemaking. FEMS Yeast Res 10: 123-133.

6. Awafo VA, Chahal DS, Simpson BK (1998) Optimization of Ethanol Production by *Saccharomyces* cerevisiae (ATCC 60868) and Pichia stipitis Y-7124: A Response Surface Model for Simultaneous Hydrolysis and Fermentation of Wheat Straw. J Food Biochemistry 22: 489-509.

7. Taniguchi M, Itaya T, Tohma T, Fujii M (1997) Ethanol Production from a Mixture of Glucose and Xylose By a Novel Co-Culture System with Two Fermentors and Two Microfiltration Modules. J Biosci Bioeng 84: 59-64.

8. Milburn M (2009) Using Metabolic Profiling Technology to Advance Cell Culture Development. Biopharm International Supplements.

9. Pham TK, Chong PK, Gan CS, Wright PC (2006) Proteomic Analysis of *Saccharomyces* Cerevisiae under High Gravity Fermentation Conditions. J Proteome Res 5: 3411-3419.

10. Washburn MP, Wolters D, Yates III JR (2001) Large-Scale Analysis of the Yeast Proteome by Multidimensional Protein Identification Technology. Nat Biotechnol 19: 242-247.

11. Laplace JM, Delgenes JP, Moletta R, Navarro JM (1992) Alcoholic Glucose and Xylose Fermentations by the Coculture Process: Compatibility and Typing of Associated Strains. Can J Microbiol 38: 654-658.

12. Thompson MR, Chourey K, Froelich JM, Erickson BK, VerBerkmoes NC (2008) Experimental Approach for Deep Proteome Measurements from Small-Scale Microbial Biomass Samples. Anal Chem 80: 9517-9525.

13. Wilmes P, Andersson AF, Lefsrud MG, Wexler M, Shah M, et.al., (2008) Community Proteogenomics Highlights Microbial Strain-Variant Protein Expression within Activated Sludge Performing Enhanced Biological Phosphorus Removal. The ISME J 2:853-864.

14. McDonald WH, Ohi R, Miyamoto DT, Mitchison TJ, Yates JR (2002) Comparison of Three Directly Coupled HPLC MS/MS Strategies for Identification of Proteins from Complex Mixtures: Single-Dimension LC-MS/MS, 2-Phase Mudpit, And 3-Phase Mudpit. Int J Mass 219: 245-251.

15. Eng JK, McCormack AL, Yates JR (1994) An Approach to Correlate Tandem Mass Spectra Data of Peptides with Amino Acid Sequences in a Protein Database. J Am Soc Mass Spectrom 5: 976-989.

16. Goltsman DSA, Denef VJ, Singer SW, VerBerkmoes NC, Lefsrud M, et.al., (2009) Community Genomic And Proteomic Analyses of Chemoautotrophic Iron-Oxidizing" Leptospirillum Rubarum"(Group II) and "Leptospirillum Ferrodiazotrophum"(Group III) Bacteria in Acid Mine Drainage Biofilms. Appl Environ Microbiol 75: 4599-4615.

17. Verberkmoes NC, Russell AL, Shah M, Godzik A, Rosenquist M, et.al., (2008) Shotgun Metaproteomics of the Human Distal Gut Microbiota. The ISME J 3:179-189.

18. Cherry JM, Adler C, Ball C, Chervitz SA, Dwight SS, et al. (1998) SGD: *Saccharomyces* Genome Database. Nucl Acids Res 26: 73-79.

19. Jeffries TW, Grigoriev IV, Grimwood J, Laplaza JM, Aerts A, et.al., (2007) Genome Sequence of the Lignocellulose-Bioconverting and Xylose-Fermenting Yeast Pichia stipitis. Nat Biotechnol 25: 319-326.

20. Kanehis, M, Goto S (2000) KEGG: Kyoto encyclopedia of Genes and Genomes. Nucleic Acids Res 28: 27-30.

21. Choi H, Fermin D, Nesvizhskii AI (2008) Significance Analysis of Spectral Count Data in Label-free Shotgun Proteomics. Mol Cell Proteomics 7: 2373-2385.

22. Davidson GS, Joe RM, Roy S, Meirelles O, Allen CP, et.al., (2011) The proteomics of quiescent and nonquiescent cell differentiation in yeast stationary-phase cultures. MBoC 22: 988-998.

23. Fernandez E, Moreno F, Rodicio R (1992) The ICL1 Gene from *Saccharomyces* cerevisiae. Eur J Biochem 204: 983-990.

24. Fu N, Peiris P, Markham J, Bavor J (2009) A Novel Co-Culture Process with Zymomonas mobilis and Pichia stipitis for Efficient Ethanol Production on Glucose/Xylose Mixtures. Enzyme Microb Tech 45: 210-217.

25. Hans MA, Heinzle E, Wittmann C (2001) Quantification of intracellular amino acids in batch cultures of *Saccharomyces* cerevisiae. Appl Microbiol Biotechnol 56: 776-779.

26. Hwang PK, Tugendreich S, Fletterick RJ (1989) Molecular analysis of GPH1, the gene encoding glycogen phosphorylase in *Saccharomyces* cerevisiae. Mol Cell Biol 9: 1659-1666.

27. Unnikrishnan I, Miller S, Meinke M, LaPorte DC (2003) Multiple positive and negative elements involved in the regulation of expression of GSY1 in *Saccharomyces* cerevisiae. J Biol Chem 278: 26450-26457.

28. Pahlman AK, Granath K, Ansell R, Hohmann, S, Adler L (2001) The yeast glycerol 3-phosphatases Gpp1p and Gpp2p are required for glycerol biosynthesis and differentially involved in the cellular responses to osmotic, anaerobic, and oxidative stress. J Biol Chem 276: 3555-3563.

29. Parrou JL, Enjalbert B, Plourde L, Bauche A, Gonzalez B, François J (1999) Dynamic responses of reserve carbohydrate metabolism under carbon and nitrogen limitations in *Saccharomyces* cerevisiae. Yeast 15: 191-203.

30. Peng J, Elias JE, Thoreen CC, Licklider LJ, Gygi SP (2003) Evaluation of multidimensional chromatography coupled with tandem mass spectrometry (LC/LC- MS/MS) for large-scale protein analysis: the yeast proteome. J Proteome Res 2: 43-50.

31. Scott EW, Baker HV (1993) Concerted Action of the Transcriptional Activators REB1, RAP1 and GCR1 in the High-level Expression of the Glycolytic Gene TPI. Mol Cell Biol 13: 543-550.

32. Sunnarborg SW, Miller SP, Unnikrishnan I, LaPorte DC (2001) Expression of the yeast glycogen phosphorylase gene is regulated by stress,Äêresponse elements and by the HOG MAP kinase pathway. Yeast 18: 1505-1514.

33. Shi NQ, Davis B, Sherman F, Cruz J, Jeffries TW (1999) Disruption of the cytochrome c gene in xylose-utilizing yeast Pichia stipitis leads to higher ethanol production. Yeast 15: 1021-1030.

34. Slininger PJ, Thompson SR, Weber S, Liu ZL, Moon J (2011) Repression of Xylose-Specific Enzymes by Ethanol in Scheffersomyces (Pichia) stipitis and Utility of Repitching Xylose-Grown Populations to Eliminate Diauxic Lag. Biotechnol Bioeng 108: 1801-1815.

35. Warner JR (2001) The Economics of Ribosome Biosynthesis in Yeast. Cold Spring Harb Symp Quant Biol 66: 567-574

28

Extraction of Chymotrypsin from Red Perch (*Sebastes marinus*) Intestine Using Reverse Micelles: Optimization of the Backward Extraction Step

Liang Zhou, Suzanne M. Budge, Abdel E. Ghaly*, Marianne S. Brooks and Deepika Dave

Process Engineering and Applied Science Department, Faculty of Engineering, Dalhousie University, Halifax, Nova Scotia, Canada

Abstract

Fish processing waste can be used to produce valuable by-products such as chymotrypsin which has applications in the food, leather, chemical and clinical industries. In this study, a reverse micelles system of AOT/isooctane was used to extract chymotrypsin from crude aqueous extract of red perch intestine. The effects of pH and KCl concentration of the backward extraction step on the total volume (TV), volume ratio (VR), total activity (TA), enzyme activity (A_E), specific activity (SA), purification fold (PF), protein concentration (Cp) and recovery yield (RY) were studied. Changing the pH from 6.5 to 8.5 and the KCl concentration from 0.5 to 2.0 M during the backward extraction step had no effects on the TV or VR. Increasing the pH from 6.5 to 7.5 increased A_E, SA, Cp, PF and RY by up to 47.06%, 30.0%, 27.0%, 26.9% and 18.47%, respectively but they all then declined with further increases in the pH. Similar trends were observed when the KCl concentration was increased from 0.5 to 1.5 M. The decreases in these parameters were due to the denaturation of protein under high pH. The highest A_E, Cp and RY were achieved with pH 7.5 and 1.0 M KCl concentration while the highest SA and PF were achieved with pH 7.5 and 1.5 M KCl concentration. Addition of isobutyl alcohol in the backward extraction step increased the TV, A_E, TA, Cp, SA, PF and RY by 13.6%, 336.4%, 342.6%, 81.1%, 146.4%, 146.2% and 345.8%, respectively. Alcohol reduced the interfacial resistance for the reverse micelles and, thus, destroyed the reverse micelles structure. The values of A_E, TA, SA, PF and RY obtained with reverse micelles methods were much higher (2.3 fold) than those obtained with the ammonium sulphate method.

Keywords: Fish waste; Chymotrypsin; Extraction; Purification; Ultrafiltration; Fractionation; Microemulsions; Enzyme activity; Protein concentration; Recovery yield

Introduction

Currently, fish waste is an approved substance for disposal at sea and the Canadian fish industry is dumping all fish waste into the sea because there is no economical way of utilizing the waste off shore and it is costly to transport the large amount of fish waste to meal plants or land-based waste disposal systems [1,2]. The decomposition of large volumes of wastes lowers the level of dissolved oxygen in the water and generates toxic by-products [3,4]. However, fish waste is a valuable resource and can be used to produce value added products such as docosahexaenoic acid (DHA) and eicosapentaenoic acid (EPA), enzymes (pepsin, trypsin, chymotrypsin), collagen and oil [5-7].

Chymotrypsin has wide applications in food, leather processing, chemical and clinical industries. Industrially, chymotrypsin is produced from fresh cattle or swine pancreas and is commonly made in either a tablet form for oral consumption or as a liquid for injection. The price of chymotrypsin is related to the cost of raw material and the purity of the products. Using fish waste, rather than fresh cattle or swine pancreas, could dramatically lower the cost of chymotrypsin production. Chymotrypsin can be extracted from fish waste using a reverse micelles method. Reverse micelles are thermodynamically stable molecules that can extract large biomolecules like proteins through electrostatic interaction that attracts soluble proteins into the inner layer of the reverse micelles [8,9]. They form amphiphilic structures in polar organic media which can be used to extract large amounts of proteins in the aqueous phase without denaturation.

The reverse micelles extraction process is divided into two steps: forward extraction and back extraction. During the forward extraction step, the aqueous and organic phases are separately prepared and homogenized. After transfer of protein from the aqueous phase to the organic phase, the phases are separated by centrifugation and the protein is measured [8-10]. In the forward extraction step, selection of surfactants and pH play significant roles in protein stabilization. Sodium di-2-ethylhexyl sulfosuccinate (AOT) is the most common surfactant used in chymotrypsin purification [8-11]. pH influences ionic molecular interactions in solution and therefore, influences the efficiency of extraction by reverse micelles [8,10]. In the backward extraction step, protein is transferred from reverse micelles to the aqueous solutions. This step is usually very slow and salt is added into the aqueous phase to assist the process [9]. However, increasing chloride ion concentration will decrease chymotrypsin yield by competing with chymotrypsin in the extraction process and the effect is particularly significant at low ionic strength [8,11]. According to Goto et al. [12] and Hu and Gulari [9], the limitations of the backward extraction step are due to the difficulty in separating proteins from AOT reverse micellar phase and the excessive time involved in the process.

Objectives

The aim of this study was to optimize the backward extraction step of the reverse micelles method while purifying chymotrypsin from fish processing waste. The specific objectives were: (a) to study the effect of pH (6.5, 7.0, 7.5, 8.0 and 8.5) and salt concentration (0.5, 1.0, 1.5 and 2.0 M) and alcohol addition in the backward extraction step of the reverse micelles method on the enzyme activity (A_E), protein concentration (Cp), specific activity (SA), purification fold (PF) and recov-

*Corresponding author: Abdel E Ghaly, Professor, Department of Process Engineering and Applied Science, Dalhousie University, Halifax, Nova Scotia, Canada

ery yield (RY)and (b) compare the effectiveness of the reverse micelles method to that of the ammonium sulphate precipitation method.

Materials and Methods

Chemicals and reagents

Tris, HCl, CaCl$_2$, NaCl, ammonium sulphate, N-benzoyl-L-tyrosine ethyl ester (BTEE) and the release, methyl alcohol and n-butyl alcohol were obtained from Sigma-Aldrich, Oakville, Ontario, Canada. AOT, isooctane (2,2,4-trimethylpentane), chymotrypsin, isobutyl alcohol and BSA (bovine serum albumin) were obtained from Fisher scientific, Ottawa, Ontario, Canada. Reagents used included 0.05M Tris-HCl buffer (0.01 M CaCl$_2$, pH 7.5), 15% v/v isobutyl alcohol, 10 μM BTEE.

Sample collection and preparation

The fish, red perch (*Sebastes marinus*), used in the experiment were collected from Clearwater Seafood's Ltd., Halifax, Nova Scotia, Canada. The intestines were separated from fish, washed with cold water and isotonic saline solution to remove the blood in the tissue according to the procedure described by Chong et al. [13] and Boeris et al. [14]. The fish intestine was chopped into small pieces (1 cm^3), weighed, marked and stored at -20°C.

Experimental design

Tables 1 and 2 show the experimental parameters investigated in this study and their levels. In the forward step pH of 8.0 and KCl concentration of 1.5M was applied [10]. In the backward extraction step, the effect of pH (6.5, 7.0, 7.5, 8.0 and 8.5) and four KCl concentration

Factors	Parameters
Forward Extraction	
AOT	20 mM
pH	7.0
Time	30 min
Temperature	4°C
Backward Extraction	
Salt concentration	0.5, 1.0, 1.5 and 2 M
pH	6.5, 7.0, 7.5, 8.0, 8.5
Time	1 hr
Temperature	4°C

No. of replicates = 3
No. of runs = 60

Table 1: Optimization of the backward extraction (changing pH and KCl concentration).

Factors	Parameters
Forward Extraction	
AOT	20 mM
pH	7.0
Time	30 min
Temperature	4°C
Backward Extraction	
Salt concentration	Optimum
pH	Optimum
Isobutyl alcohol	15% v/v alcohol or 15% v/v distilled water
Time	1 hr
Temperature	4°C

No. of replicates=3
Total no. of runs= 6

Table 2: Effect of alcohol addition during backward extraction.

(0.5, 1, 1.5 and 2 M) were studied. After optimization of the backward step, the effect of alcohol addition (15% v/v) on chymotrypsin recovery yield was studied. Finally, both the reverse micelles extraction method and the ammonium sulphate extraction method were compared on the basis of enzyme activity, specific activity purification fold and recovery yield.

Crude enzyme extraction

The extraction procedure (Figure 1) described by Heu et al. [15] and Castillo-Yáñeza et al. [16] was followed. The prepared frozen samples were thawed at 4°C overnight. A 50 g (wet basis) sample of fish gut was mixed with 150 mL isotonic saline solution and homogenized using a laboratory homogenizer (Polytron PT1035, Brinkmann Instruments, Toronto, Ontario, Canada) for 5 min, then incubated for 8 hr at 4°C to activate the chymotrypsinogen in the samples. After incubation, the sample was centrifuged at 20,000 g at 4°C for 30 min (MP4R, International Equipment Company, Needham, Massachusetts), then filtered and defatted with 50 mL CCl$_4$. The supernatant was considered a crude enzyme extract. The volume at each step was measured and the activity and concentration of crude enzyme were determined.

Ammonium sulphate extraction

The procedure (Figure 1) described by Kunitz [17] was followed. Ammonium sulphate was slowly added to the crude extract to reach 35% saturation with continuous stirring. The mixture was stirred for a further 30 minutes at 4°C and then centrifuged at 20,000 g for 15 min (MP4R, International Equipment Company, Needham, Massachusetts). The supernatant was collected and the saturation was then adjusted to 70% by addition of ammonium sulphate. After 30 min, the suspension was centrifuged at 20,000 g for 15 min. The pellets collected from the 35% and 70% saturation steps contained precipitated enzymes. Dialysis was performed on the pellets using Tris-HCl buffer. The enzyme concentration, activity and yield were then determined.

Reverse micelles extraction

The process of extracting chymotrypsin from fish waste by the reverse micelles method is shown in Figure 1. The forward extraction step was carried out usingan AOT concentration of 20 mM and at a pH of 7.0. In the forward extraction step, crude enzyme was mixed with AOT organic solution and stirring at 300 rpm using magnetic stirrer for 30 minutes and then separating the two phases by centrifugation. Aqueous solutions containing different KCl concentrations (0.5, 1, 1.5 and 2M) were prepared. Tris buffer was used as a stripping solution in the backward transfer and the pH of the solution was adjusted to 6.5, 7.0, 7.5, 8.0 or 8.5 as recommended by Hu and Gulari [9]. Equal volumes of aqueous solution and organic solution were mixed in the tube and 15% v/v isobutyl alcohol was added to the reversed micelles phase. Then the two phase systems were mixed for one hour in a beaker placed on a magnetic stirrer (Canlab NO. S8290, Atlanta GA, Georgia, USA). The mixture was then centrifuged for 15 min at 4000 rpm (MP4R, International Equipment Company, Needham, Massachusetts) in order to separate into two phases [9,12]. The enzyme activity and yield were determined. During the backward extraction step the salt concentration and pH were changed and the same procedure was followed.

Effect of alcohol

After the optimal conditions for the backward extraction were determined, the effect of alcohol addition was studied. Distilled water was used as a control. The same volumes (1.5 mL) of distilled water and isobutyl alcohol were added to the aqueous phase during backward extraction.

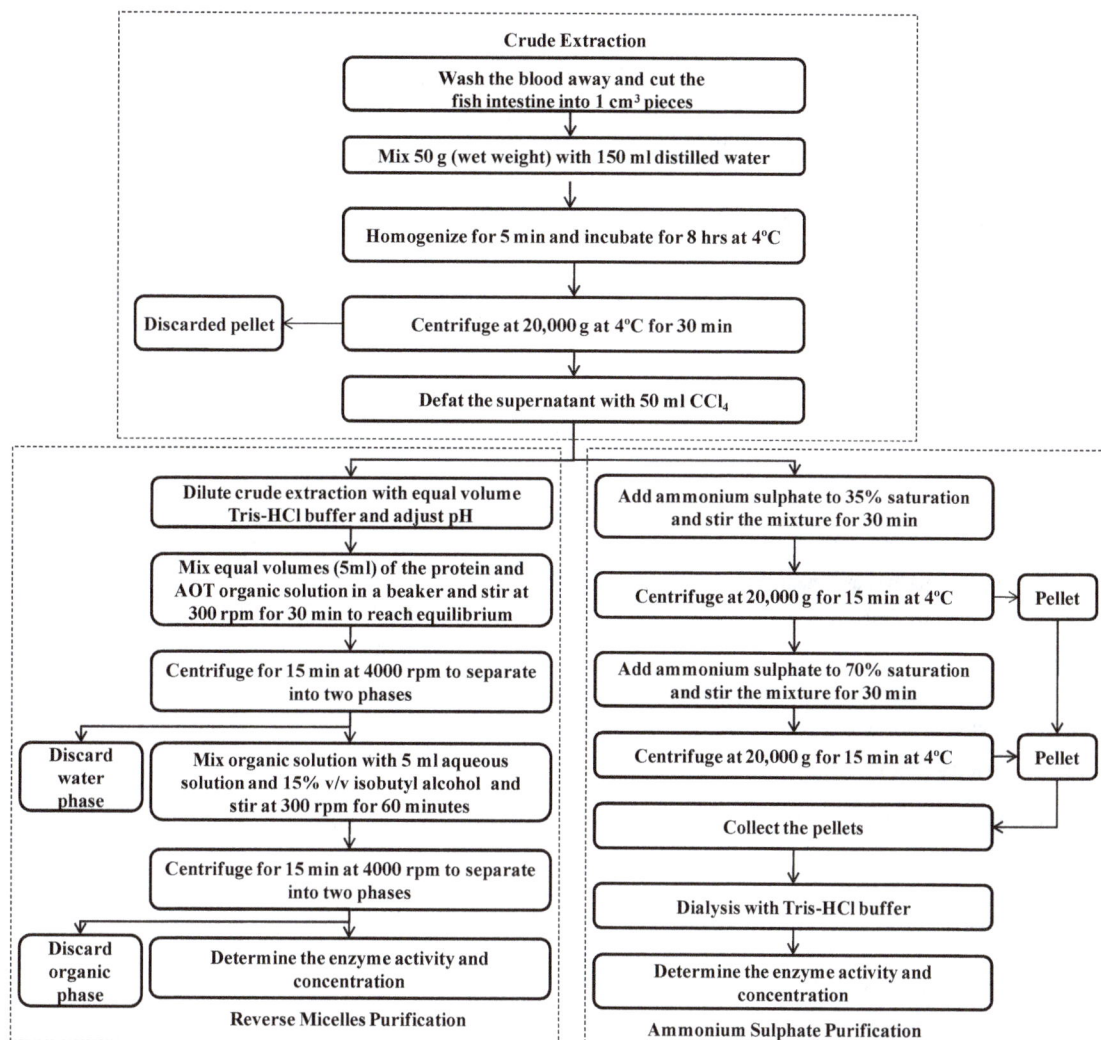

Figure 1: The extraction procedure of chymotrypsin from fish waste.

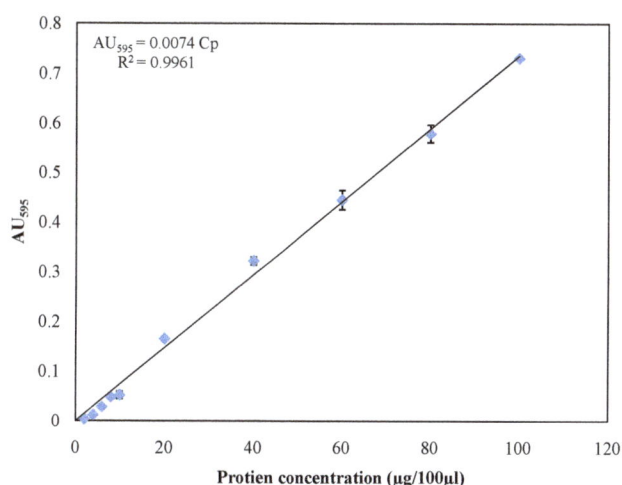

Figure 2: The standard curve for protein concentration.

Determination of protein concentration

The Bradford method was used for the determination of protein concentration according to the procedures described by Yang et al. [18] and Castillo-Yáñeza et al. [16]. Two standard curves were developed using a series of concentrations of bovine serum albumin (BSA): the standard curve (assay range 10-150 μg/mL) and a micro standard curve (assay range 1-10 μg/mL). The following solutions were prepared: (a) 0.1 g of BSA was dissolved in 10 mL of Tris-HCl buffer at room temperature (b) the stock BSA solution was diluted to span the 100- 1,500 μg/mL range and (c) BSA solution in the range of 100-900 g/mL was diluted ten times more for the micro standard curve. The two standard curves are combined together as shown in Figure 2.

10 μL of each standard was mixed with 5 mL of Bradford reagent. Each sample was allowed to incubate at room temperature for 10 minutes and the absorbance of each standard was measured at 595 nm against a blank that was composed of 10 μL of buffer and 5 mL of Bradford reagent. 0.1 mL diluted sample (concentration between 5 to 100 μg/L) was mixed with 5 mL Bradford reagent, incubated for 5

min and then the absorbance was measured at 595 nm [19]. The result was compared with the standard curve to determine the sample protein concentration.

Determination of enzyme activity

The activity of chymotrypsin was defined as the change of absorbance measured at 256 nm in one minute caused by the addition of 1 mL of chymotrypsin protein solution [20]. The substrate used in the experiment was benzoyl- tyrosine ethyl ester (BTEE). The p-nitroaniline was cleaved by BTEE and the release of N-benzoyl-L-tyrosine was followed by recording the increase in the absorbance every min for 5 min at 256 nm [13]. 1.5 mL Tris-HCl buffer (0.08 M tris, pH 7.8, 0.1M $CaCl_2$), 1.4 mL of 0.00107 M BTEE and 0.1 mL test enzyme solution were placed into cuvettes. The enzyme activity was calculated as follows.

$$A_E = \frac{\Delta U_{256} \times (3) \times (Df)}{(0.964) \times (0.10)} \quad (1)$$

Where:

A_E : Enzyme activity (Units/mL Enzyme)

ΔU_{256} : The change of the absorbance at the wave length 256 nm per minute

3 : Volume of reaction mixture (mL)

Df : Dilution factor

0.964 : Millimolar extinction coefficient of BTEE at 256 nm

0.10 : Volume of test enzyme solution used in assay (mL)

Determination of total activity (TA)

The total activity is defined as the change in the absorbance value per min for the total chymotrypsin extracted from the entire sample of red perch intestine. The total activity was calculated as follows.

$$TA = \frac{\Delta U_{256} \times (3) \times (Df)}{(0.964) \times (0.10)} \times \text{Total volume volume of crude extraction} \quad (2)$$

Determination of specific activity (SA)

Specific activity is defined as the ability of 1 mg enzyme to hydrolysis BTEE in one min at a pH of 7.5 and a temperature of 25°C. 0.1 mL enzyme solution was added into cuvettes and the change in absorbance

was measured at 256 nm every half minute for 5 minutes. The specific activity was calculated using the following equation.

$$S_A = \frac{\text{Units / mL Enzyme}}{\text{mg protein / mL Enzyme}} \quad (3)$$

Where:

S_A: Specific activity (Units/mg Protein)

Determination of purification fold (PF)

Purification fold is used to evaluate the increase in purity of the enzyme after the purification step. It was calculated using the following equation:

$$PF = \frac{\text{Units / mg Purified protein}}{\text{Units / mg Crude protein}} \quad (4)$$

Determination of recovery yield (RY)

Recovery yield is defined as the ratio of total refined enzyme activity and total crude enzyme activity. Recovery represents the chymotrypsin activity remaining in the purification process. When combined with specific activity, it can show the effectiveness of a purification method.

$$RY = \frac{\text{Units / mL Purified protein}}{\text{Units / mL Crude protein}} \times 100\% \quad (5)$$

Statistical analysis

The data for solution volume protein concentration and activity

Extraction step	TV (mL)	AE (Unit/mL)	TA (Unit)	Cp (µg/mL)	SA (Unit/mg)	PF (-)	RY (%)
After Homogenizing*	183	0.80	146.40	4486.20	0.18	-	-
AfterCentrifuging	144	0.94	135.36	1975.80	0.48	2.69	92.70
AfterDilution and pH adjustment	288	0.47	135.36	987.90	0.48	-	100.00

*Sample size: 50 g

Table 3: Crude protein extraction parameters.

Parameter	Purified chymotrypsin
A_E (Unit/mL)	0.22±0.01
TA (Unit)	1.10±0.05
Cp (µg/mL)	41.64±1.16
SA (Unit/mg)	5.31±0.19
PF (-)	11.10±0.39
RY (%)	46.72±2.23

*Sample size: 5 mL

Table 4: Extraction parameters after ammonium sulphate precipitation.

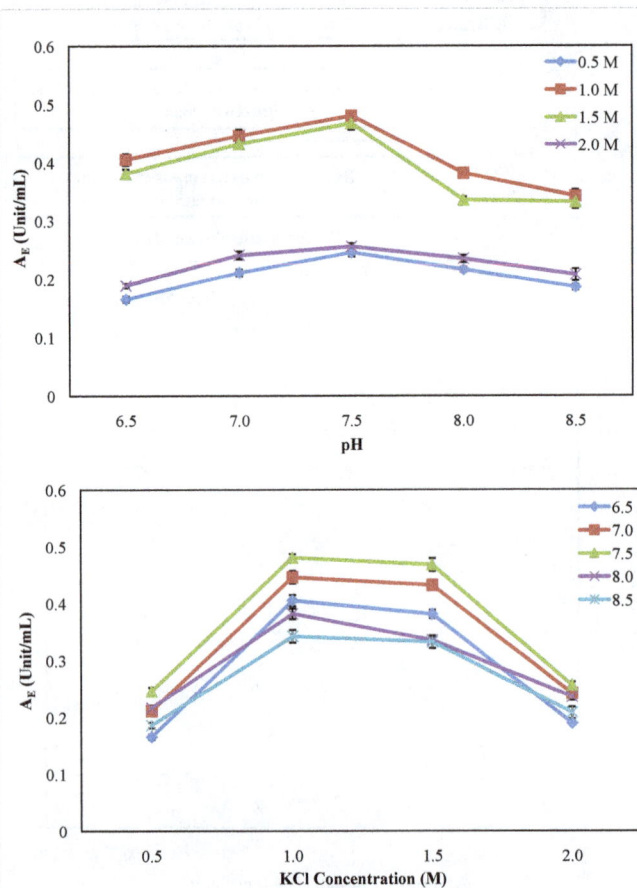

Figure 3: Effects of pH and KCl concentration on the enzyme activity.

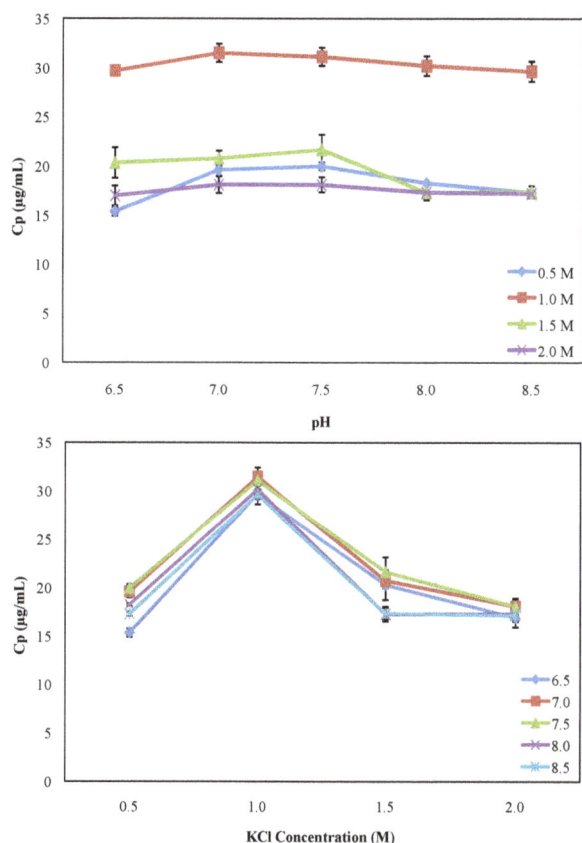

Figure 4: Effects of pH and KCl concentration on the protein concentration.

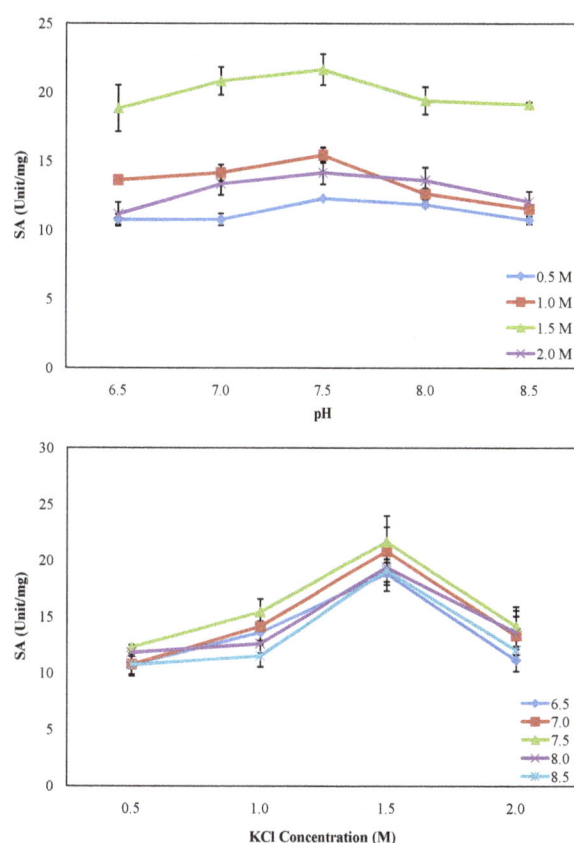

Figure 5: Effects of pH and KCl concentration on the specific activity.

were collected and total activity, recovery yield, specific activity and standard errors were calculated. The α-level was chosen as 0.05. All the statistical analysis of data was conducted using Minitab statistics software (Ver 15.1.10, Minitab Inc) to examine the coefficient data with a two-way analysis of variance (ANOVA) to determine the significant effects of single and two parameters on the results.

Results

Crude extraction

Crude protein was extracted from the intestine (50 g) of red perch and the total volume (TV) was measured after homogenization, centrifugation and dilution. The enzyme activity (A_E), total activity (TA), specific activity (SA), protein concentration (Cp), purification fold (PF) and recovery yield (RY) were determined as shown in Table 3. After centrifugation, the TV, TA and Cp decreased from 183 to 144 mL (21.32%), from 146.40 to 135.36 U (7.54%) and from 4486.2 to 1975.8 µg/mL (55.96%), while the A_E and SA increased from 0.80 to 0.94 U/mL (17.50%) and from 0.18 to 0.48 U/mg (169.1%), respectively. The purification fold and the recovery yield were 2.69 and 92.7%, respectively.

Ammonium sulphate extraction

The crude extract was purified using the ammonium sulphate precipitation method. The A_E, TA, Cp, SA, PF and RY were determined (Table 4). The A_E, TA and Cp values of the purified enzyme were lower than those of the crude enzyme. The A_E, TA and Cp decreased from 0.47 to 0.22 Unit/mL (53.3%), from 2.37 to 1.10 Unit/mL (53.3%) and from 987.90 to 41.64 µg/mL (95.8%), respectively. On the other hand, the SA and PF of purified enzyme were much higher than those of

crude enzyme. The SA increased from 0.48 to 5.31 Unit/mg (1012%) and the PF increased from 2.69 to 11.1 (313.64%). The final yield was 46.72%.

Reverse micelles extraction

The effects of salt concentration and pH on the backward extraction parameters (A_E, Cp, SA, PF and RY) were studied. Four levels of KCl salt concentrations (0.5, 1.0, 1.5 and 2.0 M) and five levels of pH (6.5, 7.0, 7.5, 8.0 and 8.5) were investigated. The results are shown in Figures 3-7. The analysis of variance performed on the data (Table 5) indicated that salt concentration and pH were highly significant at the 0.0001 level. The interaction between the salt concentration and pH was highly significant at the 0.0001 level. The highest values for the backward extraction parameters are shown in Table 6.

The volumes for water phase and organic phase remained constant regardless of the pH level and KCl concentration used. The TV was 11.5 mL and the VR was 0.77. The A_E increased initially when the pH and KCl were increased and then decreased with further increases in pH and KCl concentration (Figure 3). The maximum A_E value (0.48 unit/mL) was observed at pH 7.5 and KCl concentration of 1.0 M. The Cp showed similar trend with the maximum value (31.13 µg/mL) observed at pH 7.5 and KCl concentration of 1.0 M (Figure 4). The SA and PF showed similar behaviours with maximum values (21.67 unit/mg and 45.23) observed at pH of 7.5 and KCl concentration of 1.5 M (Figures 5 and 6). The RY had a maximum value (102.24%) at pH of 7.5 and KCl concentration of 1.0 M. These results showed that the optimum condition for the backward extraction step of chymotrypsin is at a pH of 7.5 and KCl concentration of 1.0 M and 1.5 M.

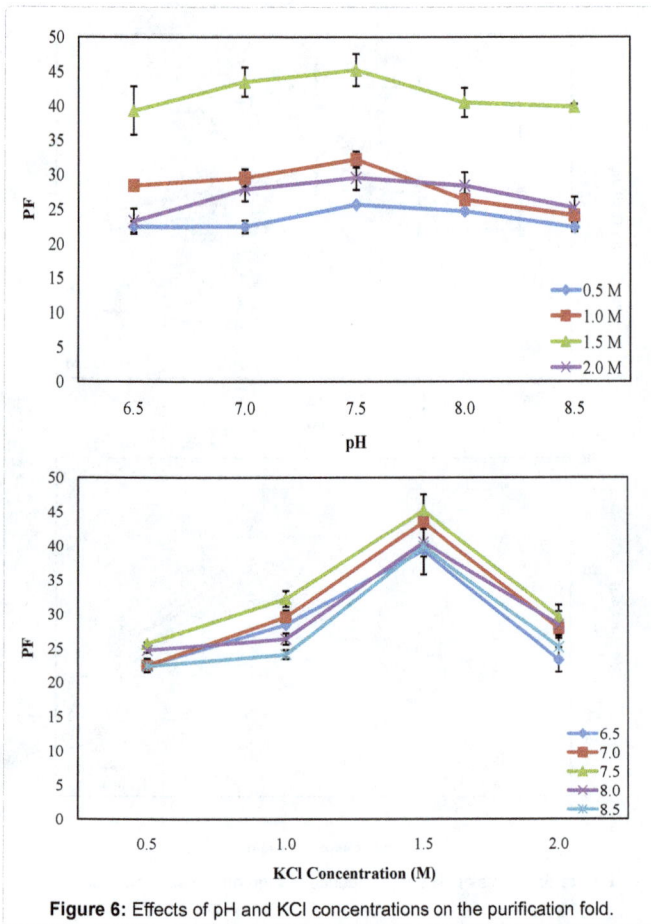

Figure 6: Effects of pH and KCl concentrations on the purification fold.

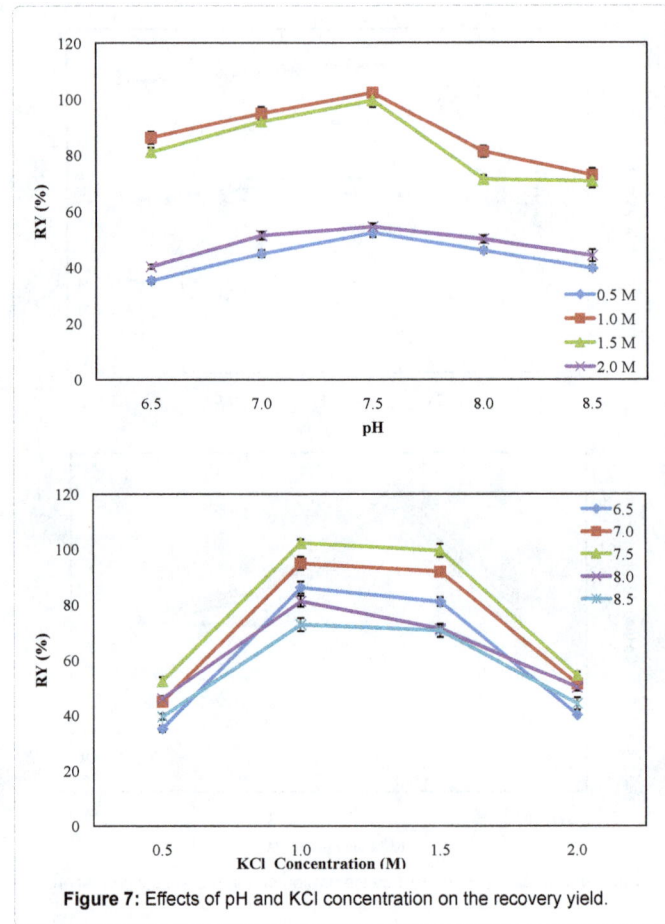

Figure 7: Effects of pH and KCl concentration on the recovery yield.

Alcohol effect

In order to determine the effect of alcohol in the backward extraction step, an experiment was carried out at the optimal conditions for forward (AOT 20 mM, and pH 7.0) and backward (KCl 1.0 M and pH 7.5) extraction in which isobutyl alcohol was added in the backward extraction step. Another experiment without alcohol was used as a control. The TV, AE, TA, Cp, PF and RY obtained form both experiments are shown in Table 7. When alcohol was added in the backward extraction step, the TV, A_E, TA, Cp, SA, PF and RY increased from 4.4 to 5.0 mL (13.6%), from 0.11 to 0.48 Unit/mL. (336.4%), from 0.54 to 2.39 U (342.6%), from 17.60 to 31.87 µg/mL (81.1%), from 6.08 to 14.98 Unit/mg (146.4%), from 12.70 to 31.27 (146.2%) and from 22.62% to 100.85% (345.8%), respectively

Comparing RM with AS

The reverse micelles (RM) method was compared to the ammonium sulphate (AS) method on the basis of their AE, TA Cp, SA, PF and RY (Table 8). The value of extraction parameters obtained with RM were much higher than those obtained with AS. The A_E, TA, and RY obtained with RM were 2 fold higher than those obtained with the AS method, while the SA and PF were 3 fold higher than those obtained with AS.

Discussion

Extraction profiles

After homogenization and centrifugation, the A_E, TA, Cp and RY

Parameters	Source	DF	SS	MS	F	P
A_E	Total	107	0.74882			
	pH	5	0.1533	0.03066	52.87	0.0001
	AOT	5	0.50872	0.10174	175.43	0.0001
	pH-AOT	25	0.04504	0.0018	3.11	0.0001
	Error	72	0.04176	0.00058		
Cp	Total	107	2496.22			
	pH	5	299.76	59.952	72.25	0.0001
	AOT	5	1930.26	386.051	465.26	0.0001
	pH-AOT	25	206.46	8.258	9.95	0.0001
	Error	72	59.74	0.83		
SA	Total	107	340.295			
	pH	5	105.573	21.1027	19.04	0.0001
	AOT	5	133.499	26.6998	24.09	0.0001
	pH-AOT	25	21.482	0.8593	0.78	0.758
	Error	72	79.801	1.1083		
PF	Total	59	3268.06			
	pH	4	213.18	53.294	21.89	0.0001
	AOT	3	2871.22	957.074	393.15	0.0001
	pH-AOT	12	86.28	7.19	2.95	0.0050
	Error	40	97.38	2.434		
RY	Total	59	28097.9			
	pH	4	3278	819.49	291.02	0.0001
	AOT	3	23562.5	7854.18	2789.23	0.0001
	pH-AOT	12	1144.7	95.39	33.88	0.0001
	Error	40	112.6	2.82		

pH = Mean effect of pH
AOT = Mean effect of AOT
pH-AOT=interaction between pH and AOT

Table 5: Analysis of variance for the various parameters.

Parameters	pH	KCl	Optimum values
A_E (Unit/mL)	7.5	1.0 M	0.48±0.06
Cp (µg/mL)	7.5	1.0 M	31.13±0.93
SA (Unit/mg)	7.5	1.5 M	21.67±1.11
PF(-)	7.5	1.5 M	45.23±2.32
RY (%)	7.5	1.0 M	102.24±1.35

*Sample size: 5 mL

Table 6: The optimum values of chymotrypsin purification parameters.

Parameters	Without alcohol	With alcohol
A_E (Unit/mL)	0.11±0.01	0.48±0.01
TA (Unit)	0.54±0.05	2.39±0.05
Cp (µg/mL)	17.60±5.48	31.87±1.31
SA (Unit/mg)	6.08±0.19	14.98±0.49
PF	12.70±0.39	31.27±1.02
RY (%)	22.62±2.23	100.85±2.23

Forward extraction: AOT concentration 20 mM, pH7.0
Backward extraction: KCl concentration 1.5 M, pH7.5
*Sample size: 5 mL

Table 7: Effect of alcohol addition on chymotrypsin purification parameters.

Parameters	AS	RM
A_E (Unit/mL)	0.22±0.01	0.48±0.01
TA (Unit)	1.10±0.05	2.39±0.05
Cp (µg/mL)	41.64±5.48	31.87±1.31
SA (Unit/mg)	5.31±0.19	14.98±0.49
PF	11.08±0.39	31.27±1.02
RY (%)	46.72±2.23	100.85±2.23

Forward extraction: AOT concentration 20 mM, pH7.0
Backward extraction: KCl concentration 1.5 M, pH7.5
*Sample size: 5 mL

Table 8: Optimum values of the AS and RM purification parameters.

decreased while the SA and PF increased which indicated that a portion of chymotrypsin was lost during the crude extraction process but the remaining portion was concentrated during the extraction.

Reverse micelles method

During the backward extraction step, salt was required to break the reverse micelles structure in the organic phase to release chymotrypsin into the aqueous phase. However, high salt concentrations will denaturate proteins [21]. The pH on the other hand affects the net charge of protein molecules and the electrostatic interaction force between the chymotrypsin and the surfactant. This in turn influences the extraction efficiency of the process [9,21-23]. In the present study, the effects of pH (6.5, 7.0, 7.5, 8.0 and 8.5) and KCl concentration (0.5, 1.0, 1.5, and 2.0 M) were investigated in order to determine the optimum conditions for the backward extraction step.

During the backward extraction step, the volumes for the organic and aqueous phases remained constant regardless of the pH and KCl concentration used. Thus, the total volume for each of the five pH levels and the four KCl concentrations was 11.5 mL and the VR of the two phases remained constant at 0.77. Several researchers reported similar results [9,12,21,22,24].

It has been reported that changes in the structure of proteins and the low rate of extraction are the main problems with the backward extraction step. Changing the pH and the concentration of salt in the aqueous phase are the most common methods applied to improve A_E, Cp, SA, PF and RY of the backward extraction [24-27]. In the study, A_E,

Cp, SA, PF and RY reached their maximum values at pH 7.5 and KCl concentration of 1.5 M.

Goto et al. [12] reported that during the backward extraction of chymotrypsin using 200 mM AOT concentration, an increase in A_E (from 1 to 6 Unit/mL) was observed when the pH was increased from 4.0 to 7.5 which was followed by a decrease (to 0.5 Unit/mL) with further increases in the pH (to 11.0). When the positive net charge on the chymotrypsin surface decreased, the electrostatic interaction between the protein and the negative AOT molecular head became weaker and chymotrypsin was released into the aqueous phase [12].

When the pH of the aqueous phase approaches the isoelectric point (*pI*) of proteins, the Cp increases because of the electrostatic interaction between the protein and AOT reverse micelles is weakened and more protein molecules are released from the reverse micelles into the aqueous phase. However, further increases in the pH result in decreasing Cp because of protein denaturation. Ono et al. [21] reported that when the pH was increased during the extraction of haemoglobin using dioleyl phosphoric acid (DOLPA), the backward extraction rate dramatically increased reaching 90% at a pH of 8.0 and then declined with further increases in the pH. Similar results had been observed by Goto et al. [12].

RY is one of the most important parameters used in evaluating the extraction process. Generally, the RY of the backward extraction is relatively lower than the forward extraction due to the strong interaction between the protein and reverse micelles [9,11,12,21,23,24,28]. Several researchers reported similar effects of pH in the backward extraction on RY [12,21,23,28]. Hebbar et al. [28] reported that during the extraction of bromelain from pineapple waste, RY increased from 68 to 100% when the aqueous phase pH was increased from 3.9 to 4.2 and then decreased with further increases in pH. Goto et al. [12] reported a 100% RY of chymotrypsin using AOT-DOLPA at pH of 7.0. The reason that RY was over 100% in this study could be due to the presence of impurities with chymotrypsin.

Chang et al. [23] stated that the KCl concentration had significant effects on the radius of reverse micelles and changing the salt concentration affected A_E, Cp, SA, PF and RY. They reported that when KCl concentration was increased from 0.2 to 0.8 M, the radius of reverse micelles decreased from 62 to 41 Å which in turn decreased protein solubility and was responsible for the release of protein to the aqueous phase. High salt concentration resulted in unstable reverse micelle structures and led to increases in the A_E, Cp, SA, PF and RY. Hebbar et al. [28] reported that during the backward extraction step of extracting bromelain from pineapple waste, the A_E increased when the KBr concentration was increased from 0.25 to 0.50 M and then decreased when the KBr concentration further increased from 0.50 to 0.75 M. Hatton et al. [29] and Dekker et al. [25] also found that a high ionic strength in the aqueous phase was not good for protein extraction in the backward extraction step. Hong et al. [24] extracted BSA, carbon anhydrase and β-lactoglubulin from AOT reverse micelles using low ionic strength conditions in the aqueous phase (0.1M KCl) and found that high ionic strength could result in protein denaturation. High ionic strength was considered to be a salt concentration >1 M. High salt concentrations can destabilize reverse micelles and release target proteins back to the aqueous phase and as such increase SA. Hebbar et al. [28] reported that when the aqueous phase pH was increased from 3.9 to 4.2 during the backward extraction of bromelain from pineapple waste, SA and PF increased from 22.6 to 56.15 CDU/mg and from 2.1 to 5.3, respectively.

Then, they decreased to 25 CDU/mg and 2.4 when the pH was further increased to 4.5. They also reported that when the KBr concentration increased from 0.25 to 0.50 M, SA increased from 20 to 56 CDU/mg and then decreased to 23 CDU/mg and PF increased from 2.1 to 5.6 and then decreased to 2.4 when the KBr concentration was further increased to 0.75 M. Hu and Gulari [9] reported a RY of 67.6% during the extraction of α-chymotrypsin using sodium bis (2-ethylhexyl) phosphate (NaDEHP) with a $CaCl_2$ concentration of 0.1 M. Hentsh et al. [11] reported a 100% RY during the backward extraction of chymotrypsin using 1.0 M KCl at pH of 8.0. In this study, SA and PF decreased when KCl concentration increased above 1.5 M, a higher concentration than that reported by others.

The optimum pH and KCl concentration that gave the highest A_E, Cp, SA, PF and RY for the backward extraction are shown in Table 4. The highest A_E, Cp and RY were reached at the pH 7.5 and KCl concentration of 1.0 M while the highest SA and PF were reached at the pH 7.5 and KCl of concentration 1.5 M. Since RY has been considered the most important parameter in evaluation of the extraction process, pH of 7.5 and KCl concentration of 1.0 M were chosen as the optimum backward extraction conditions. The optimal pH applied in the backward extractions (pH 7.5) is higher than that in forward extraction (pH 7.0). Ono et al. [21] reported pH 6.5 and 8.0 for the forward and backward extraction during the extraction steps of haemoglobin using DOLPA. Hentsch et al. [11] used pH 5.0 and 8.0 for forward and backward extraction steps during the extraction of chymotrypsin using AOT. Goto et al. [12] found the optimal pH conditions for extraction of chymotrypsin using AOT-DOLPA mixed reverse micelle system to be pH 6.8 for forward extraction and pH 7.0 for backward extraction.

Effect of alcohol on backward extraction

Adding alcohol in the backward extraction step increased TV, A_E, TA, Cp, SA, PF and RY. Hu and Gulari [9], Goto et al. [12], Ono et al. [21], Paradkar and Dordick [22] and Hong et al. [24] reported that a clear phase was quickly obtained in the presence of alcohol after stopping the stirring process and with the addition of 10-20%(v/v) alcohol in the backward extraction step, the protein transfer from reverse micelles was 10 times faster than in the absence of alcohol. Paradker and Dordick (1993) added 10%(v/v) ethyl acetate in the backward extraction step and noticed significant increases in A_E, TA and RY. Goto et al. [12] reported that without the addition of alcohol, 24 hours were required to obtain equilibrium in the back-extraction step and the extraction time was reduced to less than 2 hours by adding 10% (v/v) isobutyl alcohol. Hong et al (2000) reported that adding 10-15% isopropanol in the backward extraction step resulted in 100% extraction of pepsin and 70% extraction of chymotrypsin. Hu and Gulari [9] reported that only 10-20% RY were obtained using the NaDEHP reverse micelle system without the addition of alcohol in the backward extraction but 98% of active cytochrome-c and 67% of active chymotrypsin were recovered from the aqueous phase in the presence of alcohol.

Goto et al. [12] studied the effect of alcohol type on the RY and relative activity of recovered chymotrypsin. Their results showed that the RY of the backward extraction with isopropyl alcohol, isobutyl alcohol, isoamyl alcohol, n-hexyl alcohol, n-octyl alcohol, n-decanol and oleyl alcohol were 93.8%, 97.1%, 90.7%, 84.5%, 67.4%, 59.9% and 37.1%, respectively. The relative activity (recovered specific activity/original specific activity) for recovered chymotrypsin in the backward extraction using isopropyl alcohol, isobutyl alcohol, isoamyl alcohol, n-hexyl alcohol, n-octyl alcohol, n-decanol and oleyl alcohol were 0.03, 1.00, 0.74, 0.68, 0.24, 0.10 and 0.55 respectively.

Ono et al. [21] studied the effect of alcohol type and concentration in the backward extraction step using methanol, ethanol, isopropyl alcohol (IPA) and n-propyl alcohol (nPA) at concentrations of 0-30% (v/v). When the IPA and nPA were added in the back-extraction step, the hemoglobin recovery rate increased significantly from 0 to 60% when alcohol concentration was increased from 5 to 10% (v/v) and then decreased with further increases in the alcohol concentration. The hemoglobin recovery rate dramatically increased from 0 to 70% (v/v) when the ethanol concentration was increased from 10 to 15% (v/v) and then decreased when the concentration was further increased. The recovered hemoglobin rate dramatically increased from 0 to 70% (v/v) when the methanol concentration in the back extraction was increased from 20 to 30% (v/v). Alcohol can reduce the interfacial resistance for the reverse micelles because it promotes the fusion/fission of reverse micelles which destabilizes the structure.

Conclusions

Changing the pH from 6.5 to 8.5 and the KCl concentration from 0.5 to 2.0 M during backward extraction step (with addition of alcohol) had no effects on the total volume (TV) or the volume ratio (VR). The TV for all samples was 6.5 mL and the VR was 0.77. The reverse micelles emulsion structure was destroyed in the presence of alcohol. Increasing the pH from 6.5 to 7.5 increased the enzyme activity (A_E) by up to 47.06%, protein concentration (Cp) by up to 30.0%, specific activity (SA) by up to 27.0%, purification factor (PF) by up to 26.9% and recovery yield (RY) by up to 18.47% and they all then declined with further increases in the pH. Similarly, increases in the KCl concentration from 0.5 to 1.0 M increased A_E by up to 192.9%, Cp by up to 93.2% and RY by up to 50.97% and they all then decreased with further increases in the KCl concentration. SA and PF continued to increase up to 93.3% when the KCl concentration increases from 0.5 to 1.5 M and then decreased with further increases in the KCl concentration. The decreases in A_E, Cp, SA, PF and RY were due to the denaturation of protein under a relatively high pH and the ionic strength caused by high pH and KCl concentration in the backward extraction step. The highest A_E, Cp and RY were achieved with pH of 7.5 and 1.0 M KCl concentration. The highest SA and PF were achieved with pH of 7.5 and 1.5 M KCl concentration. The optimal conditions for the backward extraction step was pH 7.5, KCl concentration 1.0 M. Addition of alcohol in the backward extraction step increased TV by 13.6%, A_E by 336.4%, TA by 342.6%, Cp by 81.1%, SA by 146.4%, PF by 146.2% and RY by 345.8% because alcohol reduced the interfacial resistance for the reverse micelles. The values of A_E, TA, SA, PF and RY obtained with reverse micelles methods were much higher (2.3 fold) than those obtained with the AS method.

Acknowledgements

The project was funded by the Natural Sciences and Engineering Research Council (NSERC) of Canada through a Strategic Grant.

References

1. PNPPRC (1993) Pollution prevention opportunities in the fish processing industry, Pacific Northwest Pollution Prevention Research Center, Seattle, Washington, USA.

2. AMEC (2003) Management of wastes from Atlantic seafood processing operations, National programme of action Atlantic regional team, AMEC Earth & Environmental Limited, Dartmouth, Nova Scotia, Canada.

3. Bechtel PJ (2003) Properties of different fish processing by-products from pollock, cod and salmon. J Food Process Preserv 27: 101-116.

4. Gumisiriza R, Mashandete AM, Rubindamayugi MST, Kansiime F, Kivaisi AK (2009) Enhancement of anaerobic digestion of Nile perch fish processing wastewater. Afr J Biotechnol. 8: 328-333.

5. Byun HG, Park PJ, Sung NJ, Kim SK (2002) Purification and characterization of a serine proteinase from the tuna pyloric caeca. J Food Biochem 26: 479-494.

6. Swatschek D, Schatton W, Kellerman J, Muller WEG, Kreuter J (2002) Marine sponge collagen: isolation, characterization and effects on the skin parameters on pH, moisture and sebum. Eur J Pharm Biopharm 53: 107-113.

7. Kim S, Mendis E (2006) Bioactive compounds from marine processing byproducts-A review. Food Res Int 39: 383-393.

8. Jolivalt C, Minier M, Renon H (1990) Extraction of α-chymotrypsin using reverse micelles. J Colloid Interface Sci 135: 85-96.

9. Hu ZY, Gulari E (1996) Communication to the editor protein extraction using the sodium bis (2-ethylhexyl) phosphate (NaDEHP) reverse micellar system. Biotechnol Bioeng 50: 203-206.

10. Zhou L, Budge SM, Ghaly AE, Brooks MS, Dave D (2011) Extraction, purification and characterization of fish chymotrypsin: A Review. Am J Biochem Biotechnol 7: 104-123.

11. Hentsch M, Menoud P, Steiner L, Flaschel E, Renken A (1992) Optimization of the surfactant (AOT) concentration in a reverse micelle extraction process. Biotechnology Techniques 6: 359-364.

12. Goto M, Ishikawa Y, Ono T, Nakashio F, Hatton TA (1998) Extraction and activity of chymotrypsin using AOT-DOLPA mixed reversed micellar systems. Biotechnol Prog 14: 729-734.

13. Chong ASC, Hashim R, Chow-Yang L, Ali AB (2002) Partial characterization and activities of proteases from the digestive tract of discus fish Symphysodon aequifasciata. Aquaculture 203: 321-333.

14. Boeris V, Romanini D, Farruggia B, Pico G (2009) Purification of chymotrypsin from bovine pancreas using precipitation with a strong anionic polyelectrolyte. Process Biochem 44: 588-592.

15. Heu MS, Kim HR Pyeun JH (1995) Comparison of trypsin and chymotrypsin from the viscera of anchovy, Engraulis japonica. Comp Biochem Physiol B Biochem Mol Biol 112: 557-567.

16. Castillo-Yáñez FJ, Pacheco-Aguilar R, Garcia-Carreno FL (2006) Purification and biochemical characterization of chymotrypsin from the viscera of Monterey sardine (Sardinops sagax caeruleus). Food Chemistry 99: 252-259.

17. Kunitz (1948) Crystallization of salt-free chymotrypsinogen and chymotrypsin from solution in dilute ethyl alcohol. J Gen Physiol.

18. Yang F, Su WJ, Lu BJ, Wu T, Sun LC, et al. (2009) Purification and characterization of chymotrypsins from the hepatopancreas of crucian carp (Carassius auratus). Food Chemistry 116: 860-866.

19. Olson BJSC Markwell J (2007) Assays for determination of protein concentration. Current Protocols in Protein Science, John Wiley & Sons, Inc, New Jersey.

20. Chakrabarti R, Rathore RM, Mittal P, Kumar S (2006) Functional changes in digestive enzymes and characterization of proteases of silver carp (♂) and bighead carp (♀) hybrid, during early ontogeny. Aquaculture 253: 694-702.

21. Ono T, Goto M, Nakashio F, Hatton TA (1996) Extraction behaviour of haemoglobin using reverse micelles by dioleyl phosphoric acid. Biotechnol Prog 12: 793-800.

22. Paradkar VM, Dordick JS (1994) Mechanism of extraction of chymotrypsin at very low concentrations of Aerosol OT in the absence of reverse micelles. Biotechnol Bioeng 43: 529-540.

23. Chang Q, Liu H, Chen J (1994) Extraction of lysozyme, α-chymotrypsin and pepsin into reverse micelles formed using an anionic surfactant, isooctane and water. Enzyme Microb Technol 16: 970-973.

24. HongDP, Lee SS, Kuboi R (2000) Conformational transition and mass transfer in extraction of proteins by AOT-alcohol-isooctane reverse micellar systems. J Chromatogr B Biomed Sci Appl 743: 203-213.

25. Dekker M, Van't Riet K, Van der Pol JJ, Baltussen JWA, Hilhorst R, et al. (1991) Effect of temperature on the reversed micellar extraction of enzymes. Chem Eng J 46: B69-B74.

26. Nishiki T, Sato I, Muto A, Kataoka T (1998) Mass transfer characterization in forward and back extractions of lysozyme by AOT-isooctane reverse micelles across a flat liquid-liquid interface. Biochem Eng J 1: 91-97.

27. Nishiki T, Nakamura K, Kato D (2000) Forward and backward extraction rates of amino acids in reversed micellar extraction. Biochem Eng J 4: 189-195.

28. Hebbar HU, Sumana B, Raghavarao KSMS (2008) Use of reverse micellar systems for the extraction and purification of bromelain from pineapple wastes. Bioresour Technol 99: 4896-4902.

29. Hatton TA, Scamehorn JF, Harwell JH (Eds.) (1989) Surfactant-based separation processes, Marcel Dekker, New York.

The Kinetics of Alcoholic Fermentation by Two Yeast Strains in High Sugar Concentration Media

Angela Zinnai*, Francesca Venturi, Chiara Sanmartin and Gianpaolo Andrich

Department of Agriculture, Food and Environment, University of Pisa, Via del Borghetto 80, I-56124 Pisa, Italy

Abstract

Over the last two decades, most of Italian vines have produced grapes with higher sugar to total acid ratios, greater concentrations of phenols and aromatic compounds and greater potential wine quality. As a consequence, the musts obtained by these grapes are more difficult to process because of the risk of slowing or stuck of fermentation. With the aim of describing the time evolution of the sugars bioconversion during alcoholic fermentation, the kinetics of the D-glucose and D-fructose degradations, promoted by two yeast strains (*Saccharomyces cerevisiae* (strain C) e *Saccharomyces bayanus* (strain B)), was investigated using synthetic media, added or not with ethanol. The concentrations of both the substrates and the products of the sugars conversions, as well as the number of viable cells of yeasts, were determined as a function of the alcoholic fermentation time and the related kinetics constants determined.

If the reaction medium contained high concentrations of both glucose and fructose, the strains showed significant different fermentatory ability. In these conditions a stuck of fermentation occurred and the remaining sugar was only fructose (strain C) or prevailing fructose (strain B).

If the reaction medium contained only glucose as substrate, the strain C seemed more efficient while the kinetics behavior changed completely in presence of only fructose.

On the basis of the information collected using this kinetic approach, it would be possible to develop technical data sheets, specific for each yeast strain, useful to choose the optimal microbial strain as a function of the different operative conditions. Moreover the kinetic constant of hexose conversion could be adopted as bio-markers in selection and breeding of wine yeast strains having a lower tendency for sluggish fructose fermentation.

Keywords: Hexose metabolisation; Stuck of fermentation; *Saccharomyces bayanus*; *Saccharomyces cerevisiae*; Alcoholic fermentation; Kinetics parameters

Introduction

Several *Saccharomyces* species have been extensively used in wine making, sake making, and brewing processes such as in bioethanol production, despite yeasts are rather sensible to ethanol accumulation in the reaction medium [1]. In fact, in winemaking the number of stuck of fermentations is continuously increasing, particularly in countries characterized by warm climates [2].

As widely reported in literature, a number of stress factors occurring during the process, the lack of micro and macronutrients necessary for yeasts, unsuitable reaction temperatures, too low pH values, the presence of significant concentrations of inhibitors (ethanol, phenols, etc.) in the reaction medium, the development of dangerous microorganisms as well as the alteration of ionic equilibrium can induce a deep modification of the alcoholic fermentation kinetics [1,2].

Over the past two decades, wine producers aimed to produce grapes with increased sugar to total acidity ratios to obtain higher concentrations of phenols and aromatic compounds in order to increase wine quality. As a consequence, the musts obtained by these grapes are more difficult to process because they present unsuitable conditions for yeast reproduction [3,4].

The basis for the decline in fermentation rate is not fully understood. The increase in alcoholic fermentation rate by the addition of selected yeasts strains to the must could result ineffective so the residual sugars were also utilized by contaminating microorganisms able to carry out unwished metabolic pathways. Such an example, in these conditions, some heterofermentative lactic acid bacteria strains could significantly increase volatile acidity inducing a remarkable loss of quality of the alcoholic beverage. Moreover, yeasts lysis, occurring at the end of alcoholic fermentation, determines the solubilisation of cellular contents which can greatly stimulate *Brettanomyces spp.* growth [5].

Although a great number of references, providing a lot of information on the different aspects of alcoholic fermentation, are available in literature, it is still difficult to identify the possible causes of slowing or stuck fermentations even if the change of some compositional parameters (ex: D-glucose/D-fructose ratio, glycerine produced/hexoses converted) or an unusual accumulation of intermediates of sugar catabolism could be assumed as valid signals of a possible deviation from *Saccharomyces* metabolic pathways [6].

The molecular basis of the differential utilization of glucose and fructose, i.e., the glucose/fructose discrepancy in fermentation by *Saccharomyces cerevisiae*, in general, is not known. We have shown previously that different wine yeast strains have strain-specific G/F discrepancies [7,8], and the basis of these differences also is unknown.

In particular, it is interesting to find the reason why alcoholic yeasts preferably metabolise D-glucose rather than D-fructose [2,7-

---

*Corresponding author: Angela Zinnai, Department of Agriculture, Food and Environment, University of Pisa, Via del Borghetto 80, I-56124 Pisa, Italy

11], and investigate the kinetic aspects related to their conversion in order to optimize specific substrate consumption rates. Moreover, this kinetic approach appears to be potentially able to clarify many metabolic aspects related to hexoses conversion with the aim to better control alcoholic fermentation and to avoid that unwished dangerous processes take place.

To reduce the large number of variables able to influence the kinetics of alcoholic fermentation, the time evolution of different initial concentrations of D-glucose and D-fructose, dissolved in a model solution simulating a must (citrate buffer at pH=3.4 inoculated by two commercial strains of *Saccharomyces spp.*), have been investigated both in presence and in absence of ethanol in the initial reaction medium.

Materials and Methods

The experimental runs were carried out at $27.0 \pm 1.5°C$ using a 500 mL batch reactor. To ensure anaerobic and sterilized conditions the whole experimental apparatus was autoclaved and subjected to three cycles of vacuum following by replacement with nitrogen sterilized by filtration. Thus, the presence of undesired microorganisms and the aerobic utilization of sugars by yeasts were ruled out.

The characteristics of this bioreactor, realized at the Department of Agriculture, Food and Environment of the University of Pisa, were already reported in a previous paper [2]. The fermentation temperature was maintained constant by a heat exchanger, whereas the homogeneity of the reaction medium was ensured by a magnetic stirrer. The bioreactor was initially filled with 250 mL of a citrate buffer aqueous solution (pH 3.4) containing D-glucose and/or D-fructose (at five different concentrations: 556, 833, 1111, 1389, and 1667 mmol·L^{-1}) added or not with 1320 or 1450 mmol·L^{-1} of ethanol, respectively, that was sterilized by filtration. To the reaction medium containing only buffer and sugars (and ethanol when added) about 1.6 g (6.4 g·L^{-1}) of one of the two lyophilized yeasts commercial strains utilized (*S. cerevisiae Actiflore BJL souche p1* or *S. bayanus Actiflore Bayanus souche BO213*, Laffort Oenologie) were directly added to ensure a number of colony forming units (CFU) ranging from 1010 to 1011 (Table 1). This addition represented the initial time of all kinetic determinations.

The time evolution of both CFU and concentrations of both

reagents (D-glucose and/or D-fructose) and their products (glycerine and ethanol) were evaluated by a total plate count or utilizing specific commercial enzymatic kits (Megazyme), respectively [2].

The identification of the best values to be assigned to the model parameters was carried out by the specific statistical program me BURENL [9] able to identify in a space of j-dimensions (where j is equal to the number of model parameters) the minimum value of the F function, which is given by the sum of squares of differences occurring among experimental ($Y_{i,exper.}$) and calculated ($Y_{i,calc.}$) data:

$$F = \sum_{i=1}^{N} \left(Y_{i,calc.} - Y_{i,exper.} \right)^2$$

Where N represents the total number of experimental determinations. The values assumed by the model parameters at the minimum of the F function represent the best values.

To evaluate the kinetic constants related to the time evolution of the hexose under investigation (k_H, overall hexoses kinetic constants, k_F, fructose kinetic constants, k_G, glucose constants) the experimental data concerning hexose (D-glucose and or D-fructose) decrease and both ethanol and glycerine accumulations were used.

Results and Discussion

According to literature, the ability shown by *Saccharomyces spp.* to metabolise the hexoses depends on the temperature and the composition of culture media (sugars level, D-glucose to D-fructose ratio as well as ethanol concentration) [10,11]. Since the rate of a substrate conversion depends on the concentration of the microbial population present in the reaction medium (CFU/L), an high concentration of lyophilized yeasts was initially added to the reaction medium so that the great number of microbial cells could ensure a remarkable conversion of sugars in all the different operating conditions adopted (Table 1). The analytical points describing the decrease of concentrations of the two monosaccharides (D-glucose and D-fructose), and the increase of the production of ethanol and glycerol as a function of fermentation time when initial concentrations of 300 gL^{-1} (1666 mmol L^{-1}) were used, are reported in Figure 1 (Figure 1a=*Saccharomyces cerevisiae*; Figure 1b=*Saccharomyces bayanus*). The kinetic evolution of hexoses conversion was described by a first order equation and the following mathematical form introduced, in the plane (t, [H]):

Run	$[H]_{t=0}$ (mmol/L)	$[N]_{t=0}$ (CFU/L)	EtOH	$[H]_{t=end\,of\,run}$ (mmol/L)	$100 \cdot [G]_{t=end\,of\,run}/[G]_{t=0}$	$100 \cdot [F]_{t=end\,of\,run}/[F]_{t=0}$
1C	833.3 G + 833.3 F	3.20*10^{11}	no	F = 52.2	--	6.3%
2C	1111.1 G	1.25*10^{11}	no	--	--	--
3C	1111.1 F	1.10*10^{11}	no	--	--	--
4C	1666.6 G	8.92*10^{10}	no	G = 24.89	1.5%	--
5C	1666.6 F	1.38*10^{11}	no	F = 363.30	--	19.47%
6C	1111.1 G	5.75*10^{10}	yes	G = 188.0	17.2%	--
7C	1111.1 F	5.80*10^{10}	yes	F = 277.7	--	25.13%
8C	555.5 G + 555.5 F	3.10*10^{11}	yes	G = 10.3; F = 176.4	1.9%	35.0%
1B	833.3 G + 833.3 F	8.30*10^{11}	no	G = 21.5; F = 110.10	2.6%	13.3%
2B	1111.1 G	1.68*10^{11}	no	--	--	--
3B	1111.1 F	9.40*10^{10}	no	--	--	--
4B	1666.6 G	8.89*10^{10}	no	G = 60.56	3.6%	--
5B	1666.6 F	1.10*10^{11}	no	F = 73.28	--	4.4%
6B	1111.1 G	9.70*10^{10}	yes	G = 194.3	18.6%	--
7B	1111.1 F	5.10*10^{10}	yes	F = 289.06	--	27.5%
8B	555.5 G + 555.5 F	5.2*10^{10}	yes	G = 34.67; F = 170.94	6.6%	34.9%

Table 1: Values of initial concentration of hexoses (H)$_{t=0}$, yeasts (N$_{t=0}$), final concentrations of hexoses (H)$_{t=endofrun}$ and percentages of residual sugar for the experimental runs performed.

$$[H]_{t=t} = [H]_{t=0} \cdot e^{-kH}$$

Similarly, two exponential equations were used to evaluate the time evolution of glucose (G) and fructose (F) to determine the kinetic when both these hexoses are present in the reaction medium:

$$[G]_{t=t} = [G]_{t=0} \cdot e^{-kG \cdot t}$$

$$[G]_{t=t} = [G]_{t=0} \cdot e^{-kF \cdot t}$$

A good correlation among the calculated data and the experimental ones, regardless of the initial composition of the medium of reaction

(Table 2) has been successfully obtained. Therefore, the comparison of kinetic constants, related to *S. cerevisiae* and *S. bayanus* hexose conversion, has provided useful information on the fermentation efficiency of the strains examined, in the different conditions tested (Figure 2).

In the experimental runs characterized by an high concentration both of glucose and fructose initially present in the reaction medium (runs 1C, 1B) initially present in the reaction medium, D-fructose conversion became more difficult than D-glucose transformation both for the *S. cerevisiae* strain and for the *S. bayanus* strain, so that D-fructose has not been totally metabolized by yeasts and can be found partially unconverted in the reaction medium (Table 1).

Moreover, when ethanol is initially added at reaction medium (runs 6,7,8 C; 6,7,8 B), a remarkable reduction of values assumed by kinetic constant k_H could be pointed out underling the negative effect induced by this compounds on fermentative activity of both strains of yeasts so that about 20% of the initial amount remained unconverted in the reaction medium (Table 1). In particular, when the only metabolized sugar was represented by fructose (runs 7C, 7B), the percentages of unconverted hexose were the highest (Table 1) while they were nearly the same when in the reaction medium was present only D-glucose or D-glucose and D-fructose together (runs 6C, 8C and 6B, 8B).

According to the stoichiometry of alcoholic fermentation, the sum of the analytical data related to the concentrations of unconverted sugars, accumulated glycerine and half of ethanol formed did not vary significantly with time, assuming values very close to the initial concentration of sugar used (see M= molar balance; Figures 1a,1b and 3a,3b). As a consequence, a possible significant accumulation of intermediates can be ruled out and this relation can be written:

$$[H]_{t=0} - [H]_{t=t} = [0.5\ E]_{t=t} + [Gly]_{t=t}$$

Dividing both members of this equation by the amount of substrate converted ($[H]_{t=0} - [H]_{t=t}$), the following expression can be obtained:

$$1 = [0.5\ E]_{t=t}/([H]_{t=0} - [H]_{t=t}) + [Gly]_{t=t}/([H]_{t=0} - [H]_{t=t}) = R_{E,H} + R_{Gly,H}$$

Where $R_{E,H}$ and $R_{Gly,H}$ represent the fraction of hexose converted in ethanol or glycerine and ($[H]_{t=0} - [H]_{t=t}$) is the amount of sugars converted.

On this basis, the time evolutions of ethanol and glycerine

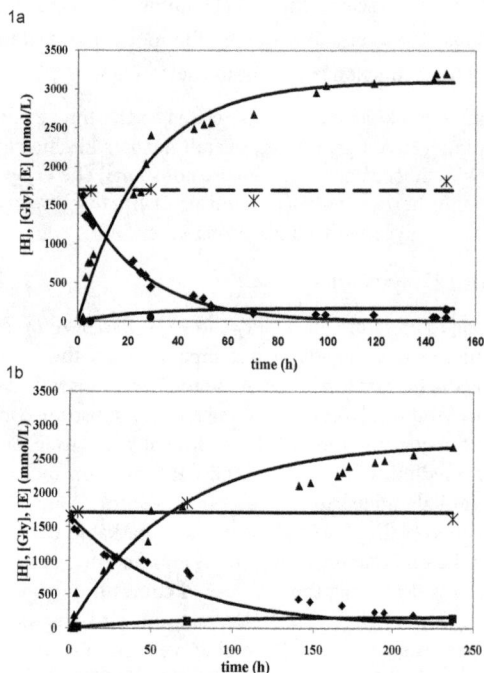

Figure 1: Experimental points (rhombs = hexoses, triangles = ethanol, circles = glycerine, stars= mass balance) and theoretic developments of hexoses conversion as a function of time of fermentation promoted by a *S. cerevisiae* strain (a) or a *S. bayanus* strain (b) in the adopted reaction medium ([D-glucose]$_{t=0}$=1667 mmol·L^{-1}, [D-fructose]$_{t=0}$=1667 mmol·L^{-1}).

Run	$k_H \pm$ i.c. (h^{-1})	r^2	$k_G \pm$ i.c. (h^{-1})	r^2	$k_F \pm$ i.c. (h^{-1})	r^2
1C	0.039 ± 0.011	0.98	0.057 ± 0.005	0.98	0.033 ± 0.002	0.98
2C	0.033 ± 0.150	0.94	--	--	--	--
3C	0.024 ± 0.021	0.98	--	--	--	--
4C	0.012 ± 0.008	0.96	--	--	--	--
5C	0.009 ± 0.001	0.98	--	--	--	--
6C	0.005 ± 0.002	0.98	--	--	--	--
7C	0.009 ± 0.003	0.98	--	--	--	--
8C	0.005 ± 0.001	0.98	0.011 ± 0.003	0.96	0.003 ± 0.001	0.98
1B	0.016 ± 0.004	0.96	0.019 ± 0.009	0.96	0.012 ± 0.003	0.96
2B	0.017 ± 0.034	0.98	--	--	--	--
3B	0.027 ± 0.021	0.96	--	--	--	--
4B	0.008 ± 0.001	0.96	--	--	--	--
5B	0.012 ± 0.001	0.98	--	--	--	--
6B	0.005 ± 0.001	0.98	--	--	--	--
7B	0.004 ± 0.002	0.98	--	--	--	--
8B	0.005 ± 0.001	0.96	0.007 ± 0.005	0.98	0.004 ± 0.001	0.96

Table 2: Values (mean ± c.i.) related to the kinetic constants of conversion of substrates (k_H) obtained for the different experimental runs performed by two strains (C = *S. cerevisiae*; B= *S. bayanus*) analyzed.

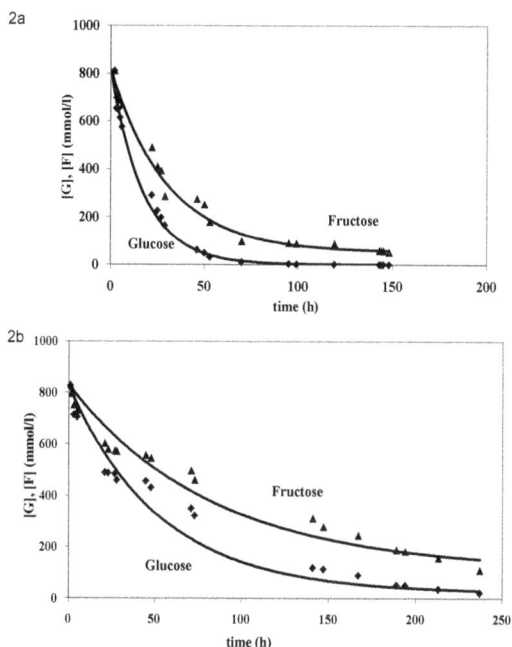

Figure 2: Experimental points (rhombs = glucose. triangles = fructose) and theoretic developments of glucose and fructose conversion as a function of time of fermentation promoted by a *S. cerevisiae* strain (a) or a *S. bayanus* strain (b) in the adopted reaction medium ([D-glucose]$_{t=0}$=1667 mmol·L^{-1}, [D-fructose]$_{t=0}$=1667 mmol·L^{-1}).

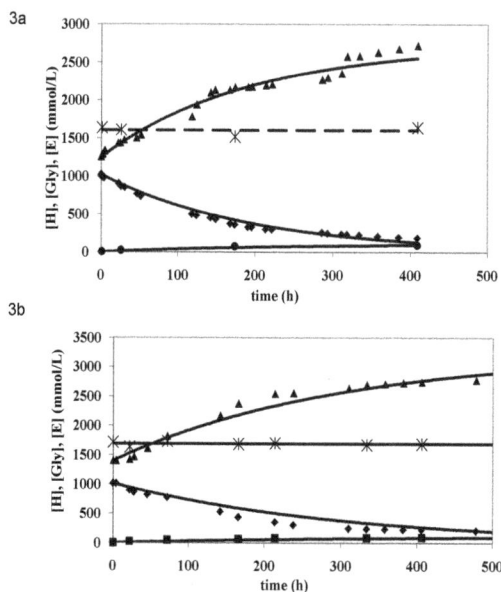

Figure 3: Experimental points (rhombs = hexoses, triangles = ethanol, circles = glycerine, stars= mass balance) and theoretic developments of hexoses conversion as a function of time of fermentation promoted by a *S. cerevisiae* strain (a) or a *S. bayanus* strain (b) in the adopted reaction medium ([D-glucose]$_{t=0}$ = 1111 mmol L^{-1} plus 255.7 ml L^{-1} ethanol; b) [D-fructose]$_{t=0}$ = 1111 mmol L^{-1} plus 306.7 ml L^{-1} ethanol.

concentrations can be described as a function of hexoses converted:

$$[E]_{t=t} = 2R_{E,H} \cdot [H]_{t=0} \cdot (1-e^{-kH \cdot t})$$

Runs	$R_{Gly,H}$	$R_{E,H}$
1C ÷ 5C	0.12 ± 0.02	0.92 ± 0.05
6C ÷ 8C	0.10 ± 0.01	0.92 ± 0.05
1B ÷ 5B	0.13 ± 0.05	0.86 ± 0.02
6B ÷ 8B	0.10 ± 0.01	0.88 ± 0.03

Table 3: Mean values and confidence intervals of the fraction of hexoses converted in glycerine ($R_{Gly,H}$) and in ethanol ($R_{E,H}$).

$$[Gly]_{t=t} = R_{Gly,H} \cdot [H]_{t=0} \cdot (1-e^{-kH \cdot t})$$

The mean values of $R_{E,H}$ and $R_{Gly,H}$ assumed similar values for runs carried out both in presence and absence of ethanol in the initial reaction medium (Table 3). This experimental evidence show that the reaction involved in the rate determining step is one of those coming first the hexose cleavage to two triose phosphates regardless the initial concentration of ethanol in the reaction medium. The ratio between the kinetic constant related to time evolution of D-glucose and D-fructose in experimental runs characterized by an high initial concentration of ethanol assumed a similar value ($k_G/k_F \sim 3:1$) both for *S. cerevisiae* strain than for *S. bayanus* one. This value was lower than that related to the runs carrying out in absence of ethanol in the initial reaction medium ($k_G/k_F \sim 1.6:1$), showing that this alcohol induced a sensible reduction of fructose conversion so that a great amount of this sugar can be found unconverted in the reaction medium (> 25%) both using strain C and strain B (Table 1).

Conclusion

The kinetics constants related to hexoses conversion (glucose and fructose) of two yeasts strains have been calculated as a function of the different operating conditions adopted. On the basis of the information collected using this kinetic approach, it would be possible to develop technical data sheets, specific for each yeast strain, useful to choose the optimal microbial strain as a function of the different operative conditions characterizing several biochemical processes (ex: wine making, sake making, brewing processes and bioethanol production). Moreover the kinetic constant of hexose conversion could be adopted as bio-markers in selection and breeding of wine yeast strains having a lower tendency for sluggish fructose fermentation.

References

1. Teixeira MC, Raposo LR, Mira NP, Lourenço AB, Sá-Correia I (2009) Genome-wide identification of Saccharomyces cerevisiae genes required for maximal tolerance to ethanol. Appl Environ Microbiol 75: 5761-5772.

2. Zinnai A, Venturi F, Sanmartin C, Quartacci MF, Andrich G (2013) Kinetics of D-glucose and D-fructose conversion during the alcoholic fermentation promoted by Saccharomyces cerevisiae. J Biosci Bioeng 115: 43-49.

3. Bauer FF, Pretorius IS (2000) Yeast Stress Response and Fermentation Efficiency: How to Survive the Making of Wine - A Review. S Afr J Enol Vitic 21: 27-51.

4. Jones GV, White MA, Cooper OR, Storchmann K (2005) Climate Change and Global Wine Quality. Climatic Change 73: 319-343.

5. Sablayrolles JM (2009) Control of alcoholic fermentation in winemaking: Current situation and prospect. Food Res Int 42: 418-424.

6. Berthels NJ, Cordero ORR, Bauer FF, Pretorius IS, Thevelein JM (2008) Correlation between glucose/fructose discrepancy and hexokinase kinetic properties in different Saccharomyces cerevisiae wine yeast strains. Appl Microbiol Biotech 77: 1083-1091.

7. Tronchoni J, Gamero A, Arroyo-López FN, Barrio E, Querol A (2009) Differences in the glucose and fructose consumption profiles in diverse Saccharomyces wine species and their hybrids during grape juice fermentation. Int J Food Microbiol 134: 237-243.

8. Loureiro V, Malfeito-Ferreira M (2003) Spoilage yeasts in the wine industry. Int J Food Microbiol 86: 23-50.

9. Buzzi FG, Manca D (1996) Politecnico, Dipartimento di Ingegneria Chimica 'G. Natta', Milan.

10. Salmon JM (1989) Effect of Sugar Transport Inactivation in Saccharomyces cerevisiae on Sluggish and Stuck Enological Fermentations. Appl Environ Microbiol 55: 953-958.

11. Guillaume C, Delobel P, Sablayrolles JM, Blondin B (2007) Molecular basis of fructose utilization by the wine yeast Saccharomyces cerevisiae: a mutated HXT3 allele enhances fructose fermentation. Appl Environ Microbiol 73: 2432-2439.

30

A Dehydrogenase Activity Test for Monitoring the Growth of *Streptomyces Venezuelae* in a Nutrient Rich Medium

T. J. Burdock, M.S. Brooks* and A.E. Ghaly

Department of Process Engineering and Applied Science, Dalhousie University, P.O. Box 1000, Halifax, Nova Scotia, Canada B3J 2X4

Abstract

Jadomycin is a novel antibiotic that has shown activities against bacteria, yeasts and fungi as well as cytotoxic properties to cancer cells. Because of the wide range of its inhibitory actions, jadomycin shows promise as a novel antibiotic and cancer treatment drug. *Streptomyces venezuelae* are aerobic bacteria that are capable of producing jadomycin when shocked by alcohol in a nutrient deprived amino acid rich medium. The size of the bacterial population that is transferred from the growth medium to the production medium can significantly affect the jadomycin yield. Therefore, the number of transferred bacteria must be accurately measured. In this study, a dehydrogenase activity measurement test was developed for *S. venezuelae* using triphenyl tetrazolium chloride (TTC) to measure the cell growth and activity in maltose-yeast extract-malt extract (MYM) broth. The dehydrogenase activity was determined by measuring the visible color changes of the TTC to triphenyl formazan (TF). The test conditions which included extraction solvent, number of extractions, incubation time, incubation temperature and medium pH were evaluated. The results showed that the triphenyl formazan was related to the number of cells. Methanol was better able to permeate the cells and extract higher amount of TF than ethanol. The amount of TF increased with the number of extractions for both solvents. A lower medium pH and/or lower temperature produced the highest amount of TF. The best test conditions that produced the highest TF yield were three extractions using methanol after an incubation time of 1 hour at a temperature of 30ºC and a medium pH of 6.

Keywords: Dehydrogenase; Triphenyl tetrazolium chloride (TTC); Jadomycin; *Streptomyces venezuelae*

Introduction

Jadomycins are a novel group of antibiotics that are produced by *Streptomyces venezuelae*. They exhibit biological activity against bacteria and yeast and demonstrate cytotoxicity against cancer cells [1]. *S. venezuelae* are typically grown in a nutrient rich medium of maltose-yeast extract-malt extract (MYM) broth prior to their introduction to a nutrient-deprived, amino acid-rich production medium [2,3]. Once transferred to the production medium, an environmental shock (ethanol or heat) is applied to induce the bacteria to produce jadomycin [4]. The size of the bacterial population that is transferred from the growth medium to the production medium can significantly affect the jadomycin yield [5]. Therefore, in order to standardize and improve the reproducibility of the jadomycin production process, the number of live bacteria that are transferred and subsequently shocked must be accurately determined.

Bacterial counts can be determined with various techniques such as plate counting [6] turbidity measurement [7], microscope enumeration [8] and dehydrogenase activity measurement (Knight et al.). The dehydrogenase activity measurement technique has the advantage over other methods in being able to quantify the number of live cells that are present in the medium [9-11] and it can be employed in a relatively short time and at a low cost [12].

The dehydrogenase activity test is based on the principle that dehydrogenase enzymes are produced by all living cells and the extent to which this enzyme group oxidizes organic matter can be related to the number of live cells present [12]. This group of enzymes transports electrons and hydrogen through a chain of intermediate electron carriers to a final electron acceptor (oxygen), resulting in the formation of water [13-16]. The activity of the co-enzymes Nicotinamide Adenine Dinucleotide (NAD) and Flavin Adenine Dinucleotide (FAD) which act as intermediate electro acceptors can be measured by the visible col-

or change of a dye such as triphenyl tetrazolium chloride (TTC) [12].

The process of measuring dehydrogenase activity involves incubating the sample in the presence of triphenyl tetrazolium chloride and an electron-donating substrate[12]. In its oxidized form, TTC is colourless, but in the presence of dehydrogenase enzymes TTC is reduced to triphenyl formazan (TF), a red water insoluble compound [12, 17-19]. The TF is retained within microbial cells and can result in highly coloured colonies when grown on agar plates [18]. The mechanism of the process is summarized in Figure 1.

Triphenyl formazan can be extracted from cells using a solvent and the concentration is determined colorimetrically by measuring the optical density at 484 nm [12]. The use of a dehydrogenase activity measurement test using TTC to measure the quantity of living cells has great potential as a quick tool for determining the optimal time to shock a population of *S. venezuelae* and start the production of jadomycin. However, the amount of TF extracted depends on number of extractions, extraction solvent, incubation time, incubation temperature, and medium pH [20].The purpose of this study was to develop a dehydrogenase activity measurement test for *S. venezuelae* that could be used to quantify the live bacterial cells. The specific objectives were to: (a) investigate the applicability of the TTC-test for measuring dehydrogenase activity in *S. venezuelae* and (b) determine the ideal test

*Corresponding authors: Su-Ling Brooks, Associate Professor, Department of Process Engineering and Applied Science, Dalhousie University, Halifax, Nova Scotia, Canada B3J 2X4

conditions (extraction solvent, number of extractions, incubation time, incubation temperature and medium pH).

Experimental Materials

Reagents

Tris (hydroxymethyl-aminomethane) buffer was used to control the pH of the samples and triphenyl tetrazolium chloride (TTC) was the tetrazolium salt used for the dehydrogenase test. A TTC-glucose reagent (1 g glucose and 2g TTC dissolved in 100 mL distilled water) was prepared and stored in the dark at 4°C until used. Triphenyl formazan (TF) was used to establish a standard curve for absorbance (OD_{484}) vs TF concentration. Alcohols (ethanol and methanol) were used to extract TF from the cells. The Tris (hydroxymethyl-aminomethane), 2, 3, 5-triphenyl tetrazolium chloride (TTC) and 2, 3, 5-triphenyl formazan (TF) were obtained from Sigma-Aldrich (Oakville, Ontario, Canada) and the glucose was obtained from BioShop (Burlington, Ontario, Canada). Ethanol and methanol were obtained from Fisher Scientific (Montreal, Quebec, Canada).

Media preparation

Maltose-yeast extract-malt extract (MYM) agar and broth were used to cultivate *Streptomyces venezuelae*. The compositions of MYM agar and broth are shown in Table 1. All media components were obtained from BioShop (Burlington, Ontario, Canada). The media components were dissolved in distilled water then autoclaved (Sterile ax, Thermo Fisher Scientific, Ottawa, Ontario, Canada) on the liquid setting (121°C and 20 Pa) for 15 minutes. The autoclaved agar was stored at 65°C to prevent solidification.

Bacteria

An initial starter plate of *Streptomyces venezuelae* ISP5230 was obtained from the Jakeman Laboratory, College of Pharmacy, Dalhousie University (Halifax, Nova Scotia, Canada) and stored at 4°C. The surface growth was used to inoculate maltose-yeast extract-malt extract (MYM) agar plates or flasks with MYM broth as needed.

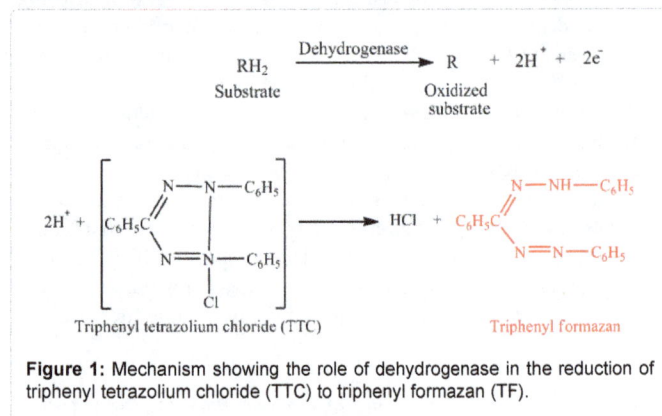

Figure 1: Mechanism showing the role of dehydrogenase in the reduction of triphenyl tetrazolium chloride (TTC) to triphenyl formazan (TF).

Component	Quantity (g/L distilled water)	
	Agar	Broth
Maltose	4.0	4.0
Yeast Extract	4.0	4.0
Malt Extract	10.0	10.0
MOPS	1.9	1.9
Agar	15.0	-

Table 1: MYM media components.

Experimental Procedure

Triphenyl formazan (tf) standard curve

A standard curve was developed to determine the concentration of TF corresponding to an absorbance measurement at 484 nm. A stock solution of 0.2 μmol/mL was prepared by dissolving 0.03 g TF in 500 mL methanol. The stock solution was diluted with methanol to produce 11 solutions with TF concentrations from 0.004 μmol/mL to 0.1 μmol/mL. The absorbance of each solution was measured using a spectrophotometer (Genesys 20, Thermo Scientific, Mississauga, Ontario, Canada) at a wavelength of 484 nm. The absorbance readings were plotted against the TF concentration of the prepared solutions as shown in Figure 2. The following linear best-fit equation ($R^2 = 0.98$) was determined:

$$AU_{484} \equiv 10.574\,TF \qquad (1)$$

where:

AU_{484} is the absorbance reading at 484 nm

TF is the concentration of triphenyl formazan (μmol/mL extraction solvent)

Microbial growth

Three 250 mL shake flasks were each filled with 175 mL of MYM broth, plugged with foam caps, covered with aluminum foil and autoclaved (SterileMax, Thermo Fisher Scientific, Ottawa, Ontario, Canada) at 121°C and 20 Pa for 15 minutes. The flasks were then inoculated with *S. venezuelae* and incubated in a controlled environment shaker (25 Incubator Shaker, New Brunswick Scientific, Edison, New Jersey, USA) at 30°C and 250 rpm. Each flask was sampled at 0, 2, 12, 14, 21, 23, 38, 40, 42, 60 and 64 hours after inoculation and the extent of cell growth was monitored over the period of 64 hour by measuring the optical density at 600 nm (OD_{600}), the number of colony forming units (CFU) and the triphenyal formazan yield (TF).

CFU determination

A series of dilutions were carried out for the determination of the number of CFU. A 1 mL aliquot of the original sample was added to an autoclaved test tube containing 9 mL of autoclaved distilled water. The test tube was capped and inverted several times to distribute the

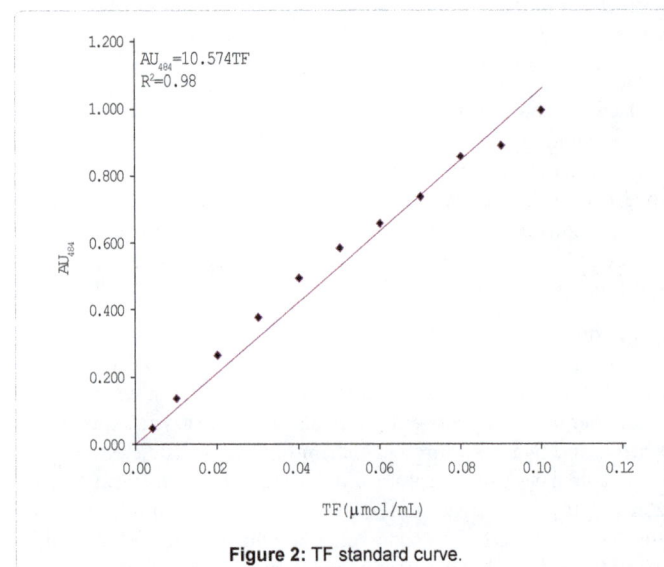

Figure 2: TF standard curve.

cells. An aliquot of 1 mL of this solution was added to a second auto-claved test tube containing 9 ml of autoclaved distilled water. This tube was capped and inverted to distribute the cells. This was carried out seven more times to a final dilution of 10^{-10}. For each of the six dilutions used (10^{-5}, 10^{-6}, 10^{-7}, 10^{-8}, 10^{-9}, 10^{-10}), 0.1 mL was added to a petri dish containing MYM agar in triplicate (given final plate dilutions of 10^{-6}, 10^{-7}, 10^{-8}, 10^{-9}, 10^{-10}, and 10^{-11}). The plates were sealed with parafilm, inverted, and incubated at 30°C in an environmentally controlled incubator (model number 2020, VWR International, Cornelius, Oregon, USA) for 24 hours. Following the incubation period, the plates were removed and the colonies were counted. The plates that had between 30-300 CFU present were used for calculating the CFU of the samples.

Dehydrogenase activity measurement

1 mL was pipetted from each sample into four test tubes. Tris buffer (2.5 mL) and TTC-glucose solution (1 mL) were added to the sample tubes (1 mL of distilled water was added to the control tube). The pH was adjusted to 7 using 1.0 N HCl and the test tubes were gently swirled to mix the content. The tubes were incubated in an environmentally controlled incubator (model number 2020, VWR International, Cornelius, Oregon, USA) at 30°C for 1 hour. The tubes were removed and centrifuged (IEC CentraCL2, Thermo Electron Corporation, Mississauga, Ontario, Canada) for 10 minutes to separate the cells from other medium components. TF extraction was carried out three times using 2.5 mL of ethanol each time. All samples were vortexed (Thermolyne Maxi Mix, Thermolyne Corporation, Hampton, New Hampshire, USA) to disrupt cell walls and leach TF from within cells followed by centrifugation to separate the cells at the bottom. Supernatants from the three extractions were combined and the absorbance of the combined supernatants was measured at 484 nm (Genesys 20, Thermo Scientific, Mississauga, Ontario, Canada).

Optimization experiments

Experiments were conducted to determine the optimum test conditions that would reduce the most TTC to TF and extract the highest amount of TF from *S. venezuelae* during growth in MYM media. The dehydrogenase activity test parameters included: solvent type, number of solvent extractions, incubation time, incubation temperature and medium pH. The values for each parameter are shown in Table 2. The study was carried out in two phases as shown in Figure 3. First, initial experiments were conducted to determine the optimal number of solvent extractions and best solvent type. Then, the best solvents and optimum number of extraction obtained from the initial experiments were used in further experiments to determine the best time, temperature and pH.

Number of extractions

To evaluate the effect of the number of extractions on the TF yield from each flask, samples were taken from the growing culture at 21, 37, 47, 62 and 64 hours after inoculation. At each sampling time, 1 mL aliquots were transferred into four test tubes (tests were carried out

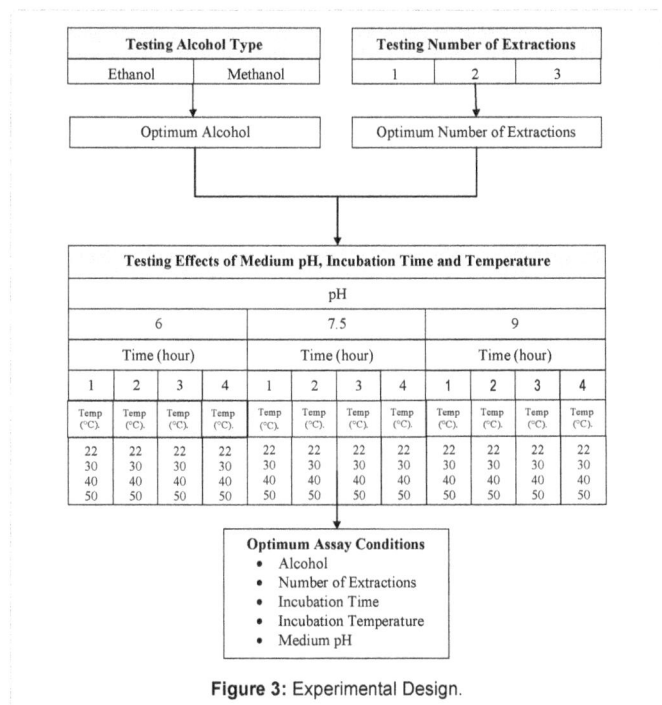

Figure 3: Experimental Design.

in triplicates with a control for each sample). Tris buffer (2.5 mL) was added to all test tubes. Then, 1 mL of TTC/glucose solution was added to each of the sample test tubes (1 ml of distilled water to control test tubes). The pH was adjusted to 6 using 1 N HCl. The test tubes were gently swirled to mix the contents, incubated in a controlled temperature oven at 50°C (Isotemp Oven, model 630 F, Fisher Scientific, Ottawa, Ontario, Canada) for 1 hour. Samples were then centrifuged (IEC CentraCL2, Thermo Electron Corporation, Mississauga, Ontario) for 10 minutes to separate the cells from the liquid media. The supernatant was discarded and 2.5 mL of ethanol was added to the cells. All samples were vortexed (Thermolyne Maxi Mix, Thermolyne Corporation, Hampton, New Hampshire, USA) to aid in the extraction of TF (red colour) from the cells. Samples were centrifuged again, the supernatant decanted and absorbance measured at 484 nm using the control to zero the spectrophotometer (Genesys 20, Thermo Scientific, Mississauga, Ontario, Canada). A second extraction with ethanol was carried out, the supernatant was combined with that from the first extraction and the absorbance was measured. A third extraction with ethanol was carried out, the supernatant combined with those from the previous two extractions and the absorbance was measured.

Type of extraction solvent

To evaluate the effectiveness of solvent in extracting TF, samples were taken from the growing culture at 21, 37, 47, 62 and 64 hours. At each sampling time, aliquots (1 mL) were transferred from each flask into the two groups of test tubes (ethanol and methanol). Tests were carried out in triplicates and a control for each flask sampled and solvent tested. Tris buffer (2.5 mL) was added to all the test tubes and 1 mL of the TTC/glucose solution was added to the sample test tubes (or 1 mL of distilled water to control test tubes). The pH was adjusted to 6 using 1 N HCl. The test tubes were gently swirled to mix the contents and incubated at 50°C for 1 hour in a temperature controlled oven (Isotemp Oven, model 630F, Fisher Scientific, Ottawa, Ontario, Canada). Samples were then centrifuged (IEC CentraCL2, Thermo Electron Cor-

Extraction Solvent	Number of Extractions	Incubation Time (hour)	Incubation Temperature (°C)	Medium pH
Methanol	1	1	22	6
Ethanol	2	2	30	7.5
	3	3	40	9
		4	50	

Table 2: Dehydrogenase activity assay conditions.

poration, Mississauga, Ontario) for 10 minutes and the supernatant discarded. Extraction of TF was carried out three times using 2.5 mL of either methanol or ethanol. After each addition of solvent to the cells, the samples were vortexed, centrifuged and the supernatant decanted. The absorbance of the supernatants collected from three extractions for each solvent for each solvent was measured at 484 nm using a spectrophotometer (Genesys 20, Thermo Scientific, Mississauga, Ontario, Canada).

Incubation time, temperature and medium ph

After 64 hours of growth, the contents of all flasks were combined into a 1 L flask and refrigerated at 4°C until required. For each incubation temperature investigated (22°C, 30°C, 40°C and 50°C), three medium pH values (6, 7.5 and 9) and four incubation times (1, 2, 3 and 4 hours) were tested. The resulting 48 tests were carried out in triplicate with a control. For all tests, 1 mL aliquots of the MYM broth with *S. venezuelae* growth were added to test tubes. The pH was adjusted to 6, 7.5 or 9 using 1 N HCl or NaOH as needed. Tris buffer (2.5 mL) and TTC/glucose solution (1 mL) were added to each tube. Tubes were manually swirled to mix contents and incubated for either 1, 2, 3, or 4 hours. Tubes were incubated at 22°C or 30°C in controlled environment incubator (Model 2020, VWR International, Cornelius, Oregon, USA), and at 40°C or 50°C in temperature controlled oven (Isotemp Oven, model 630F, Fisher Scientific, Ottawa, Ontario, Canada). Samples were then centrifuged (IEC CentraCL2, Thermo Electron Corporation, Mississauga, Ontario) for 10 minutes and the supernatant discarded. Extraction of TF was carried out three times using 2.5 mL methanol each time. After each addition of solvent to the cells, the samples were vortexed, centrifuged and the supernatant decanted. The supernatants obtained from the three extractions were combined after each extraction and the absorbance was measured at 484 nm using a spectrophotometer (Genesys 20, Thermo Scientific, Mississauga, Ontario, Canada).

Results

Microbial growth

The microbial growth as determined by measuring optical density at 600 nm (OD_{600}), the number of colony forming units (CFU) and the triphenyl fromazan (TF) yield. The results are presented in Figure 4. There was an initial lag period followed by exponential growth phase. The lag period and specific growth phase were determined graphically according to the procedure described by [15] as shown in Figure 5. The lag period and specific growth were 10.3 hour and 0.3 h^{-1}, respectively.

Number of extractions

In order to test the effect of the number of extractions on the final TF yield from *S. venezualae* cells grown in MYM broth, one, two and three extractions were carried out at a medium pH of 6, an incubation time of 1 hour and an incubation temperature of 50°C. The results presented in Figure 6 showed higher TF yields with increasing number of extractions at all sampling times (different population sizes). The results also showed that the stationary growth phase was reached after the 60 hours of growth.

Type of extraction solvent

Ethanol and methanol were used to extract TF from samples taken during the growth of *S. venezuelae* cells. The test was carried out using three extractions at a medium pH of 6, an incubation temperature of 50°C and an incubation time of 1 hour. The TF results shown in Figure

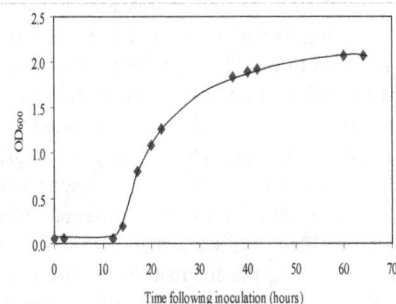

(a) Optical density

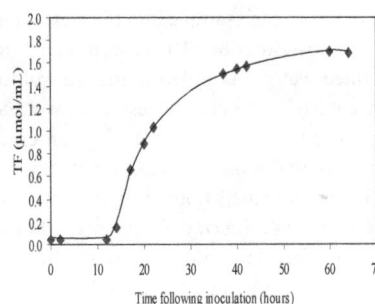

(b) TF yield

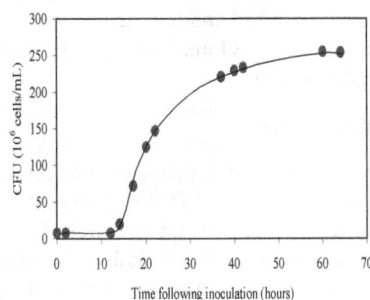

(c) CFU

Figure 4: *S. venezuelae* growth measured by optical density, TF yield and CFU.

7 indicated that both solvents showed an increase in TF yield over the time as the number of bacteria increased. The results also showed that the TF extracted by ethanol and methanol started to decline after 60 hours of growth indicating the start of the stationary growth phase.

Medium pH

Figures 8 and 9 show the effect of pH on TF yield at varying incubation temperatures and incubation times, respectively. All plots display a similar concave shape with the pH value of 7.5 resulting in the lowest TF yield and the pH of 6 resulting in the greatest TF yield at all incubation times and temperatures. However, longer incubation time and lower temperature resulted in slightly higher TF yields.

Incubation time

Figures 10 and 11 show the effect of incubation time on TF yield

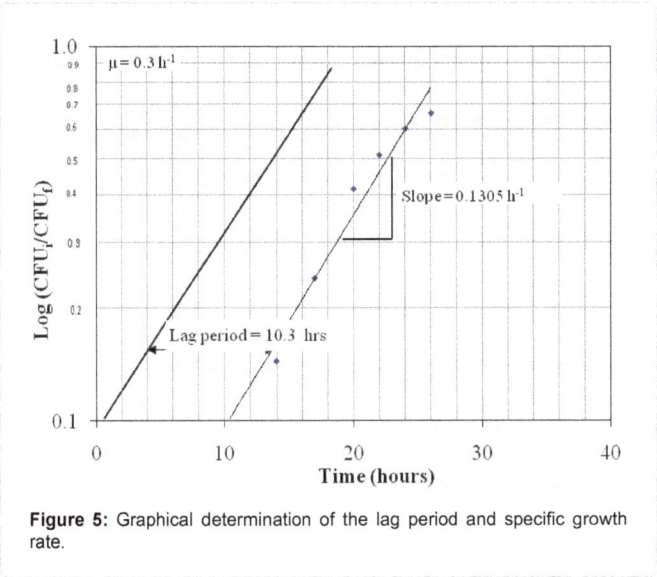

Figure 5: Graphical determination of the lag period and specific growth rate.

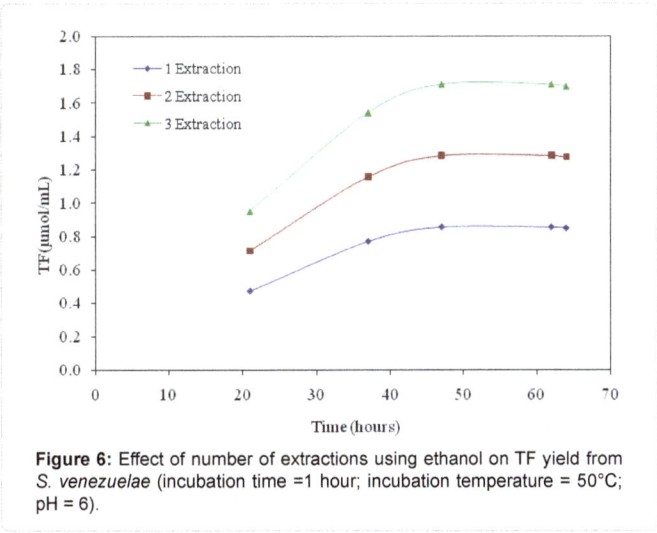

Figure 6: Effect of number of extractions using ethanol on TF yield from *S. venezuelae* (incubation time =1 hour; incubation temperature = 50°C; pH = 6).

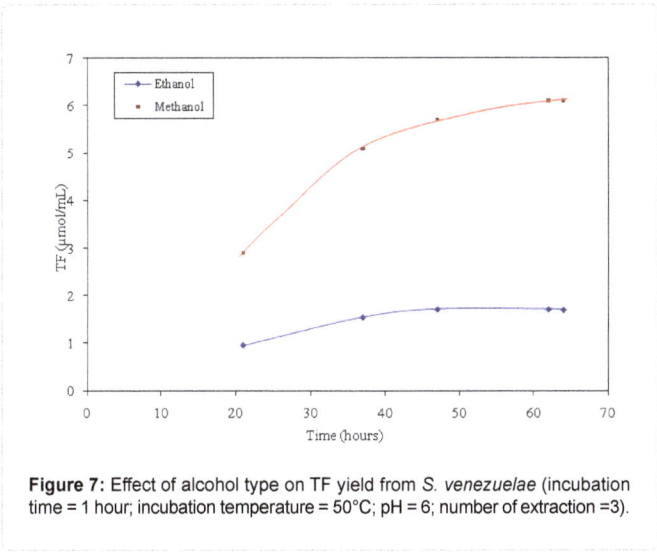

Figure 7: Effect of alcohol type on TF yield from *S. venezuelae* (incubation time = 1 hour; incubation temperature = 50°C; pH = 6; number of extraction =3).

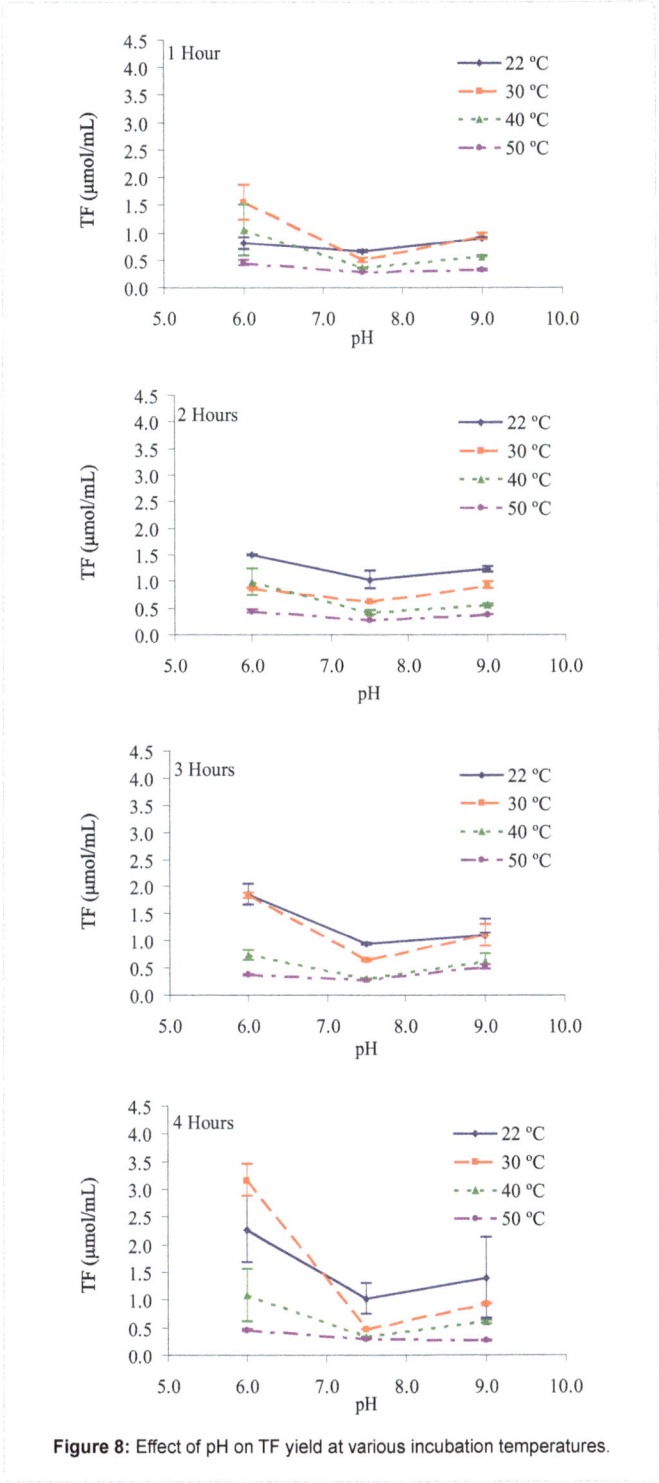

Figure 8: Effect of pH on TF yield at various incubation temperatures.

at different medium pH values and incubation temperatures, respectively. At all incubation times, higher incubation temperatures (40°C and 50°C) resulted in lower TF yields than those obtained at the lower temperatures (22°C and 30°C) but the highest TF yields were always achieved at a pH of 6. It seems that the effect of incubation time was dependent on the temperature. At higher temperatures (40°C and 50°C), 1 hour incubation time produced the higher TF yield while at lower temperature (22°C and 30°C), 4 hours incubation time produced the highest TF yield.

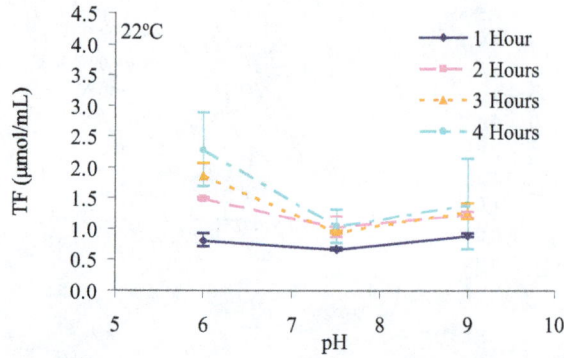

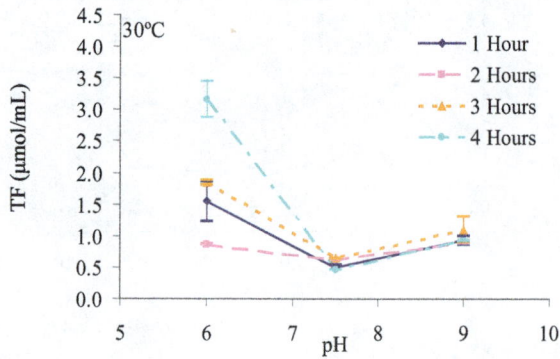

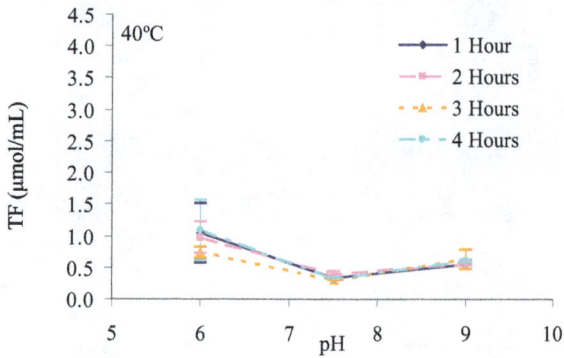

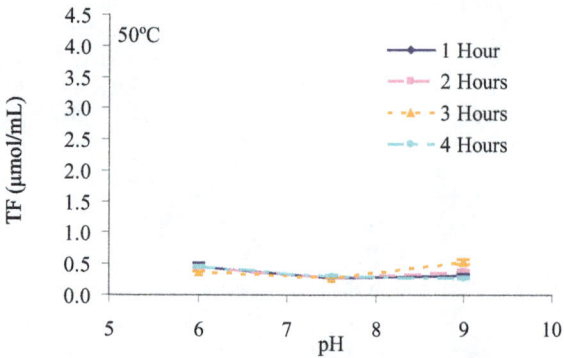

Figure 9: Effect of pH on TF yield at various incubation times.

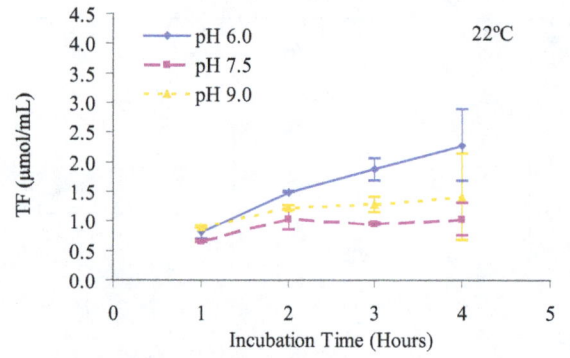

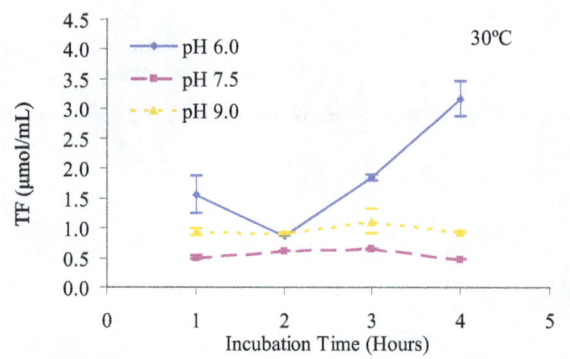

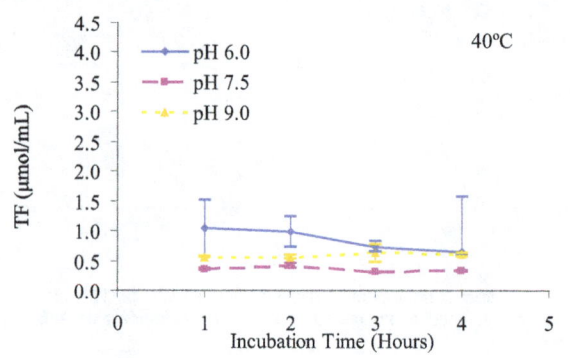

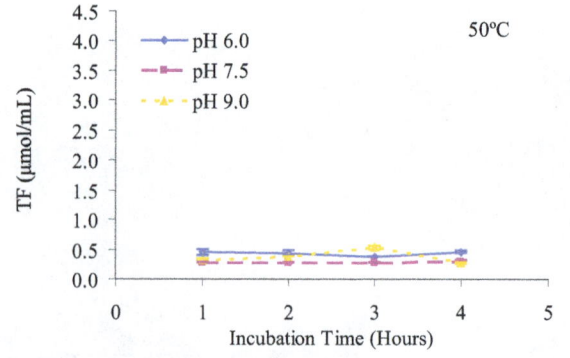

Figure 10: Effect of incubation time on TF yield at various pHs.

Incubation temperature

Figures 12 and 13 show the effect of incubation temperature on TF yield at different pH values and incubation times, respectively. For all incubation temperatures, the highest TF yield was obtained at a pH of 6. However, the incubation temperatures of 22°C and 30°C appeared to result in higher TF yields than those observed at higher incubation temperatures (40°C and 50°C).

Discussion

Jakeman et al. (2006) monitored *S. venezuelae* population during

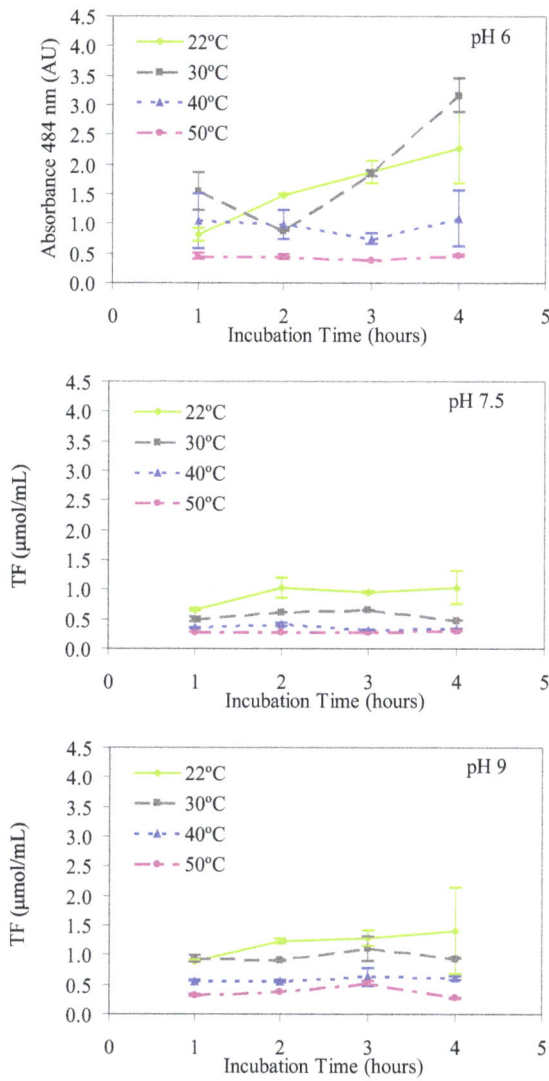

Figure 11: Effect of incubation time on TF yield at various incubation temperatures.

the growth period by measuring the optical density at 600 nm (OD_{600}). In this study, the change of *S. venezuelae* population during the growth period was monitored by measuring the optical density at 600 nm (OD_{600}), the number of colony forming units (CFU) and the triphenyl formazan yield (TF). The relationships between CFU, OD_{600} and TF are presented in Figure 14. The amount of TF extracted had a much better correlation with CFU than that observed between the CFU and OD_{600}. The results clearly indicated the effectiveness of dehydrogenase activity as an accurate measure of cell growth.

The OD_{600}, TF yield and CFU curves showed a lag period of approximately 10.3 hours, during which *S. venezuelae* adjusted to the new growth medium and environmental conditions. After the initial lag period, the bacteria grew exponentially before reaching the stationary phase at approximately 60 hours. The specific growth measured in this study was 0.3 h^{-1}. [21] reported maximum specific growth rate of 0.23 h^{-1} for *S. venezuelae* grown in media containing soluble starch at 30° C. [22] reported a maximum specific growth rate of 0.14 h^{-1} for *S. venezuelae* grown in MYM medium at 27°C.

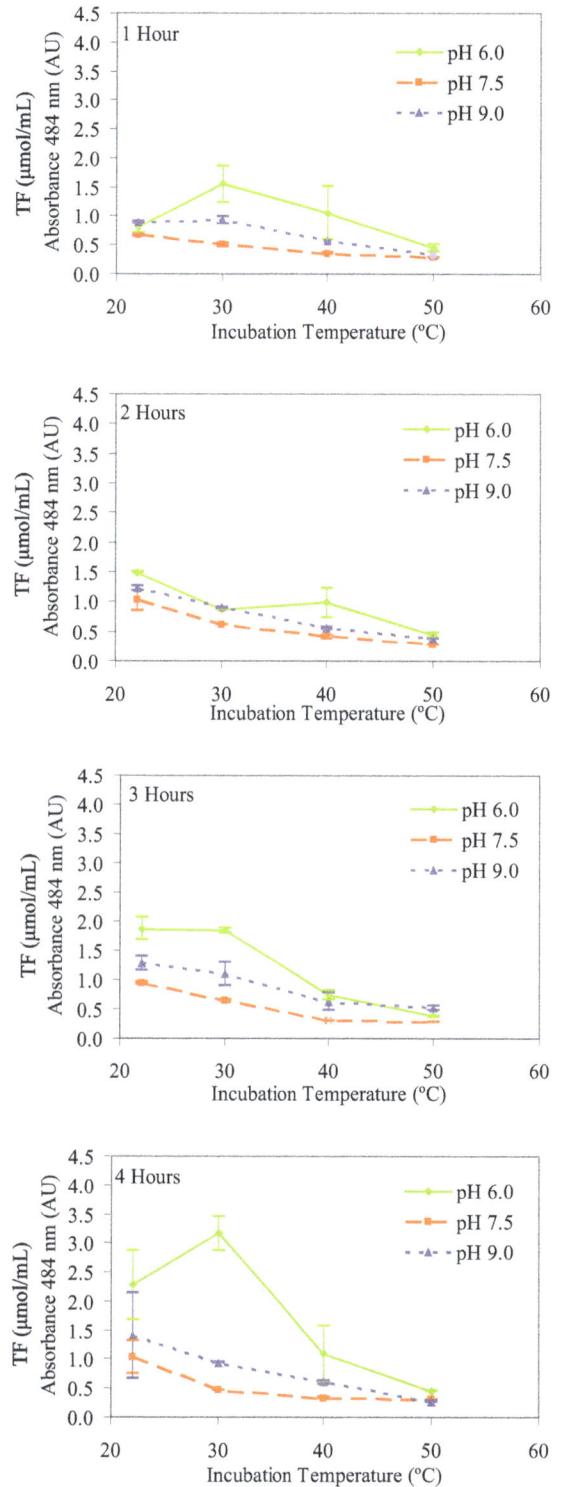

Figure 12: Effect of incubation temperature on TF yield at various pHs.

The specific TF yield (μmol/CFU) was calculated by dividing the TF yield by the CFU in order to assess the cell activity during the growth period. Figure 15 indicated that the cell activities during the lag phase (0.65 x 10^{-8} μmol/CFU) and stationary phase (0.67 x 10^{-8}μmol/CFU) were lower than that observed during the exponential growth period

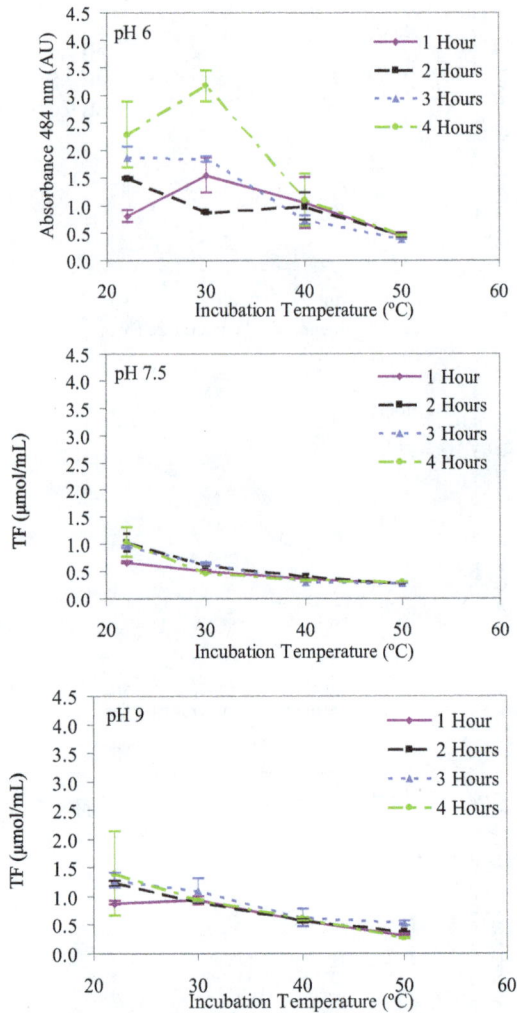

Figure 13: Effect of incubation temperature on TF yield at different incubation times.

(a) OD₆₀₀ vs TF

(b) CFU vs TF

Figure 14: The relationships between CFU, OD600 and TF.

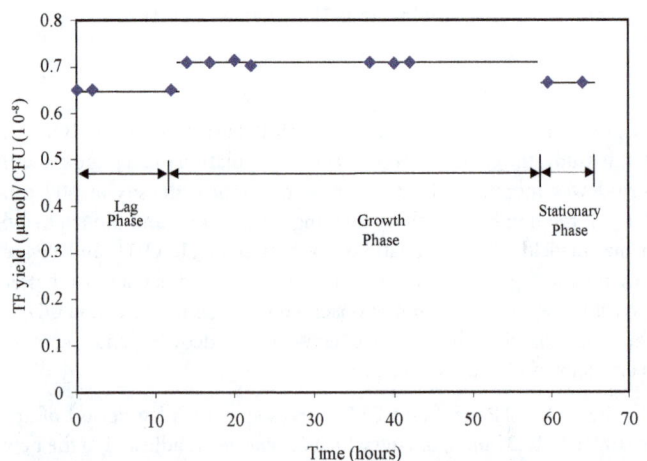

Figure 15: The activity of S. venezuelae as measured by specific TF yield during growth in MYM broth.

$(0.7 \times 10^{-8}$ μmol/CFU). However, the specific TF yield remained constant during the entire growth period indicating the accuracy of TF as a measure of the cell growth.

Number of extractions

It was observed that for most cases, all of the red colour was removed from the pelletized cells after three extractions. Thus, the dehydrogenase activity test for S. venezuelae using TTC should be carried out with at least three extractions in order to achieve the highest TF yield. Other researchers observed a higher recovery of formazan using sequential extractions [17, 23-25]. Green and Nahara (1980) found that extraction with ethanol followed by ethyl acetate achieved the highest yield from muscle cells, but lower yields were observed when the solvents were reversed. They also observed lower TF yields when one and two extractions were done with the same solvent and concluded that the number of extractions was a critical factor that must be considered when developing tests involving solubilization of formazan. In the study by Ghaly and Mahmoud (2006), two extractions with ethanol were found sufficient to extract TF from *Aspergillus niger* vegetative cells. This indicates that the number of extractions may depend on the type of cells.

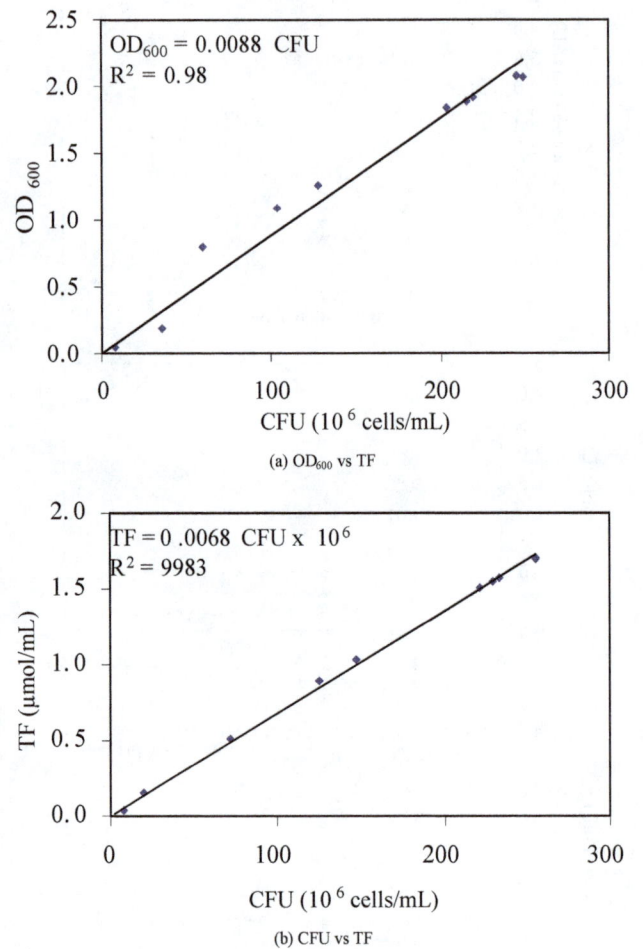

The specific TF yield (μmol/CFU) of the three extractions from the samples taken at 21, 37, 47, 62 and 64 hours was calculated Table 3. The results indicated that 2ⁿᵈ and 3ʳᵈ extractions increased the TF yield by

51% and 100% respectively. Also, the cell activity remained constant during the exponential growth period (at the 21, 37 and 47 hours) and slightly decreased during the stationary growth period (after 62 and 64 hours).

Type of solvent

It was noticed that after three washes with ethanol, the cell pellets were still red, which may explain why the TF yield from ethanol was lower. Thus, methanol is the better solvent to use for measuring dehydrogenase activity in *S. venezuelae* using TTC. In the study by Burton and Lanza (1986), the dehydrogenase activities of microbial consortia from sediment slurry samples were measured with TTC using different solvents. They ranked the solvents based on TF yield from highest to lowest as follows: tetrachloroethylene, acetone, propanol, ethanol and methanol. However, the differences between ethanol and methanol were not statistically significant at the 95% confidence level. Tayler and May (2000) [26] found that for the tetrazolium salt INT (2-(4-iodophenyl)-3-(4nitrophenyl)-5-phenyl tetrazolium chloride), methanol was found to be a better solvent than 95% ethanol for the extraction of INT-formazan from bacterial cells. Lee et al. (1988) [27] investigated the extraction of INT-formazan in activated sludge and filamentous bacteria and found methanol to be less effective than dimethylsulfoxide (DMSO) and tetrachloroethylene/acetone as solvents. However, by increasing the permeability of the cells with lysozyme or Triton X-100 prior to using methanol, the same level of formazan yield was achieved as with the other solvents. This indicates that the ability of the solvent to permeate the cells is a critical factor in the extraction of formazan. In the present study, methanol was better able to permeate the cells than ethanol.

The specific TF yield ($\mu mol/CFU$) after three extractions with ethanol and methanol at 21, 37, 47, 62 and 64 hours was calculated Table 4. The results showed that the TF yield for methanol was 259 % higher than that of ethanol. Also, the specific TF yield remained constant during the exponential growth period (at the 21, 37 and 47 hours) and slightly decreased during the stationary growth periods (at the 62 and 64 hours).

Medium pH

Several researchers reported high TF yields at different pH values. For example, Mahmoud and Ghaly (2004) found that at pH less than 7, no reduction of TTC occurred for both cheese whey (*Kluyveromyces fragilis*) and compost materials (mixed culture). Ghaly and Ben-Hassan (1993) [28] found that maximum dehydrogenase activities for both *Kluyveromyces fragilis* and *Candida pseudotropicalis* yeasts grown in cheese whey were at a pH of 7 and the activities were reduced at the acidic and basic levels of pH. Backor and Fahselt (2005) [29] reported that significantly acidic pHs (1.5 - 3) resulted in lower TTC reduction in lichens. Ghaly and Mahmoud (2006) [30] observed higher TF yield at a pH of 9 for *A. niger* grown in chitin. However, Mahmoud and Ghaly (2004) reported non-enzymatic reduction of TTC to TF at high pH values. In this study, a pH of 6 was the most appropriate value for measuring dehydrogenase activity in *S. venezuelae* during growth in MYM broth. It is not clear however if non-enzymatic reduction of TTC to TF occurred at pH 9.

Incubation time

The results showed slight increase in TF yield as when the incubation time increased from 1 hour to 4 hours. Several investigators reported that incubating samples for longer times increased the extent of TTC reduction to TF. Mahmoud and Ghaly (2006) reported that TF yield

Time (hours)	Specific TF Yield (10^{-8} µmol/CFU)		
	1st Extraction	2nd Extraction	3rd Extraction
21	0.35	0.53	0.70
37	0.35	0.52	0.70
47	0.35	0.53	0.70
62	0.33	0.50	0.67
64	0.33	0.50	0.67

Table 3: Specific TF yield for various ethanol extractions at various sampling time

Time (hours)	Specific TF Yield (10^{-8} µmol/CFU)	
	Ethanol	Methanol
21	0.70	2.51
37	0.70	2.51
47	0.70	2.51
62	0.67	2.49
64	0.67	2.49

Table 4: Specific TF yield for ethanol and methanol at various sampling time.

for *A. niger* grown in chitin increased exponentially when incubation time was increased form 1.5 hours to 4.5 hours. Ghaly and Ben-Hassan (1993) reported increased TF yield with increased incubation time for both *Kluyveromyces fragilis* and *Candida pseudotropicalis* yeasts grown in cheese whey, but the TF yield started to plateau after 80 hours in both cases. Mathew and Obbard (2001) [31] reported increased INT-formazan yield with increased incubation time for petroleum-contaminated beach sediments, but the TF yield started to level off after 22 hours of incubation. Although the TF yield obtained at an incubation period of 4 hours was slightly higher than the TF yield obtained at an incubation time of 1 hour, it is more practical to use 1 hour since the resulting TF yield is measurable and can provide good representation of cell growth and activity.

Incubation temperature

In the literature, there have been reports of higher TF yields as a result of increasing incubation temperature for microbial populations [30,32]. In this study, higher temperature (40°C and 50°C) showed a negative effect on the activity of the bacteria probably due to enzymatic inhibition at higher temperature [34]. According to Breed (1957) [33] *S. venezuelae* are soil bacteria, and therefore, achieve optimal growth with the lower temperature ranges investigated (22°C-30°C) in this study. Doull et al. (1993) reported a decreased growth for *S. venezuelae* at temperatures of 37° C and 42° C, compared to that at a control temperature of 27° C. Therefore, 30° C is an optimum temperature for measuring the dehydrogenase activity of *S. venezuelae* with TTC test.

Conclusions

A dehydrogenase activity measurement test using triphenyl tetrazolium chloride (TTC) was successfully developed for *Streptomyces venezuelae* growth in MYM broth. TF yield ($\mu mol/mL$) was related to the number of cells was measured by optical density (OD_{600}). It was found that methanol was able to extract a greater yield of the red triphenyl formazan (TF) than ethanol and that the TF yield increased with the number of extractions. High TF yields were observed at low pH value and/or low temperatures. Lower temperatures (22-30°C) required longer incubation time compared to higher temperature (40-50° C). Based on the results obtained from this study, the optimum conditions for measuring the dehydrogenase activity of *S. venezuelae* (reducing TTC to TF and extracting highest amount of TF) are three extractions with methanol after incubation time of 1 hour at a medium pH of 6 and incubation temperature of 30°C.

Acknowledgements

The research was funded by the National Science and Engineering Research Council (NSERC) of Canada.

References

1. Zheng J, Rix U, Mattingly C, Adams V, Chen Q, et al. (2005) Cytotoxic activities of new jadomycin derivatives. J. Antibi 58: 405-408.

2. Doull JL, Singh AK, Hoare M, Ayer SW (1994) Conditions for the production of jadomycin B by Streptomyces venezuelae ISP5230: effects of heat shock, ethanol treatment and phage infection. J. Ind Microbio 13: 120-125.

3. Jakeman DL, Farrell S, Young W, Doucet RJ, Timmons SC (2005) Novel jadomycins: incorporation of non-natural and natural amino acids. Bioorg Med Chem Lett 15: 1447–1449.

4. Burdock TJ, Giffin AH, Brooks MS, Ghaly AE (2008) Heat Balance analysis during the production of jadomycin C. Amer J Biochem Biotech 4: 7-18.

5. Jakeman DL, Graham CL, Young W, Vining LC (2006) Culture conditions improving the production of jadomycin B. J Ind Microbio Biotech 33: 767-772.

6. Balestra GM and Misaghi IJ (1997) Increasing the efficiency of the plate counting method for estimating bacterial diversity. J. Microbio Method 30: 111-117 .

7. Lindqvist R (2006) Estimation of staphylococcus aureus growth parameters from turbidity data: characterization of strain variation and comparison of methods. Applied and Environmental Microbiology 72: 4862-4870.

8. Lebaron P, Troussellier M, Got P (1994) Accucary of epifluorescence microscopy counts for direct estimates of bacterial numbers. J. Microbio Method 19: 89-94.

9. Beyer L, Waehendor C, Eisner DC, Knabe R (1992) Suitability of dehydrogenase activity assay as an index of soil biological activity. Biology and Fertility of Soils 16: 52-56.

10. Alisi CS, Nwanyanwu CE, Akujobi and Ibegbulem CO (2008) Inhibition of dehydrogenase activity in pathogenic bacteria isolates by aqueous extracts of Musa paradisiaca (Var Sapientum). African Journal of Biotechnology 7: 1821-1825.

11. Zhao X, Li H, Wu Q, Li Y, Zhao C, et al. (2010) Analysis of dehydrogenase activity in phytoremediation of composite pollution sediment. 2nd Conference on Environmental Science and Information Application Technology.

12. Mahmoud NS, Ghaly AE (2004) Influence of temperature and pH on the nonenzymatic reduction of triphenyltetrazolium chloride. Biotech Progr 20: 346-353.

13. Ruhlin A, Tyler G (1973) Heavy metal pollution and decomposition of spruce needle litter. Oikos 24: 402-416.

14. Rogers JE, Li SW (1985) Effect of metals and other inorganic ions on soil microbial activity: Soil dehydrogenase assay as a simple toxicity test. Bull Environ Contam Toxicol 34: 858-865.

15. Ghaly AE, Kok R, Ingrahm JM (1989) Growth rate determination of heterogeneous microbial population in swine manure. Appl Biochem Biotech 22:59-78.

16. Chander K, Brookes PC (1991) Is the dehydrogenase assay invalid as a method to estimate microbial activity in copper contaminated soils? Soil Biol Biochem 23: 909-915.

17. Friedel JK, Molter K, Fischer WR (1994) Comparison and improvement of methods for determining soil dehydrogenase activity by using triphenyltetrazolium chloride and iodonitrotetrazolium chloride. Biol Fert Soils 18: 291-296.

18. Beloti V, Barros M, de Freitas JC, Nero LA, de Souza JA, et al (1998) Frequency of 2, 3, 5-triphenyltetrazolium chloride (TTC) non-reducing bacteria in pasteurized milk. Rev Microbiol 30: 137-140.

19. Olga P, Peta K, Jelena M, Srdjan R (2008) Screening method for detection of hydrocarbon-oxidizing bacteria in oil-contaminated water and soil specimens. J. Microbio Meth 74: 110-113.

20. Mahmoud NS (2005) Novel biotechnological approach for the production of chitin and de-icing agents. Ph.D Thesis, Dalhousie University.

21. Abdel-Fattah YR (2007) Application of fractional factorial design for the development of production media for the pikromycin macrolide family by Streptomyces venezuelae. Trends Applied Sci. Res 2: 472-482.

22. Glazebrook M, Doull JL, Stuttardan C, Vining LC (1990) Sporulation of Streptomyces venezuelae in submerged cultures. J. General Microbio 136: 581-588.

23. Green JD, Narahara HT (1980) Assay of succinate dehydrogenase activity by the tetrazolium method: Evaluation of an improved technique in skeletal muscle fractions. J Histochem Cytochem 28: 408-412.

24. Burton GA, Lanza GR (1986) Variables affecting two electron transport system assays. Appl. Environ. Microbiol 51: 931-937

25. Mahmoud NS, Ghaly AE (2006) Optimum condition for measuring dehydrogenase activity of Aspergillus niger using TTC. Ame J Bioche Biotech 2: 186-194.

26. Tayler S, May E (2000) Investigation of the localization of bacterial activity on sandstone from ancient monuments. Int Biodeterior Biodegrad 46: 327-333.

27. Lee CW, Koopman B, Bitton G (1988) Evaluation of the formazan extraction step of INT-dehydrogenase assay. Toxic Assess 3: 41-54.

28. Ghaly AE, Ben-Hassan RM (1993) Dehydrogenase activity measurement in yeast fermentation. App Biochem Biotech 43: 77-91.

29. Backor M, Fahselt D (2005) Tetrazolium reduction as an indicator of environmental stress in lichens and isolated bionts. Environ Exper Bot 53: 125-133.

30. Ghaly AE, Mahmoud NS (2006) Optimum conditions for measuring dehydrogenase activity of Aspergillus niger using TTC. Ame J Biochem Biotech 2: 186-194.

31. Mathew M, Obbard JP (2001) Optimization of dehydrogenase activity measurements in beach sediments contaminated with petroleum hydrocarbons. Biotech Lett 23: 227-230.

32. Trevors JT (1984) Effect of substrate concentration, inorganic nitrogen, O_2 concentration, temperature and pH on dehydrogenase activity in soil. Plan Soi 77:285-293.

33. Breed RS (1957) Bergey's Manual of Determinative Bacteriology, 7th edn, Wiliams and Wilkins Company, Baltimore, MD.

34. Doull JL, Ayer SW, Singh AK, Thibault P (1993) Production of a novel polyketide antibiotic, jadomycin B, by Streptomyces venezuelae following heat-shock. J Antibiotics 46: 869-871.

31

The Kinetics of Fermentations in Sourdough Bread Stored at Different Temperature and Influence on Bread Quality

Venturi F*, Andrich G, Sanmartin C and Zinnai A

Department of Agriculture, Food and Environment (DAFE), University of Pisa, Italy

Abstract

The fermentations induced by the utilization of sourdough in bread-making, are able to enhance the qualitative properties of the final dough, improving its volume, texture and flavor, so to obtain a bread characterized by high qualitative properties and able to retard its staling process.

In particular the working conditions adopted can deeply affect the ratio occurring between the populations of lactic acid bacteria and alcoholic yeasts of the sourdough and then also the productions of the related metabolites, which deeply affect the organoleptic characteristics and then also the quality of the final bread.

The effect induced on the microbial and chemical composition of the sourdough by different values of the storage temperature utilized (13, 19 and 27°C) between two successive refreshes (~ 24 hours), was evaluated to put in evidence the different sensory characteristics assumed by the corresponding breads. The sensory profiles of the obtained breads evaluated by the descriptive analyses, were carried on by a panel of trained assessors using a sensorial sheet specifically developed for this purpose and characterized by unstructured graphical intensity scales; the reliability of judgments obtained was evaluated by statistical analysis.

So it was possible to put in evidence the high degree of correlation occurring among microbiological and chemical data of the sourdoughs and the sensorial characteristics of the corresponding bread.

Among the three storage temperatures of the sourdough, 19°C appears to be able to ensure the best organoleptic characteristics to this particular bread.

Keywords: Sourdough; Bread; Storage temperature; Lactic bacteria; Yeasts

Introduction

Worldwide bread consumption accounts for one of the largest consumed foodstuffs, with over 20 billion pounds (9 billion kg) of bread being produced annually. This demand has been driven by consumers' seeking convenient fresh products that provide a source of nutritional value. Consequently, freshness is a key component in consumer acceptability and choice of bread [1,2]. The use of sourdough is a technology of expanding interest for improvement of flavor, structure and stability of baked goods [3], because it can actively retard starch digestibility leading to low glycemic responses, modulate levels and bioaccessibility of bioactive compounds, and improve mineral bioavailability [4]. Sourdough is a mixture of flour and water fermented with lactic acid bacteria (LAB) and yeasts, which determine its characteristics in terms of acid production, aroma and leavening [5]. A slack flour dough is inoculated with microbial starter, "mother culture", which is constantly renewed in a cyclical way, using specified recipes and ripening conditions [6]. As a consequence, sourdough is a unique food ecosystem: it selects for LAB and yeasts which are adapted to the environment, and hosts highly specific microbial communities [7,8]. When used in optimized proportions, sourdough can improve volume, texture, flavor, nutritional value of bread and increase the shelf life by retarding the staling process and by protecting bread from mould and bacterial spoilage [7]. The changes in cereal matrix, potentially improving nutritional quality, are numerous. They include acid production, suggested to retard starch digestibility, and to adjust pH to a range which support the action of certain endogenous enzymes, thus changing the bioavailability pattern of minerals and potentially protective compounds in the blood circulation [9]. The action of enzymes during fermentation also causes hydrolysis and solubilisation of grain macromolecules, such as proteins (i.e.: gluten) and cell wall polysaccharides. It has also been suggested that degradation of gluten may render bread better suitable for celiac people [4]. Some strains of lactobacillus bacteria, involved in the process of souring of dough, produce an enzyme that breaks down a protein to be toxic to people with celiac disease [10].

Several types of traditional Italian bread have in common a long-time sourdough fermentation step and are known for their peculiar nutritional and qualitative traits in comparison with bakery products obtained with bread making protocols based on the use of selected yeasts and shorter fermentation step [11] so that many of them has been awarded the designation POD (i.e. bread of Altamura) or PGI (i.e. breads of Genzano, Matera and Coppia bread from Ferrara) recognized by the EU agricultural product quality policy.

As with other food processing, the challenge in fermenting cereal raw materials lies in the ability to combine good sensory quality with demonstrated nutritional and health benefits. Some of the mechanisms to improve and enhance the nutritional effects of fermented cereal systems are dependent on adjustment of the acidity for optimal action

Corresponding author: Venturi F, Department of Agriculture, Food and Environment (DAFE), University of Pisa, Italy

of the enzyme system present. Other mechanisms may be directly linked to other metabolites produced by yeast and lactic acid bacteria (l.a.b.), and then the control of different metabolic routes in the fermenting organisms becomes a key issue [9,4].

The aim of this research activity was to investigate on lactic bacteria and yeasts metabolic activity when different storage temperatures were adopted during the sourdough leavening, to determine how they affect the chemical composition of the sourdough and the sensory characteristics of the corresponding bread. So it would be possible to ensure a constant quality with elements of tipicity and recognisability that make a product different from each other similar.

Materials and Methods

Sourdough bread production

Every day the sourdough utilized was refreshed by following the methodology generally adopted in Tuscan bakeries. In particular it was utilized soft wheat flour type (Giambastiani mill, Lucca, Italy, "Consortium of Promotion and Protection of Tuscan Sourdough Bread"), water (60% of used flour) and a part (30% of flour) of sourdough coming from the previous baking [12]. According to the methodology described in the EU Regulation for PDO (Protected Designation of Origin) "Pane Toscano" [13,14] sourdough consist of a portion of dough from a previous preparation which, kept in a suitable environment, undergoes a gradual process of fermentation and acidification.

When suitably refreshed this sourdough, called "the starter" (biga), is able to initiate the rising process when combined with new dough [13].

In detail, at each cycle of refresh, the following amounts of the involved components are used to prepare the starter (biga): 1 kg of soft wheat flour type 0, at least 500 mL of sterilized water, at least 200 g of sourdough coming from the previous back-sloping process. After the mixing, an aliquot of the biga is stored in controlled conditions for at least 8 hours to be utilized in the refresh process of following day, while the remaining portion is maintained at room temperature for at least 2, 30 hours before to be cooked [14].

In the laboratory tests, the following amounts of the three involved components were used in order to refresh the sourdough utilized: sourdough coming from the previous back-sloping process (75 g), soft wheat flour type 0 (250 g) and sterilized water (150 g). These ingredients were mixed for 20 min in a kneading machine. A sample (50 g) of this dough was collected to be analyzed, while the remaining aliquot was further divided into two different portions, the first (150 g) was stored inside a temperature controlled cell for about 24 hours to be utilized in the refresh process of following day, while a part (200 g) of the remaining portion was maintained at 30°C for 4 hours before to be cooked (30 min at 230°C) inside an automatic oven to produce the wished bread [12]. Three different values of storage temperatures (13°C, 19°C and 27°C) were tested, while, for every temperature analyzed, the refresh procedure and the related sample collection was carried on for about two weeks.

Microbial analysis of sourdough

The sourdough utilized in this research was initially obtained by the Consortium for the Promotion and Protection of PDO "Pane Toscano" sourdough bread [13,14].

In order to follow the time evolution of microbial populations in the sourdough during the storage as a function of the working conditions adopted, during each experimental run several samples of sourdough were aseptically collected into sterilized jars and viable and cultivable populations of yeasts and lactic acid bacteria (CFU/g d.m.) were evaluated following ICC standard methods [15]: n° 146 for Yeasts (WL agar) and n° 147 for Bacteria (MRS agar modified) [12].

As reported in a previous paper [16], the microbial composition of the sourdough was already investigated: the Yeasts population was represented mainly by *Saccharomyces cereviasiae strain*, while several strains of Lactic Acid Bacteria were identified by DGGE analysis (*Lb sanfranciscensis, Lb brevis, Lb curvatus, Lb plantarum and Weissella confusa*).

Chemical characterisation: sourdough and cooked bread

Concentrations of the main fermentative metabolites (ethanol, D/L - lactic acid) produced in the sourdough during the storage time and in the bread samples after cooking, were determined by using specific Enzymic Kits (Megazyme), after pre-extraction with HCl 0,1N [12].

Sensorial analysis of bread (crust and crumb)

Descriptive analyses was used to determine the sensory profiles of the bread samples. A panel of trained assessors evaluated bread samples 2 hours after to be taken out of the oven. The crumb separately from bread crust were tasted to better identify the specific contribution of these two bread fractions.

A 20 g portion of each sample, including 10g of crust and 10g of crumb, was presented to assessors in 3-digit coded glasses covered with a glass cover; 10 min intervals were allowed between each sample. All samples were assessed in duplicate. For evaluation, each assessor was provided with filtered water and un-salted crackers and asked to cleanse their palate between tastings. In addition, assessors received a list of attributes that included definitions to aid in their assessments [1,17]. Sample attributes were scored on unstructured 100 mm line scales labeled from low at 5 mm to high at 95 mm intervals. For each attribute (Table 1), ratings on the unstructured line scale were measured geometrically to produce intensity values [18].

Sensorial evaluation of bread crumb			
Aspect	**Smell**	**Taste**	**Structure**
Intensity of color	Grain	Sweet	Elasticity
Percentage of white	Acetic Acid	Salty	Compressibility
Density	Hay	Acid	Deformability
Porosity	Yeast	Bitter	Resistance to chewing
Structure regularity	Rancid	Grain Aroma	Surface Moistness
Homogeneity	Frank	Hay Aroma	Compactness
		Yeast Aroma	Cohesiveness
		Astringent	Juiceness
		Aftertaste	
Sensorial evaluation of bread crust			
Aspect	**Smell**	**Taste**	**Structure**
Intensity of color	Intensity of smell	Sweet	Structure regularity
Regularity of color	Cereals	Salty	Hardness
Tonality of color	Roasted	Acid	Friability
(yellow/brown)	Burned	Bitter	Crispness
	Fragrant	Cereal Aroma	Resistance to
		Toasted Aroma	detachment crust/crumb
		Burned Aroma	Amount of residual
		Astringent	crumb after detachment
		Aftertaste	

Table 1: Sensorial descriptors utilized to describe separately the crumb and the crust of the bread.

Statistical analysis

To evaluate the statistical significance of the data obtained, the chemical and sensorial evaluations were performed in duplicate, the microbiological ones in triplicate. Statistical analysis of data was performed by one-way ANOVA (CoStat, Cohort 6.0), and means separation by the Tukey's HSD test at $P \leq 0.05$ of significance.

Results and Discussion

The effect of the storage temperature adopted on microbial evolution in sourdough

To evaluate the effect induced by the main working variables (temperature, number of refreshes, gas composition of surrounding atmosphere), it was adopted a rigid experimental protocol changing only a parameter one at a time. So, according to the protocol previously reported (see Materials and Methods), the effect of the storage temperature of sourdough employed in the refresh process, was evaluated in laboratory scale so that all the other working variables could be more efficiently controlled. Figures 1a and 1b report the evolutions of the concentrations (CFU/g d.m.) of yeasts and lactic acid bacteria as a function of storage time of the sourdough at the analyzed temperatures (13, 19 and 27°C).

While for yeasts (Figure 1a), the use of the lowest temperature (13°C) determines an insufficient rate of growth, at the intermediate temperature (19°C) the concentration of viable cells increases and reaches the maximum value at the highest temperature (27°C). On the contrary, bacteria (Figure 1b) appear less sensible to the lowest temperature than yeasts, and over time they reach very similar values at all the temperatures adopted.

The influence of storage temperature adopted on the production of the main microbial metabolites in the sourdough and in the cooked bread

Sourdough: By changing the storage temperature of sourdough, the yeasts concentrations varied such as those of their metabolites,

which are able to deeply affect the characteristics of the final bread. In particular, the concentration of ethanol in the sourdough (Figure 2a) was too low at the storage temperature of 13°C, so that the leavening process, due to carbon dioxide, which is produced at equimolar amount of ethanol by yeasts, could be considered not well developed. Although the concentration of lactic acid bacteria was similar over time at the different temperatures studied, the lactic acid production seems to be higher when temperature increases. So at 27°C, the specific activity of l.a.b. appears higher than at the other temperatures adopted (Figure 2b).

Cooked bread: As a consequence of the influence of storage temperature on the microbial and chemical composition of sourdough, it was possible to put in evidence that when the sourdough was stored at 19°C also the corresponding bread (crust and crumb fractions) was characterized by lower concentrations of acidic compounds (D+L-lactic acid and acetic acid) and higher concentration of ethanol (Figure 3)

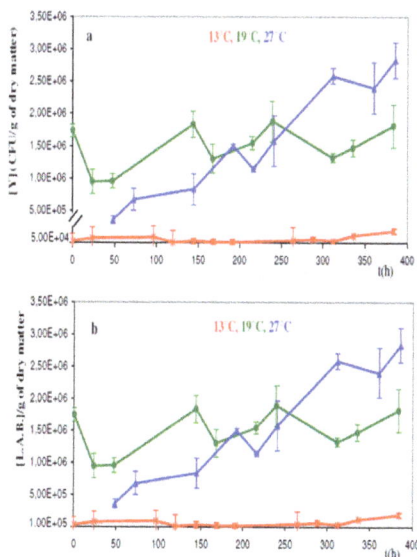

Figure 2: Time evolution of experimental points related to the concentrations (mmol/g d.m.) of Acetic Acid (a) and D+L-lactic acid (b) found in the sourdough as a function of temperatures adopted.

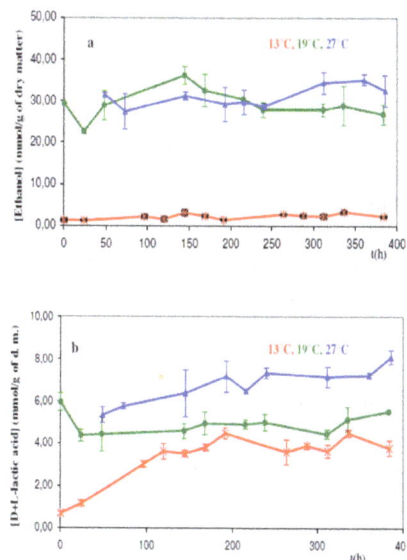

Figure 1: Time evolution of experimental points related to the concentrations (CFU/g d.m.) of yeasts (a) and lactic acid bacteria (b) found in the sourdough as a function of temperatures adopted.

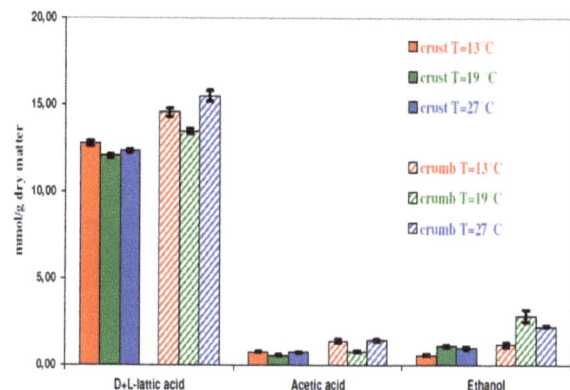

Figure 3: Concentration of the main fermentative metabolites in crust and crumb fractions as a function of temperature adopted during the storage of the corresponding sourdough.

than the breads obtained by the other sourdoughs. The differences are more evident in the crumb fraction than in the crust one.

The influence of storage temperature adopted on sensorial expression of bread

By changing the storage temperature of the sourdough, yeasts and lactic acid bacteria metabolites varied greatly, affecting deeply the organoleptic characteristics of the cooked product. Because the attributes of the view and the smell are greatly influenced by the working conditions adopted during cooking process, which were the same for all the bread samples tasted, the differences in sensorial expression of the breads obtained by the three different sourdoughs were mainly related to the attributes of the taste.

In Figures 4a and 4b, the spider plots of the attributes of the taste, respectively related to crust and crumb of the breads obtained by the sourdoughs stored at different temperatures, are reported: when lactic acid bacteria and yeasts were present in a good concentration (bacteria/yeasts ≈ 10/1), the bread obtained was well leavened with a concentration of acidic compounds more tolerable by consumers. In fact, as shown in Figures 4a,4b, the sensorial analysis of the breads obtained in the different experimental runs of baking, produced the best results when sourdough was maintained at 19°C (sweet, low acidity and aftertaste, more elevated wheat and yeast flavours of crumb fraction as well as higher expression of aromatic components in crust fraction). On the contrary, when the temperature of sourdough storage was 13°C, yeasts were present at too low concentration, so that the bread obtained was not enough leavened and characterised by an unpleasant acid perception due to the reduced value assumed by the ratio lactic bacteria/yeasts (≈ 100/1).

In synthesis, the evaluation of the overall pleasantness (Figure 5) associated to the three different bread produced, indicated that the best

sensorial profile was attributed to the sample of bread obtained by the sourdough stored at 19°C.

Conclusions

Although some studies was already carried out to correlate microbial metabolites of many sourdoughs to the working conditions adopted [5-9], no information were still available concerning the typical Tuscan bread production.

Because the storage temperature adopted during the sourdough leavening deeply affects not only the chemical composition of sourdough but also the sensorial characteristics of the bread, some chemical indexes (Ethanol/D+L-lactic acid; Ethanol/ (D+L-lactic acid + Acetic acid); D+L-lactic acid/Acetic acid) were calculated by the ratio of the concentration of the main microbial metabolites determined in both crust and crumb fractions of Tuscan sourdough bread. Among the indexes which was calculated, the ratio D+L-Lactic acid/Acetic acid (Table 2), that could represent a good marker of activity of homo/ hetero lactic bacteria fermenting strains, has been the best index able to provide some reliable indications about the quality of the bread. In particular, when the temperature utilized for the storage of sourdough was too low (13°C) or too high (27°C) this index showed values very close to each other and significantly lower than that showed by the crumb and the crust of the bread obtained by the sourdough stored at 19°C and characterized by the best sensorial profile.

The use of this new approach based on the integration of the information coming from both kinetic, chemical and sensorial data, has made possible the identification of the best operative conditions which could be adopted during Tuscan sourdough storage. The same experimental procedure make it possible to evaluate the effect induced by the other main working variables (number of refreshes, gas composition of surrounding atmosphere, etc.) on the properties of the

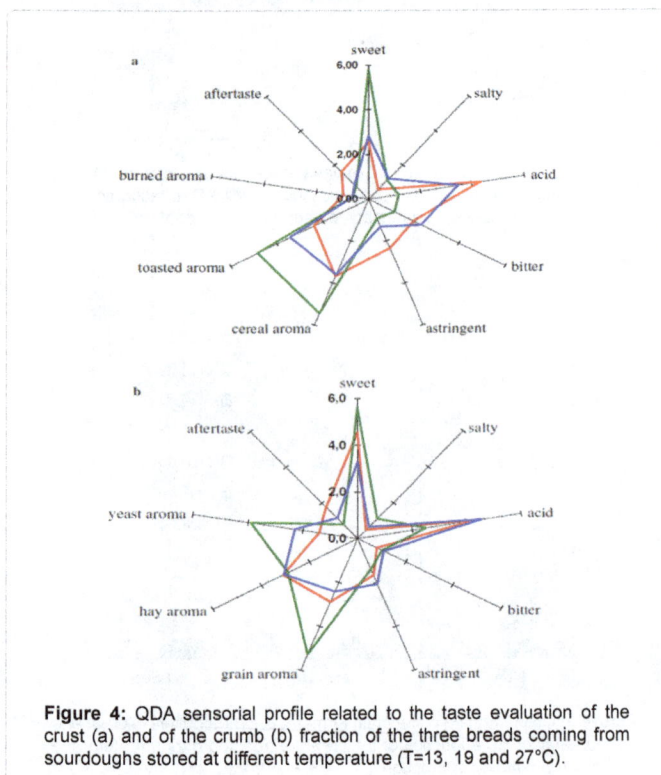

Figure 4: QDA sensorial profile related to the taste evaluation of the crust (a) and of the crumb (b) fraction of the three breads coming from sourdoughs stored at different temperature (T=13, 19 and 27°C).

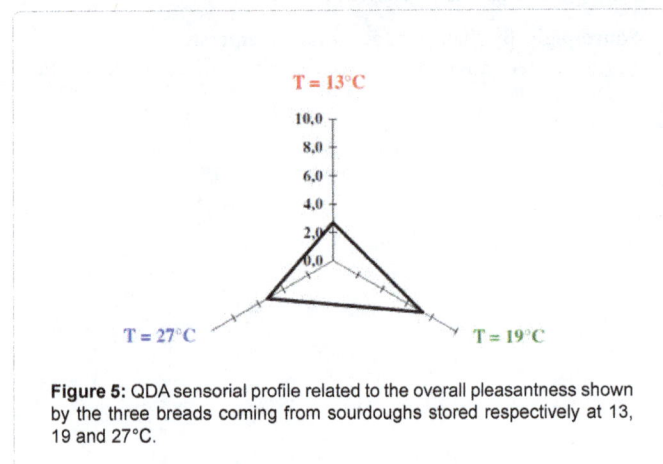

Figure 5: QDA sensorial profile related to the overall pleasantness shown by the three breads coming from sourdoughs stored respectively at 13, 19 and 27°C.

Sample	Ethanol / D+L−lactic acid	Ethanol / D+L−Lactic ac + Acetic ac	D+L−Lactic ac / Acetic ac
Crust (T=13°C)	0.05	0.003	15.97
Crust (T=19°C)	0.09	0.007	20.25
Crust (T=27°C)	0.08	0.006	15.87
Crumb (T=13°C)	0.08	0.005	10.16
Crumb (T=19°C)	0.21	0.015	16.57
Crumb (T=27°C)	0.15	0.009	10.50

Table 2: Chemical indexes obtained by the chemical markers determined in crust and crumb fractions of the different breads.

final product. So it would be possible not only to obtain the sourdough bread characterised by the most favourable chemical and sensory properties but also to identify markers/parameters able to protect this product against fraud and imitation.

This experimental study represent part of a research project which was developed with the aim to obtain the PDO appellation for Tuscan sourdough bread [14].

References

1. Heenan SP, Dufour JP, Hamid N, Harvey W, Delahunty CM (2008) The sensory quality of fresh bread: Descriptive attributes and consumer perceptions. Food Res Int 41: 989-997.

2. Lambert JL, Le-Bail A, Zuniga R, Van-Haesendonck I, Van-Zeveren E, et al. (2009) The attitudes of European consumers toward innovation in bread; interest of the consumers toward selected quality attributes. J Sens Stud 24: 204-219.

3. Rehman S, Paterson A, Piggott JR (2006) Flavour in sourdough breads: a review. Trends Food Sci Tech 17: 557-566.

4. Poutanen K, Flander L, Katina K (2009) Sourdough and cereal fermentation in a nutritional perspective. Food Microbiol 26: 693-699.

5. Gül H, Özçelik S, Saˇgdıç O, Certel M (2005) Sourdough bread production with lactobacilli and S. cerevisiae isolated from sourdoughs. Process Biochem 40: 691-697.

6. Mondal A, Datta AK (2008) Bread baking - A review. J Food Eng 86: 465-474.

7. De Vuyst L, Vancanneyt M (2007) Biodiversity and identification of sourdough lactic acid bacteria. Food Microbiol 24: 120-127.

8. De Vuyst L, Van Kerrebroeck S, Harth H, Huys G, Daniel HM, et al. (2013) Microbial ecology of sourdough fermentations: Diverse or uniform? Food Microbiol, in press.

9. Katina K, Heinio RL, Autio K, Poutanen K (2006) Optimization of sourdough process for improved sensory profile and texture of wheat bread. Food Sci Technol-LEB 39: 1189-1202.

10. Moroni AV, Dal Bello F, Arendt EK (2009) Sourdough in gluten-free breadmaking: an ancient technology to solve a novel issue? Food Microbiol 26: 676-684.

11. Brescia MA, Sacco D, Sgaramella A, Pasqualone A, Simeone R, et al. (2007) Characterisation of different typical Italian breads by means of traditional, spectroscopic and image analyses. Food Chem 104: 429-438.

12. Andreotti A (2010) I principali aspetti chimico composizionali, tecnologici e microbiologici coinvolti nella produzione del Pane Toscano a lievitazione naturale. PhD Thesis, University of Pisa, DAFE.

13. Disciplinare di Produzione della denominazione di origine protetta «PANE TOSCANO», Ministero per le Politiche Agricole, Alimentari e Forestali.

14. Council Regulation (EC) No 510/2006 on the protection of geographical indications and designations of origin for agricultural products and foodstuffs 'PANE TOSCANO' EC No: IT-PDO-0005-01016-18.07.2012, C235/19-C235/22.

15. ICC (International association for cereal science and technology) (1994) General principles of the available ICC standard methods. International Association for Cereal Science and Technology.

16. Andreotti A, Agnolucci M, Zinnai A, Venturi F, Andrich G, et al. (2012) Tracciabilità e qualità del "pane quotidiano". In: Alimentazione e qualità della vita. Donne e cibo: risorse, salute, immagine. Ed. Plus, Pisa University Press.

17. Elia M (2011) A procedure for sensory evaluation of bread: Protocol developed by a trained panel. J Sens Stud 26: 269-277.

18. Heenan SP, Dufour JP, Hamid N, Harvey W, Delahunty CM (2009) Characterisation of fresh bread flavour: Relationships between sensory characteristics and volatile composition. Food Chem 116: 249-257.

32

Structural Elucidation and Antioxidant Activity of a Polysaccharide from Mycelia Fermentation of *Hirsutella sinensis* Isolated from *Ophiocordyceps sinensis*

Jin-Hua Liu[1], Ze-Jian Wang[1]*, Yu-hua Wang[2], Ju Chu[1]*, Ying-Ping, Zhuang[1] and Si-Liang Zhang[1]

[1]*State Key Laboratory ofBioreactor Engineering, East China University of Science and Technology, Shanghai 200237, China*

[2]*Zhu Feng Ophiocordyceps Pharmacy Corporation, Ltd. P R Qinghai, China*

Abstract

The structure and antioxidant activity of a polysaccharide from mycelia fermentation of *Hirsutella sinensis* were analyzed. The natural active component water-soluble polysaccharides was isolated from mycelia, and three polysaccharide fractions HSP-1, HSP-2, and HSP-3 were purified with chromatography and the structures were identified. The structural characteristics determination with a combination of chemical and instrumental analysis methods showed that the mainly component HSP-1 was about 1.7×10^4 Da, and composed of glucose, mannose and galactose at a molar ratio of 4.5:1.0:1.4. Further researches revealed that HSP-1 was a branched polysaccharide possessing a backbone of $(1{\rightarrow}4)$-α-D-glucose residues (~70%), $(1{\rightarrow}4)$-α-D-mannose residues (~15%) and $(1{\rightarrow}4)$-α-D-galactose residues (~15%). The branches were at the $(1,2,4,6{\rightarrow})$-α-D-glucose residues (~8%) of the backbone, mainly composed of $(1{\rightarrow}4)$-α-D-glucose residues, $(1{\rightarrow}4)$-α-D-galcatose residues, $(1{\rightarrow}4)$-α-D-mannose residues, and terminated with α-D-galactose residues. The *in vitro* antioxidant assay proved HSP-1 possessed the hydroxyl radical-scavenging activity with an IC_{50} value of 0.834 mg/mL.

Keywords: *Ophiocordyceps sinensis*; *Hirsutella sinensis*; Mycelia fermentation; Polysaccharide structure; Antioxidant activity

Nomenclature: HSP: *Hirsutella sinensis* polysaccharide; IC_{50}: The polysaccharide concentration for hemi-inhibitable hydroxyl radical; IPS: Intracellular polysaccharides

Introduction

In recent years, many polysaccharides and polysaccharide-protein complexes isolated from fungi have been used as a source of therapeutic agents [1,2]. Many studies showed that fungi polysaccharide have series of pharmacological action, including anti-oxidation, hypoglycemic, boost immunity, anti-fatigue and anti-cancer [3-7]. Therefore, it is significant to discover and extract the valuable polysaccharides from fungi as safe compounds for functional foods or medicine.

O. sinensis, called *Cordyceps* or *Dong Chong Xia Cao* in China, is one of the most valuable traditional Chinese medicinal fungi. It is generally used to nourish the kidney, moisten the lung, fight fatigue and enhance immunity [8]. Furthermore, the wild *O. sinensis* is exiguity and expensive in the market, so the mycelia fermentation has become to an economical method to meet large requirement of the market [5]. Several Intracellular Polysaccharides (IPS) have been purified from the mycelia of *O. sinensis*, and the molecular structures have been elucidated [9-11]. In this study, the mycelia we used are called *Hirsutella sinensis*, which is a novel fungus isolated from the fruiting body of the wild *O. sinensis* on the Tibetan Plateau. It has been identified as an anamorphic fungus by the Chinese Academy of Sciences. Numerous liquid fermentations have been conducted to optimize the production of mycelia biomass. However, the IPS purified from *H. sinensis* has not been reported yet.

Since the structure of IPS is closely related with its functions, it would be of interest for an in-depth research. The aim of this study is to characterize the molecular structure and antioxidant activity of the polysaccharide, HSP-1, which was isolated and purified from the crude IPS produced by the *H.sinensis* liquid fermentation.

Materials and Methods

Fungus and mycelia fermentation

The strain used in this research was *Hirsutella sinensis*, which was identified as the anamorph of *Ophiocordyceps sinensis* [12,13]. 0.8 L culture was pre-cultivated in a 1 L flask for 10 days at 16°C, 180 rpm and natural pH. The medium consists of, in (g/L): glucose (30.0), yeast extract (22.0), KH_2PO_4(0.1), $MgSO_4.7H_2O$ (0.05). Then it was transferred to a 50 L fermentation tank by 10% (v/v) inoculums size in a liquid medium, which containing, in (g/L): glucose (40.0), yeast extract (33.0), KH_2PO_4(0.1), $MgSO_4.7H_2O$ (0.05). The total volume of the fermentation medium was 30 L, cultivated for 10 days at 16°C, 200rpm. During the fermentation process, the pHwas natural. After the fermentation finished, the mycelium was collected and dried at 60°C for 24 h.

Isolation and purification of polysaccharide

The dried *H. sinensis* mycelium was extracted by deionized water at 100°C for 90 min, and the ratio of solid to liquid was 1:10 (w/v), which was repeated three times. The supernatant was collected

*Corresponding authors: Ze-Jian Wang, State Key Laboratory of Bioreactor Engineering, East China University of Science and Technology P.O. Box 329, 130 Meilong Road, Shanghai 200237 People's Republic of China

Ju Chu, State Key Laboratory of Bioreactor Engineering, East China University of Science and Technology P.O. Box 329, 130 Meilong Road, Shanghai 200237 People's Republic of China

after and concentrated under reduced pressure. It was mixed with 3 volume of ethanol and precipitated for 24 h at 4°C to obtain the crude polysaccharide. The protein was removed by Sevag method [14], and dialyzed against running water and deionized water for 48 h. After the non-dialyzable phase was precipitated with 3 volume of ethanol, the precipitation was collected by centrifugation. The precipitation was washed with absolute ethyl alcohol, acetone and diethyl ether, and finally was dried under vacuum.

The obtained crude polysaccharide was purified by DEAE-cellulose column eluted with gradient NaCl aqueous solution (0-1 M). The fractions were collected and detected by phenol-sulfuric acid [15]. The resulting fractions were further purified by Sephadex G-100 column eluted with deionized water. Three polysaccharide fractions were detected after purification process, which were termed HSP-1, HSP-2, and HSP-3 respectively. The main fraction HSP-1 was used for the further structure elucidation and antioxidant activity assay.

Homogeneity and molecular weight measurement of HSP-1

The homogeneity of HSP-1 was measured by Sephacryl S-300 HR column chromatography and Ultraviolet (UV) spectroscopy scanning. Gel chromatographic method [9,16] was used to measure the molecular weight of HSP-1. The blue dextran 2000 and different weight-average molecular weights standard dextrans T-500, T-70, T-40 and T-10 were passed through the Sephacryl S-300 HR column, eluted with deionized water at a flow rate of 0.2 mL/min. The standard curve was established using the elution volumes plotted against the negative logarithms of their known molecular weights. HSP-1 (5 mg) dissolved in deionized water (0.5 mL)passed through the column, so the molecular weight of HSP-1 was obtained by plotting the elution volume with the standard curve.

Analysis of monosaccharide composition

Dried HSP-1 (5 mg) was hydrolyzed with 2 mL trifluoroacetic acid (TFA) (2 M) at 100°C for 8 h. After the hydrolysis finished, excess acid evaporated under reduced pressure and then washed three times with absolute ethyl alcohol. The hydrolysate with 20 mg methoxylamine hydrochloride was dissolved in 1 mL pyridine at 70°C for 2 h, then 150 μL of the sample was mixed with 100 μL of bis(trimethylsilyl) trifluoroacetamide (trimethylchlorosilane 1%) derivatized at 70°C for 1 h. The silylation derivatized sample was ready for Gas Chromatograph-Mass Spectrometer (GC-MS) analysis. GC-MS analysis was conducted with an Agilent Technologies 7890A/5975C instrument, using a HP-5MS capillary column (30 m × 0.25 mm × 0.25 nm). The initial column temperature was kept at 70°Cfor 4 min, first increased to 200°C at 3°C/min, kept for 0 min, and then increased to 300°C at 10°C/min, kept for 5 min. The ionization potential was 70 eV and the temperature of the ion source was 280°C. Similarly, the standard monosaccharides D-glucose, D-mannose, D-galactose, D-arabinose D-xylose and D-inositol were derivatized.

Periodate oxidation-Smith degradation

The HSP-1 (20 mg) was oxidized with 15 mM $NaIO_4$ (20 mL) and kept in the dark at 4°C. 100 μL aliquots were withdrawn for every 12 h, the aliquots were diluted to 25 mL and tested using a spectrophometer at 223 nm [17]. The oxidation was stopped by adding 1 mL glycol until the absorbance did not change any more.The production of HCOOH was measured by titration with 0.005 M NaOH and the consumption of $NaIO_4$ was determined by spectrophotometric methods [18]. The solution was dialyzed against running water and deionized water each

for 24 h, reduced by $NaBH_4$ (70 mg) overnight, neutralized with 50% acetic acid, dialyzed against running water and deionized water each for 24 h, and vacuum dried. The product was hydrolyzed with 2 M trifluoroacetic acid (TFA) at 100°C for 8 h in a sealed tube and analyzed by GC-MS with the same method as mentioned above for the analysis of monosaccharide composition.

Methylation analysis

The HSP-1 (20 mg) was methylated by the method of Needs and Selvendran [19]. The methylated HSP-1 was depolymerized with 90% HCOOH for 6 h at 100°C. The residues were hydrolyzed with 2 M TFA (2 mL) for 8 h after removal of the HCOOH. The resulting products were silylation derivatized and analyzed by GC-MS as the method mentioned for the analysis of monosaccharide composition. The methylated sugar linkages were identified on the basis of the retention time and fragmentation patterns [19,20].

Partial hydrolysis of HSP-1

The HSP-1 (100 mg) was hydrolyzed by 0.05 M TFA (4 mL) at 100°C for 12 h, After dilution the products with deionized water and dialyzed against deionized water for 24 h in a dialysis bag (cut off 3, 500 Da), the solution in the bag was diluted with ethanol. The fraction out of the bag and the precipitate and supernatant in the bag were collected, driedand hydrolyzed. GC-MS analysis was carried out to understand the monosaccharide composition.

IR analysis

1mg HSP-1 was ground and mixed with KBr before being flaked. TheInfra-Red (IR) spectrums werescanned in the range of 400-4000 cm^{-1}on a Nicolet 6700 Fourier transformed IR spectrophotometer.

NMR spectroscopy

20 mg HSP-1 was dried under vacuum over P_2O_5 for several days and then put it into a 5-mm Nuclear Magnetic Resonance (NMR) tube with 1 mL of D_2O. NMR (^1H, ^{13}C) spectra were accomplished with a Bruker 400 spectrometer.

Hydroxyl radical-scavenging activity

Hydroxyl radical-scavenging assay was carried out by Fenton's reaction method described by He et al. [21] with a slight modification. Briefly, the reaction mixture included 1.0 mL of brilliant green (0.435 mM), 2.0 mL of $FeSO_4$ (0.5 mM), 1.5 mL of H_2O_2 (3.0%), 0.5 mL samples of different concentrations and was made up to 5.0 mL with deionized water. The absorbance of the reaction mixture was measured at 624 nm after incubating at room temperature for 20 min.

The hydroxyl radicals can eliminate the brilliant green, so the scavenging ability for hydroxyl radical can be characterized by the absorbance variation of the reaction mixture. The hydroxyl radical-scavenging activity can be expressed as:

$$\text{Scavenging rate (\%)} = (A_O - A_S)/(A - A_O) \times 100\% \quad (1)$$

Where, A_S is the absorbance of the mixture with the sample, A_O is the absorbance of the control without the sample and A is the absorbance in the absence of the sample and Fenton reaction system. Vc was used as a control.

Statistical analysis

The data were expressed as means ± SD. A statistical analysis of

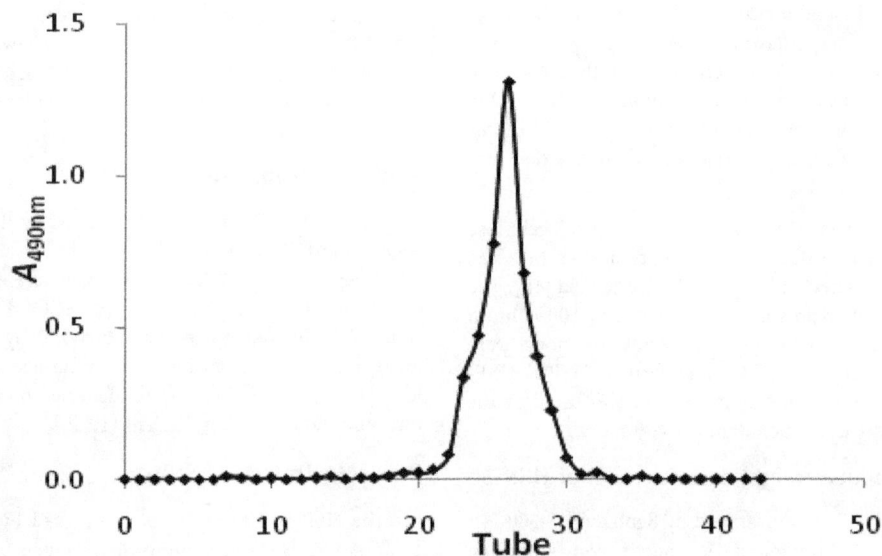

Figure 1: Profile of HSP-1 tested with Sephacryl S-300 HR column chromatography.

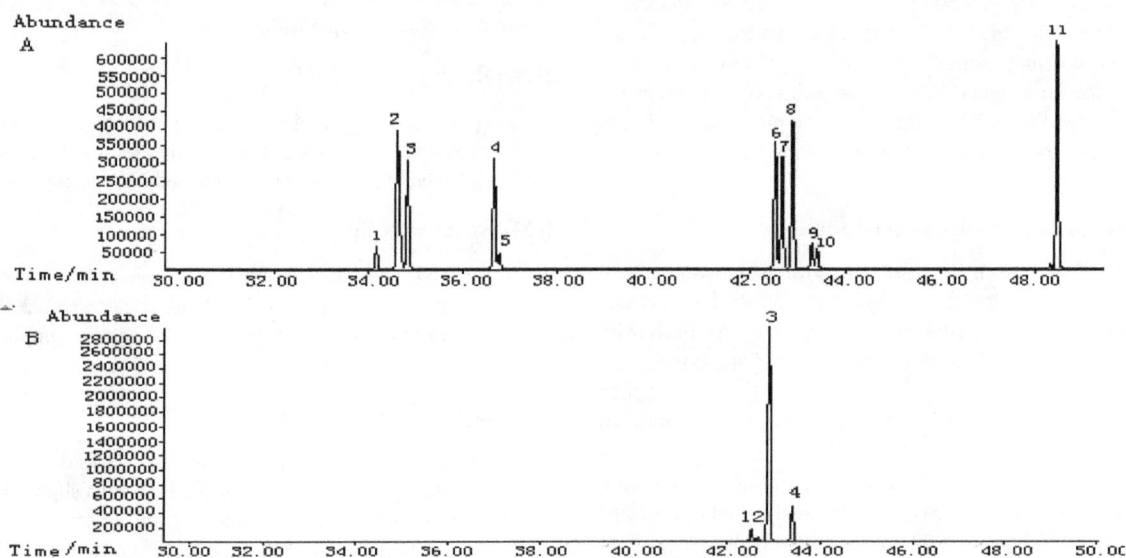

Figure 2: GC-MS profiles of silylation derivatized (A) monosaccharide standards (Peaks: Xyl (1,3); Ara (2); Gal (4, 5, 9, 10); Man (6, 7); Glu (8); Ino (11)) and (B) HSP-1 constituents (Peaks: Man (1, 2); Glu (3); Gal (4)). Three monosaccharides were identified in the hydrolysate of HSP-1, and the composition was D-glucose, D-mannose and D-galactose with a molar ratio of 4.5:1.0:1.4.

the data was assessed by Student's T-test. $P < 0.05$ was considered as a statistically significant difference.

Results and Discussion

Isolation, purification and composition of polysaccharide

Crude polysaccharide was extracted from the mycelium of *Hirsutella sinensis* with a yield of 8.239%. After the polysaccharidefractionated by DEAE-cellulose and Sephadex G-100 column with phenol sulfuric acid to detect the polysaccharide distribution, three main fractions were obtained and termed HSP-1, HSP-2 and HSP-3, respectively. The main fraction HSP-1 was chosen for subsequent analysis.

The purification of HSP-1 was tested with the Sephacryl S-300 HR column using phenol sulfuric acid to detect the polysaccharide distribution. A single and symmetric sharp peak was obtained (Figure 1). The UV spectrum of the HSP-1 solution showed no absorption at 260 nm and 280 nm, indicating that HSP-1 did not contain either nucleic acid or protein; therefore it is a homogeneous polysaccharide. Using different dextran markers passed through a Sephacryl S-300 HR column, the average molecular weight of HSP-1 was 1.7×10^4 Da. The total carbohydrate content was 92.4% determined by the phenol-sulfuric acid method.

The monosaccharide composition of HSP-1 was measured by silylation derivatization and GC-MS analysis (Figure 2). Three

Methylated sugar	Molar ratio	Linkages
2,3,6-Me_3-Gal	2.97	1,4-
2,3,4,6-Me_4-Gal	1.89	T-
2,3,6-Me_3-Man	4.12	1,4-
2,3,6-Me_3-Glu	17.88	1,4-
3-Me-Glu	1.00	1,2,4,6-

Table 1: GC-MS results from the methylated product of HSP-1.

Fraction	Molar ratio		
	D-glucose	D-mannose	D-galactose
A	4.835	1	1.422
B	5.724	1	1.463
C	2.944	1	6.665

A: precipitation in the dialysis bag
B: supernatant in the dialysis bag
C: fraction out of the dialysis bag

Table 2: GC-MS analysis results of fractions from partial acid hydrolysis of HSP-1.

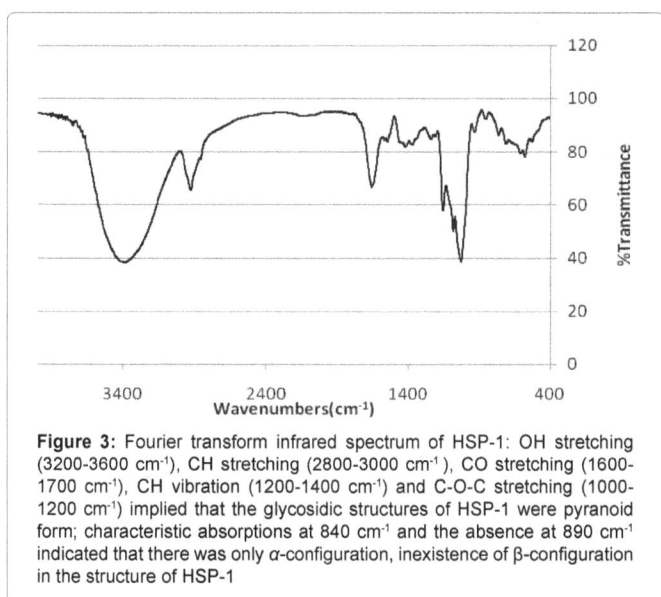

Figure 3: Fourier transform infrared spectrum of HSP-1: OH stretching (3200-3600 cm^{-1}), CH stretching (2800-3000 cm^{-1}), CO stretching (1600-1700 cm^{-1}), CH vibration (1200-1400 cm^{-1}) and C-O-C stretching (1000-1200 cm^{-1}) implied that the glycosidic structures of HSP-1 were pyranoid form; characteristic absorptions at 840 cm^{-1} and the absence at 890 cm^{-1} indicated that there was only α-configuration, inexistence of β-configuration in the structure of HSP-1

monosaccharides were identified in the hydrolysates of HSP-1, and the composition was D-glucose, D-mannose and D-galactose with a molar ratio of 4.5:1.0:1.4.

Structure characterization of HSP-1

The results of periodate oxidation which showed the ratio of HIO$_4$ consumption to formic acid production was 2.644 (larger than 2). It meant there were 1,2- or 1,2,6- or 1,4- or 1,4,6-linkages in the HSP-1 structure which was not able to produce formic acid when being oxidized. Furthermore, Glycerin, erythritol and D-glucose were detected from the periodate oxidation product of HSP-1 by GC-MS. The presence of glycerin suggested that there were 1- or 1,6- or 1,2- or 1,2,6-linkages in the structure of HSP-1; the presence of erythritol suggested that there should have 1,4- or 1,4,6-linkage; the presence of glucose suggested that a part of glucose in the HSP-1 should be in 1,3- or 1,3,6- or 1,2,3- or 1,2,4- or 1,3,4- or 1,2,3,4-linkage which could not be oxidized by periodate.

The methylated HSP-1 was analyzed by GC-MS (Table 1), and the result showed five components, namely 2,3,6-Me_3-Gal, 2,3,4,6-Me_4-Gal, 2,3,6-Me_3-Man, 2,3,6-Me_3-Glu, 3-Me-Glu in a molar ratio of 2.97: 1.89: 4.12: 17.88: 1.00 (about 3:2:4:18:1). This pattern of linkage was in good agreement with the results by periodate oxidation and Smith

degradation, which showed a good correlation between terminal and the branched residues. Furthermore, the molar ratio was fitted well with the monosaccharide composition and the ratio of HSP-1 measured above.

The monosaccharide compositions of HSP-1 fractions derived from partial acid hydrolysis were subjected to GC-MS analysis (Table 2). The results showed that the precipitation in the dialysis bag (fraction A) which mainly composed of the backbone structure of HSP-1, was mainly consisted of D-glucose with a little amount of D-mannose and D-galactose in a molar ratio of 4.835:1:1.422; the supernatant in the dialysis bag (fraction B) which mainly composed of the branch chains of HSP-1 was consisted of D-glucose, D-mannose and D-galactose in a molar ratio of 5.724:1:1.463; D-glucose, D-mannose and D-galactose in a molar ratio of 2.944:1:6.665 were presented out of the dialysis bag (fraction C), which indicated its existence in the terminal position of the branch chains. The molar ratio from the partial acid hydrolysis matched well with the monosaccharide composition and ratio of HSP-1 and the methylation results measured above.

The Fourier Transform Infrared Spectroscopy (FT-IR) spectrum of HSP-1 was presented in Figure 3. In the spectrum, the attributions of the main absorptions were the characteristic of glycosidic structures and related to OH stretching (3200-3600 cm^{-1}); CH stretching (2800-3000 cm^{-1}); CO stretching (1600-1700 cm^{-1}); CH vibration (1200-1400 cm^{-1}); C-O-C stretching (1000-1200 cm^{-1}). It implied that the glycosidic structures of HSP-1 were pyranoid form. Moreover, the characteristic absorptions at 840 cm^{-1} and the absence at 890 cm^{-1} indicated that there was only α-configuration, inexistence of β-configuration in the structure of HSP-1.

In the ^{1}H NMR spectrum (Figure 4A), the chemical shifted from 4.9 to 5.6 ppm corresponding to α configuration [22]. The region shifted from 3.3 to 4.2 ppm was assigned to protons of carbons C2 to C6 of glycosidic ring [23]. In the ^{13}C NMR spectrum (Figure 4B), six strong signals between 60 to 100 ppm were attributed to C1, C2, C3,

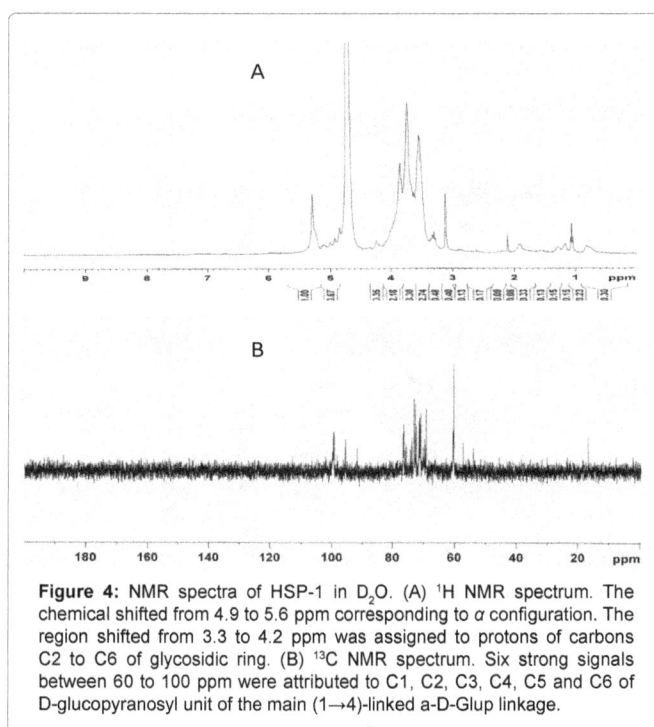

Figure 4: NMR spectra of HSP-1 in D$_2$O. (A) ^{1}H NMR spectrum. The chemical shifted from 4.9 to 5.6 ppm corresponding to α configuration. The region shifted from 3.3 to 4.2 ppm was assigned to protons of carbons C2 to C6 of glycosidic ring. (B) ^{13}C NMR spectrum. Six strong signals between 60 to 100 ppm were attributed to C1, C2, C3, C4, C5 and C6 of D-glucopyranosyl unit of the main (1→4)-linked a-D-Glup linkage.

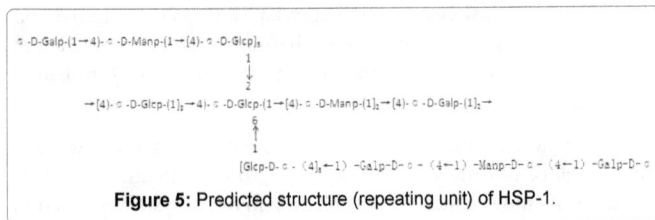

Figure 5: Predicted structure (repeating unit) of HSP-1.

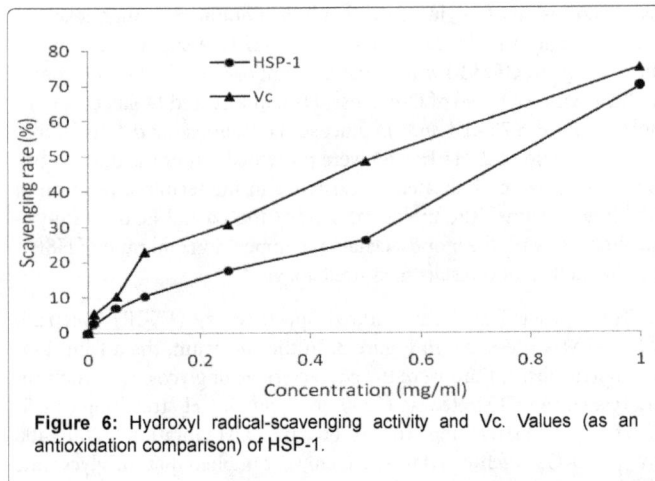

Figure 6: Hydroxyl radical-scavenging activity and Vc. Values (as an antioxidation comparison) of HSP-1.

C4, C5 and C6 of D-glucopyranosyl unit of the main (1→4)-linked a-D-Glu$p$ linkage [22]. In the anomeric carbon region of 95-105 ppm, five main signals were detected, namely δ 95.70, 95.83, 99.44, 99.53, and 99.66. Based on the data obtained in the literatures [6,24-26], in the anomeric carbon region, signal at δ 95.70 was attributed to C-1 of (1→4)- linked α-Gal$p$, δ 95.83 to C-1 of (1→4)-linked α-Man$p$, δ 99.44 to C-1 of (1→2,4,6)-linked α-Glu$p$, δ 99.53 to C-1 of T-linked α-Gal$p$ and δ 99.66 to C-1 of (1→4)-linked α-Glu$p$, respectively. The presence of C-1 signal demonstrated that all monosaccharides should be in pyran ring, because of the resonance of furan ring should be in 107-109 ppm [27]. The results of NMR analysis were in agreement with the results of GC-MS and FT-IR data.

On the basis of the results gained above, the structure of the polysaccharide HSP-1 was demonstrated that HSP-1 was a branched polysaccharide possessing a backbone of (1→4)-α-D-glucose residues (~70%), (1→4)-α-D-mannose residues (~15%) and (1→4)-α-D-galactose residues (~15%). The branches are at the (1,2,4,6→)-α-D-glucose residues (~8%) of the backbone, mainly composed of (1→4)-α-D-glucose residues, (1→4)-α-D-galactose residues, (1→4)-α-D-mannose residues, and mainly terminated with α-D-galactose residues. One of the possible repeating units of HSP-1 was shown in Figure 5.

Hydroxyl radical-scavenging activity of HSP-1

As shown in Figure 6, HSP-1 was found to have a high hydroxyl radical-scavenging activity in a concentration-dependent manner. In a concentration range from 0 to 1 mg/mL, the scavenging rate was strengthened with the concentration increasing, with an IC_{50} value of 0.834 mg/mL. On the same conditions, Vitamin C (Vc) showed a slightly higher scavenging rate on the hydroxyl radical, with an IC_{50} value of 0.590 mg/mL. Three repetitions were conducted and the RSD values were not more than 3.0%.

The Reactive Oxygen Species (ROS) like hydroxyl radicals, superoxide anion and hydrogen peroxide, are bound up with the pathogenesis of various diseases [11,28]. Hydroxyl radical can cause

severe damage to the biomolecules, so the antioxidant activity of the HSP-1 polysaccharide to hydroxyl radical-scavenging rate in vitro was measured in the present study.

The hydroxyl radical-scavenging activity of the polysaccharides might be influenced by the protein content, Molecular Weight (MW) or monosaccharide composition [29-31]. However, the relationship between the structure and antioxidant activity was still uncertain, therefore the further study was focused on the scavenging activity mechanism of the hydroxyl radicals, which was beneficial to understand the bioactivity of the polysaccharide HSP-1.

Hirsutella sinensis is a kind of hyphomycete; there were few reports about the structure and antioxidant activity of its polysaccharides. One exopolysaccharide produced by *Hirsutella sp.* showed antibacterial activity against gram-positive bacterium and the monosaccharide component of EPS was mannose, galactose and glucose with a molar ratio of 4.0:8.2:1.0. Its molecular weight was 23 kDa [32]. A novel polysaccharide designated EPS-1A with an average molecular weight around 40 kDa was fractionated and purified from the crude exopolysaccharide (EPS) isolated from fermentation broth of Cs-HK1, a *Tolypocladium s.* Fungus isolated from wild *Cordyceps sinensis*. EPS-1A was composed of glucose, mannose and galactose at 15.2:3.6:1.0 M ratio [11]. Recently, studies could also be found in the research of the polysaccharides isolated from cultured *Cordyceps* mycelia and *Cordyceps militaris*. By using anti-oxidation activity-guided fractionation, a 210 kDa polysaccharide was isolated from cultured *Cordyceps* mycelia. The polysaccharide, containing glucose, mannose and galactose in a ratio of 1:0.6:0.75, had a strong anti-oxidation activity. The pretreatment of isolated polysaccharide on the cultured rat pheochromocytoma $PC1_2$ cells showed strong protective effect against hydrogen peroxide (H_2O_2)-induced insult [3].The water-soluble crude polysaccharides were obtained and purified from the fruiting bodies of cultured *Cordyceps militaris*, giving main three polysaccharide fractions termed P_{50}-1, P_{70}-1, and P_{70}-2, structural features of P_{70}-1 were investigated. In the *in vitro* antioxidant assay, P_{70}-1 was found to possess hydroxyl radical-scavenging activity with an IC_{50} value of 0.548 mg/mL [33]. A novel polysaccharide named CBP-1 was isolated from the fruiting body of cultured *Cordyceps militaris*; its structural features were investigated. In the *in vitro* antioxidant assay, CBP-1 was found to possess the hydroxyl radical-scavenging activity with an IC_{50} value of0.638 mg/mL [34].

Conclusion

Previous researches on *Hirsutella sinensis* showed that fungi polysaccharides have series of pharmacological action and the structure of polysaccharide is closely related with its functions. However, hardly any research focus on the polysaccharide purification from *H. sinensis*, it would be of interest for an in-depth research on the molecular structure and bioactivity of the polysaccharide isolated from *Hirsutella sinensis*.

In this research, three polysaccharide fractions named HSP-1, HSP-2, and HSP-3 were purified from mycelia fermentation of *Hirsutella sinensis*. The polysaccharide HSP-1 isolated was identified as a heteropolysaccharide, which was composed of glucose, mannose and galactose in a molar ratio of 4.5:1.0:1.4. The HSP-1 was a branched polysaccharide possessing a backbone of (1→4)-α-D-glucose residues (~70%), (1→4)-α-D-mannose residues (~15%) and (1→4)-α-D-galactose residues (~15%). Anti-oxidation tests showed that HSP-1 could sweep the hydroxyl radical with an IC_{50} value of 0.834 mg/mL.

Acknowledgement

This work was financially supported by Grant from the Major State Basic Research Development Program of China (973 Program), No. 2013CB733600, and

2012CB72100x, and Science and technology support plan of Hebei (13272803D). We also thank Zhu Feng Cordyceps Pharmacy Corporation, LTD. P R (Qinghai, China) for donating the industrial strain.

References

1. Carbonero ER, Smiderle FR, Gracher AHP, Mellinger CG, Torri G, et al. (2006) Structure of two glucans and a galactofuranomannan from the lichen *Umbilicaria mammulata*. Carbohydr Polym 63: 13-18.

2. Novak M, Vetvicka V (2008) Beta-glucans, history, and the present: immunomodulatory aspects and mechanisms of action. J Immunotoxicol 5: 47-57.

3. Yu R, Yang W, Song L, Yan C, Zhang Z, et al. (2007) Structural characterization and antioxidant activity of a polysaccharide from the fruiting bodies of cultured *Cordyceps militaris*. Carbohydr Polym 70: 430-436.

4. Ye M, Qiu T, Peng W, Chen WX, Ye YW, et al. (2011) Purification, characterization and hypoglycemic activity of extracellular polysaccharides from *Lachnum calyculiforme*. Carbohydr Polym 86: 285-290.

5. Lee JS, Kwon JS, Yun JS, Pahk JW, Shin WC, et al. (2010) Structural characterization of immunostimulating polysaccharide from cultured mycelia of *Cordyceps militaris*. Carbohydr Polym 80: 1011-1017.

6. Chi AP, Chen JP, Wang ZZ, Xiong ZY, Li QX (2008) Morphological and structural characterization of a polysaccharide from *Gynostemma pentaphyllum* Makino and its anti-exercise fatigue activity. Carbohydr Polym 74: 868-874.

7. Yan JK, Wang WQ, Li L, Wu JY (2011) Physiochemical properties and antitumor activities of two a-glucans isolated from hot water and alkaline extracts of *Cordyceps* (Cs-HK1) fungal mycelia. Carbohydr Polym 85: 753-758.

8. Zhu JS, Halpern GM, Jones K (1998) The scientific rediscovery of an ancient Chinese herbal medicine: *Cordyceps sinensis*: part I. J Altern Complement Med 4: 289-303.

9. Yalin W, Cuirong S, Yuanjiang P (2006) Studies on isolation and structural features of a polysaccharide from the mycelium of a Chinese edible fungus (*Cordyceps sinensis*). Carbohydr Polym 63: 251-256.

10. Wu Y, Hu N, Pan Y, Zhou L, Zhou X (2007) Isolation and characterization of a mannoglucan from edible *Cordyceps sinensis* mycelium. Carbohydr Res 342: 870-875.

11. Yan JK, Li L, Wang ZM, Wu JY (2010) Structural elucidation of an exopolysaccharide from mycelial fermentation of a *Tolypocladium* sp. fungus isolated from wild *Cordyceps sinensis*. Carbohydr Polym 79: 125-130.

12. Chen YQ, Wang N, Qu L, Li T, Zhang W (2001) Determination of the anamorph of *Cordyceps sinensis* inferred from the analysis of the ribosomal DNA internal transcribed spacers and 5.8S rDNA. Biochem Syst Ecol 29: 597-607.

13. Liu ZY, Yao YJ, Liang ZQ, Liu AY, Pegler DN, et al. (2001) Molecular evidence for the anamorph-teleomorph connection in *Cordyceps sinensis*. Mycol Res 105: 827-832.

14. Staub AM (1965) Removal of protein-Sevag method. Method Carbohyd Chem 5: 5-6.

15. Dubois M, Gilles K, Hamilton JK, Rebers PA, Smith F (1956) Colorimetric method for determination of sugars and related substances. Anal Chem 28: 350-356.

16. Wang GY, Liang ZY, Zhang LP (2001) Studies on the structure of JS$_1$–The water soluble polysaccharide isolated by alkaline from *Hippophae rhamnoides* L. Chem J Chin Univ 22: 1688-1690.

17. Linker A, Evans LR, Impallomeni G (2001) The structure of a polysaccharide from infectious strains of *Burkholderia cepacia*. Carbohydr Res 335: 45-54.

18. Aspinall GO, Ferrier RJ (1957) A spectrophotometric method for the determination of periodate consumed during the oxidation of carbohydrates. Chem Ind 7: 1216-1221.

19. Needs PW, Selvendran R (1993) Avoiding oxidative degradation during sodium hydroxide/methyl iodide-mediated carbohydrate methylation in dimethyl sulfoxide. Carbohydr Res 245: 1-10.

20. Bjorndal H, Lindberg B, Svenndon S (1967) Mass spectrometry of partially methylated alditol acetates. Carbohydr Res 5: 433-440.

21. He ZS, Luo H, Cao CH, Cui ZW (2004) Photometric determination of hydroxyl free radical in Fenton system by brilliant green. Amer J Chin Clin Med 6: 236-237.

22. Kawagishi H, Kanao T, Mizuno T, Shimura K, Ito H, et al. (1990) Formolysis of a potent antitumor (1 → 6)-β-d-glucan-protein complex from *Agaricus blazei* fruiting bodies and antitumor activity of the resulting products. Carbohydr Polym 12: 393-403.

23. Chauveau C, Talaga P, Wieruszeski JM, Strecker G, Chavant L (1996) A water-soluble beta-D-glucan from *Boletus erythropus*. Phytochemistry 43: 413-415.

24. Gulin S, Kussak A, Jansson PE, Widmalm G (2001) Structural studies of S-7, another exocellular polysaccharide containing 2-deoxy-arabino-hexuronic acid. Carbohydr Res 331: 285-290.

25. Xu C, Chen YL, Zhang M (2004) Structural characterization of the polysaccharide DMP2a-1 from *Dendrobium moniliforme*. Chin Pharm J 39: 900-902.

26. Zhang WJ (1999) Technology of biochemical research on compound polysaccharide. Hangzhou: Zhejiang University Press.

27. Wang Y, Yin H, Lv X, Wang Y, Gao H, et al. (2010) Protection of chronic renal failure by a polysaccharide from *Cordyceps sinensis*. Fitoterapia 81: 397-402.

28. Busciglio J, Yankner BA (1995) Apoptosis and increased generation of reactive oxygen species in Down's syndrome neurons *in vitro*. Nature 378: 776-779.

29. Chen HX, Zhang M, Qu ZS, Xie BJ (2008) Antioxidant activities of different fractions of polysaccharide conjugates from green tea (*Camellia sinensis*). Food Chem 106: 559-563.

30. Lin CL, Wang CC, Chang SC, Inbaraj BS, Chen BH (2009) Antioxidative activity of polysaccharide fractions isolated from *Lycium barbarum* Linnaeus. Int J Biol Macromol 45: 146-151.

31. Zhu BW, Wang LS, Zhou DY, Li DM, Sun LM, et al. (2008) Antioxidant activity of sulphated polysaccharide conjugates from abalone (*Haliotis discus hannai Ino*). Eur Food Res Technol 227: 1663-1668.

32. Li R, Jiang XL, Guan HS (2010) Optimization of mycelium biomass and exopolysaccharides production by *Hirsutella* sp. In submerged fermentation and evaluation of exopolysaccharides antibacterial activity. African Journal of Biotechnology 9: 195-202.

33. Li SP, Zhao KJ, Ji ZN, Song ZH, Dong TT, et al. (2003) A polysaccharide isolated from *Cordyceps sinensis*, a traditional Chinese medicine, protects PC12 cells against hydrogen peroxide-induced injury. Life Sci 73: 2503-2513.

34. Yu RM, Yin Y, Yang W, Ma WL, Yang L, et al. (2009) Structural elucidation and biological activity of a novel polysaccharide by alkaline extraction from cultured *Cordyceps militaris*. Carbohydr Polym 75: 166-171.

33

Use of Pectin Rich Fruit Wastes for Polygalacturonase Production by *Aspergillus awamori* MTCC 9166 in Solid State Fermentation

P.Naga Padma[2], K.Anuradha[2], B. Nagaraju[1], V.Selva Kumar[2] and Gopal Reddy[1]*

[1]*Department of Microbiology, Osmania University, Hyderabad 7, India*
[2]*BVB Vivekananda College, Secunderabad 94, India*

Abstract

Polygalacturonase enzyme has industrial application in extraction and clarification of fruit juices. Production of polygalacturonase (PGU) by *Aspergillus awamori* MTCC 9166 was studied in solid state fermentation using different pectin-rich fruit wastes like apple peel, banana peel, citrus (orange) peel, jackfruit rind, mango peel, and pine apple peel. Sugar free fruit peels were prepared by water treatment and dried materials were used as substrates for PGU production. Highest enzyme production was with jack fruit rind and mango peel at 65% moisture content, 28°C, pH 5.2, 10^6 spores/gm inoculum size for jack fruit rind and 10^8 spores/gm for mango peel and 96 h incubation period. These studies indicate that locally available waste raw materials, jack fruit rind and mango peel, have good potential as substrates for PGU production. Use of such waste raw material is not only cost effective but also caters to the cause of disposing of waste at no cost, which is important for developing Indian economy. This is the first report on use of fruit wastes as substrates for production of polygalacturonases by *Aspergillus awamori* MTCC 9166 in solid state fermentation. The enzyme production by this strain is more than the reported strains.

Keywords: Solid state fermentation; Solid substrates; Polygalacturonase; Fruit wastes; Aspergillus awamori

Introduction

Pectinases are a group of enzymes that degrade pectins present in middle lamella and primary cell walls of plant tissues [1]. These have wide applications in the food industry for clarification of fruit juices, wines [2,3], coffee and tea fermentations [3] and extraction of essential oils [4] etc. Pectinases produced by different microbes are divided into depolymerizing enzymes and saponifying enzymes. Depolymerizing enzymes are polymethylgalacturonases, pectin lyases, polygalacturonases and pectate lyases and saponifying enzymes are pectinesterases [3,5]. They have significant commercial value with a share of about 25% in global sales of food enzymes [6].

The production of pectinolytic enzymes has been widely reported in bacteria and filamentous fungi [7]. Fungal polygalacturonases are very significant for clarification of fruit juices, wines and for extraction of vegetable oils [8]. Their significance in clarification of fruit juices is due to the fact that their optimal pH closer to that of many fruit juices.

Solid state fermentation (SSF) is the process of cultivation of microbes on solid material in the near absence of free water. Polygalacturonase (PGU) production is reported to be significantly higher in solid state fermentation than in submerged fermentation (SMF) [9].

In the present study pectin rich solid wastes were used as substrates for production of PGU by *Aspergillus awamori* MTCC 9166. Various fruit wastes used were apple peel, banana peel, orange peel, jackfruit rind, mango peel, and pine apple peel. Different fermentation conditions like moisture content, temperature, pH, inoculums size and incubation period were optimized using the best substrates, jackfruit rind and mango peel, for production of PGU in SSF.

Methodology

Microorganism

Aspergillus awamori MTCC 9166 strain was isolated from vegetable dumpyard soil and maintained on PDA slants in refrigerator [10].

Pretreatment of solid substrates

Peels of various fruits like apple, banana, citrus (orange), jackfruit rind, mango and pine apple were subjected to water treatment till sugar free. These were dried and cut into small pieces in the size range of 0.5- 0.75cm. Their pectin content was determined by carbazole method [10,11].

Media and their moisture

Different fruit wastes as solid support and pectin substrate sources were taken and various volumes of Czapek nutrient solution was added to optimize moisture content.

Inoculum optimization

Fungal spores were scrapped from PDA slants to water suspension and added in various concentrations to media flasks to optimize inoculum size. One ml of spore suspension with a spore count in the range of 105-109spores/gm was inoculated to each 10gm of substrate.

Cultural conditions

Experiments were carried out in 100 ml flasks with 10gm solid substrate. Fermentation parameters like pH, temperature and incubation period were tested for optimization. The ranges were pH 4.5 - 6, temperature 25 - 40°C and incubation period 2 - 10 days.

---

*Corresponding author: Gopal Reddy, Department of Microbiology, Osmania University, Hyderabad 7, India

Enzyme extraction and assay

Enzyme was extracted using acetate buffer at pH 5.2 and assayed by measuring the D-galacturonic acid realeased from polygalacturonic acid as substrate by Miller's method [12]. One unit of enzyme activity is defined as the amount of enzyme required to produce 1μ mole of galacturonic acid per minute at room temperature.

Results

Pectin rich fruit wastes were studied for selection of solid substrates for polygalacturonase (PGU) production (Figure 1). Figure 1 indicates shows polygalacturonase production using different pectin rich substrates at various moisture levels. It also indicates that 65% moisture was best for all substrates studied. Two substrates, jack fruit rind and mango peel, showed comparatively higher enzyme production. Therefore these were used as solid substrates for further study on optimization of fermentation conditions for PGU production.

Inoculum size for polygalacturonase production was optimized and the results are presented in Figure 2. The optimized inoculum size was 106 spores/gm for jack fruit rind and 108 spores/gm for mango peel. The pH range was tested from 4.5 to 6 and optimum pH was 5.25 (Figure 3). The effect of temperature was tested in the range of 25° - 40°C and the optimum temperature was 28° C (Figure 4).

The fermentation cycle by solid state fermentation (SSF) for poly-

galacturonase was studied for a period of 240 hours or 10 days (Figure 5). Peak production was at 96 hrs or 4th day for both jack fruit rind and mango peel as substrates. The overall polygalacturonase production by *Aspergillus awamori* MTCC 9166 was good by using jack fruit rind (412 U /gm) than mango peel (217 U /gm) as solid substrate.

Discussion

Solid state fermentation (SSF) is advantageous compared to submerged fermentation as microbe-substrate interactions are efficient in this providing scope for reaching high product concentration, and also there is less liquid effluent that makes product recovery more efficient [13]. SSF was selected as there are reports of highest PGU production by this method [14-16]. Moisture content plays a significant role in SSF as it gives a good compromise between water availability and substrate swelling. It is also significant as both oxygen availability and its diffusion depend on this [14,17]. Therefore it was determined for different raw pectin substrates selected for study and all showed good response at 65% moisture level as shown in figure1. Comparitive study of 65% moisture level for PGU production in different pectin substrates used both as solid support and pectin rich inducers indicated (Figure 2) that jack fruit rind and mango peel are good substrates as they are also rich in pectin [11,10]. Inoculum size is important as it influences efficient substrate microbe interaction and 106 spores/gm was optimum for jack fruit rind and 10^8 spores/gm for mango peel (Figure 3). Optimum fer-

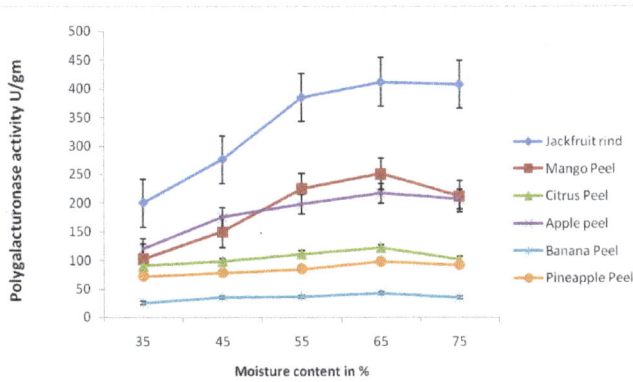

Figure 1: Effect of moisture content on polygalacturonase production by *Aspergillus awamori* MTCC 9166 in solid state fermentation using different pectin rich substrates.

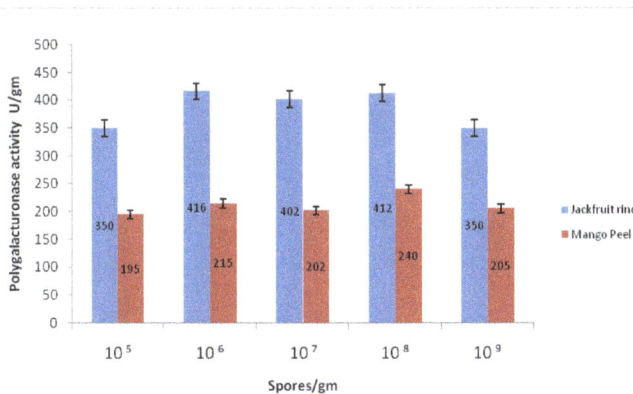

Figure 2 : Effect of inoculum size on polygalacturonase production by *Aspergillus awamori* MTCC 9166 in solid state fermentation. The p value is 0.0000 .

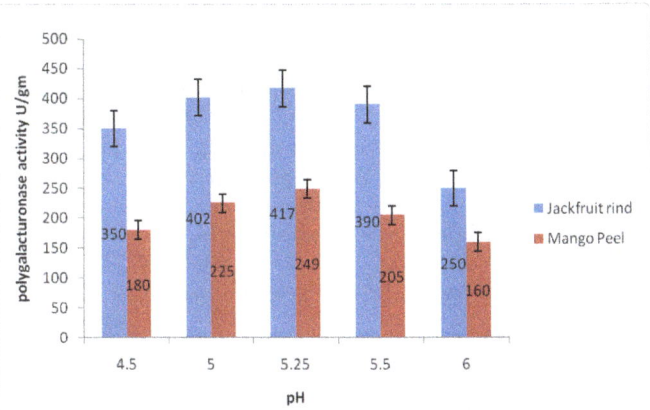

Figure 3 : Effect of pH   on polygalacturonase production by *Aspergillus awamori* MTCC 9166 in solid state fermentation. The p value is 0.001641.

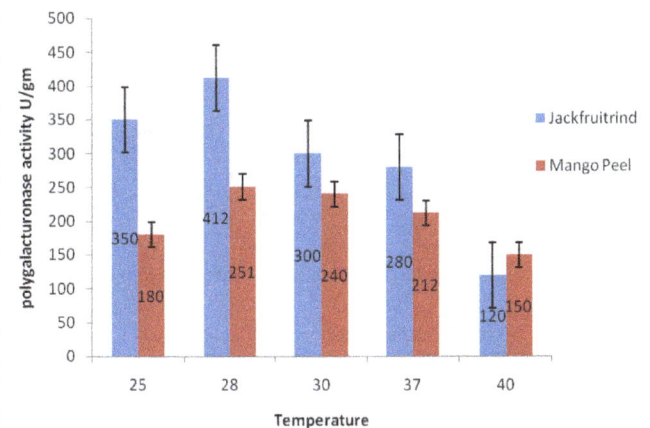

Figure 4 : Effect of temperature on polygalacturonase production by *Aspergillus awamori* MTCC 9166 in solid state fermentation. The p value is 0.139135.

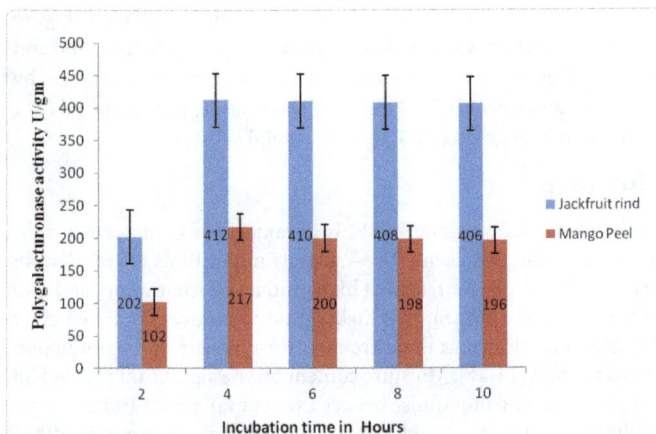

Figure 5 : Effect of incubation time for polygalacturonase production by *Aspergillus awamori* MTCC 9166 in solid state fermentation. The p value is 0.00392845.

mentation conditions bring out the highest product from any producer strain [18]. When these were tested as indicated in Figures 4 and 5, the response was good for an acidic pH 5.25 and a mesophilic temperature 28° C as it is a fungal organism and the result is similar to reported observations [19-21]. Producer strains need to have shorter fermentation cycles as this would be cost effective. In the present study the strain showed peak production at 4th day or 96 hours for both jack fruit rind and mango peel which is less than the reported for the same species. The organism studied could be an efficient producer strain as it responded well to a cheaper locally available solid substrates like jack fruit rind which is a waste that if not disposed could cause waste disposal problems. Response of this strain to submerged fermentation with jack fruit rind was also significant as reported earlier [10]. The shorter fermentation cycle also makes the strain cost effective for commercial exploitation. For a developing and comparatively poorer economy like that of India, use of such waste raw material is not only cost effective but also caters to the cause of disposing off waste at no cost.

Acknowledgements

The authors (PNP and KA) are grateful to the management of BVB Vivekananda College for encouraging to carryout this work.

References

1. Ismail AS (1996) Utilization of orange peels for the production of multienzyme complexes by some fungal strains. Proc Biochem 31: 645-650.

2. Alkorta I, Garbisu C, Llama MJ, Serra JL (1998) Industrial applications of pectic enzymes: A review. Proc Biochem 33: 21-28.

3. Whitaker JR (1984) Pectic substances, pectic enzymes and haze formation in fruit juices. Enzyme Microb Technol 6: 341-349.

4. Jayani RS, Saxena S, Gupta R (2005) Microbial pectinolytic enzymes: A review. Proc Biochem 40: 2931-2944.

5. Whitaker JR (1990) Microbial Pectinolytic enzymes. WM Fogarty and CT Kelly Eds. Microbial Enzymes and biotechnology 2nd edition, Elsiever Science Ltd, London.

6. Jarvis MC (1984) Structure and properties of pectin gels in plant cell walls. Plant cell Env 7: 153-164.

7. Pedrolli DB, Gomes E, Monti R, Carmona EC (2008) Studies on productivity and charecterisation of polygalacturonase from *Aspergillus giganteus* submerged culture using citrus pectin and orange waste. Appl Biochem Biotechnol 144: 191-200.

8. Naidu GSN, Panda T (1998) Production of pectinolytic enzymes-A review. Bioprocess Biosyst Eng 19 : 355-361.

9. Castilho LR, Medronho RA, Alves TLM (2000) Production and extraction of pectinases obtained by solid state fermentation of agroindustrial residues with *Aspergillus niger*. Bioresour Technol 71: 45-50.

10. Anuradha K, Naga Padma P, Venkateshwar S, Gopal Reddy (2010) Fungal isolates from natural pectic substrates for polygalaturonase and multienzyme production. Ind Microbiol 50: 339-344.

11. Ranganna S (1979) Handbook of Analysis and Quality control for fruit and vegetable products. Tata McGraw-Hill Pub Co Ltd. New Delhi.

12. Miller GL (1959) Use of dinitrosalicylic acid reagent for determination of reducing sugar. Anal Chem 31: 426-428.

13. Pandey A (2001) Production of enzymes by solid-state fermentation. In : Pandey A, Soccol CR, Rodriguez-Leon JA and Nigam P (Eds) Solid-State fermentation in biotech nology. Fundamentals and applications. Asiatech Publishers, New Delhi.

14. Blandino A, Iqbalsyah T, Pandiella S, Cantero D, Webb C (2002) Polygalacturonase production by *Aspergillus awamori* on wheat bran in solid-state fermentation. Appl Microbiol Biotechnol 58: 164-169.

15. Dartora AB, Bertolin TE, Bilibio D, Silveira MM, Costa JAV (2002) Evaluation of filamentous fungi and inducers for the production of endo-polygalacturonase by solid state fermentation. Z Naturforsch 57: 666-670.

16. Dayanand A, Patil SR (2006) Production of pectinase from deseeded dried sunflower head by *Aspergillums niger* in submerged and solid-state conditions. Bioresour Technol 97: 2054-2058.

17. Acuna-Arguelles M, Gutierrez-Rojas M, Viniegra-Gonzalez G, Favela-Torres E (1994) Effect of water activity on exo-pectinase production by *Aspergillus niger* CH4 in solid state fermentation. Biotechnol Lett 16: 23-28.

18. Baracat MC, Vanetti MCD, Araujo EF, Silva DO (1991) Growth conditions of a pectinolytic *Aspergillus fumigatus* for degumming of natural fibres. Biotechnol Lett 13: 693-696.

19. Cavalitto FS, Arcas JA, Hours RA (1996) Pectinase production profile of *Aspergillus foetidus* in solid-state cultures at different acidities. Biotechnol Lett 18: 251-256.

20. Hours RA, Katsuragi T, Sakai T (1994) Growth and protopectinase production of *Aspergillus awamori* in solid-state culture at different acidities. J Biosci Bioeng 78: 426-430.

21. Sudheer Kumar Y, Varakumar S, Reddy OVS (2010) Production and optimization of polygalacturonase from mango (*Mangifera indica* L.) peel using *Fusarium moniliforme* in solid state fermentation. World J Microbiol Biotechnol 26: 1973-1980.

34

Study on the Effect of pH, Temperature and Aeration on the Cellular Growth and Xanthan Production by *Xanthomonas campestris* Using Waste Residual Molasses

P. Mudoi*, P. Bharali and B. K. Konwar

Department of Molecular Biology and Biotechnology, Tezpur University, Tezpur, Assam-784028, India

Abstract

Waste residual molasses, a non-edible portion produced during the processing of sugarcane juice for the preparation of molasses, may be an alternative low-cost renewable substrate to the pricey food-grade molasses for xanthan production. Systematic strategies were applied to improve xanthan production with a newly isolated indigenous strain *Xanthomonas campestris* originated from Tezpur, Assam. Analyses with TLC, HPLC and FTIR show that the polymer consisted mainly of glucose, galactose and glucornic acid but showed no evidence of xylose, arabinose or glycoprotein in the polysaccharide. The isolated xanthan exhibited all the required physico-chemical characteristics and were examined by using TGA, DSC, XRD and SEM. Maximum concentration of xanthan was observed after 24h of incubation of the culture media, pH 7 at 28°C with 200 rpm. The viscosity of xanthan was found to be stable over a wide range of pH, reduced with the increase in temperature and raised at the higher xanthan concentration. The results obtained in the present investigation are noteworthy for the possible xanthan production from low-cost waste residual molasses at an industrial level.

Keywords: Xanthan; pH; Temperature; Rheological properties; Stable

Introduction

Xanthan is a type of microbial polysaccharide, produced by specific bacterial sp. such as *Xanthomonas campestris*. These are generally regarded as safe (GRS) and approved for its use in food by the Food and Drug Administration (FDA) [1]. The molecular weight of xanthan is approximately around 2×10^6, but it can reach up to 13×10^6 to 50×10^6 [2]. The polymer is acidic in nature and made up of pentasaccharide subunits, forming a cellulose backbone with trisaccharide side-chains composed of mannose (β 1,4), glucuronic acid (β 1,2) and mannose attached to alternate glucose residues in the backbone by α-1,3 linkages [3]. Xanthan exhibits high pseudo-plastic character in aqueous solutions with high viscosity at low shear forces, resistant to enzymatic degradation and extremely stable over a wide range of pH (2-11), temperature (up to 9°C) and salinity (up to 15% g/l NaCl) [4]. Due to such characteristic behaviors xanthan has numerous applications such as stabilizing, viscosifying, emulsifying, thickening, and suspending agents in various fields [3-5].

The cost involvement with the fermentation medium represents a critical aspect for the commercial production of xanthan. Searching for cheaper carbon sources, in place of glucose or sucrose, might lower the cost for the xanthan production. In such circumstances, non-edible crude molasses, a waste by-product of sugarcane based industries is an interesting substrate for xanthan gum production, since it is renewable source and the by-product is produced in a large quantity during the processing of sugarcane juice. The present work aims at providing relevant scientific information about the direct use of such residual waste molasses, an abundant agro-industrial residue of sugarcane based industries, to produce xanthan. The objective of the present investigation was to produce cost effective xanthan using waste residual molasses as sole feedstock carbon source. Physico-chemical characterization of xanthan gum was studied and the effect of supplementation of nutrients, pH, temperature and agitation was evaluated to optimize the production of xanthan.

Material and Methods

Microorganisms and culture conditions

The waste residual molasses were collected from the sugarcane processing unit of Tezpur, Assam, India. The total soluble solids of sugarcane molasses were adjusted by adding water at a ratio of 3:1 (v/w). The diluted molasses was preheated at 90°C in a water bath for 15 min with continuous stirring and centrifuged at 10000 rpm for 25 min. An indigenous strain of *Xanthomonas campestris* was used throughout the study. The inoculum was prepared by transferring a single colony from the slant culture to the 100 ml of Erlenmeyer conical flasks containing 25 ml of the sterile yeast mould broth (YMB) and subsequently incubated in an orbital incubator shaker at 28 ± 1°C for 48 h at 180 rpm. Fermentation was carried out in 250 ml Erlenmeyer flask with a working volume of 100 ml consisted of waste residual molasses 175g/l; yeast extract, 5.0 gm; peptone, 10.0 gm; NaCl, 10.0 gm; and K_2HPO_4, 4.0 gm. The pH of the culture medium was adjusted to 7.0 and sterilization was done by autoclaving at 121°C for 15 min and incubated for 48 h at 28°C at 180 rpm. To determine the effect of each of the component of xanthan production, different sources of carbon, nitrogen and salts were added to the basic medium. Important parameters (pH, agitation and temperature) were also studied separately. All experiments were repeated thrice.

*Corresponding author: P Mudoi, Department of Molecular Biology and Biotechnology, Tezpur University, Tezpur, Assam-784028, India

Xanthan recovery

The fermented culture broth was centrifuged at 12,000 rpm for 30 min at 4°C. The culture supernatant was added with three volume of ice cold ethanol (95%, v/v) and kept overnight at 4°C to precipitate. The mixture was then centrifuged at 10,000 rpm for 30 min at 4°C to recover the precipitate. The precipitate was then washed with 95% (v/v) ethyl alcohol and dried in a hot air oven for overnight period at 45°C. The production of the biopolymers by the bacterial strain was determined gravimetrically and the average was expressed in g/l.

Analytical methods

The qualitative analysis of the isolated xanthan was done by thin layer chromatography (TLC) [6]. TLC was performed for the monosaccharide components using a mobile system consist of butanol-ethanol-water (5:5:4, v/v/v). The resultant spots on the TLC plates were visualized by spraying with a solution of 5% (v/v) sulphuric acid in ethanol followed by heating at 120°C. The hydrolyzed xanthan was dissolved in acetonitrile: water solution (75:25, v/v), filtered and analyzed by HPLC (waters 2489 UV/Visible detector). The separation was performed in isocratic mode (acetonitrile: water; 75:25, v/v), with a photodiode array detector at 254 nm using a flow rate of 1.0 ml/min on an analytical column (C18, 250×4.6 mm). The element content of the isolated xanthan was determined by an elemental analyzer (Perkin Elmer; Model PR 2400 Series II). IR spectrum of the isolated xanthan gum was recorded on a Perkin Elmer- Spectrum100 using KBr. The isolated xanthan was further characterized by ^{1}H NMR and ^{13}C spectroscopy on a Jeol ECS-400 NMR model [7].

Physical characterization

Thermo-stability of the isolated xanthan was determined using a thermo gravimetric analyzer Shimadzu TG50, Japan. The dynamic differential scanning calorimetry (DSC) experiments were conducted in N$_2$ atmosphere, using a Shimadzu differential scanning calorimeter (model DSC-60). X-ray diffraction study was carried out on Model-Miniflex, Rigaku at room temperature over the range of 2Θ = 2-60°. A thin xanthan film was prepared and mounted on to a stub using double sided carbon tape, coated with a thin layer of platinum using ion sputter JFC 1100 and examined under JSM 6390LV, JEOL scanning electron microscope at a magnification of 2000-10, 000 X.

Rheological characterization

The rheological behavior was determined using a Brookfield LVT Viscometer spindle no. 1 at varying shear rate range of 3-60 rpm. All rheological measurements have been carried out at 25°C. The viscosity of the aqueous solutions of xanthan samples were studied in the temperature range of 25-80°C, and at different pH values. The rpm values were converted to shear rate (sec^{-1}) using a factor (0.34) provided by the manufacturer connected to LVT Model.

Results and Discussion

Chemical characterization

The carbohydrate composition of the xanthan extracted from waste residual molasses supplemented medium composed of four neutral sugars that were separated at different relative mobility (Rf) (Figures 1a and b). The isolated xanthan contains glucose, galactose, and glucuronic acid. Since, the relative mobility of glucose and arabinose were very close as shown by TLC; further verification was done by HPLC. HPLC analysis confirmed the presence of glucose, galactose, arabinose,

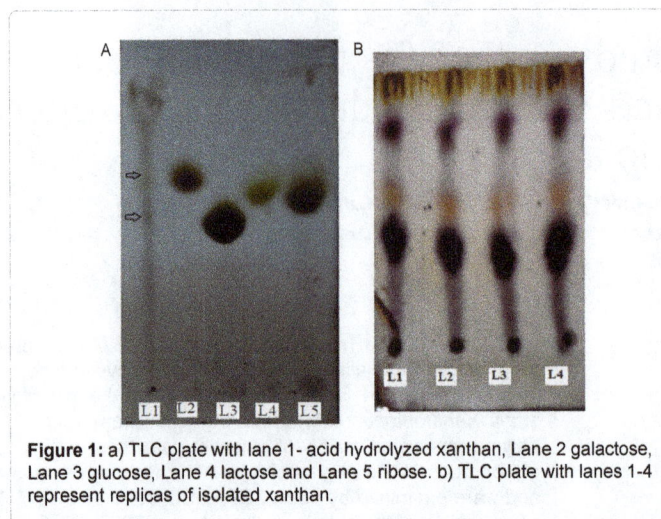

Figure 1: a) TLC plate with lane 1- acid hydrolyzed xanthan, Lane 2 galactose, Lane 3 glucose, Lane 4 lactose and Lane 5 ribose. b) TLC plate with lanes 1-4 represent replicas of isolated xanthan.

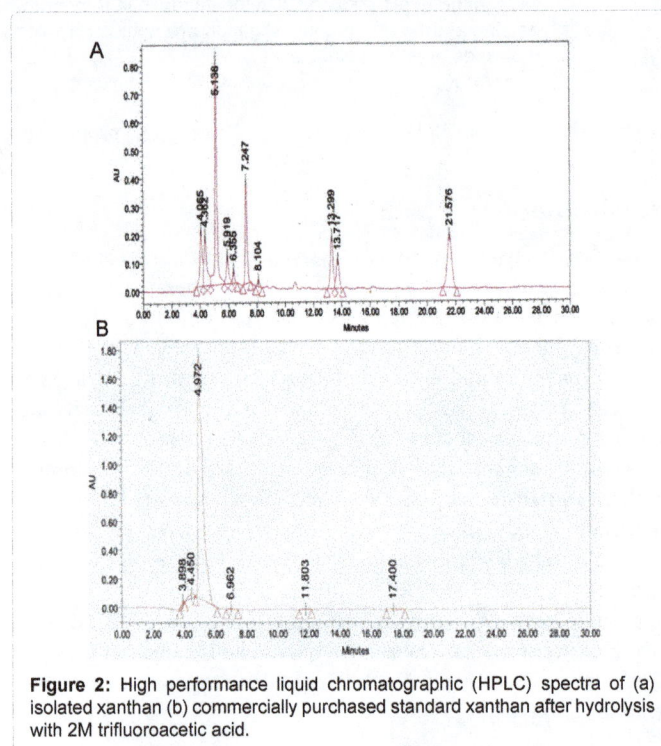

Figure 2: High performance liquid chromatographic (HPLC) spectra of (a) isolated xanthan (b) commercially purchased standard xanthan after hydrolysis with 2M trifluoroacetic acid.

glucuronic acid, and absence of xylose and rhamnose sugar (Figures 2a and 2b). Further, HPLC analysis indicated the presence of around 98.4% of glucose in the acid hydrolyzed. Lowson and Symes, Souza and Venduscolo and Silva et al. [8-10] reported the presence of rhamnose in the xanthan. Silva et al. [10] reported another type of xanthan that devoid of rhamnose. Heyraud et al. [11] reported a type of xanthan that devoid of glucornic acid. Hence, such variation in the chemical composition of xanthan depends directly on the bacteria. The results of FTIR (Figures 3a and 3b) and NMR (Figures 4a and 4b) analysis are presented in Table 1.

Physical characterization

The thermal properties of the isolated xanthan were studied at room temperature (28 ± 2°C) and are shown in Figure 5a. The first reduction in the weight occurred over the temperature range of 30-91.0°C, which

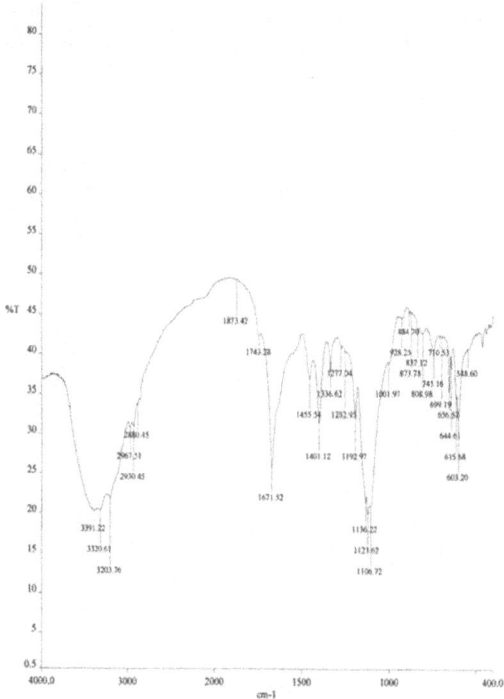

Figure 3: Fourier transforms infrared (FT-IR) spectra of (a) isolated xanthan (b) commercially purchased standard xanthan.

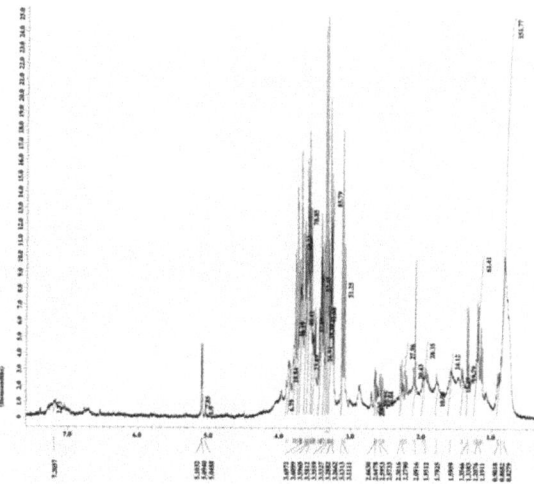

Figure 4: a) ¹H and b) ¹³C NMR spectra of isolated xanthan.

compound was converted from its crystalline state to rubbery state. The melting point of the polymer was 178°C (Tm). Xanthan was found to be almost similar like that of a typical semi-crystalline amorphous material (Figure 6) and confirms the findings of DSC study. The

Spectroscopic and analytical analyses	Observation	Interference/Remark
FTIR	Band at 3391.22 cm⁻¹	-OH stretching of the hydroxyl group
	peaks between 2880.45 and 2967.51 cm⁻¹	-CH stretching of methyl and methylene groups
	1743.28cm⁻¹	-C=O stretching of the acetate group
	peaks at 615.68 cm-1 and 1455.54 cm⁻¹	-COO groups
	strong signals at 1401.12 cm-1 and 1252.95 cm⁻¹	carboxylate asymmetric stretching and -C=O acetate deformation
	absorption peaks between 928.25-745.16 cm⁻¹	β- glycoside linkages
	absorption peaks at 1671.52 cm⁻¹	pyruvate group
	absorption peaks at 1743.28 cm⁻¹	acetyl group
¹H NMR	1.44	pyruvate
	2.12 ppm	acetate groups
¹³C NMR	107 to 95 ppm	quaternary C_2 of pyruvate
		C_1 carbon of glucose, galactose and glucuronic acid
	175.04 ppm	carbonyl carbons of proteins, fatty acids and exopolysaccharide
	107.8 and 75.9 ppm	carbons of the glucose and ketal group of pyruvic acid and polymeric chain of glucose

Table 1: FTIR and NMR analysis of the isolated xanthan.

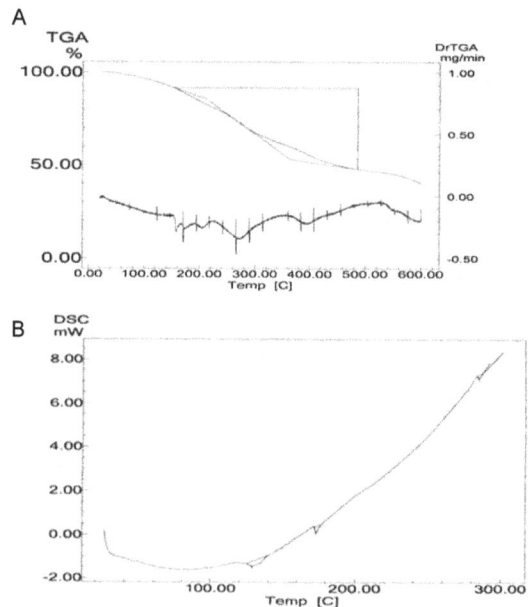

Figure 5: Typical thermograph of (a) Thermo-gravimetric analysis (TGA) and (b) Differential scanning calorimetry (DSC) of isolated xanthan.

might be due to the loss of residual water present in the sample. The second reduction between the temperature ranges of 215.1 and 303.2°C described the dehydration and decarboxylation of the xanthan, leads to the formation of inter and intra molecular anhydride. About 62.0% of weight loss occurred within this section of thermogram. The third decomposition stage was in the temperature range of 365.2 - 470.1°C and due to the degradation of the residual polymer. The result of differential scanning calorimetric analysis is shown in Figure 5b, the glass transition temperature (Tg) appeared at 29°C. Temperature at 130°C was determined as crystalline melting point where the

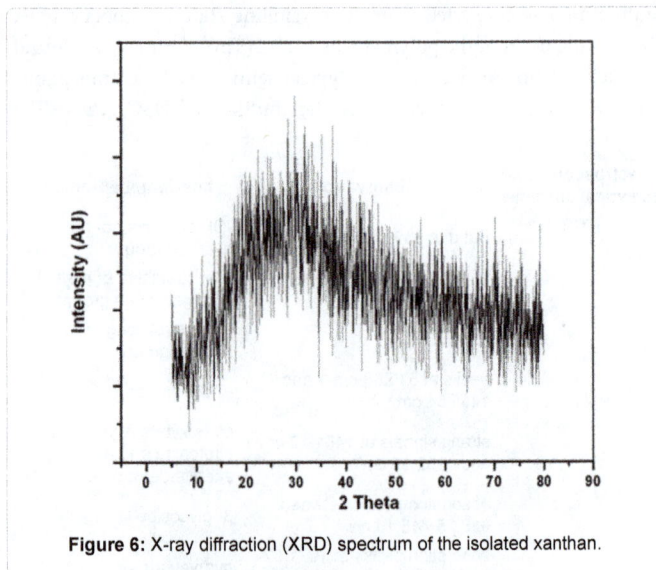

Figure 6: X-ray diffraction (XRD) spectrum of the isolated xanthan.

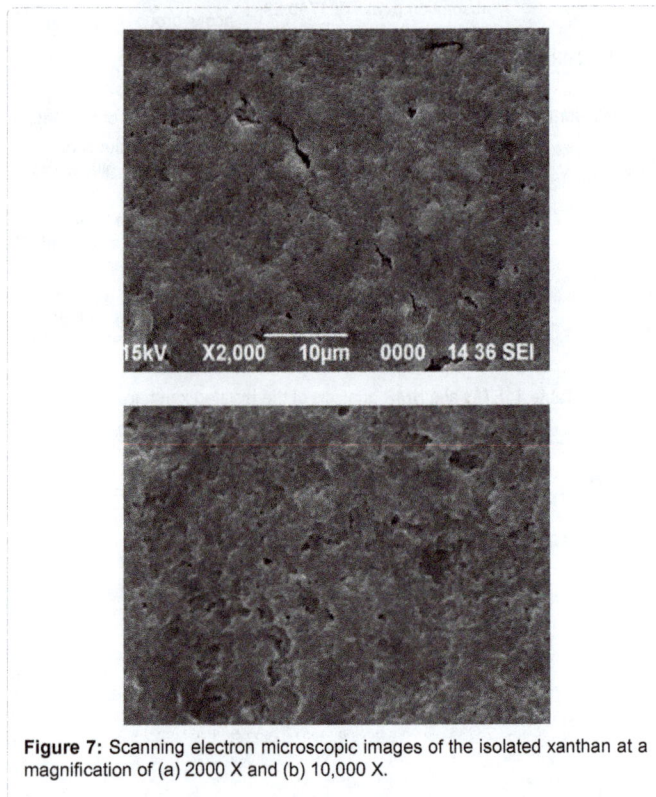

Figure 7: Scanning electron microscopic images of the isolated xanthan at a magnification of (a) 2000 X and (b) 10,000 X.

by Kalogiannis et al. [17] was found to be efficient on the basis of yield per liter of the medium. The results showed that the amount of xanthan produced on waste residual molasses was comparatively different with that of produce from glucose, sucrose, lactose, galactose and maltose (Figure 8). The effects of different nutritional parameters (include concentration of K_2HPO_4, NaCl, and molasses) and culture conditions (pH, temperature and aeration) on xanthan gum production are presented in Figures 9a, 9b and 9c and Figures 10a, 10b and 10c. An

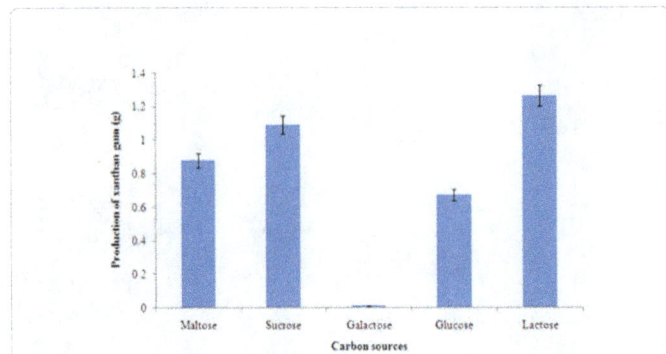

Figure 8: Effect of different carbon sources on the production of xanthan by *Xanthomonas campestris*.

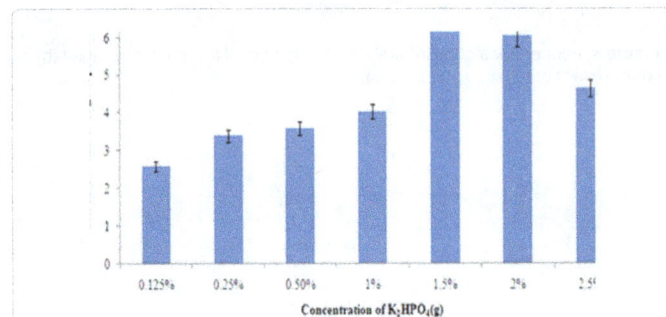

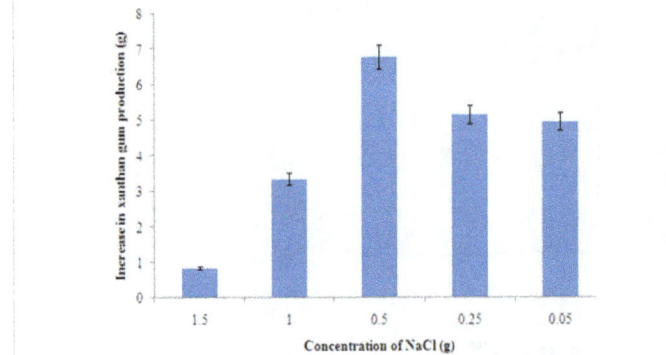

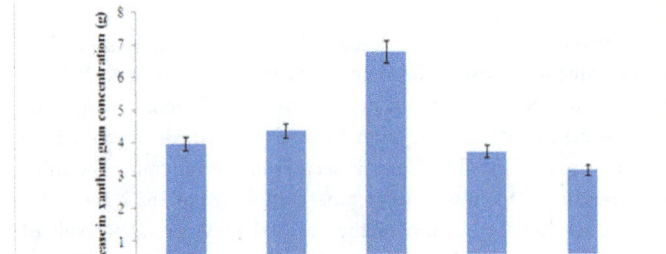

Figure 9: Effect of (a) K_2HPO_4 (b) NaCl and (c) concentration of waste molasses on the production of xanthan by *Xanthomonas campestris*.

theoretical values for xanthan elemental analysis are 34.35% C, 6.45% H, and 54.62% O, respectively. The presence of nitrogen in the xanthan samples might be the due to the presence of impurities (proteins). The cross-sectional view of the dry xanthan exhibited the homogeneity of the biopolymer and shown in Figures 7a and 7b. The appearances of pores in the prepared film are due to the evaporation of water on drying. Formation of pores played a crucial role in fast swelling and de-swelling kinetics of the xanthan [12].

Optimization of xanthan production

Out of five different fermentation media [13-17], media described

increase on the concentration of NaCl on the medium was found to improve the xanthan production. The effect of K_2HPO_4 showed that the concentration was not significant in the range investigated. Therefore additional experiments varying the K_2HPO_4 concentrations viz. 0.125%, 0.25%, 0.5% and 1.0% (w/v) were carried out. The maximum production was 5.23 g/l at 1.5% of K_2HPO_4. Previous reports indicated that K_2HPO_4 can be useful as a buffering agent as well as a nutrient for the *X. campestris* [17,18]. Various authors have reported the use of agro-industrial wastes such as citrus waste [19], olive mill wastewaters [20], cheese whey [10] and date juice by-products [21] as carbon source for xanthan production. Hence, on the basis xanthan production per liter, use of waste residual molasses appears to be favorable

Rheological studies

The effect of xanthan concentration, pH and temperature on the rheological behavior of xanthan produced by the indigenous strain *X. campestris* are shown in Table 2. The differences in apparent viscosity (AV) values were more evident at low concentrations of xanthan. At concentration above 0.5% (w/v) of xanthan gum attains relatively high viscosity indicating the pseudoplastic behavior in the solutions and the apparent viscosity decreased with the increase in shear rate. Such behavior is a characteristics feature shown by polymeric solutions of microbial polysaccharides with large molecular weight [22,23]. When shear stress was increased gradually, viscosity was progressively

getting reduced because the aggregates of xanthan molecules bound through hydrogen bonding and polymer entanglement are gradually disrupted under the pressure of applied shear. However, on removal of shearing stress, the initial viscosity of the solution is recovered almost immediately [24]. Xanthan solutions are specific in their ability to retain their viscosity until it reaches its melting temperature. At such temperature, the viscosity sharply decreases due to a reversible molecular conformation change [24]. The effects of different temperatures on xanthan gum viscosity are shown in Table 2. Decrease in the viscosity of xanthan solution was observed with the increase in the temperature up to 80°C, though it was only a transient phenomenon but the solution returned to their original viscosity upon cooling. The appeared viscosity of the solutions decreased with increasing temperature at all pH values. To determine the function of pH, the stability of xanthan solution was checked over the pH range of 3.5-7.0 and results are presented in Table 2. The viscosity values, however, were not influenced by pH changes and did not show any linear correlation with pH values. Since xanthan is a neutral and nonionic polymer, its viscosity was independent of pH and remained stable from a pH range of 2 to 10. Such stability of the xanthan at various pH ranges may find its application in a number of food products.

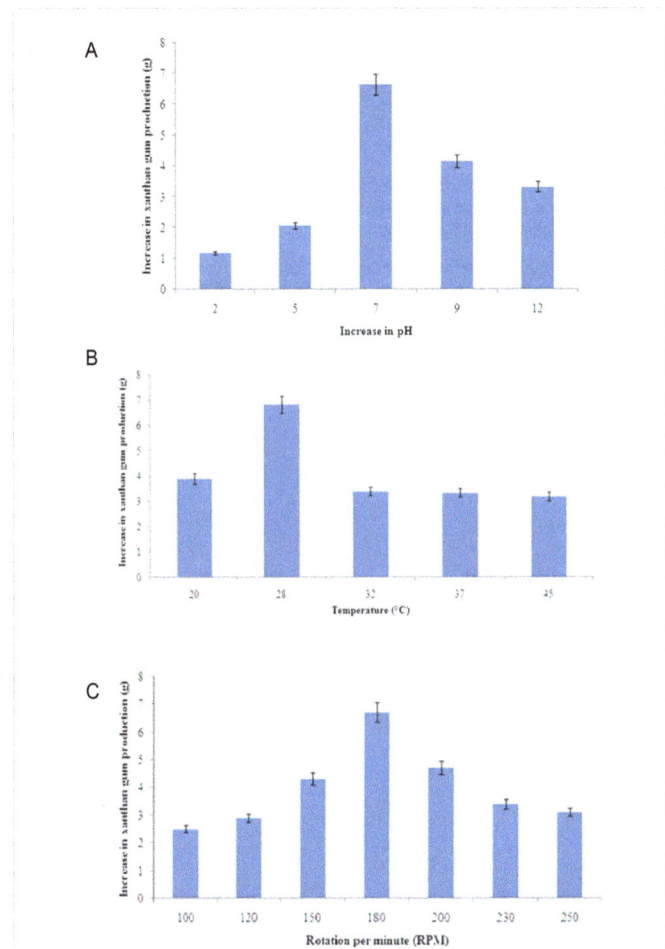

Figure 10: Effect of (a) pH (b) temperature and (c) aeration on the production of xanthan by *Xanthomonas campestris*.

Concentration of xanthan gum (%)	pH	Temperature (°C)	Viscosity (mPa. sec)
0.75	3.5	25	119 ± 0.7
		40	123 ± 0.5
		60	99 ± 0.8
		80	93 ± 0.4
	5	25	120 ± 0.8
		40	112 ± 0.6
		60	102 ± 0.7
		80	97 ± 0.0
	7	25	123.6 ± 0.5
		40	120 ± 0.6
		60	108 ± 0.4
		80	96 ± 0.2
1.5	3.5	25	137 ± 0.5
		40	133 ± 0.7
		60	127 ± 0.3
		80	100 ± 0.4
	5	25	139 ± 0.1
		40	131 ± 0.2
		60	120 ± 0.5
		80	102 ± 0.4
	7	25	150 ±0.6
		40	96 ± 0.5
		60	84 ± 0.8
		80	102 ± 0.7
3.0	3.5	25	159 ± 0.6
		40	148 ± 0.5
		60	136 ± 0.4
		80	123 ± 0.3
	5	25	154 ± 0.3
		40	152 ± 0.2
		60	138 ± 0.6
		80	119 ± 0.5
	7	25	162 ± 0.2
		40	154 ± 0.4
		60	141 ± 0.8
		80	124 ± 0.5

Table 2: Effect of xanthan gum concentration, pH and temperature on the viscosity of xanthan.

Conclusion

The potentiality of using a low-cost renewable residual molasses as a substrate for xanthan production by *X. campestris* seems to be advantageous from the view point of economy. Under optimal conditions a maximum of 5.23 g/l xanthan could be produced using such feedstock. The isolated xanthan exhibited a greater degree of pseudo plasticity as shown by the progressive reduction in the viscosity in direct response to shear. The isolated gum was nonionic in nature and stable to a wide range temperature and pH. The present study suggests the possibility of using residual waste molasses as carbon feedstock and the xanthan produced on such substrates could be useful in a variety of industrial applications.

References

1. De Vuyst L, Vermeire A (1994) Use of industrial medium components for xanthan production by Xanthomonas campestris NRRL-B-1459. Appl Microbiol Biotechnol 42: 187-191.

2. Katzen F, Becker A, Zorreguieta A, Pühler A, Ielpi L (1996) Promoter analysis of the Xanthomonas campestris pv. campestris gum operon directing biosynthesis of the xanthan polysaccharide. J Bacteriol 178: 4313-4318.

3. Rosalam S, England R (2006) Review of xanthan gum production from unmodified starches by Xanthomonas campestris sp. Enzyme Microb Tech 39: 197-207.

4. Lachke A (2004) Xanthan - A versatile gum. Resonance 9: 25-33.

5. Margaritis A, Pace GW (1985) Microbial polysaccharides. In: Comprehensive Biotechnology Moo-Young M (ed). Pergamon Press, Oxford, UK, pp 1005-1104.

6. Talaga P, Stahl B, Wieruszeski JM, Hillenkamp F, Tsuyumu S, et al. (1996) Cell-associated glucans of Burkholderia solanacearum and Xanthomonas campestris pv. citri: a new family of periplasmic glucans. J Bacteriol 178: 2263-2271.

7. Hamcerencu M, Desbrieres J, Popa M, Khoukh A, Riess G (2007) New unsaturated derivatives of Xanthan gum: Synthesis and characterization. Polymer 48: 1921-1929.

8. Lowson CJ, Symes KC (1977) Oligosaccharides produced by partial acetolysis of xanthan gum. Carbohyd Res 58: 433-438.

9. Souza AS, Vendruscolo CT (1999) Produção e caracterização dos biopolímeros sintetizados por Xanthomonas campestris pv. pruni cepas 24 e58. Ciência e Engenharia 8: 115-123.

10. Silva MF, Fornari RCG, Mazutti MA, Oliveira D, de Padilha FF, et al. (2009) Production and characterization of xantham gum by Xanthomonas campestris using cheese whey as sole carbon source. J Food Eng 90: 119-123.

11. Heyraud A, Sayah B, Vojnov A, Colin-Morel P, Gey C, et al. (1998) Structure of an extracellular mannosylated cellulose produced by a mutant strain of Xanthomonas campestris. Cell Mol Biol (Noisy-le-grand) 44: 447-454.

12. Shu CH, Yang ST (1990) Effects of temperature on cell growth and xanthan production in batch cultures of Xanthomonas campestris. Biotechnol Bioeng 35: 454-468.

13. Esgalhado ME, Roseiro JC, Collaco MTA (1995) Interactive Effects of pH and Temperature on Cell Growth and Polymer Production by Xanthomonas campestris. Process Biochem 30: 667-671.

14. Honma T, Nagura S, Mutofushi K (1996) Method for fermentation production of Xanthan gum, USP No: US5580763 A.

15. Letisse F, Chevallereau P, Simon JL, Lindley ND (2001) Kinetic analysis of growth and xanthan gum production with Xanthomonas campestris on sucrose, using sequentially consumed nitrogen sources. Appl Microbiol Biotechnol 55: 417-422.

16. Kalogiannis S, Lakovidou G, Liakopoulou-Kyriakides M, Kyriakidis DA, Skaracis GN (2003) Optimization of xanthan gum production by Xanthomonas campestris grown in molasses. Process Biochem 39: 249-256.

17. Demain AL (2000) Small bugs, big business: the economic power of the microbe. Biotechnol Adv 18: 499-514.

18. Bilanovic D, Shelef G, Green M (1994) Xanthan fermentation of citrus waste. Biores Technol 48: 169-172.

19. López MJ, Moreno J, Ramos-Cormenzana A (2001) Xanthomonas campestris strain selection for xanthan production from olive mill wastewaters. Water Res 35: 1828-1830.

20. Salah RB, Chaari K, Besbes S, Ktari N, Blecker C, et al. (2010) Optimisation of xanthan gum production by palm date (Phoenix dactylifera L.) juice by-products using response surface methodology. Food Chem 121: 627-633.

21. Cacik F, Dondo RG, Marqués D (2001) Optimal control of a batch bioreactor for the production of xanthan gum. Comput Chem Eng 25: 409-418.

22. Rao YM, Suresh AK, Suraishkumar GK (2003) Free radical aspects of Xanthomonas campestris cultivation with liquid phase oxygen supply strategy. Process Biochem 38: 1301-1310.

23. Sworn GM (2000) Xanthan Gum. In: Handbook of Hydrocolloids, Phillips GO, Williams PA (eds). CRC Press LLC, Boca Raton, pp 103-115.

24. Dawkins NL, Nnanna IA (1995) Studies on oat gum [(1→3, 1→4)-β-D-glucan]: composition, molecular weight estimation and rheological properties. Food Hydrocolloid 9: 1-7.

Permissions

The contributors of this book come from diverse backgrounds, making this book a truly international effort. This book will bring forth new frontiers with its revolutionizing research information and detailed analysis of the nascent developments around the world.

We would like to thank all the contributing authors for lending their expertise to make the book truly unique. They have played a crucial role in the development of this book. Without their invaluable contributions this book wouldn't have been possible. They have made vital efforts to compile up to date information on the varied aspects of this subject to make this book a valuable addition to the collection of many professionals and students.

This book was conceptualized with the vision of imparting up-to-date information and advanced data in this field. To ensure the same, a matchless editorial board was set up. Every individual on the board went through rigorous rounds of assessment to prove their worth. After which they invested a large part of their time researching and compiling the most relevant data for our readers.

The editorial board has been involved in producing this book since its inception. They have spent rigorous hours researching and exploring the diverse topics which have resulted in the successful publishing of this book. They have passed on their knowledge of decades through this book. To expedite this challenging task, the publisher supported the team at every step. A small team of assistant editors was also appointed to further simplify the editing procedure and attain best results for the readers.

Apart from the editorial board, the designing team has also invested a significant amount of their time in understanding the subject and creating the most relevant covers. They scrutinized every image to scout for the most suitable representation of the subject and create an appropriate cover for the book.

The publishing team has been an ardent support to the editorial, designing and production team. Their endless efforts to recruit the best for this project, has resulted in the accomplishment of this book. They are a veteran in the field of academics and their pool of knowledge is as vast as their experience in printing. Their expertise and guidance has proved useful at every step. Their uncompromising quality standards have made this book an exceptional effort. Their encouragement from time to time has been an inspiration for everyone.

The publisher and the editorial board hope that this book will prove to be a valuable piece of knowledge for researchers, students, practitioners and scholars across the globe.

List of Contributors

Sari Metsämuuronen
Lappeenranta University of Technology, PO Box 20, FI-53851 Lappeenranta, Finland

Heli Siren
Department of Chemistry, University of Helsinki, PO Box 55, FI-00014 University of Helsinki, Finland

Stefano Piccoli
Applied Bioinformatics Group, Dept. of Biotechnology, University of Verona, Italy

Alejandro Giorgetti
Applied Bioinformatics Group, Dept. of Biotechnology, University of Verona, Italy
German Research School for Simulation Sciences, Jülich, Germany

Gunjan Gautam
Department of Biotechnology, Motilal Nehru National Institute of Technology, Allahabad, India

Vishwas Mishra
Department of Biotechnology, Motilal Nehru National Institute of Technology, Allahabad, India

Payal Verma
Department of Biotechnology, Motilal Nehru National Institute of Technology, Allahabad, India

Ajay Kumar Pandey
Department of Biotechnology, Motilal Nehru National Institute of Technology, Allahabad, India

Sangeeta Negi
Department of Biotechnology, Motilal Nehru National Institute of Technology, Allahabad, India

Sudhish Mishra
Department of Biology, Edward Waters College, 1658 Kings Road, Jacksonville, FL 32209, USA

Anita Mandal
Department of Biology, Edward Waters College, 1658 Kings Road, Jacksonville, FL 32209, USA

Prabir K. Mandal
Department of Biology, Edward Waters College, 1658 Kings Road, Jacksonville, FL 32209, USA

Cristiane Fernandes de Assis
Department of Chemical Engineering, Federal University of Rio Grande do Norte, Natal, Rio Grande do Norte, Brazil

Raniere Fagundes Melo-Silveira
Department of Biochemistry, Federal University of Rio Grande do Norte, Natal, Rio Grande do Norte, Brazil

Ruth Medeiros de Oliveira
Department of Biochemistry, Federal University of Rio Grande do Norte, Natal, Rio Grande do Norte, Brazil

Leandro Silva Costa
Department of Biochemistry, Federal University of Rio Grande do Norte, Natal, Rio Grande do Norte, Brazil

Hugo Alexandre de Oliveira Rocha
Department of Biochemistry, Federal University of Rio Grande do Norte, Natal, Rio Grande do Norte, Brazil

Gorete Ribeiro de Macedo
Department of Chemical Engineering, Federal University of Rio Grande do Norte, Natal, Rio Grande do Norte, Brazil

Everaldo Silvino dos Santos
Department of Chemical Engineering, Federal University of Rio Grande do Norte, Natal, Rio Grande do Norte, Brazil

Xiaohe Chu
State Key Laboratory of Bioreactor Engineering, East China University of Science and Technology, Shanghai 200237, P.R China
Zhejiang Shenghua Biok Biology Co., Zhejiang 313220, P.R. China

Zijing Zhen
State Key Laboratory of Bioreactor Engineering, East China University of Science and Technology, Shanghai 200237, P.R China

Zhenyu Tang
State Key Laboratory of Bioreactor Engineering, East China University of Science and Technology, Shanghai 200237, P.R China

Yingping Zhuang
State Key Laboratory of Bioreactor Engineering, East China University of Science and Technology, Shanghai 200237, P.R China

Ju Chu
State Key Laboratory of Bioreactor Engineering, East China University of Science and Technology, Shanghai 200237, P.R China

Siliang Zhang
State Key Laboratory of Bioreactor Engineering, East China University of Science and Technology, Shanghai 200237, P.R China

Meijin Guo
State Key Laboratory of Bioreactor Engineering, East China University of Science and Technology, Shanghai 200237, P.R China

JM Araya-Garay
Department of Microbiology, University of Santiago de Compostela, Spain

L. Feijoo-Siota
Department of Microbiology, University of Santiago de Compostela, Spain

P. Veiga-Crespo
Department of Microbiology, University of Santiago de Compostela, Spain
School of Biotechnology, University of Santiago de Compostela, 15782 Santiago de Compostela, Spain

A.Sánchez-Pérez
Discipline of Physiology, Bosch Institute, University of Sydney, NSW 2006, Australia

T. González Villa
Department of Microbiology, University of Santiago de Compostela, Spain
School of Biotechnology, University of Santiago de Compostela, 15782 Santiago de Compostela, Spain

Sreeahila Retnadhas
Applied and Industrial Microbiology Laboratory, Department of Biotechnology, Bhupat and Jyoti Mehta School of Biosciences, Indian Institute of Technology Madras, Chennai, India

Sathyanarayana N Gummadi
Applied and Industrial Microbiology Laboratory, Department of Biotechnology, Bhupat and Jyoti Mehta School of Biosciences, Indian Institute of Technology Madras, Chennai, India

Daniel E. Rodríguez Fernández
Department of Bioprocess Engineering and Biotechnology, Federal University of Paraná, P.O. Box 19011, Zip Code 81.531-970, Curitiba, PR, Brazil

José A. Rodríguez León
Department of Biology Science, Positivo University, Prof. Pedro Viriato Parigot de Souza Street, 5300, Zip Code 81.280-330, Curitiba, PR, Brazil

Julio C. de Carvalho
Department of Bioprocess Engineering and Biotechnology, Federal University of Paraná, P.O. Box 19011, Zip Code 81.531-970, Curitiba, PR, Brazil

Susan G. Karp
Department of Bioprocess Engineering and Biotechnology, Federal University of Paraná, P.O. Box 19011, Zip Code 81.531-970, Curitiba, PR, Brazil

José L. Parada
Department of Biology Science, Positivo University, Prof. Pedro Viriato Parigot de Souza Street, 5300, Zip Code 81.280-330, Curitiba, PR, Brazil

Carlos R. Soccol
Department of Bioprocess Engineering and Biotechnology, Federal University of Paraná, P.O. Box 19011, Zip Code 81.531-970, Curitiba, PR, Brazil

Yu Ji
Department of Biochemical Engineering, University College London, Torrington Place, London WC1E 7JE, UK

Yuhong Zhou
Department of Biochemical Engineering, University College London, Torrington Place, London WC1E 7JE, UK

Dessy Ariyanti
Department of Chemical Engineering, Faculty of Engineering, Diponegoro University, Prof Soedarto, SH Kampus Tembalang, Semarang, Indonesia

Noer Abyor Handayani
Department of Chemical Engineering, Faculty of Engineering, Diponegoro University, Prof Soedarto, SH Kampus Tembalang, Semarang, Indonesia

Hadiyanto
Department of Chemical Engineering, Faculty of Engineering, Diponegoro University, Prof Soedarto, SH Kampus Tembalang, Semarang, Indonesia

Jenny Dussán
Centro de Investigaciones Microbiológicas-CIMIC, Departamento de Ciencias Biológicas, Universidad de los Andes, Bogotá, Colombia

Mónica Numpaque
Centro de Investigaciones Microbiológicas-CIMIC, Departamento de Ciencias Biológicas, Universidad de los Andes, Bogotá, Colombia

Di Lu
Department of Pharmacy, General Hospital of Guangzhou Military Command, Guangzhou, People's Republic of China

Wei Zhao
Department of Pharmacy, General Hospital of Guangzhou Military Command, Guangzhou, People's Republic of China

Kuanpeng Zhu
Department of Pharmacy, General Hospital of Guangzhou Military Command, Guangzhou, People's Republic of China

Shu-jin Zhao
Department of Pharmacy, General Hospital of Guangzhou Military Command, Guangzhou, People's Republic of China

Ekta Tiwary
Department of Microbiology, University of Delhi, South Campus, New Delhi, India-110021

Rani Gupta
Department of Microbiology, University of Delhi, South Campus, New Delhi, India-110021

Antonio Carlos Augusto da Costa
Museum of Astronomy and Related Sciences, Department of Documentation and Archives, 586 Rua General Bruce, S. Christopher, Rio de Janeiro, Brazil

Lucia Alves da Silva Lino
Museum of Astronomy and Related Sciences, Department of Documentation and Archives, 586 Rua General Bruce, S. Christopher, Rio de Janeiro, Brazil

Ozana Hannesch
Museum of Astronomy and Related Sciences, Department of Documentation and Archives, 586 Rua General Bruce, S. Christopher, Rio de Janeiro, Brazil

Nadia Abdi
Ecole Nationale Polytechnique d'Alger, B.P. 182-16200, El Harrach, Alger, Algeria

Lila Bensaadallah
Ecole Nationale Polytechnique d'Alger, B.P. 182-16200, El Harrach, Alger, Algeria

Nadjib Drouiche
Ecole Nationale Polytechnique d'Alger, B.P. 182-16200, El Harrach, Alger, Algeria
University of Technology of Compiegne, Departement Genie chimique, B.P. 20.509, 60205 Compiègne cedex, France

Hocine Grib
Ecole Nationale Polytechnique d'Alger, B.P. 182-16200, El Harrach, Alger, Algeria

Hakim Lounici
Ecole Nationale Polytechnique d'Alger, B.P. 182-16200, El Harrach, Alger, Algeria

Andre Pauss
Centre de Recherche en Technologie des Semi-conducteurs pour l'energetique (CRTSE), 2, Bd Frantz Fanon BP140, Alger-7 Merveilles, 16000, Algeria

Nabil Mameri
Centre de Recherche en Technologie des Semi-conducteurs pour l'energetique (CRTSE), 2, Bd Frantz Fanon BP140, Alger-7 Merveilles, 16000, Algeria

Arafat M Goja
College of Food Science and Technology, Huazhong Agricultural University, Wuhan, Hubei 430070, China
Department of Food Science and Technology, Faculty of Agriculture and Natural Resources, Bakht Alruda University, ED Dueim, 1311, Sudan

Hong Yang
College of Food Science and Technology, Huazhong Agricultural University, Wuhan, Hubei 430070, China
Key Laboratory of Environment Correlative Dietology, Huazhong Agricultural University, Ministry of Education, Wuhan, Hubei 430070, China
National R&D Branch Center for Conventional Freshwater Fish Processing (Wuhan), Wuhan, Hubei 430070, China
Aquatic Product Engineering and Technology Research Center of Hubei Province, Wuhan, Hubei 430070, China

Min Cui
State Key Laboratory of Agricultural Microbiology, Huazhong Agricultural University, Wuhan, Hubei 430070, China
Laboratory of Animal Virology, College of Veterinary Medicine, Huazhong Agricultural University, Wuhan, Hubei 430070, China

Charles Li
College of Food Science, Fujian Agricultural and Forestry University, Fuzhou, China

Soltana Fellahi
Laboratory of Microbiology and Plant Biology, Department of Biotechnology Faculty of Sciences of Nature and Life, Mostaganem University, Mostaganem, Algeria
Swedish Centre for Resource Recovery, University of Borås, Borås, Sweden

Taha I Zaghloul
Institute of Graduate Studies and Research, University of Alexandria, Alexandria, Egypt

Elisabeth Feuk-Lagerstedt
Swedish Centre for Resource Recovery, University of Borås, Borås, Sweden

Mohammad J Taherzadeh
Swedish Centre for Resource Recovery, University of Borås, Borås, Sweden

Tianyin Huang
School of Environmental Science and Engineering, Suzhou University of Science and Technology, Suzhou, Jiangsu, 215011, P.R. China

Yongxin Gui
School of Environmental Science and Engineering, Suzhou University of Science and Technology, Suzhou, Jiangsu, 215011, P.R. China

Wei Wu
School of Environmental Science and Engineering, Suzhou University of Science and Technology, Suzhou, Jiangsu, 215011, P.R. China

Kai Feng
School of Environmental Science and Engineering, Suzhou University of Science and Technology, Suzhou, Jiangsu, 215011, P.R. China

Feng Liu
School of Environmental Science and Engineering, Suzhou University of Science and Technology, Suzhou, Jiangsu, 215011, P.R. China

VV Ramakrishnan
Department of Process Engineering and Applied Science, Dalhousie University, Halifax, Nova Scotia, Canada

AE Ghaly
Department of Process Engineering and Applied Science, Dalhousie University, Halifax, Nova Scotia, Canada

MS Brooks
Department of Process Engineering and Applied Science, Dalhousie University, Halifax, Nova Scotia, Canada

Budge SM
Department of Process Engineering and Applied Science, Dalhousie University, Halifax, Nova Scotia, Canada

Mahdi Karimi
Khorasan Razavi Agricultural and Natural Resources Research Center, Iran

Milad Fathi
Khorasan Razavi Agricultural and Natural Resources Research Center, Iran
Department of Food Science and Technology, Isfahan University of Technology (IUT), Isfahan, 84156-83111, Iran

Zahra Sheykholeslam
Khorasan Razavi Agricultural and Natural Resources Research Center, Iran

Bahareh Sahraiyan
Department of Food Science and Technology, Ferdowsi University of Mashhad (FUM), Mashhad, Iran

Fariba Naghipoor
Department of Food Science and Technology, Ferdowsi University of Mashhad (FUM), Mashhad, Iran

Kohji Nakazawa
Department of Life and Environment Engineering, The University of Kitakyushu 1-1 Hibikino, Wakamatsu-ku, Kitakyushu, Fukuoka 808-0135, Japan

Yukako Shinmura
Department of Life and Environment Engineering, The University of Kitakyushu 1-1 Hibikino, Wakamatsu-ku, Kitakyushu, Fukuoka 808-0135, Japan

Ami Higuchi
Department of Life and Environment Engineering, The University of Kitakyushu 1-1 Hibikino, Wakamatsu-ku, Kitakyushu, Fukuoka 808-0135, Japan

Yusuke Sakai
Department of Life and Environment Engineering, The University of Kitakyushu 1-1 Hibikino, Wakamatsu-ku, Kitakyushu, Fukuoka 808-0135, Japan

Fengqiang Wang
Bioprocess Development, Merck Research Laboratories, Union, NJ, USA

Dennis Driscoll
Bioprocess Development, Merck Research Laboratories, Union, NJ, USA

Daisy Richardson
Bioprocess Development, Merck Research Laboratories, Union, NJ, USA

Alexandre Ambrogelly
Bioprocess Development, Merck Research Laboratories, Union, NJ, USA

Feifei Wang
The Research Center of China Hemp Materials, Beijing 100027, P. R. China
Key Laboratory Industrial Fermentation Microbiology, University of Science & Technology, Tianjin 300457, P. R. China

Jianchun Zhang
The Research Center of China Hemp Materials, Beijing 100027, P. R. China
The Quartermaster Equipment Institute of GLD of PLA, Beijing 100010, P. R. China

Limin Hao
The Research Center of China Hemp Materials, Beijing 100027, P. R. China
The Quartermaster Equipment Institute of GLD of PLA, Beijing 100010, P. R. China

Shiru Jia
Key Laboratory Industrial Fermentation Microbiology, University of Science & Technology, Tianjin 300457, P. R. China

Jianming Ba
Department of Endocrinology, Chinese PLA General Hospital, Beijing 100853, P. R. China

Shuang Niu
The Research Center of China Hemp Materials, Beijing 100027, P. R. China
Key Laboratory Industrial Fermentation Microbiology, University of Science & Technology, Tianjin 300457, P. R. China

Fumimasa Nomura
Department of Biomedical Information, Institute of Biomaterials and Bioengineering, Tokyo Medical and Dental University, 2-3-10 Kanda-Surugadai, Chiyoda, Tokyo 101- 0062, Japan

Tomoyuki Kaneko
Department of Biomedical Information, Institute of Biomaterials and Bioengineering, Tokyo Medical and Dental University, 2-3-10 Kanda-Surugadai, Chiyoda, Tokyo 101- 0062, Japan

Akihiro Hattori
Department of Biomedical Information, Institute of Biomaterials and Bioengineering, Tokyo Medical and Dental University, 2-3-10 Kanda-Surugadai, Chiyoda, Tokyo 101- 0062, Japan

Kenji Yasuda
Department of Biomedical Information, Institute of Biomaterials and Bioengineering, Tokyo Medical and Dental University, 2-3-10 Kanda-Surugadai, Chiyoda, Tokyo 101- 0062, Japan

Ying Zha
Microbiology & Systems Biology, TNO, Utrechtsweg 48, 3704 HE, Zeist, The Netherlands
Netherlands Metabolomics Centre (NMC), Einsteinweg 55, 2333 CC Leiden, The Netherlands

Bas Muilwijk
TNO Triskelion BV, Utrechtseweg 48, 3700 AV, Zeist, The Netherlands

Leon Coulier
Netherlands Metabolomics Centre (NMC), Einsteinweg 55, 2333 CC Leiden, The Netherlands
Research group Quality & Safety, TNO, Utrechtsweg 48, 3704 HE, Zeist, The Netherlands

Peter J. Punt
Microbiology & Systems Biology, TNO, Utrechtsweg 48, 3704 HE, Zeist, The Netherlands
Netherlands Metabolomics Centre (NMC), Einsteinweg 55, 2333 CC Leiden, The Netherlands

Eric L Huang
Bioresource Engineering, McGill University, 21111 Lakeshore Road, Ste-Anne-de-Bellevue, Quebec, H9X 3V9, Canada

Mark G Lefsrud
Bioresource Engineering, McGill University, 21111 Lakeshore Road, Ste-Anne-de-Bellevue, Quebec, H9X 3V9, Canada

Liang Zhou
Process Engineering and Applied Science Department, Faculty of Engineering, Dalhousie University, Halifax, Nova Scotia, Canada

Suzanne M. Budge
Process Engineering and Applied Science Department, Faculty of Engineering, Dalhousie University, Halifax, Nova Scotia, Canada

Abdel E. Ghaly
Process Engineering and Applied Science Department, Faculty of Engineering, Dalhousie University, Halifax, Nova Scotia, Canada

Marianne S. Brooks
Process Engineering and Applied Science Department, Faculty of Engineering, Dalhousie University, Halifax, Nova Scotia, Canada

Deepika Dave
Process Engineering and Applied Science Department, Faculty of Engineering, Dalhousie University, Halifax, Nova Scotia, Canada

Angela Zinnai
Department of Agriculture, Food and Environment, University of Pisa, Via del Borghetto 80, I-56124 Pisa, Italy

Francesca Venturi
Department of Agriculture, Food and Environment, University of Pisa, Via del Borghetto 80, I-56124 Pisa, Italy

Chiara Sanmartin
Department of Agriculture, Food and Environment, University of Pisa, Via del Borghetto 80, I-56124 Pisa, Italy

Gianpaolo Andrich
Department of Agriculture, Food and Environment, University of Pisa, Via del Borghetto 80, I-56124 Pisa, Italy

T. J. Burdock
Department of Process Engineering and Applied Science, Dalhousie University, P.O. Box 1000, Halifax, Nova Scotia, Canada B3J 2X4

M. S. Brooks
Department of Process Engineering and Applied Science, Dalhousie University, P.O. Box 1000, Halifax, Nova Scotia, Canada B3J 2X4

A. E. Ghaly
Department of Process Engineering and Applied Science, Dalhousie University, P.O. Box 1000, Halifax, Nova Scotia, Canada B3J 2X4

F. Venturi
Department of Agriculture, Food and Environment (DAFE), University of Pisa, Italy

G. Andrich
Department of Agriculture, Food and Environment (DAFE), University of Pisa, Italy

C. Sanmartin
Department of Agriculture, Food and Environment (DAFE), University of Pisa, Italy

A. Zinnai
Department of Agriculture, Food and Environment (DAFE), University of Pisa, Italy

Jin-Hua Liu
State Key Laboratory ofBioreactor Engineering, East China University of Science and Technology, Shanghai 200237, China

Ze-Jian Wang
State Key Laboratory ofBioreactor Engineering, East China University of Science and Technology, Shanghai 200237, China

Yu-hua Wang
Zhu Feng Ophiocordyceps Pharmacy Corporation, Ltd. P R Qinghai, China

Ju Chu
State Key Laboratory ofBioreactor Engineering, East China University of Science and Technology, Shanghai 200237, China

Ying-Ping, Zhuang
State Key Laboratory ofBioreactor Engineering, East China University of Science and Technology, Shanghai 200237, China

Si-Liang Zhang
State Key Laboratory ofBioreactor Engineering, East China University of Science and Technology, Shanghai 200237, China

P. Naga Padma
BVB Vivekananda College, Secunderabad 94, India

K. Anuradha
BVB Vivekananda College, Secunderabad 94, India

B. Nagaraju
Department of Microbiology, Osmania University, Hyderabad 7, India

V.Selva Kumar
BVB Vivekananda College, Secunderabad 94, India

Gopal Reddy
Department of Microbiology, Osmania University, Hyderabad 7, India

P. Mudoi
Department of Molecular Biology and Biotechnology, Tezpur University, Tezpur, Assam-784028, India

P. Bharali
Department of Molecular Biology and Biotechnology, Tezpur University, Tezpur, Assam-784028, India

B. K. Konwar
Department of Molecular Biology and Biotechnology, Tezpur University, Tezpur, Assam-784028, India

www.ingramcontent.com/pod-product-compliance
Lightning Source LLC
Chambersburg PA
CBHW080623200326
41458CB00013B/4481

9 781682 861905